ENVIRONMENTAL CHEMISTRY

FIFTH EDITION

STANLEY E. MANAHAN

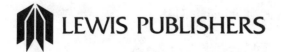

LEWIS PUBLISHERS

Library of Congress Cataloging-in-Publication Data

Manahan, Stanley E.
    Environmental chemistry / Stanley E. Manahan -- 5th ed.
        p. cm.
    Includes biographical references and index.
    1. Environmental Chemistry.   I. Title
QD31.2.M35 1991
628.5'01'54--dc20   91-188
ISBN 0-87371-425-3

LEWIS PUBLISHERS, INC
121 South Main Street, Chelsea, Michigan, 48118

PRINTED IN THE UNITED STATES OF AMERICA

Stanley E. Manahan is Professor of Chemistry at the University of Missouri – Columbia, where he has been on the faculty since 1965. He received his A.B. in chemistry from Emporia State University in 1960 and his Ph.D. in analytical chemistry from the University of Kansas in 1965. Since 1968 his primary research and professional activities have been in environmental chemistry and have included development of methods for the chemical analysis of pollutant species, environmental aspects of coal conversion processes, development of coal products useful for pollutant control, hazardous waste treatment, and toxicological chemistry. He teaches courses on environmental chemistry, hazardous wastes, toxicological chemistry, and analytical chemistry and has lectured on these topics throughout the U.S. as an American Chemical Society Local Section tour speaker. He is also President of ChemChar Research, Inc., a firm working with the development of non-incinerative thermochemical and electrothermochemical treatment of mixed hazardous substances containing refractory organic compounds and heavy metals.

Professor Manahan has written books on hazardous wastes (*Hazardous Waste Chemistry, Toxicology and Treatment*, 1990, Lewis Publishers Inc.), toxicological chemistry (*Toxicological Chemistry*, 1989, Lewis Publishers Inc.), applied chemistry, and quantitative chemical analysis. He has been the author or co-author of approximately 70 research articles.

Cover illustration: Artist's rendition of the reactor chamber of the ChemChar Electrothermochemical Hazardous Waste Treatment Process, courtesy of Robert Anderton and ChemChar Research, Inc. This process uses a combination of high temperature, microplasma discharges, and photolysis under chemically reducing conditions to destroy especially refractory organic compounds in mixed wastes and wastes containing heavy metals. Radionuclides and heavy metals are retained in the carbonaceous reaction matrix.

ACKNOWLEDGMENTS

The author would like to acknowledge the assistance of Vivian Collier, Kathy Walters, and the rest of the staff of Lewis Publishers who have been outstanding to work with in preparing this book. The author would also like to express his thanks to the graduate students in his group — Audrey McGowin, Robert Anderton, Robert Welch, Laura Kinner, and Rhys Thomas — for their excellent research work in the area of environmental chemistry and waste treatment. The greatest satisfaction of a university professorship is the privilege of working with outstanding graduate students such as these. The assistance of Anne F. Manahan in producing this work is gratefully acknowledged.

The author appreciates the efforts of those reviewers who have carefully and constructively critiqued the book and would appreciate receiving copies of their reviews. Those most well qualified to review a book are, of course, those individuals who have used it. Feedback in the form of comments and suggestions from users are welcomed and may be directed to the author at 123 Chemistry Building, University of Missouri, Columbia, Missouri, 65211 U.S.A.

PREFACE

Environmental chemistry is now a mature, viable discipline. It is an exciting area that combines the application of chemical principles to the biggest challenge facing humankind today — the maintenance and enhancement of environmental quality.

Chemists and their profession have a special role to play in the environmental arena. In past decades, chemical processes producing massive amounts of a wide variety of chemicals have given humankind an unprecedented standard of living and quality of life. However, this has exacted a price of pollution and environmental degradation. On the other hand, it is only through the enlightened applications of chemistry that environmental quality can be improved. It is essential, therefore, that anyone entering the chemical profession now — as well as those actively engaged in it — have a basic understanding of environmental chemistry and its applications. It is toward that end that this book has been written.

Individuals in other professions need to have some knowledge of chemistry and environmental chemistry if they are to make a contribution to environmental improvement. Without such a background, non-chemists are sometimes prone to chemophobia, inclined to avoid all chemicals at any costs. To do so, of course, would be to avoid food, air, water, and all other substances — an obvious impossibility. It is one of the more important missions of today's chemical educators to teach non-chemists — including political office-holders, the legal profession, and journalists — about the fascinating and essential discipline of chemistry. In order to assist in that effort, *Environmental Chemistry* has been written at a level such that, with a little extra study of organic and analytical chemistry, an individual with a background in beginning chemistry can understand the material presented.

This book begins with an introduction to environmental chemistry in Chapter 1. Included in this chapter is a brief review of several aspects of solutions and chemical equilibria useful in mastering the next several chapters that follow. In addition, Chapter 1 contains a brief overview of organic chemistry to assist readers who may not have a background in this discipline. Chapters 2–8 deal with aquatic chemistry, first addressing the basic principles of this topic and finishing with a coverage of water pollution (Chapter 7) and water treatment (Chapter 8). Included also is a discussion of the essential role played by microorganisms in aquatic chemical phenomena ("Aquatic Microbial Biochemistry," Chapter 6). Chapters 9–14 cover atmospheric chemistry, concluding with a discussion of major threats to the global atmosphere, particularly from greenhouse gases and ozone-depleting chemicals (Chapter 14, "The Endangered Global Atmosphere"). Chapters 15–19 deal with the geosphere and hazardous substances, which often end up as discarded materials in the geosphere. Chapter 15 covers the fundamental aspects of the geosphere and Chapter 16 emphasizes soil chemistry. The nature and sources of hazardous wastes are discussed in Chapter 17, their environmental chemistry in Chapter 18, and their treatment, minimization, and recycling in 19. Chapter 20 covers the effects of pollutants and hazardous substances on living organisms as presented from the view of toxicological chemistry. Resources and energy are reviewed in Chapter 21.

As the year 2000 approaches, evidence abounds of the stresses that human activities are placing on Planet Earth that may threaten its existence as a place hospitable to life — as examples, production of chemicals that may destroy protective stratospheric ozone, release of greenhouse gases to the atmosphere, excessive demands upon limited water supplies, and destruction of tropical rain forests. Environmental chemistry can play a key role in alleviating problems such as these and it is the author's hope that this book will help its readers to understand the ways in which this discipline can contribute to environmental preservation and improvement.

CONTENTS

ENVIRONMENTAL CHEMISTRY

FIFTH EDITION

Chemistry and Environmental Chemistry

1.1. CHEMICAL SCIENCE AND THE ENVIRONMENT

The importance of protecting our planet from degradation and ruin is a top priority of thinking human beings all over the Earth. It is now generally realized that our environment is endangered by a broad range of human activities.[1] Many urban dwellers would agree that a breath of fresh air is a rare commodity in the city. Some underground sources of drinking water are being threatened by the insidious movement of hazardous waste chemical leachates through groundwater aquifers. The exhausts of fossil fuel combustion and chlorofluorocarbons used for air conditioning and other purposes may cause irreversible damage to the atmosphere. The multiplicative effects of increased human population and increased environmental burdens per person continue to strain Earth's capacity to sustain life.

As 1991 gets underway, warnings of a stressed global environment abound. As a result of the Persian Gulf war, large quantities of soot from burning petroleum have been released to the atmosphere. Huge oil slicks on the Gulf have destroyed wildlife and threatened the operation of water desalination plants. Life-threatening hunger resulting from long-term drought continues to plague parts of Africa. A prolonged drought in California has forced unprecedented water restrictions in that state.

In order to combat threats to our environment, it is necessary to understand the nature and magnitude of the problems involved. Before discussing these problems further, it is essential to recognize the fact that science and technology must play key roles in solving environmental problems. Only through the proper application of science and technology, under the direction of people with a strong environmental consciousness and a basic knowledge of the environmental sciences, can humankind survive on the limited resources of this planet.

Chemistry is often portrayed as the villain in environmental degradation. It is true that chemical products, produced in large quantities and badly misused, have caused vast environmental harm. But it is nonsense to think of a chemical-free environment. All matter — the air we breathe, the food we eat, the life substances that make up our own bodies — consists of chemicals. Some "natural" environmental contaminants can be just as toxic and damaging as synthetic ones. Any meaningful efforts to solve environmental problems require an understanding of chemical processes that occur in water, air, and soil as well as environmental chemical processes that occur in living systems. These are the topics addressed by environmental chemistry.

1.2. ENVIRONMENTAL CHEMISTRY AND ENVIRONMENTAL BIOCHEMISTRY

What is environmental chemistry? This question is a little difficult to answer because environmental chemistry encompasses many different topics. It may involve a study of Freon reactions in the stratosphere or an analysis of toxic Kepone deposits in ocean sediments. It also covers the chemistry and biochemistry of volatile and soluble organometallic compounds biosynthesized by anaerobic bacteria. **Environmental chemistry** is *the study of the sources, reactions, transport, effects, and fates of chemical species in water, soil, and air environments.*

Environmental chemistry is not a new discipline. Excellent work has been done in this field for the greater part of this century. Until about 1970, most of this work was done in academic departments or industrial groups other than chemistry groups. Much of it was performed by people whose primary training was not in chemistry. Thus, when pesticides were synthesized, biologists observed firsthand some of the less desirable consequences of their use. When detergents were formulated, sanitary engineers were startled to see sewage treatment plant aeration tanks vanish under meter-thick blankets of foam, while limnologists wondered why previously normal lakes suddenly became choked with stinking blue-green algae. But even today, relatively few chemists are exposed to material dealing with environmental chemistry as part of their training.

At an accelerating rate in recent years, however, many chemists have become deeply involved with the investigation of environmental problems. Academic chemistry departments have found that environmental chemistry courses appeal to students, and many graduate students are attracted to environmental chemistry research. Help-wanted ads have included increasing numbers of openings for environmental chemists among those of the more traditional chemical subdisciplines. Industries have found that well-trained environmental chemists at least help avoid difficulties with regulatory agencies and at best are instrumental in developing profitable pollution-control products and processes.

Some background in environmental chemistry should be part of the training of every chemistry student. The ecologically illiterate chemist can be a very dangerous species. Chemists must be aware of the possible effects their products and processes might have upon the environment. Furthermore, any serious attempt to solve environmental problems must involve the extensive use of chemicals and chemical processes.

There are some things that environmental chemistry is not. It is not just the same old chemistry with a different cover and title. Because it deals with natural systems, it is more complicated and difficult than "pure" chemistry. Students sometimes find this hard to grasp. Accustomed to the clear-cut concepts of relatively simple, well-defined systems, they may find environmental chemistry to be poorly defined, vague, and confusing. More often than not, it is impossible to come up with a simple answer to an environmental chemistry problem. But, building on an ever-increasing body of knowledge, the environmental chemist can make educated guesses as to how environmental systems will behave.

One of environmental chemistry's major challenges is the determination of the nature and quantity of specific pollutants in the environment. Thus, chemical analysis is a vital first step in environmental chemistry research. The difficulty of analyzing for many environmental pollutants can be awesome. Significant levels of air pollutants may consist of less than a microgram per cubic meter of air. For many water pollutants, one part per million by weight (essentially 1 milligram per liter) is a very high value. Environmentally significant levels of some pollutants may be only a few parts per trillion. Thus, it is obvious that the chemical analyses used to study some environmental systems require a very low limit of detection.

However, environmental chemistry is not the same as analytical chemistry, which is only one of the many subdisciplines that are involved in the study of the chemistry of the environment. A "brute-force" approach to environmental control, involving the monitoring of each environmental niche for every possible pollutant, increases employment for chemists and raises sales of chemical instruments but is a wasteful way to detect and solve environmental problems, degenerating into a mindless exercise in the collection of marginally useful numbers. We must be smarter than that. In order for chemistry to make a maximum contribution to the solution of environmental problems, the chemist must work toward an understanding of the nature, reactions, and transport of chemical species in the environment. Analytical chemistry is a fundamental and crucial part of that endeavor but is by no means all of it.

The ultimate environmental concern is that of life itself. The discipline that deals specifically with the effects of environmental chemical species on life is **environmental biochemistry**. A related area, **toxicological chemistry** is *the chemistry of toxic substances with emphasis upon their interactions with biologic tissue and living organisms*.[2] Toxicological chemistry, which is discussed in detail in Chapter 20, deals with the chemical nature and reactions of toxic substances and involves their origins, uses, and chemical aspects of exposure, fates, and disposal.

Strong environmental interactions occur among water, air, soil, and living systems — the hydrosphere, atmosphere, lithosphere (geosphere), and biosphere, respectively. These are shown in Figure 1.1 for environmental pollutants. Environmental chemistry and environmental biochemistry must consider such interactions.

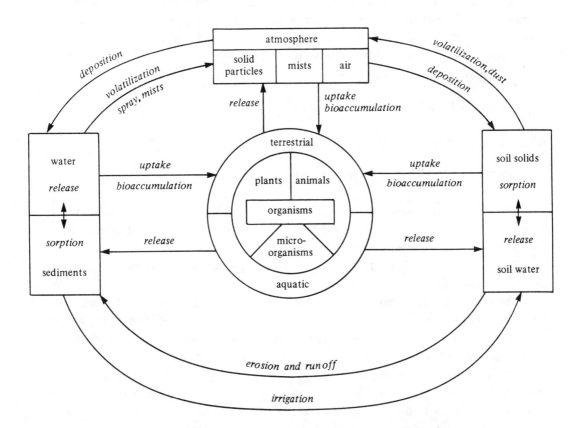

Figure 1.1. Interactions of pollutants among the hydrosphere, atmosphere, lithosphere, and biosphere.

1.3. SOME DEFINITIONS

In some cases pollution is a clear-cut phenomenon, whereas in others it lies largely in the eyes of the beholder. Toxic organochlorine solvent residues leached into water supplies from a hazardous waste chemical dump are pollutants in anybody's view. However, loud rock music amplified to a high decibel level by the sometimes questionable miracle of modern electronics is pleasant to some people and a very definite form of noise pollution to others. Frequently, time and place determine what may be called a pollutant. The phosphate that the sewage treatment plant operator has to remove from wastewater is chemically the same as the phosphate that the farmer a few miles away has to buy at high prices for fertilizer. Most pollutants are, in fact, resources gone to waste; as resources become more scarce and expensive, economic pressure will almost automatically force solutions to many pollution problems.

A reasonable definition of a **pollutant** is a substance present in greater than natural concentration as a result of human activity and having a net detrimental effect upon its environment or upon something of value in that environment. **Contaminants**, which are not classified as pollutants unless they have some detrimental effect, cause deviations from the normal composition of an environment.

Every pollutant originates from a **source**. The source is particularly important, because it is generally the logical place to eliminate pollution. After a pollutant is released from a source, it may act upon a receptor. The **receptor** is anything that is affected by the pollutant. Humans whose eyes smart from oxidants in the atmosphere are receptors. Trout fingerlings that may die after exposure to dieldrin in water are also receptors. Eventually, if the pollutant is long-lived, it may be deposited in a **sink**, a long-time repository of the pollutant. Here it will remain for a long time, though not necessarily permanently. Thus, a limestone wall may be a sink for atmospheric sulfuric acid, through the reaction,

$$CaCO_3 + H_2SO_4 \rightarrow CaSO_4 + H_2O + CO_2 \tag{1.3.1}$$

which fixes the sulfate as part of the wall composition.

1.4. WATER, AIR, SOIL, AND LIFE

As noted above, environmental chemistry and environmental biochemistry may be subdivided into areas involving the chemistry of the hydrosphere, the lithosphere, the atmosphere, and the biosphere. All matter, from minerals in the outer layers of the Earth's crust to relatively stable ions in the upper reaches of the atmosphere, may be included in one of these categories.

The **hydrosphere** refers to water in its many forms. It includes the oceans, lakes, streams, reservoirs, snowpack, glaciers, the polar ice caps, and water under the ground (groundwater). For the study of environmental chemistry, however, liquid water and the reactions of the chemical species in it are of predominant importance.

The **lithosphere** includes the outer parts of the solid Earth. In general, the term refers to minerals encountered in the Earth's crust and to the complex and variable mixture of minerals, organic matter, water, and air making up soil. Insofar as environmental chemistry is concerned, the soil is the most significant part of the lithosphere.

The **atmosphere** is the envelope of gases surrounding the Earth. The atmosphere is subdivided into different regions depending on altitude. Atmospheric chemistry varies a great deal with altitude, exposure to solar radiation, pollution load, and other factors.

The term **biosphere** refers to life. It includes living organisms and their immediate surroundings. The biosphere is influenced tremendously by the chemistry of the environment and, in turn, exerts a powerful influence upon the chemistry of most environments, particularly the lithosphere and hydrosphere. Moreover, biological activity is responsible for the present composition of the atmosphere (specifically, high oxygen level and low carbon dioxide level), and plants still influence the atmosphere, for example, by emitting terpenes which form a sort of smog, such as that observed in the U.S. Great Smoky or Blue Ridge Mountains.

1.5. CHEMISTRY AND CHEMICAL REACTIONS

Ideally, the user of this book should have a background in organic and analytical chemistry, in addition to fundamentals of general chemistry. Some familiarity with applied physical chemistry is also helpful. However, most individuals who need to use environmental chemistry will not have this much background in chemical science. Therefore, the assumption is made that readers have had at least one semester of general chemistry. Those who do not are referred to texts dealing with this subject at a fundamental level[3,4].

The remainder of this chapter deals with some basic points of general chemistry and outlines the fundamentals of organic chemistry. Readers who are familiar with these areas may want to omit these sections.

Solutions

Because much of the material in this book deals with substances dissolved in solution, a brief review of solutions is given here. A **solution** is formed when a solid, gas, or another liquid in contact with a liquid becomes dispersed homogeneously throughout the liquid in a molecular form. The substance, called a **solute**, is said to **dissolve**. The liquid is called a **solvent**. There may be no readily visible evidence that a solute is present in the solvent; for example, a deadly poisonous solution of sodium cyanide in water looks like pure water! The solution may have a strong color, as is the case for intensely purple solutions of potassium permanganate, $KMnO_4$. It may have a strong odor, such as that of ammonia, NH_3, dissolved in water.

Solution Concentration

The quantity of solute relative to that of solvent or solution is called the **solution concentration**. Concentrations are expressed in numerous ways. Very high concentrations are often given as percent by weight. For example commercial concentrated hydrochloric acid is 36% HCl, meaning that 36% of the weight has come from dissolved HCl and 64% from water solvent. Concentrations of very dilute solutions, such as those of hazardous waste leachate containing low levels of contaminants, are expressed as weight of solute per unit volume of solution. Common units are milligrams per liter (mg/L) or micrograms per liter (µg/L). Since a liter of water weighs essentially 1,000 grams, a concentration of 1 mg/L is equal to 1 part per million (ppm) and a concentration of 1 µg/L is equal to 1 part per billion (ppb).

Chemists often express concentrations in moles per liter, or **molarity, M**. Molarity is given by the relationship,

$$M = \frac{\text{Number of moles of solute}}{\text{Number of liters of solution}} \qquad (1.5.1)$$

where the number of moles of a substance is its mass in grams divided by its molar mass.

Concentration of H^+ Ion and pH

Acids, such as HCl and H_2SO_4, produce H^+ ion, whereas bases, such as sodium hydroxide and calcium hydroxide (NaOH and $Ca(OH)_2$, respectively), produce hydroxide ion, OH^-. Molar concentrations of hydrogen ion, $[H^+]$, range over many orders of magnitude and are conveniently expressed by pH defined as

$$pH = -\log[H^+] \qquad (1.5.2)$$

In absolutely pure water the value of $[H^+]$ is exactly 1×10^{-7} mole/L, the pH is 7.00, and the solution is **neutral** (neither acidic nor basic). **Acidic** solutions have pH values of less than 7 and **basic** solutions have pH values of greater than 7.

Solution Equilibria

Many of the phenomena in aquatic chemistry (Chapters 2–8) and geochemistry (Chapters 15 and 16) involve solution equilibrium. In a general sense solution equilibrium deals with the extent to which **reversible** acid-base, solubilization (precipitation), complexation, or oxidation-reduction reactions proceed in a forward or backward direction.[5] This is expressed for a generalized equilibrium reaction,

$$aA + bB \rightarrow cC + dD \qquad (1.5.3)$$

by the following **equilibrium constant expression:**

$$\frac{[C]^c[D]^d}{[A]^a[B]^b} = K \qquad (1.5.4)$$

where K is the **equilibrium constant.**

A reversible reaction may approach equilibrium from either direction. In the example above, if A were mixed with B, or C were mixed with D, the reaction would proceed in a forward or reverse direction such that the concentrations of species — [A], [B], [C], and [D] — substituted into the equilibrium expression gave a value equal to K.

As expressed by **Le Châtelier's principle**, a stress placed upon a system in equilibrium will shift the equilibrium to relieve the stress. For example, adding product "D" to a system in equilibrium will cause Reaction 1.5.3 to shift to the left, consuming "C" and producing "A" and "B," until the equilibrium constant expression is again satisfied. This **mass action effect** is the driving force behind many environmental chemical phenomena.

In most cases this book uses concentrations and pressures in equilibrium constant expression calculations. When this is done, K is not exactly constant with varying concentrations and pressures; it is an *approximate equilibrium constant* that applies only to limited conditions. **Thermodynamic equilibrium constants** are more exact forms derived from thermodynamic data that make use of *activities* in place of concentrations. At a specified temperature, the value of a thermodynamic equilibrium constant is applicable over a wide concentration range. The activity of a species, commonly denoted as a_X for species "X" expresses how effectively it interacts with its surroundings, such as other solutes or electrodes in solution. (The analogy may be drawn of an environmental chemistry class with a "concentration" of 20 students per classroom. Their "activity" in relating to the subject is likely to be much higher on a cold, rainy day than on a balmy, sunny day in springtime.) Activities approach concentrations at low values of concentration. The thermodynamic equilibrium constant expression for Reaction 1.5.3 is expressed as

$$\frac{a_C^c \, a_D^d}{a_A^a \, a_B^b} = K \tag{1.5.5}$$

There are several major kinds of equilibria in aqueous solution. One of these is acid-base equilibrium (see Chapter 3) as exemplified by the ionization of acetic acid, HAc,

$$HAc \xleftarrow{} \rightarrow H^+ + Ac^- \tag{1.5.6}$$

for which the acid dissociation constant is

$$\frac{[H^+][Ac^-]}{[HAc]} = K = 1.75 \times 10^{-5} \quad \text{(at 25°C)} \tag{1.5.7}$$

Very similar expressions are obtained for the formation and dissociation of metal **complexes** or **complex ions** (Chapter 3), formed by the reaction of a metal ion in solution with a **complexing agent** or **ligand**, both of which are capable of independent existence in solution. This can be shown by the reaction of iron(III) ion and thiocyanate ligand

$$Fe^{3+} + SCN^- \xleftarrow{} \rightarrow FeSCN^{2+} \tag{1.5.8}$$

for which the **formation constant expression** is:

$$\frac{[FeSCN^{2+}]}{[Fe^{3+}][SCN^-]} = K_f = 1.07 \times 10^3 \quad \text{(at 25°C)} \tag{1.5.9}$$

The bright red color of the $FeSCN^{2+}$ complex formed could be used to test for the presence of iron(III) in acid mine water (Chapter 7).

An example of an **oxidation-reduction reaction**, those that involve the transfer of electrons between species, is

$$MnO_4^- + 5Fe^{2+} + 8H^+ \xleftarrow{} \rightarrow Mn^{2+} + 5Fe^{3+} + 4H_2O \tag{1.5.10}$$

for which the equilibrium expression is:

$$\frac{[Mn^{2+}][Fe^{3+}]^5}{[MnO_4^-][Fe^{2+}]^5[H^+]^8} = K = 3 \times 10^{62} \quad \text{(at 25°C)} \tag{1.5.11}$$

The value of K is calculated from the Nernst equation, as explained in Chapter 4.

Distribution between Phases

As discussed in Chapter 5, many important environmental chemical phenomena involve distribution of species between phases. This most commonly involves the equilibria between species in solution and in a solid phase. **Solubility equilibria** (see Chapter 5) deal with reactions such as,

$$AgCl(s) \rightarrow Ag^+ + Cl^- \tag{1.5.12}$$

in which one of the participants is a slightly soluble (virtually insoluble) salt and for which the equilibrium constant is,

$$[Ag^+][Cl^-] = K_{sp} = 1.82 \times 10^{-10} \quad \text{(at 25°C)} \tag{1.5.13}$$

a **solubility product**. Note that in the equilibrium constant expression there is not a value given for the solid AgCl. This is because the activity of a solid is constant at a specific temperature and is contained in the value of K_{sp}.

An important example of distribution between phases is that of a hazardous waste species partitioned between water and a body of immiscible organic liquid in a hazardous waste site. The equilibrium for such a reaction,

$$X(aq) \longleftrightarrow X(org) \tag{1.5.14}$$

is described by the **distribution law** expressed by a **distribution coefficient** or **partition coefficient** in the following form:

$$\frac{[X(org)]}{[X(aq)]} = K_d \tag{1.5.15}$$

1.6. ORGANIC CHEMISTRY

Most carbon-containing compounds are **organic chemicals** and are discussed under the heading of **organic chemistry**. Organic chemistry is a vast, diverse, discipline because of the enormous number of organic compounds that exist as a consequence of the versatile bonding capabilities of carbon. Such diversity is due to the ability of carbon atoms to bond to each other in a limitless variety of straight chains, branched chains, and rings as illustrated in Figure 1.2. This figure shows several **alkanes**, which are **hydrocarbon compounds** containing only carbon and hydrogen and no double bonds (see alkenes later in this section), triple bonds, or aromatic entities (see Section 1.8). The four hydrocarbon molecules in Figure 1.2 contain 8 carbon atoms each. In one of the molecules, all of the carbon atoms are in a straight chain and in two they are in branched chains, whereas in a fourth, 6 of the carbon atoms are in a ring.

Among organic chemicals are included the majority of important industrial compounds, synthetic polymers, agricultural chemicals, biological materials, and most substances that are of concern because of their toxicities and other hazards. These few pages present only a few of the most fundamental concepts and definitions of organic chemistry needed to understand organic substances in the environment and do not pretend to teach this vast discipline to a reader unfamiliar with it. The reader needing a more complete background in the subject is referred to one of the briefer organic chemistry texts,[6,7] or full-semester texts.[8]

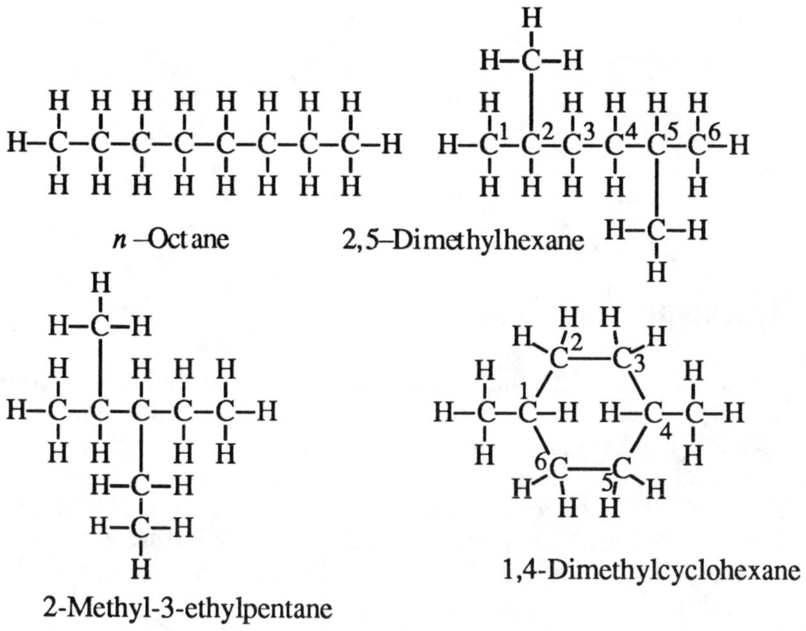

Figure 1.2. Structural formulas of four hydrocarbons, each containing 8 carbon atoms, that illustrate the structural diversity possible with organic compounds. Numbers used to denote locations of atoms for purposes of naming are shown on two of the compounds.

1.7. ORGANIC FORMULAS, STRUCTURES, AND NAMES

Formulas of organic compounds present information at several different levels of sophistication. **Molecular formulas**, such as that of octane (C_8H_{18}), give the number of each kind of atom in a molecule of a compound. As shown in Figure 1.2, however, the molecular formula of C_8H_{18}, may apply to several alkanes, each one of which has unique chemical, physical, and toxicological properties. These different compounds are designated by **structural formulas** showing how the atoms in a molecule are arranged. Compounds that have the same molecular, but different structural, formulas are called **structural isomers**. Of the compounds shown in Figure 1.2, n-octane, 2,5-dimethylhexane, and 2-methyl-3-ethylpentane are structural isomers, all having the formula C_8H_{18}, whereas 1,4-dimethylcyclohexane is not a structural isomer of the other three compounds because its molecular formula is C_8H_{16}.

Figure 1.3 illustrates another kind of isomerism, called **cis-trans**, isomerism, that is possible for alkenes. Unlike the alkanes shown in Figure 1.3, **alkenes**, sometimes still called **olefins**, are hydrocarbons with a **double bond** composed of four shared electrons. The double bond is represented in structural formulas by a double line, =. The two carbon atoms connected by the double bond cannot rotate relative to each other, so that some alkenes are *cis-trans* isomers that have different parts of the molecule oriented differently in space, although these parts occur in the same order. Both alkenes illustrated in Figure 1.3 have a molecular formula of C_4H_8. In the case of *cis*-2-butene, the two CH_3 groups attached to the C=C carbon atoms are on the same side of the molecule, whereas in *trans*-2-butene they are on opposite sides.

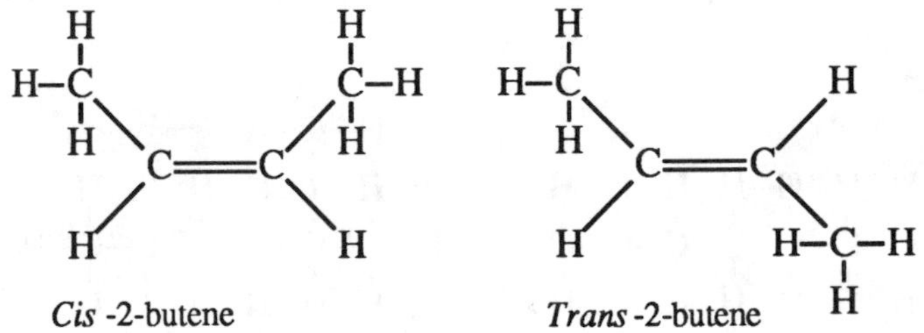

Figure 1.3. *Cis* and *trans* isomers of the alkene, 2-butene.

Condensed Structural Formulas

To save space, structural formulas are conveniently abbreviated as **condensed structural formulas**. The condensed structural formula of 2-methyl-3-ethylpentane is $CH_3CH(CH_3)CH(C_2H_5)CH_2CH_3$ where the CH_3 (methyl) and C_2H_5 (ethyl) groups are placed in parentheses to show that they are branches attached to the longest continuous chain of carbon atoms, which contains 5 carbon atoms. It is understood that each of the methyl and ethyl groups is attached to the carbon immediately preceding it in the condensed structural formula (methyl attached to the second carbon atom, ethyl to the third).

As illustrated by the examples in Figure 1.4, the structural formulas of organic molecules may be represented in a very compact form by lines and by figures such as hexagons. The ends and intersections of straight line segments in these formulas indicate the locations of carbon atoms. Carbon atoms at the terminal ends of lines are understood to have three H atoms attached, C atoms at the intersections of two lines are understood to have *two* H atoms attached to each, *one* H atom is attached to a carbon represented by the intersection of three lines, and *no* hydrogen atoms are bonded to C atoms where four lines intersect. Other atoms or groups of atoms, such as the Cl atom or OH group, that are substituted for H atoms are shown by their symbols attached to a C atom with a line.

Organic Nomenclature

Systematic names, from which the structures of organic molecules can be deduced, have been assigned to all known organic compounds. The more common organic compounds, including hazardous organic sustances, likewise have **common names** that have no structural implications. Although it is not possible to cover organic nomenclature in any detail in this chapter, the basic approach to nomenclature is presented here along with some pertinent examples.

Consider the alkanes shown in Figure 1.2. The fact that *n*-octane has no side chains is denoted by "*n*", that it has 8 carbon atoms is denoted by "oct," and that it is an alkane is indicated by "ane." The names of compounds with branched chains or atoms other than H or C attached make use of numbers that stand for positions on the longest continuous chain of carbon atoms in the molecule. This convention is illustrated by the second compound in Figure 1.2. It gets the hexane part of the name from the fact that it is an alkane with 6 carbon atoms in its longest continuous chain ("hex" stands for 6). However, it has a methyl group (CH₃) attached on the second carbon atom of the chain and another on the fifth. Hence the full systematic name of the compound is 2,5-dimethylhexane, where "di" indicates two methyl groups. In the case of 2-methyl-3-ethylpentane, the longest continuous chain of carbon atoms contains 5 carbon atoms, denoted by pentane, a methyl group is attached to the second carbon atom, and an ethyl group, C₂H₅, on the third carbon atom The last compound shown in the figure has 6 carbon atoms in a ring, indicated by the prefix "cyclo," so it is a cyclohexane compound. Furthermore, the carbon in the ring to which one of the methyl groups is attached is designated by "1" and another methyl group is attached to the fourth carbon atom around the ring. Therefore, the full name of the compound is 1,4-dimethylcyclohexane.

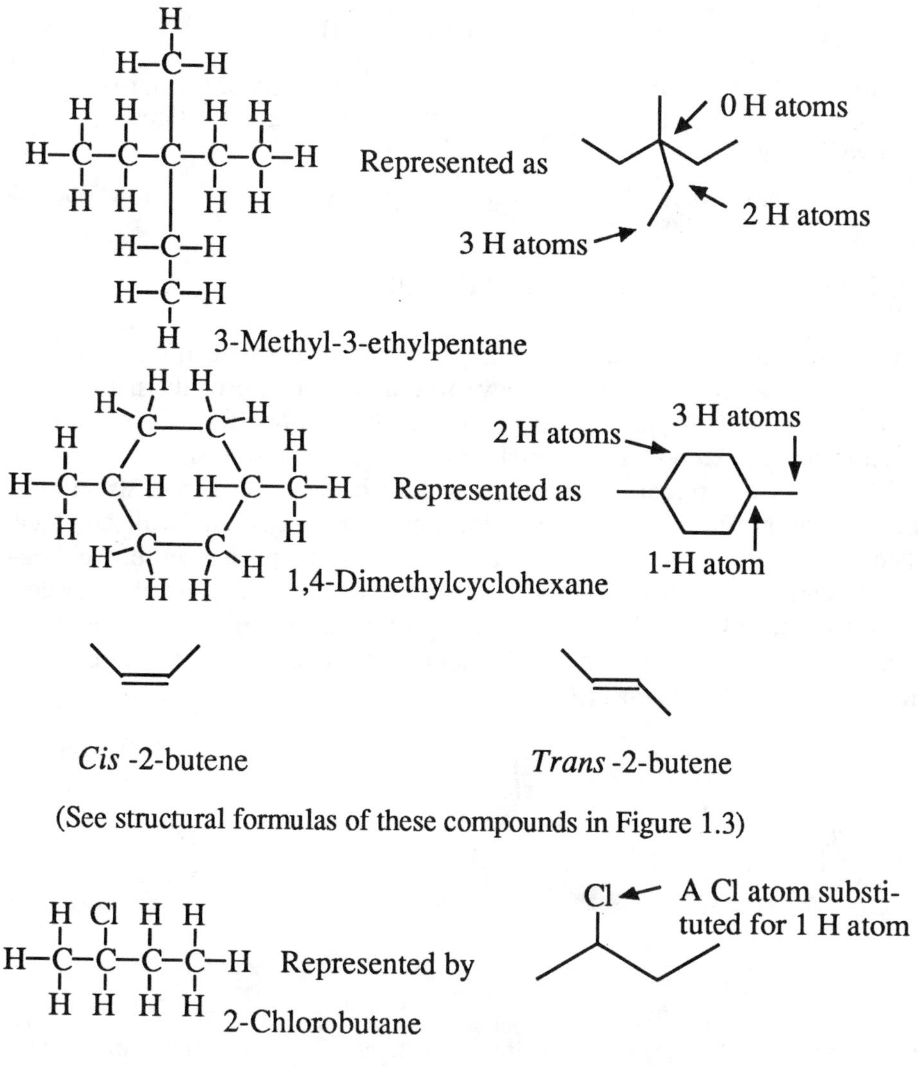

Figure 1.4. Representation of structural formulas with lines. A carbon atom is understood to be at each corner and at the end of each line. The numbers of hydrogen atoms attached to carbons at several specific locations are shown by arrows.

Molecular Geometry

The three-dimensional shape of a molecule, that is, its **molecular geometry,** determines in part its properties, particularly its interactions with biological systems. Shapes of molecules are represented in drawings by lines of normal, uniform thickness for bonds in the plane of the paper, broken lines for bonds extending away from the viewer, and heavy lines for bonds extending toward the viewer. These conventions are shown by the example of dichloromethane, CH_2Cl_2, an important organochloride solvent and extractant, as illustrated in Figure 1.5.

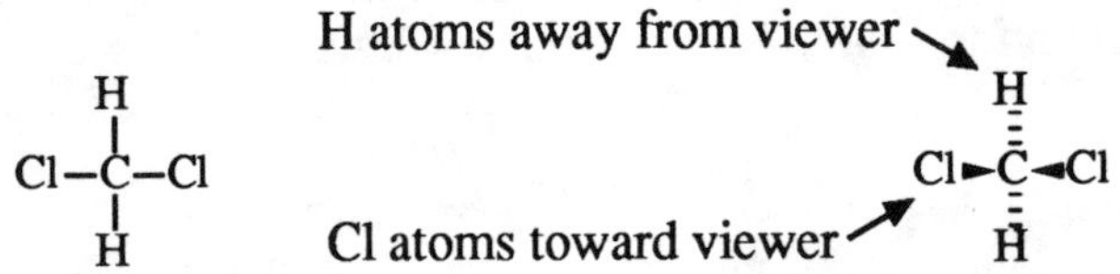

H atoms away from viewer

Cl atoms toward viewer

Structural formula of dichloromethane in two dimensions

Structural formula of dichloro- methane represented in three dimensions

Figure 1.5. Structural formulas of dichloromethane, CH_2Cl_2; the formula on the right provides a three-dimensional representation.

1.8. AROMATIC ORGANIC COMPOUNDS

Aromatic compounds (arenes, or **aryl compounds)** constitute an important special class of organic substances. They are unique because of the nature of their bonds, a subject that is beyond the scope of this chapter. Most aromatic compounds discussed in this book contain 6-carbon-atom benzene rings as shown for benzene, C_6H_6, in Figure 1.6. Aromatic compounds such as benzene have ring structures and are held together in part by particularly stable bonds that contain delocalized clouds of so-called π (pi, pronounced "pie") electrons. In an oversimplified sense, the structure of benzene can be visualized as resonating between the two equivalent structures shown on the left in Figure 1.6 by the shifting of electrons in chemical bonds. This structure can be shown more simply and accurately by a hexagon with a circle in it, as illustrated on the right in Figure 1.6.

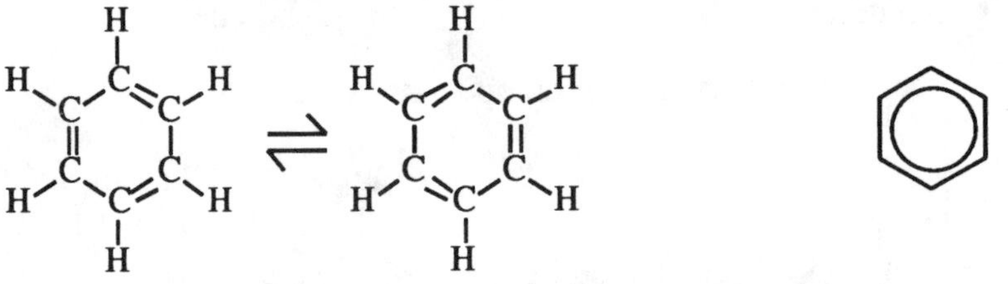

Figure 1.6. Representation of the aromatic benzene molecule with two resonance structures (left) and, more accurately, as a hexagon with a circle in it (right). Unless shown by symbols of other atoms, it is understood that a C atom is at each corner and that one H atom is bonded to each C atom.

As shown in Figure 1.7, some arenes, such as naphthalene and the polycyclic aromatic compound, benzo(a)pyrene, contain fused rings.

Naming Aromatic Compounds

The chlorinated derivative of phenol, 3-chlorophenol, shown in Figure 1.7, may be used to illustrate the numbering of positions in a 6-membered aromatic ring for the purpose of naming of arenes. Phenol has an -OH group attached to the benzene ring, the carbon to which this group is attached is assigned the number 1, and the other carbon atoms are numbered clockwise around the ring. When chlorine is substituted on the third carbon atom from the -OH group, the compound is 3-chlorophenol. This example also illustrates an older system of nomenclature in which the prefixes, *ortho*, *meta*, and *para*, are employed to show substituent groups on carbons 2 (or 6), 3 (or 5), and 4, respectively.

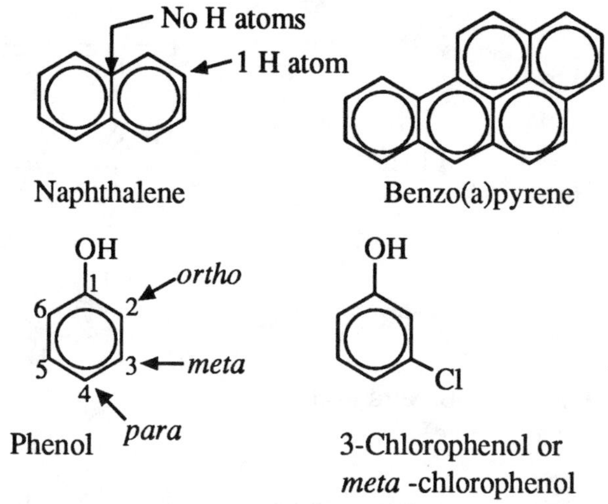

Naphthalene

Benzo(a)pyrene

Phenol

3-Chlorophenol or
meta -chlorophenol

Figure 1.7. Aromatic compounds containing fused rings (top) and showing the numbering of carbon atoms in an aromatic ring for purposes of nomenclature.

1.9. ORGANIC FUNCTIONAL GROUPS

The discussion of organic chemistry so far in this chapter has emphasized hydrocarbon compounds that contain only hydrogen and carbon and the influence of molecular structure upon chemical behavior. It has been shown that hydrocarbons may exist as alkanes, alkenes, and arenes, depending upon the kinds of bonds between carbon atoms. The presence of elements other than hydrogen and carbon in organic molecules greatly increases the diversity of their chemical behavior. **Functional groups** consist of specific bonding configurations of atoms in organic molecules. Most functional groups contain at least one element other than carbon or hydrogen, although two carbon atoms joined by a double bond (alkenes) or triple bond (alkynes) are likewise considered to be functional groups. Table 1.1 shows some of the major functional groups that determine the nature of organic compounds.

Organic Acids, Alcohols, and Esters

Various functional groups in organic molecules may undergo reactions to form other kinds of functional groups. Among the more important such groups formed in life processes (see the discussion of biochemistry in Chapter 20) are **esters** produced by the reaction of alcohols and carboxylic acids with the loss of water. The formation of methyl ester is shown in Figure 1.8. An ester can also undergo the reverse reaction, the addition of water by hydrolysis, to regenerate the original alcohol and acid. In addition to their occurrence as fats, oils, and waxes in biological systems, some esters are important industrial chemicals and occur in hazardous wastes.

Table 1.1. Examples of Some Important Functional Groups

Type of functional group	Example compound	Structural formula of functional group[1]
Alkene (olefin)	Propene (propylene)	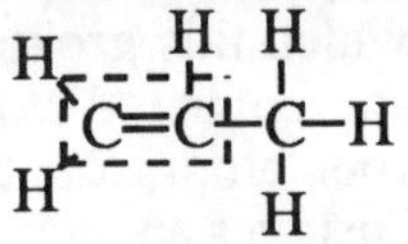
Alkyne	Acetylene	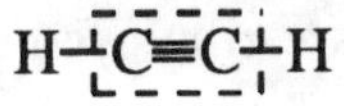
Alcohol (-OH attached to alkyl group)	2-Propanol	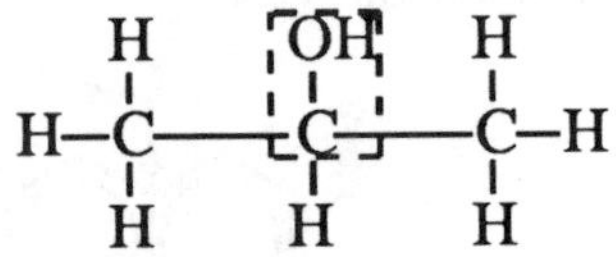
Phenol (-OH attached to aryl group)	Phenol	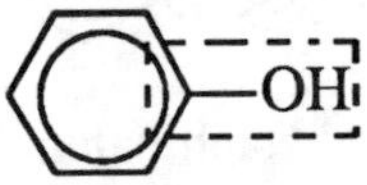
Ketone	Acetone	

(When $-\overset{O}{\underset{}{\overset{\|}{C}}}-H$ group is on end carbon, compound is an aldehyde)

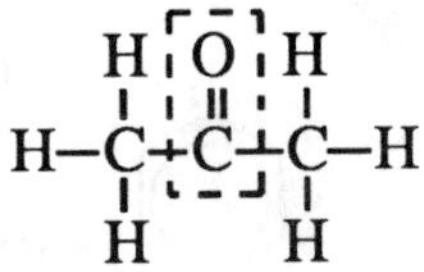

Amine	Methylamine	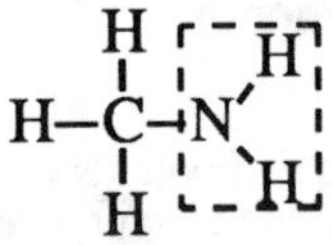
Nitro compounds	Nitromethane	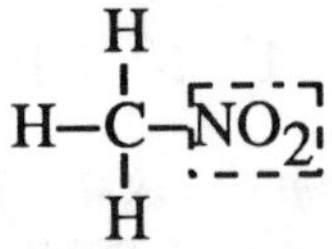
Sulfonic acids	Benzenesulfonic acid	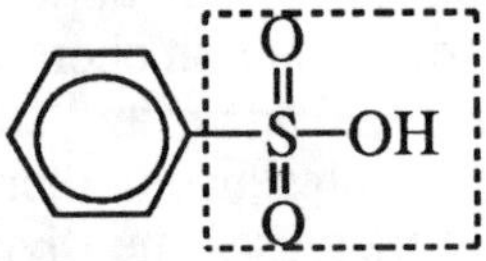
Organohalides	1,1–Dichloro-ethane	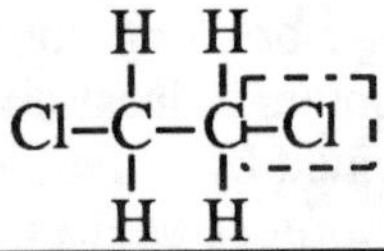

[1] Functional group outlined by dashed line

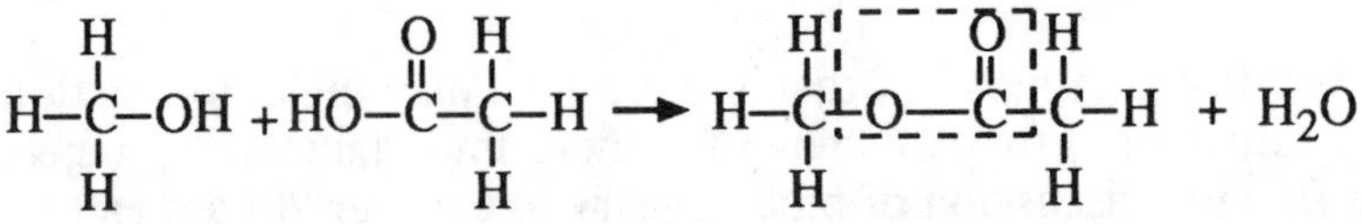

Methyl   Acetic acid (a car-   Methyl acetate
alcohol   boxylic acid)   ester

Figure 1.8. Formation of a simple ester with the loss of water. The functional groups characteristic of carboxylic acids and esters formed from alcohols and carboxylic acids are outlined by dashed rectangles.

1.10. SYNTHETIC POLYMERS

A large fraction of the chemical industry worldwide is devoted to polymer man-ufacture, which is very important to environmental chemistry, both for the pollutants it produces and for the uses of polymeric materials to mitigate pollution. Synthetic **polymers**, such as polyvinylchloride (Figure 1.9), are produced when small molecules called **monomers** (Figure 1.10) bond together to form a much smaller number of very large molecules. Many natural products are polymers; for example, cellulose in wood, paper, and many other materials is a polymer of the sugar glucose. Synthetic polymers form the basis of many industries, such as rubber, plastics, and textiles manufacture.

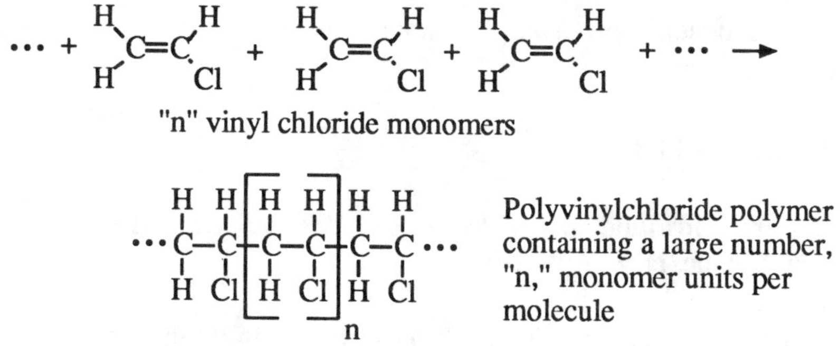

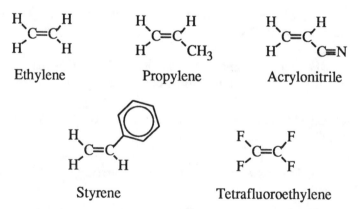

Figure 1.9. Polyvinylchloride polymer.

Figure 1.10. Monomers from which commonly used polymers are synthesized.

An important example of a polymer is that of polyvinylchloride, shown in Figure 1.9. This polymer is synthesized in large quantities for the manufacture of water and sewer pipe, water-repellant liners, and other plastic materials. Other major polymers include polyethylene (plastic bags, milk cartons), polypropylene, (impact-resistant plastics, indoor-outdoor carpeting), polyacrylonitrile (Orlon, carpets), polystyrene (foam insulation), and polytetrafluoroethylene (Teflon coatings, bearings); the monomers from which these substances are made are shown in Figure 1.10.

Many of the hazards from the polymer industry arise from the monomers used as raw materials. Many monomers are reactive and flammable, with a tendency to form explosive vapor mixtures with air. All have a certain degree of toxicity; vinyl chloride is a known human carcinogen. The combustion of many polymers may result in the evolution of toxic gases, such as hydrogen cyanide (HCN) from polyacrylonitrile or hydrogen chloride (HCl) from polyvinylchloride. Another hazard presented by plastics results from the presence of **plasticizers** added to provide essential properties, such as flexibility. The most widely used plasticizers are phthalates, such as dimethyl-phthalate,

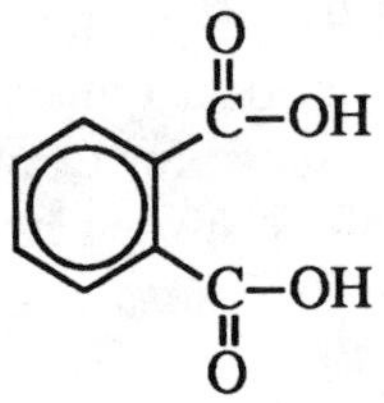

Dimethylphthalate

which are environmentally persistent, resistant to treatment processes, and prone to undergo bioaccumulation.

LITERATURE CITED

1. World Resources Institute, *World Resources 1990-91*, Oxford University Press, 20 Madison Avenue, New York, NY.

2. Manahan, Stanley E., *Toxicological Chemistry*, Lewis Publishers, Inc., Chelsea, Michigan, 1989.

3. Manahan, Stanley E., *General Applied Chemistry*, 2nd ed., Brooks/Cole Publishing Co., Pacific Grove, CA, 1982.

4. Rosenberg, Jerome L., and Lawrence M. Epstein, *College Chemistry*, 7th ed., Schaum's Outline Series, McGraw-Hill Publishing Company, New York, 1990.

5. Manahan, Stanley E., *Quantitative Chemical Analysis*, Brooks/Cole Publishing Co., Pacific Grove, CA, 1986.

6. Atkins, Robert C., and Francis A. Carey, *Organic Chemistry — A Brief Course*, McGraw-Hill Publishing Company, New York, 1990.

7. Fessenden, Ralph J., and Joan S. Fessenden, *Fundamentals of Organic Chemistry*, Harper and Row, Publishers, Scranton, Pennsylvania, 1990.

8. Wade, L. G., Jr., *Organic Chemistry*, 2nd ed., Prentice Hall, Englewood Cliffs, New Jersey, 1991.

SUPPLEMENTARY REFERENCES

Bradley, Raymond, and Stephen Duguid, Eds.,*Environmental Ethics*, Vol. 2., Institute for the Humanities, Simon Fraser University, Burnaby, BC, Canada, 1989.

Whelan, Elizabeth M., *Toxic Terror*, Jameson Books, Ottawa, Illinois, 1985.

Paustenbach, Dennis J., Ed., *The Risk Assessment of Environmental and Human Health Hazards: A Textbook of Case Studies*, John Wiley and Sons, New York, NY, 1989.

Richardson, Mervyn L., Ed., *Risk Assessment of Chemicals in the Environment.* Royal Society of Chemistry, Letchworth, England, 1990.

Krimsky, Sheldon and Alonzo Plough, *Environmental Hazards: Communicating Risks as a Social Process*, Auburn House Publishing Co., Dover, Massachusetts, 1988.

Benarde, Melvin A., *Our Precarious Habitat: Fifteen Years Later*, John Wiley and Sons, New York, NY, 1989.

McKibben, Bill, *The End of Nature*, Random House, New York, NY, 1989.

Begon, Michael, John L. Harper, and Colin R. Townsend, *Ecology. Individuals, Populations and Communities*, 2nd ed. Blackwell Scientific, Boston, 1990.

Cohen, Gary, and John O'Connor, *Fighting Toxics. A Manual for Protecting your Family, Community, and Workplace*, Island Press, Washington, DC, 1990.

Mathews, Christopher K., and K. E. van Holde, *Biochemistry*, Benjamin/Cummings, Redwood City, CA, 1990.

Jeffery, G. H., Vogel's, *Textbook of Quantitative Chemical Analysis*, 5th ed. Longman Scientific, Harlow, U.K., 1989.

Landy, Marc K., Marc J. Roberts, and Stephen R. Thomas, *The Environmental Protection Agency. Asking the Wrong Questions*, Oxford University Press, New York, 1990.

Marks, Dawn B., *Biochemistry*, Williams and Wilkins, Baltimore, 1990.

Holum, John R., *Fundamentals of General, Organic, and Biological Chemistry*, 4th ed. Wiley, New York, 1990.

Hynes, H. Patricia, *The Recurring Silent Spring*, Pergamon, Elmsford, NY, 1989.

Friday, Laurie, and Ronald Laskey, Eds., *The Fragile Environment. The Darwin College Lectures*, Cambridge University Press, New York, 1989.

Goudie, Andrew, *The Human Impact on the Natural Environment*, 3rd ed., MIT Press, Cambridge, MA, 1990.

Silver, Cheryl Simon, and Ruth S. DeFries, *One Earth, One Future. Our Changing Global Environment*, National Academy Press, Washington, DC, 1990.

Ray, Dixy Lee, and Lou Guzzo, *Trashing the Planet. How Science Can Help Us Deal with Acid Rain, Depletion of the Ozone, and Nuclear Waste (Among Other Things)*. Regnery Gateway, Washington, DC, 1990.

Johnson, Richard G., Ed., *Global Environmental Change. The Role of Space in Understanding Earth*, Published for the American Astronautical Society by Univelt, San Diego, CA, 1990.

The International Geosphere-Biosphere Programme. A Study of Global Change: The Initial Core Projects, IGBP Secretariat, Stockholm, 1990.

Hutzinger, Otto, Ed., 1988, *The Handbook of Environmental Chemistry*. Vol. 2, *Reactions and Processes*.

World Resources Institute and International Institute for Environment and Development, *World Resources, 1988-89*, Basic Books, New York, 1988.

Buchner, W., *Industrial Inorganic Chemistry*, VCH, New York, 1989.

Clarke, Lee, *Acceptable Risk? Making Decisions in a Toxic Environment*, University of California Press, Berkeley, 1989.

QUESTIONS AND PROBLEMS

1. Under what circumstances does a contaminant become a pollutant?

2. Calculate the molar concentration and pH of a solution that is 1 part per thousand in HCl.

3. What characteristic must a reaction have in order for meaningful equilibrium calculations to be performed on it?

4. What is the distinction between approximate and thermodynamic equilibrium constants?

5. Having read sections 1.6–1.10, how would you define organic chemistry?

6. Both 2,5-dimethylhexane and 1,4-dimethylcyclohexane are 8-carbon hydrocarbons containing only single bonds. Why are they not considered isomers of each other?

7. Why are there *cis/trans* isomers of 2-butene, but not of 1-butene, which has the same molecular formula, but has the double bond between the first and second carbon atom?

8. Of what general class of compound is the structure,

characteristic?

9. Give systematic names to each of the compounds shown below:

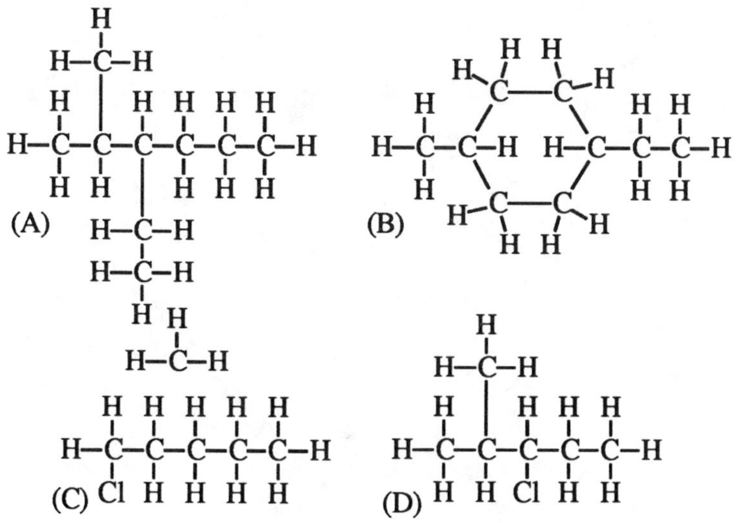

(A)

(B)

(C)

(D)

10. Name each of the kinds of functional groups illustrated below:

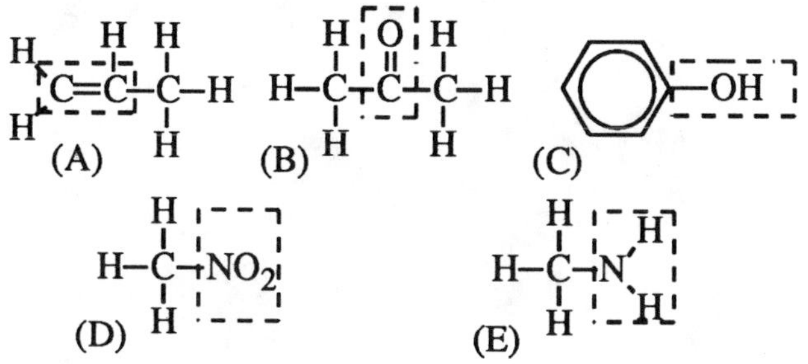

(A) (B) (C)

(D) (E)

11. What is the definition of a functional group?

Properties of Water and Bodies of Water

2.1. WATER: QUALITY, QUANTITY, AND CHEMISTRY

Throughout history, the quality and quantity of water available to humans have been vital factors in determining their well-being. Whole civilizations have disappeared because of water shortages resulting from changes in climate. Even in temperate climates, fluctuations in precipitation cause problems. Devastating droughts in Africa during the 1980s resulted in catastrophic crop failures and starvation. In 1990 floods in northern Texas and southern Oklahoma caused widespread destruction. By the end of 1990, the fifth consecutive year of drought in California had resulted in the imposition of water rationing in that state.

Waterborne diseases such as cholera and typhoid killed millions of people in the past. Some of these diseases still cause great misery in less developed countries. Ambitious programs of dam and dike construction have reduced flood damage, but they have had a number of undesirable side effects in some areas, such as inundation of farmland by reservoirs, and unsafe dams prone to failure. Problems with water supply quantity and quality remain and in some respects are becoming more serious. These problems include increased water use due to population growth, contamination of drinking water by improperly discarded hazardous wastes (see Chapters 17 and 18), and destruction of wildlife by water pollution.

This chapter is concerned with the properties of water and bodies of water. It considers groundwater and water in rivers, lakes, estuaries, and oceans as well as the phenomena that determine the distribution and circulation of chemical species in natural waters. Any consideration of aquatic chemistry requires some understanding of the sources, transport, characteristics, and composition of water. The chemical reactions that occur in water and the chemical species found in it are strongly influenced by the environment in which the water is found. The chemistry of water exposed to the atmosphere is quite different from that of water at the bottom of a lake. Microorganisms play an essential role in determining the chemical composition of water. Thus, in discussing water chemistry, it is necessary to consider the many general factors that influence this chemistry.

2.2. SOURCES, USES, AND CYCLES OF WATER:

The world's water supply is found in the five parts of the **hydrologic cycle** (Figure 2.1). A large portion of the water is found in the oceans. Another fraction is present as water vapor in the atmosphere (clouds). Some water is contained in the solid state as ice and snow in snowpacks, glaciers, and the polar ice caps. Surface water is found in lakes, streams, and reservoirs. Groundwater is located in aquifers underground.

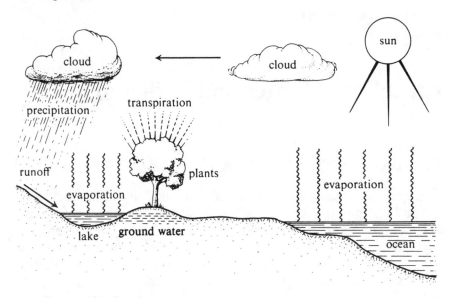

Figure 2.1. The hydrologic cycle.

There is a strong connection between the *hydrosphere,* where water is found, and the *lithosphere*, or land; human activities affect both. For example, disturbance of land by conversion of grasslands or forests to agricultural land or intensification of agricultural production may reduce vegetation cover, decreasing **transpiration** (loss of water vapor by plants) and affecting the microclimate. The result is increased rain runoff, erosion, and accumulation of silt in bodies of water. The nutrient cycles may be accelerated, leading to nutrient enrichment of surface waters. This, in turn, can profoundly affect the chemical and biological characteristics of bodies of water.

The water that humans use is primarily fresh surface water and groundwater. In arid regions, a small fraction of the water supply comes from the ocean, a source that is likely to become more important as the world's supply of fresh water dwindles relative to demand. Saline or brackish groundwaters may also be utilized in some areas.

Groundwater and surface water have appreciably different characteristics. Many substances either dissolve in surface water or become suspended in it on its way to the ocean. Surface water in a lake or reservoir that contains the mineral nutrients essential for algal growth may support a heavy growth of algae. Surface water with a high level of biodegradable organic material, used as food by bacteria, normally contains a large population of bacteria. All these factors have a profound effect upon the quality of surface water.

Groundwater may dissolve minerals from the formations through which it passes. Most microorganisms originally present in groundwater are gradually filtered out as it seeps through mineral formations. Occasionally, the content of undesirable salts may become excessively high in groundwater, although it is generally superior to surface water as a domestic water source.

In the continental United States, an average of approximately 1.48×10^{13} liters of water fall as precipitation each day, an average of 76 cm per year. Of that amount, approximately 1.02×10^{13} liters per day, or 53 cm per year, are lost by evaporation and transpiration. Thus, the water theoretically available for use is approximately 4.4 x 10^{12} liters per day, or only 23 centimeters per year. At present, the U.S. uses 1.6 x 10^{12} liters per day, or 8 centimeters of the average annual precipitation. This amounts to an almost 10-fold increase from a usage of 1.66×10^{11} liters per day at the turn of the century. Even more striking is the per capita increase from about 40 liters per day in 1900 to around 600 liters per day now. Much of this increase is accounted for by high agricultural and industrial use, each of which accounts for approximately 46% of total consumption. Municipal use consumes the remaining 8%.

A major problem with water supply is its non-uniform distribution with location and time. As shown in Figure 2.2, precipitation falls unevenly in the continental U.S. This is a problem because people in areas with low precipitation often consume more water than people in regions with more rainfall. Rapid population growth in the more arid southwestern states of the U.S. during the last four decades has further aggravated the problem. Water shortages are becoming more acute in southwestern U.S., which contains six of the nation's eleven largest cities (Los Angeles, Houston, Dallas, San Diego, Phoenix, and San Antonio). Other problem areas include Florida, where overdevelopment of coastal areas threatens Lake Okeechobee; the Northeast, plagued by deteriorating water systems; and the High Plains, ranging from the Texas panhandle to Nebraska, where irrigation demands on the Ogalla aquifer are dropping the water table steadily, with no hope of recharge. These problems are minor, however, in comparison to those in some parts of Africa where water shortages are contributing to real famine conditions.

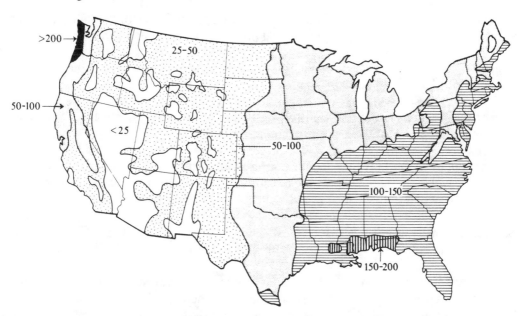

Figure 2.2. Distribution of precipitation in the continental U.S., showing average annual rainfall in centimeters.

Some drastic changes in the pattern of water use appear inevitable. The impact may be particularly severe upon agriculture, as illustrated, for example, by drastically curtailment of irrigation water to California agriculture in 1991. In the southwestern U.S., for example, agriculture accounts for the bulk of total water usage — 85 percent in California, 90 percent in New Mexico, 89 percent in Arizona, and 68 percent in Texas. In some areas industries and municipalities are willing to buy their water at prices up to ten times that paid for irrigation water. The increased cost of water could have marked effects on food prices and availability in the U.S.

Water continues to be the subject of heated disputes among land owners and governmental agencies. The state of South Dakota has protested the release of water from reservoirs in the state to maintain barge traffic on the lower Missouri River. Suggestions to transfer water from Washington, Oregon, or Northern California to meet growing demand in Southern California generate heated discussion.

2.3 . THE PROPERTIES OF WATER, A UNIQUE SUBSTANCE

The study of water is known as **hydrology,** and is divided into a number of subcategories. **Limnology** is the branch of the science dealing with the characteristics of fresh water, including biological properties as well as chemical and physical properties. **Oceanography** is the science of the ocean and its physical and chemical characteristics.

Water has a number of unique properties that are essential to life, many of which are due to water's ability to form hydrogen bonds. These characteristics are summarized in Table 2.1.

Table 2.1. Important Properties of Water

Property	Effects and Significance
Excellent solvent	Transport of nutrients and waste products, making biological processes possible in an aqueous medium
Highest dielectric constant in solution	High solubility of ionic substances and their ionization in solution
Higher surface tension than any other liquid	Controlling factor in physiology; governs drop and surface phenomena
Transparent to visible and longer-wavelength fraction of ultraviolet light	Colorless, allowing light required for photosynthesis to reach considerable depths in bodies of water
Maximum density as a liquid at 4°C	Ice floats; vertical circulation restricted in stratified bodies of water
Higher heat of evaporation than any other material	Determines transfer of heat and water molecules between the atmosphere and bodies of water
Higher latent heat of fusion than any other liquid except ammonia	Temperature stabilized at the freezing point of water
Higher heat capacity than any other liquid except ammonia	Stabilization of temperatures of organisms and geographical regions

Water is an excellent solvent for many materials; thus it is the basic transport medium for nutrients and waste products in life processes. The extremely high dielectric constant of water relative to other liquids has a profound effect upon its solvent properties, in that most ionic materials are dissociated in water. With the exception of liquid ammonia, water has the highest heat capacity of any liquid or solid, 1 cal x g^{-1} x deg^{-1}. Because of this high heat capacity, a relatively large amount of heat is required to change appreciably the temperature of a mass of water; hence, a body of water can have a stabilizing effect upon the temperature of nearby geographic regions. In addition, this property prevents sudden large changes of temperature in large bodies of water and thereby protects aquatic organisms from the shock of abrupt temperature variations. The extremely high heat of vaporization of water, 585 cal/g at 20°C, likewise stabilizes the temperature of bodies of water and the surrounding geographic regions. It also influences the transfer of heat and water vapor between bodies of water and the atmosphere. Water has its maximum density at 4°C, a temperature above its freezing point. The fortunate consequence of this fact is that ice floats, so that few large bodies of water ever freeze solid. Furthermore, the pattern of vertical circulation of water in lakes, a determining factor in their chemistry and biology, is governed largely by the unique temperature-density relationship of water.

2.4. THE CHARACTERISTICS OF BODIES OF WATER

The physical condition of a body of water strongly influences the chemical and biological processes that occur in water. **Surface water** occurs primarily in streams, lakes, and reservoirs. Lakes may be classified as oligotrophic, eutrophic, or dystrophic, an order that often parallels the life of the lake. **Oligotrophic** lakes are deep, generally clear, deficient in nutrients, and without much biological activity. **Eutrophic** lakes have more nutrients, support more life, and are more turbid. **Dystrophic** lakes are shallow, clogged with plant life, and normally contain colored water with a low pH. **Wetlands** are flooded areas in which the water is shallow enough to enable growth of bottom-rooted plants.

Some constructed reservoirs are very similar to lakes, while others differ a great deal from them. Reservoirs with a large volume relative to their inflow and outflow are called **storage reservoirs.** Reservoirs with a large rate of flow-through compared to their volume are called **run-of-the-river reservoirs**. The physical, chemical, and biological properties of water in the two types of reservoirs may vary appreciably. Water in storage reservoirs more closely resembles lake water, whereas water in run-of-the-river reservoirs is much like river water.

Impounding water in reservoirs may have some profound effects upon water quality. These changes result from factors such as different velocities, changed detention time, and altered surface-to-volume ratios relative to the streams that were impounded. Some resulting beneficial changes due to impoundment are a decrease in the level of organic matter, a reduction in turbidity, and a decrease in hardness (calcium and magnesium content). Some detrimental changes are lower oxygen levels due to decreased reaeration, decreased mixing, accumulation of pollutants, lack of bottom scour produced by flowing water scrubbing a stream bottom, and increased growth of algae. Algal growth may be enhanced when suspended solids settle from impounded water, causing increased exposure of the algae to sunlight. Stagnant water in the bottom of a reservoir may be of low quality. Oxygen levels frequently go to almost zero near the bottom, and odorous hydrogen sulfide is produced by the reduction of sulfur compounds in the low oxygen environment. Insoluble iron(III) and manganese(IV) species are reduced to soluble iron(II) and manganese(II) ions which must be removed prior to using the water.

Estuaries constitute another type of body of water, consisting of arms of the ocean into which streams flow. The mixing of fresh and salt water gives estuaries unique chemical and biological properties. Estuaries are the breeding grounds of much marine life, which makes their preservation very important.

Water's unique temperature-density relationship results in the formation of distinct layers within nonflowing bodies of water, as shown in Figure 2.3. During the summer a surface layer (**epilimnion**) is heated by solar radiation and, because of its lower density, floats upon the bottom layer, or **hypolimnion**. This phenomenon is called **thermal stratification.** When an appreciable temperature difference exists between the two layers, they do not mix but behave independently and have very different chemical and biological properties. The epilimnion, which is exposed to light, may have a heavy growth of algae. As a result of exposure to the atmosphere and (during daylight hours) because of the photosynthetic activity of algae, the epilimnion contains relatively higher levels of dissolved oxygen and generally is aerobic. In the hypolimnion, bacterial action on biodegradable organic material may cause the water to become anaerobic. As a consequence, chemical species in a relatively reduced form tend to predominate in the hypolimnion.

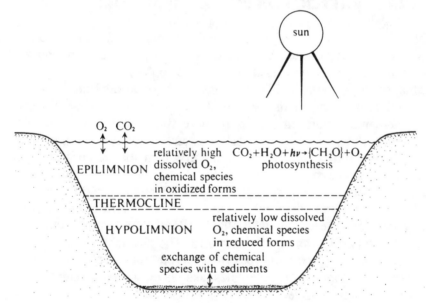

Figure 2.3. Stratification of a lake.

The shear-plane, or layer between epilimnion and hypolimnion, is called the **thermocline**. During the autumn, when the epilimnion cools, a point is reached at which the temperatures of the epilimnion and hypolimnion are equal. This disappearance of thermal stratification causes the entire body of water to behave as a hydrological unit, and the resultant mixing is known as **overturn**. An overturn also generally occurs in the spring. During the overturn, the chemical and physical characteristics of the body of water become much more uniform, and a number of chemical, physical, and biological changes may result. Biological activity may increase from the mixing of nutrients. Changes in water composition during overturn may cause disruption in water-treatment processes.

The chemistry and biology of the Earth's vast oceans are unique because of the ocean's high salt content, great depth, and other factors. Oceanographic chemistry is a discipline in its own right. The environmental problems of the oceans have increased greatly in recent years because of ocean dumping of pollutants, oil spills, and increased utilization of natural resources from the oceans.

2.5. AQUATIC LIFE

The living organisms (**biota**) in an aquatic ecosystem may be classified as either autotrophic or heterotrophic. **Autotrophic** biota utilize solar or chemical energy to fix elements from simple, nonliving inorganic material into complex life molecules that compose living organisms. Algae are typical autotrophic aquatic organisms. Generally, CO_2, NO_3^-, and $H_2PO_4^-$/HPO_4^{2-} are sources of C, N, and P, respectively, for autotrophic organisms. Organisms that utilize solar energy to synthesize organic matter from inorganic materials are called **producers**.

Heterotrophic organisms utilize the organic substances produced by autotrophic organisms as energy sources and as the raw materials for the synthesis of their own biomass. **Decomposers** (or **reducers**) are a subclass of the heterotrophic organisms and consist of chiefly bacteria and fungi, which ultimately break down material of biological origin to the simple compounds originally fixed by the autotrophic organisms.

The ability of a body of water to produce living material is known as its **productivity.** Productivity results from a combination of physical and chemical factors. Water of low productivity generally is desirable for water supply or for swimming. Relatively high productivity is required for the support of fish. Excessive productivity can result in choking by weeds and can cause odor problems. The growth of algae may become quite high in very productive waters, with the result that the concurrent decomposition of dead algae reduces oxygen levels in the water to very low values. This set of conditions is commonly called **eutrophication**.

Life forms higher than algae and bacteria — fish, for example — comprise a comparatively small fraction of the biomass in most aquatic systems. The influence of these higher life forms upon aquatic chemistry is minimal. However, aquatic life is strongly influenced by the physical and chemical properties of the body of water in which it lives. *Temperature, transparency,* and *turbulence* are the three main physical properties affecting aquatic life. Very low water temperatures result in very slow biological processes, whereas very high temperatures are fatal to most organisms. A difference of only a few degrees can produce large differences in the kinds of organisms present. Thermal discharges of hot water from power plants (*cooling water*) frequently kill temperature-sensitive fish while increasing the growth of fish and other species that are adapted to higher temperatures. The transparency of water is particularly important in determining the growth of algae. Thus, turbid water may not be very productive of biomass, even though it has the nutrients, optimum temperature, and other conditions needed. Turbulence is an important factor in mixing and transport processes in water. Some small organisms (**plankton**) depend upon water currents for their own mobility. Water turbulence is largely responsible for the transport of nutrients to living organisms and of waste products away from them. It plays a role in the transport of oxygen, carbon dioxide, and other gases through a body of water and in the exchange of these gases at the water-atmosphere interface. Moderate turbulence is generally beneficial to aquatic life.

Dissolved oxygen (DO) frequently is the key substance in determining the extent and kinds of life in a body of water. Oxygen deficiency is fatal to many aquatic animals such as fish. The presence of oxygen can be equally fatal to many kinds of anaerobic bacteria.

Biochemical oxygen demand, BOD, is another important water-quality parameter. It refers to the amount of oxygen utilized when the organic matter in a given volume of water is degraded biologically. A body of water with a high biochemical oxygen demand, and no means of rapidly replenishing the oxygen, obviously cannot sustain organisms that require oxygen. The importance of BOD is discussed in more detail in Chapter 7.

Carbon dioxide is produced by respiratory processes in waters and sediments and can also enter water from the atmosphere. Carbon dioxide is required for the photosynthetic production of biomass by algae and in some cases is a limiting factor. High levels of carbon dioxide produced by the degradation of organic matter in water can cause excessive algal growth and productivity.

The levels of nutrients in water frequently determine its productivity. Aquatic plant life requires an adequate supply of carbon (CO_2), nitrogen (nitrate), phosphorus (orthophosphate), and trace elements such as iron. In many cases, phosphorus is the limiting nutrient and is generally controlled in attempts to limit excess productivity.

The salinity of water also determines the kinds of life forms present. Irrigation waters may pick up harmful levels of salt. Marine life obviously requires or tolerates salt water, whereas many fresh-water organisms are intolerant of salt.

2.6. AQUATIC CHEMISTRY

This section introduces some aspects of the environmental chemistry of water (aquatic chemistry) that will be expanded upon in subsequent chapters. The study of chemical processes in water is rarely easy. Even under carefully controlled conditions in a laboratory, the investigation of chemical species in water can be very difficult. Generally, the media used in laboratory investigations are made up to constant ionic strength with a relatively inert electrolyte. This minimizes effects arising from differences in factors such as activity coefficients or, when potentiometric measurements are made, liquid junction potentials at the solution-reference electrode interface. The temperature of the medium may be regulated to one hundredth of a degree, or even more accurately if necessary. If an equilibrium constant is to be measured, the system may be allowed to reach equilibrium over a very long period. However, the wide variation among values given for the same constants in the chemical literature attests to the difficulty of describing even relatively simple chemical systems under the most carefully controlled conditions.

Compared to the laboratory, it is much more difficult to describe chemical phenomena in natural water systems. Such systems are very complex and a description of their chemistry must take many variables into consideration. In addition to water, these systems contain mineral phases, gas phases, and organisms. As open, dynamic systems, they have variable inputs and outputs of energy and mass. Therefore, except under unusual circumstances, a true equilibrium condition is not obtained, although an approximately steady-state aquatic system frequently exists. Most metals found in natural waters do not exist as simple hydrated cations in the water, and oxyanions often are found as polynuclear species, rather than as simple monomers. The nature of chemical species in water containing bacteria or algae is strongly influenced by the action of these organisms. Thus, an exact description of the chemistry of a natural water system based upon acid-base, solubility, and complexation equilibrium constants, redox potential, pH, and other chemical parameters is not possible. Therefore, the systems must be described by simplified **models**, often based around equilibrium chemical concepts. Though not exact, nor entirely realistic, such models can yield useful generalizations and insights pertaining to the nature of aquatic chemical processes and provide guidelines for the description and measurement of natural water systems. Though greatly simplified, such models are very helpful in visualizing the conditions that determine chemical species and their reactions in natural waters and wastewaters.

2.7. GASES IN WATER

Dissolved gases — O_2 for fish and CO_2 for photosynthetic algae — are crucial to the welfare of living species in water. Some gases in water can also cause problems, such as the death of fish from bubbles of nitrogen formed in the blood of fish exposed to water supersaturated with N_2. Volcanic carbon dioxide evolved from Lake Nyos in the African country of Cameroon asphyxiated 1,700 people in 1986, and an estimated 300 million cubic meters of the gas had accumulated in the lake as of 1991.[1]

The solubilities of gases in water are calculated with **Henry's Law**, which states that *the solubility of a gas in a liquid is proportional to the partial pressure of that gas in contact with the liquid.* These calculations are discussed in some detail in Chapter 5.

Oxygen in Water

Without an appreciable level of dissolved oxygen, many kinds of aquatic organisms cannot exist in water. Dissolved oxygen is consumed by the degradation of organic matter in water. Many fish kills are caused not from the direct toxicity of pollutants but a deficiency of oxygen because of its consumption in the biodegradation of pollutants.

Most elemental oxygen comes from the atmosphere, which is 20.95% oxygen by volume of dry air. Therefore, the ability of a body of water to reoxygenate itself by contact with the atmosphere is an important characteristic. Oxygen is produced by the photosynthetic action of algae, but this process is really not an efficient means of oxygenating water because some of the oxygen formed by photosynthesis during the daylight hours is lost at night when the algae consume oxygen as part of their metabolic processes. When the algae die, the degradation of their biomass also consumes oxygen.

The solubility of oxygen in water depends upon water temperature, the partial pressure of oxygen in the atmosphere, and the salt content of the water. The calculation of oxygen solubility as a function of partial pressure is discussed in Section 5.3, where it is shown that the concentration of oxygen in water at 25°C in equilibrium with air at atmospheric pressure is only 8.32 mg/L. Thus, water in equilibrium with air cannot contain a high level of dissolved oxygen compared to many other solute species. If oxygen-consuming processes are occurring in the water, the dissolved oxygen level may rapidly approach zero unless some efficient mechanism for the reaeration of water is operative, such as turbulent flow in a shallow stream or air pumped into the aeration tank of an activated sludge secondary waste treatment facility (see Chapter 8). The problem becomes largely one of kinetics, in which there is a limit to the rate at which oxygen is transferred across the air-water interface. This rate depends upon turbulence, air bubble size, temperature, and other factors. It is important to distinguish between *solubility*, which is the maximum dissolved oxygen concentration at equilibrium, and dissolved oxygen *concentration*, which is generally not the equilibrium concentration and is limited by the rate at which oxygen dissolves.

If organic matter of biological origin is represented by the formula $\{CH_2O\}$, the consumption of oxygen in water by the degradation of organic matter may be expressed by the following reaction:

$$\{CH_2O\} + O_2 \rightarrow CO_2 + H_2O \tag{2.7.1}$$

The weight of organic material required to consume the 8.3 mg of O_2 in a liter of water in equilibrium with the atmosphere at 25°C is given by using these figures in a simple stoichiometric calculation based on Equation 2.7.1, which yields a value of 7.8 mg of $\{CH_2O\}$. Thus, the microorganism-mediated degradation of only 7 or 8 mg of organic material can completely consume the oxygen in one liter of water initially saturated with air at 25°C. The depletion of oxygen to levels below those that will sustain aerobic organisms requires the degradation of even less organic matter at higher temperatures (where the solubility of oxygen is less) or in water not initially saturated with atmospheric oxygen. Furthermore, there are no common aquatic chemical reactions that replenish dissolved oxygen; except for oxygen provided by photosynthesis, it must come from the atmosphere.

The temperature effect on the solubility of gases in water is especially important in the case of oxygen. The solubility of oxygen in water decreases from 14.74 mg/L at 0°C to 7.03 mg/L at 35°C. At higher temperatures, the decreased solubility of oxygen, combined with the increased respiration rate of aquatic organisms, frequently causes a condition in which a higher demand for oxygen accompanied by lower solubility of the gas in water results in severe oxygen depletion.

2.8. CARBON DIOXIDE AND CARBONATE SPECIES

Because of carbon dioxide's acidic character, it is much more complicated to calculate the solubility of CO_2 in water than to calculate the solubility of a nonreactive gas like O_2 or N_2. Consideration of the chemical interactions of carbon dioxide in water is required. Calculations of CO_2 solubility are discussed in Sections 3.2 and 5.3.

Carbon dioxide, bicarbonate ion, and carbonate ion have an extremely important influence upon the chemistry of water. Many minerals are deposited as salts of the carbonate ion, CO_3^{2-}. Algae in water utilize dissolved CO_2 in the synthesis of biomass. The equilibrium of dissolved CO_2 with the atmosphere,

$$CO_2(\text{water}) \leftarrow \rightarrow CO_2(\text{atmosphere}) \tag{2.8.1}$$

and equilibrium of CO_3^{2-} ion between aquatic solution and solid carbonate minerals,

$$MCO_3(\text{slightly soluble carbonate salt}) \leftarrow \rightarrow M^{2+} + CO_3^{2+} \tag{2.8.2}$$

have a strong buffering effect upon the pH of water.

Carbon dioxide is only about 0.035% by volume of normal dry air. As a consequence of the low level of atmospheric CO_2, water totally lacking in alkalinity in equilibrium with the atmosphere contains only a very low level of carbon dioxide.

However, the formation of HCO_3^- and CO_3^{2-} greatly increases the solubility of carbon dioxide. High concentrations of free carbon dioxide in water may adversely affect respiration and gas exchange of aquatic animals. It may even cause death and should not exceed levels of 25 mg/L in water.

A large share of the carbon dioxide found in water is a product of the breakdown of organic matter by bacteria. Even algae, which utilize CO_2 in photosynthesis, produce it through their metabolic processes in the absence of light. As water seeps through layers of decaying organic matter while infiltrating the ground, it may dissolve a great deal of CO_2 produced by the respiration of organisms in the soil. Later, as water goes through limestone formations, it dissolves calcium carbonate because of the presence of the dissolved CO_2:

$$CaCO_3(s) + CO_2(aq) + H_2O \longleftrightarrow Ca^{2+} + 2HCO_3^- \tag{2.8.3}$$

This process is the one by which limestone caves are formed.

Although CO_2 in water is often represented as H_2CO_3, the equilibrium constant for the reaction

$$CO_2(aq) + H_2O \longleftrightarrow H_2CO_3(aq) \tag{2.8.4}$$

is only around 2×10^{-3} at 25°C, so just a small fraction of the dissolved carbon dioxide is actually present as the species H_2CO_3. In this text, nonionized carbon dioxide in water will be designated simply as CO_2, which in subsequent discussions will stand for the total of dissolved molecular CO_2 and undissociated H_2CO_3.

Depending on pH, different species predominate in the $CO_2/HCO_3^-/CO_3^{2-}$ system in water. Unionized CO_2 predominates below pH 6, HCO_3^- ion in the pH range from about 6–10 (most natural water), and CO_3^{2-} in highly basic water above pH 10. This concept is discussed in greater detail in Section 3.3 and is illustrated in Figure 3.1.

2.9. ALKALINITY

The capacity of water to accept H^+ ions (protons) is called **alkalinity**. Alkalinity is important in water treatment and in the chemistry and biology of natural waters. Frequently, the alkalinity of water must be known to calculate the quantities of chemicals to be added in treating the water. Highly alkaline water often has a high pH and generally contains elevated levels of dissolved solids. These characteristics may be detrimental for water to be used in boilers, food processing, and municipal water systems. Alkalinity serves as a pH buffer and reservoir for inorganic carbon, thus helping to determine the ability of a water to support algal growth and other aquatic life. It is used by biologists as a measure of water fertility. Generally, the basic species responsible for alkalinity in water are bicarbonate ion, carbonate ion, and hydroxide ion:

$$HCO_3^- + H^+ \rightarrow CO_2 + H_2O \tag{2.9.1}$$

$$CO_3^{2-} + H^+ \rightarrow HCO_3^- \tag{2.9.2}$$

$$OH^- + H^+ \rightarrow H_2O \tag{2.9.3}$$

Other, usually minor, contributors to alkalinity are ammonia and the conjugate bases of phosphoric, silicic, boric, and organic acids.

Alkalinity generally is expressed as *phenolphthalein alkalinity*, corresponding to titration with acid to the pH at which HCO_3^- is the predominant carbonate species (pH 8.3), or *total alkalinity*, corresponding to titration with acid to the methyl orange endpoint (pH 4.3), where both bicarbonate and carbonate species have been converted to CO_2.

It is important to distinguish between high *basicity*, manifested by an elevated pH, and high *alkalinity*, the capacity to accept H^+. Whereas pH is an *intensity* factor, alkalinity is a *capacity* factor. This may be illustrated by comparing a solution of 1.00×10^{-3} M NaOH with a solution of 0.100 M HCO_3^-. The sodium hydroxide solution is quite basic, with a pH of 11, but a liter of it will neutralize only 1.00×10^{-3} mole of acid. The pH of the sodium bicarbonate solution is 8.34, much lower than that of the NaOH. However, a liter of the sodium bicarbonate solution will neutralize 0.100 mole of acid; therefore, its alkalinity is 100 times that of the more basic NaOH solution.

As an example of a water-treatment process in which water alkalinity is important, consider the use of *filter alum*, $Al_2(SO_4)_3 \cdot 18H_2O$ as a coagulant. The hydrated aluminum ion is acidic, and when added to water it reacts with base to form gelatinous aluminum hydroxide,

$$Al(H_2O)_6^{3+} + 3OH^- \rightarrow Al(OH)_3(s) + 6H_2O \qquad (2.9.4)$$

which settles and carries suspended matter with it. This reaction removes alkalinity from the water. Sometimes the addition of more alkalinity is required to prevent the water from becoming too acidic.

In engineering terms, alkalinity frequently is expressed in units of mg/L of $CaCO_3$, based upon the following acid-neutralizing reaction:

$$CaCO_3 + 2H^+ \rightarrow Ca^{2+} + CO_2 + H_2O \qquad (2.9.5)$$

The equivalent weight of calcium carbonate is one-half its formula weight. Expressing alkalinity in terms of mg/L of $CaCO_3$ can, however, lead to confusion, and equivalents/L is preferable notation for the chemist.

2.10. ACIDITY

Acidity as applied to natural water systems is the capacity of the water to neutralize OH^-. Acidic water is not frequently encountered, except in cases of severe pollution. Acidity generally results from the presence of weak acids such as $H_2PO_4^-$, CO_2, H_2S, proteins, fatty acids, and acidic metal ions, particularly Fe^{3+}. Acidity is more difficult to determine than is alkalinity. One reason for the difficulty in determining acidity is that two of the major contributors are CO_2 and H_2S, both volatile solutes which are readily lost from the sample. The acquisition and preservation of representative samples of water to be analyzed for these gases is difficult.

The term *free mineral acid* is applied to strong acids such as H_2SO_4 and HCl in water. Pollutant acid mine water contains an appreciable concentration of free mineral acid. Whereas total acidity is determined by titration with base to the phenolphthalein endpoint (pH 8.2), free mineral acid is determined by titration with base to the methyl orange endpoint (pH 4.3).

The acidic character of some hydrated metal ions may contribute to acidity, for example:

$$Al(H_2O)_6^{3+} + H_2O \leftarrow\rightarrow Al(H_2O)_5OH^{2+} + H_3O^+ \qquad (2.10.1)$$

For brevity in this book, the hydronium ion, H_3O^+, is abbreviated simply as H^+ and proton-accepting water is omitted so that the above equation becomes

$$Al(H_2O)_6^{3+} \leftarrow\rightarrow Al(H_2O)_5OH^{2+} + H^+ \qquad (2.10.2)$$

Some industrial wastes, for example pickling liquor used to remove corrosion from steel, contain acidic metal ions and often some excess strong acid. For such wastes the determination of acidity is important in calculating the amount of lime, or other chemicals, that must be added to neutralize the acid.

2.11. METAL IONS AND CALCIUM IN WATER

The formula of a metal ion in aqueous solution usually is written M^{n+}, which signifies the simple hydrated metal cation $M(H_2O)_x^{n+}$. A bare metal ion, Mg^{2+} for example, cannot exist as a separate entity in water. In order to secure the highest stability of their outer electron shells, metal ions in water are bonded, or *coordinated*, to water molecules or other stronger bases (electron-donor partners) that might be present.

Metal ions in aqueous solution seek to reach a state of maximum stability through chemical reactions. Acid-base, precipitation, complexation, and oxidation-reduction reactions all provide means through which metal ions in water are transformed to more stable forms. The ways in which these reactions affect water chemistry are discussed in later chapters.

Hydrated Metal Ions as Acids

Hydrated metal ions, particularly those with a charge of +3 or more, tend to lose protons in aqueous solution, and fit the definition of Brönsted acids. (Recall that according to the Brönsted definition, acids are proton donors and bases are proton acceptors.) The acidity of a metal ion increases with charge and decreases with increasing radius. As shown by the reaction,

$$Fe(H_2O)_6^{3+} \leftarrow\rightarrow Fe(H_2O)_5OH^{2+} + H^+ \qquad (2.11.1)$$

hydrated iron(III) ion is a relatively strong acid, with K_{a1} of 8.9×10^{-4}. Hydrated trivalent metal ions, such as iron(III), generally are minus at least one hydrogen ion at neutral pH values or above. For tetravalent metal ions, the completely protonated forms, $M(H_2O)_x^{4+}$, is rare even at very low pH values. Commonly, O^{2-} is coordinated to tetravalent metal ions; an example is the vanadium(IV) species, VO^{2+}. Generally, divalent metal ions do not lose a hydrogen ion at pH values below 6, whereas monovalent metal ions such as Na^+ do not act as acids at all in this pH range and exist in water solution as simple hydrated ions.

The tendency of hydrated metal ions to behave as acids may have a profound effect upon the aquatic environment. A good example is *acid mine water* (see Chapter 7), which derives part of its acidic character from the character of hydrated iron(III):

$$Fe(H_2O)_6^{3+} \longleftrightarrow Fe(OH)_3(s) + 3H^+ + 3H_2O \qquad (2.11.2)$$

Hydroxide, OH$^-$, bonded to a metal ion, may function as a bridging group to join two or more metals together through the following dehydration–dimerization:

$$2Fe(H_2O)_5OH^{2+} \longrightarrow (H_2O)_4Fe \underset{\underset{H}{O}}{\overset{\overset{H}{O}}{<}} Fe(H_2O)_4^{4+} + 2H_2O \qquad (2.11.3)$$

Among the metals other than iron(III) forming polymeric species with OH$^-$ as a bridging group are Al(III), Be(II), Bi(III), Ce(IV), Co(III), Cu(II), Ga(III), Mo(V), Pb(II), Sc(II), Sn(IV), and U(VI). Additional hydrogen ions may be lost from water molecules bonded to the dimers, furnishing OH$^-$ groups for further bonding and leading to the formation of polymeric hydrolytic species. If the process continues, colloidal hydroxy polymers are formed, and finally precipitates are produced. This process is thought to be the general one by which hydrated iron(III) oxide, $Fe_2O_3 \cdot x(H_2O)$, or ferric hydroxide, $Fe(OH)_3$, is precipitated from solutions containing iron(III).

Calcium and Hardness

Of the cations found in most fresh-water systems, calcium generally has the highest concentration. The chemistry of calcium, although complicated enough, is simpler than that of the transition metal ions found in water. Calcium is a key element in many geochemical processes, and minerals constitute the primary sources of calcium ion in waters. Among the primary contributing minerals are gypsum, $CaSO_4 \cdot 2H_2O$; anhydrite, $CaSO_4$; dolomite, $CaMg(CO_3)_2$; and calcite and aragonite, which are different mineral forms of $CaCO_3$.

Water containing a high level of carbon dioxide readily dissolves calcium from its carbonate minerals:

$$CaCO_3(s) + CO_2(aq) + H_2O \longleftrightarrow Ca^{2+} + 2HCO_3^- \qquad (2.11.4)$$

When the above equation is reversed and CO_2 is lost from the water, calcium carbonate deposits are formed. The concentration of CO_2 in water determines the extent of dissolution of calcium carbonate. The carbon dioxide that water may gain by equilibration with the atmosphere is not sufficient to account for the levels of calcium dissolved in natural waters, especially groundwaters. Rather, the respiration of microorganisms degrading organic matter in water, sediments, and soil accounts for the very high levels of CO_2 and HCO_3^- observed in water. This is an extremely important factor in aquatic chemical processes and geochemical transformations.

Calcium ion, along with magnesium and sometimes iron(II) ion, accounts for water hardness. The most common manifestation of water hardness is the curdy precipitate formed by soap in hard water. *Temporary hardness* is due to the presence of calcium and bicarbonate ions in water and may be eliminated by boiling the water, thus causing the reversal of Equation 2.11.4:

$$Ca^{2+} + 2HCO_3^- \longleftrightarrow CaCO_3(s) + CO_2(g) + H_2O \tag{2.11.5}$$

Increased temperature may force this reaction to the right by evolving CO_2 gas and a white precipitate of calcium carbonate may form in boiling water having temporary hardness.

The ionic radius of calcium, 0.99 Å, is relatively large compared to some other divalent metal ions found in water. For example, the ionic radius of Mn^{2+} is 0.80 Å and that of Fe^{2+} is 0.75 Å. Thus, the charge density of the Ca^{2+} ion is less than that of these other divalent metal ions. As a consequence, the waters of coordination around Ca^{2+} are less strongly bound, and there is less of a tendency for hydrated Ca^{2+} ions to lose protons. Furthermore, there is relatively little tendency for Ca^{2+} to form complex ions. Under the conditions obtaining in most fresh-water systems, the primary soluble calcium species present is Ca^{2+}. However, at very high levels of HCO_3^-, the ion pair $Ca^{2+}HCO_3^-$ may be present in appreciable amounts. Similarly, in waters of high sulfate content, the soluble ion pair $Ca^{2+}SO_4^{2-}$ is present.

2.12. OTHER CHEMICAL SPECIES IN WATER

Much of the preceding discussion has dealt with dissolved species that are particularly abundant or important in water. A number of other chemical species are present naturally, and are significant, in water. Some of these can also be pollutants and are discussed in later chapters. These chemical species and their sources, behavior, and significance in natural waters are summarized in Table 2.2.

Table 2.2. Chemical Species Commonly Occurring in Water

Substance	Sources	Behavior and significance in water
Aluminum	Aluminum-containing minerals	Occurs as $Al(H_2O)_6^{3+}$ below pH 4.0; loses H^+ to yield $Al(OH)(H_2O)_5^{2+}$ from pH 4.5 to 6.5; forms hydroxy-bridged polymers (see Section 2.8); precipitates as gibbsite, $Al_2O_3 \cdot 3\,H_2O$; amphoteric, forming $Al(OH)_4^-$ above pH 10; forms strong complexes with F^-; precipitates with silica and orthophosphate ions.
Chloride, Cl^-	Minerals, pollution	Does not react chemically with many species in water; harmless at relatively low concentrations; major anion associated with excess salinity at higher levels.
Fluoride, F^-	Minerals, water additive	Forms HF, $pK_a = 3.13$, at low pH; forms insoluble salts with Ca(II), Ba(II), Sr(II), Pb(II); commonly substitutes for OH^-; harmful to bones and teeth above approximately 10 mg/L; prevents tooth decay at levels around 1 mg/L, and is commonly added to water for that purpose.

Substance	Sources	Behavior and significance in water
Iron	Minerals, acid mine water	Occurs as soluble Fe^{2+} under reducing conditions, such as occur in ground water or lake bottom waters; iron(III) must be present as particulate matter or organically bound iron at normal pH's because of the very low solubility of $Fe(OH)_3$ (see Section 2.8); very undesirable solute in water because of formation of $Fe(OH)_3$ deposits; commonly found at levels of 1–10 mg/L in ground waters.
Magnesium	Minerals, such as dolomite, $CaMg(CO_3)_2$	Occurs as Mg^{2+} ion; properties similar to Ca^{2+}, except Mg^{2+} has a much smaller atomic radius of 0.65 Å, holding waters of hydration more strongly; concentrations usually lower than Ca^{2+}, typically 10 mg/L.
Manganese	Minerals	Present as MnO_2 in the presence of oxygen; reduced to soluble Mn^{2+} in ground water and other oxygen-deficient water; low toxicity, but staining tendency of MnO_2 formed by oxidation of Mn^{2+} requires very low levels in municipal water; often precipitates as $MnCO_3$.
Nitrogen	Minerals, decay of nitrogenous organic matter, pollution	Nitrogen species are among the most important species in water. Inorganic nitrogen is present as NO_3^- in the presence, and NH_4^+ in the absence, of oxygen, whereas toxic nitrite, NO_2^-, may be an intermediate form. Nitrate is an algal nutrient that may contribute to excess algal growth. NH_4^+ ion is a weak acid with $pK_a = 9.26$. Unlike NO_3^-, it is strongly bound to soil. Microorganisms catalyze interconversions among various oxidation states of N in water. Organic nitrogen in water is bound to various pollutant organic compounds and biological compounds. (Aspects of nitrogen in water are discussed in Sections 3.12, 4.20, 5.14–5.18, 6.12, 7.11, 7.15, 8.10, 8.19, 10.8, 10.12, and elsewhere in the text.)
Potassium, K^+	Mineral matter, fertilizer runoff, forest fire runoff	May be leached from minerals such as feldspar, $KAlSi_3O_8$; essential plant nutrient; usually occurs at levels of several mg/L, at which it is not a pollutant.
Phosphorus	Minerals, fertilizer runoff, domestic wastes (from detergents)	Occurs in natural waters as anions of orthophosphoric acid, H_3PO_4, $pK_{a1} = 2.17$, $pK_{a2} = 7.31$, $pK_{a3} = 12.36$; the anions $H_2PO_4^-$ and HPO_4^{2-} are predominant in normal water pH ranges; may be present as organic phosphorus; algal nutrient often contributing to excessive algal growth; occurs in natural waters at levels of a few hundredths or tenths of a mg/L.
Silicon	Minerals, such as sodium feldspar albite, $NaAlSi_3O_8$, pollutants	Present in water at normal levels of 1–30 mg/L; occurs as colloidal SiO_2, polynuclear silicate species, such as $Si_4O_6(OH)_6^{2-}$, or silicic acid, H_4SiO_4, $pK_a = 9.46$
Sulfur	Minerals, pollutants, acid mine water, acid rain	Sulfate ion, SO_4^{2-}, predominates under aerobic conditions; hydrogen sulfide, H_2S, is produced in anaerobic waters; H_2S is toxic, but SO_4^{2-} is harmless at moderate levels.
Sodium, Na^+	Minerals, pollution	There are very few reactions by which Na^+ is precipitated or absorbed; no direct harm at lower levels, but higher levels are associated with salt-water pollution, which kills plants; normal levels are several mg/L.

LITERATURE CITED

1. "Warning on Gas in Cameroon Lake," *New York Times*, January 22, 1991, p. B7.

SUPPLEMENTARY REFERENCES

Hammer, Donald A., Ed., *Constructed Wetlands for Wastewater Treatment*, Lewis Publishers, Inc., Chelsea, MI, 1989.

Averett, Robert C., and Diane M. Mcknight, *Chemical Quality of Water and the Hydrologic Cycle*, Lewis Publishers, Chelsea, MI, 1987.

Clesceri, Lenore S., Arnold E. Greenberg, and R. Rhodes Trussell, Eds., *Standard Methods for the Examination of Water and Wastewater*, 17th Edition, American Public Health Association, Washington, DC, 1989.

Hites, Ronald A., and Steven J. Eisenreich, Eds., *Sources and Fates of Aquatic Pollutants*, American Chemical Society, Washington, DC, 1987.

Van der Leeden, Frits, Fred L. Troise, and David Keith Todd, *The Water Encyclopedia*, 2nd ed. Lewis, Chelsea, MI, 1990.

Werner Stumm, Ed., *Aquatic Chemical Kinetics. Reaction Rates of Processes in Natural Waters*, Wiley, New York, 1990.

Laboratory Manual for the Examination of Water, Waste Water, and Soil. H. H. Rump and H. Krist. VCH, New York, 1989.

Faust, S. D., and M. A. Osman, *Chemistry of Natural Waters*, Ann Arbor Science Publishers, Inc., Ann Arbor, Mich, 1981.

Hem, J. D., *Study and Interpretation of the Chemical Characteristics of Natural Water*, 2nd ed., U. S. Geological Survey Paper **1473**, U. S. Geological Survey, Washington, DC, 1970.

Höll, K., *Water: Examination, Assessment, Conditioning, Chemistry, Bacteriology, Biology*, Waltery de Gruyter, Inc., Berlin, 1972.

Jenne, E. A., *Chemical Modeling in Aqueous Systems*, ACS Symposium Series **93**, American Chemical Society, Washington, DC, 1979.

Riley, J. P., and R. Chester, Eds., *Chemical Oceanography*, Academic Press, New York, 1978.

Ross, D. A., *Introduction to Oceanography*, 2nd ed., Prentice-Hall, Inc., Englewood Cliffs, NJ, 1977.

Snoeyink, V. L., and D. Jenkins, *Water Chemistry*, John Wiley and Sons, Inc., New York, 1980.

Stumm, Werner, and James J. Morgan, *Aquatic Chemistry*, 2nd ed., Wiley-Interscience, New York, 1981.

QUESTIONS AND PROBLEMS

1. A sample of groundwater heavily contaminated with soluble inorganic iron is brought to the surface and the alkalinity is determined without exposing the sample to the atmosphere. Why does a portion of such a sample exposed to the atmosphere for some time exhibit a decreased alkalinity?

2. What indirect role is played by bacteria in the formation of limestone caves?

3. Over the long term, irrigation must be carried out so that there is an appreciable amount of runoff, although a much smaller quantity of water would be sufficient to wet the ground. Why must there be some runoff?

4. Of the following, the true statement is that the specific species H_2CO_3 (a) is the predominant form of CO_2 dissolved in water; (b) exists only at pH values above 9; (c) makes up only a small fraction of CO_2 dissolved in water, even at low pH; (d) is not known to exist at all; (e) is formed by the reaction between CO_2 and OH^-.

5. Alkalinity is **not** (a) a measure of the degree to which water can support algal growth, (b) the capacity of water to neutralize acid, (c) a measure of the capacity of water to resist a decrease in pH, (d) a measure of pH, (e) important in considerations of water treatment.

6. Why may oxygen levels become rather low at night in water supporting a heavy growth of algae?

7. What is the molar concentration of O_2 in water in equilibrium with atmospheric air at 25°C?

8. Explain why a solution of $Fe_2(SO_4^{2-})_3$ in water is acidic.

9. How does the ionic radius of ions such as Ca^{2+}, Fe^{2+}, and Mn^{2+} correlate with their relative tendencies to be acidic?

10. An individual measured the pH of a water sample as 11.2 and reported it to be "highly alkaline." Is that statement necessarily true?

Fundamentals of Aquatic Chemistry

3.1. INTRODUCTION

The topic of aquatic chemistry was introduced in Section 2.6. The current chapter expands upon the fundamentals of aquatic chemistry with emphasis upon acid-base and complexation phenomena. Oxidation-reduction reactions and equilibria are discussed in Chapter 4 and details of solubility calculations are given in Chapter 5.

To understand water pollution, it is first necessary to have an appreciation of chemical phenomena that occur in water. These phenomena, in turn, are strongly influenced by the nature of water — including its polar character, tendency to form hydrogen bonds, and ability to hydrate metal ions — as discussed in Chapter 2.

Aquatic environmental chemical phenomena involve processes familiar to chemists, including acid-base, solubility, oxidation–reduction, and complexation reactions. Although most aquatic chemical phenomena are discussed here from the thermodynamic (equilibrium) viewpoint, it is important to keep in mind that kinetics — rates of reactions — are very important in aquatic chemistry. Biological processes play a key role in aquatic chemistry. For example, algae undergoing photosynthesis can raise the pH of water by removing aqueous CO_2, thereby converting HCO_3^- ion to CO_3^{2-} ion, which reacts with Ca^{2+} in water to precipitate $CaCO_3$.

This chapter discusses several kinds of species that are natural to water, but which can be pollutants in excess. For example, water alkalinity is a normal constituent of natural waters required to maintain water fertility, but is toxic to plants and detrimental to water quality at higher levels.

3.2. WATER ACIDITY AND CARBON DIOXIDE IN WATER

Acidity as applied to natural water and wastewater is the capacity of the water to neutralize OH^-; it is analogous to alkalinity, the capacity to neutralize H^+, which is discussed in the next section. Although virtually all water has some alkalinity, acidic water is not frequently encountered, except in cases of severe pollution. Acidity generally results from the presence of weak acids such as $H_2PO_4^-$, CO_2, H_2S, proteins, fatty acids, and acidic metal ions, particularly Fe^{3+}. Acidity is more difficult to determine than is alkalinity. One reason for the difficulty in determining acidity is that two of the major contributors are CO_2, and H_2S, both volatile solutes which are readily lost from the sample so that the acquisition and preservation of representative samples of water to be analyzed for these gases is difficult.

From the pollution standpoint, strong acids are the most important contributors to acidity. The term *free mineral acid* is applied to strong acids such as H_2SO_4 and HCl in water. Acid mine water is a common water pollutant that contains an appreciable concentration of free mineral acid. Whereas total acidity is determined by titration with base to the phenolphthalein endpoint (pH 8.2), free mineral acid is determined by titration with base to the methyl orange endpoint (pH 4.3).

The acidic character of some hydrated metal ions may contribute to acidity, for example:

$$Al(H_2O)_6^{3+} \longleftrightarrow Al(H_2O)_5OH^{2+} + H^+ \tag{3.2.1}$$

Some industrial wastes, such as spent steel pickling liquor, contain acidic metal ions and often some excess strong acid. For such wastes the determination of acidity is important in calculating the amount of lime, or other chemicals, that must be added to neutralize the acid.

Carbon Dioxide in Water

Carbon dioxide, CO_2, is a weak acid in water. Because of the presence of carbon dioxide in air and its production from microbial decay of organic matter, dissolved CO_2 is present in virtually all natural waters and wastewaters. Carbon dioxide is a weak acid so that rainfall from even an absolutely unpolluted atmosphere is slightly acidic due to the presence of dissolved CO_2.

The concentration of gaseous CO_2 in the atmosphere varies with location and season; it is increasing by about 1 part per million (ppm) by volume per year. For purposes of calculation here, the concentration of atmospheric CO_2 will be taken as 350 ppm (0.0350%) in dry air. At 25°C water in equilibrium with unpolluted air containing 350 ppm carbon dioxide has a $CO_2(aq)$ concentration of 1.146×10^{-5} M (see Henry's law calculation of gas solubility in Section 5.3), and this value will be used for subsequent calculations.

Although CO_2 in water is often represented as H_2CO_3, the equilibrium constant for the reaction

$$CO_2(aq) + H_2O \longleftrightarrow H_2CO_3 \tag{3.2.2}$$

is only around 2×10^{-3} at 25°C, so just a small fraction of the dissolved carbon dioxide is actually present as H_2CO_3. In this text, nonionized carbon dioxide in water will be designated simply as CO_2, which in subsequent discussions will stand for the total of dissolved molecular CO_2 and undissociated H_2CO_3.

The CO_2–HCO_3^-–CO_3^{2-} system in water may be described by the equations,

$$CO_2 + H_2O \longleftrightarrow HCO_3^- + H^+ \tag{3.2.3}$$

$$K_{a1} = \frac{[H^+][HCO_3^-]}{[CO_2]} = 4.45 \times 10^{-7} \quad pK_{a1} = 6.35 \tag{3.2.4}$$

$$HCO_3^- + \longleftrightarrow CO_3^{2-} + H^+ \tag{3.2.5}$$

$$K_{a2} = \frac{[H^+][CO_3^{2-}]}{[HCO_3^-]} = 4.69 \times 10^{-11} \quad pK_{a2} = 10.33 \tag{3.2.6}$$

where $pK_a = -\log K_a$. The predominant species formed by CO_2 dissolved in water depends upon pH. This is best shown by a **distribution of species diagram** with pH as a master variable as illustrated in Figure 3.1. Such a diagram shows the major

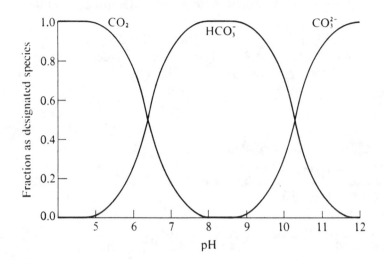

Figure 3.1. Distribution of species diagram for the $CO_2–HCO_3^- –CO_3^{2-}$ system in water.

species present in solution as a function of pH. For CO_2 in aqueous solution, the diagram is a series of plots of the fractions present as CO_2, HCO_3^-, and CO_3^{2-} as a function of pH. These fractions, designated as α_x, are given by the following expressions:

$$\alpha_{CO_2} = \frac{[CO_2]}{[CO_2] + [HCO_3^-] + [CO_3^{2-}]} \tag{3.2.7}$$

$$\alpha_{HCO_3^-} = \frac{[HCO_3^-]}{[CO_2] + [HCO_3^-] + [CO_3^{2-}]} \tag{3.2.8}$$

$$\alpha_{CO_3^{2-}} = \frac{[CO_3^{2-}]}{[CO_2] + [HCO_3^-] + [CO_3^{2-}]} \tag{3.2.9}$$

Substitution of the expressions for K_{a1} and K_{a2} into the α expressions gives the fractions of species as a function of acid dissociation constants and hydrogen ion concentration:

$$\alpha_{CO_2} = \frac{[H^+]^2}{[H^+]^2 + K_{a1}[H^+] + K_{a1}K_{a2}} \tag{3.2.10}$$

$$\alpha_{HCO_3^-} = \frac{K_{a1}[H^+]}{[H^+]^2 + K_{a1}[H^+] + K_{a1}K_{a2}} \tag{3.2.11}$$

$$\alpha_{CO_3^{2-}} = \frac{K_{a1}K_{a2}}{[H^+]^2 + K_{a1}[H^+] + K_{a1}K_{a2}} \tag{3.2.12}$$

Calculations from these expressions show that for pH significantly below pK_{a1}, α_{CO_2} is essentially 1, when $pH = pK_{a1}$, $\alpha_{CO_2} = \alpha_{HCO_3^-}$; when $pH = \frac{1}{2}(pK_{a1} + pK_{a2})$, $\alpha_{HCO_3^-}$ is at its maximum value of 0.98; when $pH = pK_{a2}$, $\alpha_{HCO_3^-} = \alpha_{CO_3^{2-}}$; and for pH significantly below pK_{a2}, $\alpha_{CO_3^{2-}}$ is essentially 1. The distribution of species diagram in Figure 3.1 shows that hydrogen carbonate (bicarbonate) ion (HCO_3^-) is the predominant species in the pH range found in most waters, with CO_2 predominating in more acidic waters.

As mentioned above, the value of $[CO_2(aq)]$ in water at 25°C in equilibrium with air that is 350 ppm CO_2 is 1.146×10^{-5} M. The carbon dioxide dissociates partially in water to produce equal concentrations of H^+ and HCO_3^-:

$$CO_2 + H_2O \xleftarrow{\longrightarrow} HCO_3^- + H^+ \tag{3.2.3}$$

The concentrations of H^+ and HCO_3^- are calculated from K_{a1}:

$$K_{a1} = \frac{[H^+][HCO_3^-]}{[CO_2]} = \frac{[H^+]^2}{1.146 \times 10^{-5}} = 4.45 \times 10^{-7} \tag{3.2.13}$$

$$[H^+] = [HCO_3^-] = (1.146 \times 10^{-5} \times 4.45 \times 10^{-7})^{1/2} = 2.25 \times 10^{-6}$$

$$pH = 5.65$$

This calculation explains why pure water that has equilibrated with the unpolluted atmosphere is slightly acidic with a pH somewhat less than 7.

3.3. ALKALINITY

Alkalinity, the capacity of water to accept H^+ ions, was discussed in Section 2.9, where it was noted that alkalinity in most water is due to the presence of HCO_3^- ion. Based on a knowledge of the acid-base behavior of the CO_2-HCO_3^--CO_3^{2-} system in water, alkalinity may be discussed in more detail. In this discussion it will be assumed that the only contributors to alkalinity are HCO_3^-, CO_3^{2-}, and OH^-.

Natural water typically has an alkalinity, designated here as "[alk]" of 1.00×10^{-3} equivalents per liter (eq/L) meaning that 1 liter of the water will neutralize 1.00×10^{-3} moles of acid. The contributions made by different species to alkalinity depend upon pH. This is shown here by calculation of the relative contributions to alkalinity of HCO_3^-, CO_3^{2-}, and OH^- at pH 7.00 and pH 10.00. First, for water at pH 7.00, $[OH^-]$ is too low to make any significant contribution to the alkalinity. Furthermore, as shown by Figure 3.1, at pH 7.00 $[HCO_3^-] \gg [CO_3^{2-}]$. Therefore, the alkalinity is due to HCO_3^- and $[HCO_3^-] = 1.00 \times 10^{-3}$ M. Substitution into the expression for K_{a1} shows that at pH 7.00 and $[HCO_3^-] = 1.00 \times 10^{-3}$ M, the value of $[CO(aq)]$ is 2.25×10^{-4} M, somewhat higher than the value that arises from water in equilibrium with atmospheric air, but readily reached due to the presence of carbon dioxide from bacterial decay in water and sediments.

Consider next the case of water with the same alkalinity, 1.00×10^{-3} eq/L that has a pH of 10.00. At this higher pH both OH^- and CO_3^{2-} are present at significant concentrations compared to HCO_3^- and the following may be calculated:

$$[alk] = [HCO_3^-] + 2[CO_3^{2-}] + [OH^-] = 1.00 \times 10^{-3} \qquad (3.3.1)$$

The concentration of CO_3^{2-} is multiplied by 2 because each CO_3^{2-} ion can neutralize 2 H^+ ions. The other two equations that must be solved to get the concentrations of HCO_3^-, CO_3^{2-}, and OH^- are

$$[OH^-] = \frac{K_w}{[H^+]} = \frac{1.00 \times 10^{-14}}{1.00 \times 10^{-10}} = 1.00 \times 10^{-4} \qquad (3.3.2)$$

and

$$[CO_3^{2-}] = \frac{K_{a2}[HCO_3^-]}{[H^+]} \qquad (3.3.3)$$

Solving these three equations gives $[HCO_3^-] = 4.64 \times 10^{-4}$ M and $[CO_3^{2-}] = 2.18 \times 10^{-4}$ M. Therefore, the contributions to the alkalinity of this solution are the following:

$$4.64 \times 10^{-4} \text{ eq/L from } HCO_3^-$$
$$2 \times 2.18 \times 10^{-4} = 4.36 \times 10^{-4} \text{ eq/L from } CO_3^{2-}$$
$$\underline{1.00 \times 10^{-4} \text{ eq/L from } OH^-}$$
$$alk = 1.00 \times 10^{-3} \text{ eq/L}$$

The values given above can be used to show that at the same alkalinity value the concentration of total dissolved inorganic carbon, [C],

$$[C] = [CO_2] + [HCO_3^-] + [CO_3^{2-}] \qquad (3.3.4)$$

varies with pH. At pH 7.00,

$$[C]_{pH\ 7} = 2.25 \times 10^{-4} + 1.00 \times 10^{-3} + 0 = 1.22 \times 10^{-3} \qquad (3.3.5)$$

whereas at pH 10.00,

$$[C]_{pH\ 10} = 0 + 2.18 \times 10^{-4} + 4.64 \times 10^{-4} = 6.82 \times 10^{-4} \qquad (3.3.6)$$

The calculation above shows that the dissolved inorganic carbon concentration at pH 10.00 is only about half that at pH 7.00. This is because at pH 10 major contributions to alkalinity are made by CO_3^{2-} ion, each of which has twice the alkalinity of each HCO_3^- ion, and by OH^-, which does not contain any carbon. The lower inorganic carbon concentration at pH 10 shows that the aquatic system can donate dissolved inorganic carbon for use in photosynthesis with a change in pH but none in alkalinity. This pH-dependent difference in dissolved inorganic carbon concentration represents a significant potential source of carbon for algae growing in water which fix carbon by the overall reactions

$$CO_2 + H_2O + hv \rightarrow \{CH_2O\} + O_2 \qquad (3.3.7)$$

and

$$HCO_3^- + H_2O + h\nu \rightarrow \{CH_2O\} + OH^- + O_2 \tag{3.3.8}$$

As dissolved inorganic carbon is used up to synthesize biomass, $\{CH_2O\}$, the water becomes more basic. The amount of inorganic carbon that can be consumed before the water becomes too basic to allow algal reproduction is proportional to the alkalinity. In going from pH 7.00 to pH 10.00 the amount of inorganic carbon consumed from 1.00 L of water having an alkalinity of 1.00×10^{-3} eq/L is

$$[C]_{pH\,7} \times 1\,L - [C]_{pH\,10} \times 1\,L =$$

$$1.22 \times 10^{-3}\,mol - 6.82 \times 10^{-4}\,mol = 5.4 \times 10^{-4}\,mol \tag{3.3.9}$$

This translates to an increase of 5.4×10^{-4} mol/L of biomass. Since the formula mass of $\{CH_2O\}$ is 30, the weight of biomass produced amounts to 16 mg/L. Assuming no input of additional CO_2, at higher alkalinity more biomass is produced for the same change in pH, whereas at lower alkalinity less is produced. Because of this effect, biologists use alkalinity as a measure of water fertility.

The increased solubility of carbon dioxide in water with an elevated alkalinity can be illustrated by comparing its solubility in pure water (alkalinity 0) to its solubility in water initially containing 1.00×10^{-3} M NaOH (alkalinity 1.00×10^{-3} eq/L). The number of moles of CO_2 that will dissolve in a liter of pure water from the atmosphere containing 350 ppm carbon dioxide is

$$Solubility = [CO_2(aq)] + [HCO_3^-] \tag{3.3.10}$$

Substituting values calculated in Section 3.2 gives

$$Solubility = 1.146 \times 10^{-5} + 2.25 \times 10^{-6} = 1.371 \times 10^{-5}\,M$$

The solubility of CO_2 in water initially 1.00×10^{-3} M in NaOH is about 100-fold higher because of uptake of CO_2 by the reaction

$$CO_2(ag) + OH^- \leftrightarrow HCO_3^- \tag{3.3.11}$$

so that

$$Solubility = [CO_2(aq)] + [HCO_3^-]$$
$$= 1.146 \times 10^{-5} + 1.00 \times 10^{-3} = 1.01 \times 10^{-3}\,M \tag{3.3.12}$$

At pH values below 7, $[H^+]$ in water detracts significantly from alkalinity, and its concentration must be subtracted in computing the total alkalinity. Therefore, the following equation is the complete equation for alkalinity in a medium where the only contributors to it are HCO_3^-, CO_3^{2-}, and OH^-:

$$[alk] = [HCO_3^-] + 2[CO_3^{2-}] + [OH^-] - [H^+] \tag{3.3.13}$$

3.4. CALCIUM AND OTHER METALS IN WATER

Of the cations found in most fresh-water systems, calcium generally has the highest concentration. Calcium in water is commonly called **water hardness**. The chemistry of calcium, although complicated enough, is simpler than that of the transition metal ions found in water. Calcium is a key element in many geochemical processes, and minerals constitute the primary sources of calcium ion in waters. Among the major minerals contributing to dissolved calcium in water are gypsum, $CaSO_4 \cdot 2H_2O$; anhydrite, $CaSO_4$; dolomite, $CaCO_3 \cdot MgCO_3$; and calcite and aragonite, which are different mineral forms of $CaCO_3$.

Water containing a high level of carbon dioxide readily dissolves calcium from its carbonate minerals:

$$CaCO_3(s) + CO_2(aq) + H_2O \longleftrightarrow Ca^{2+} + 2HCO_3^- \qquad (3.4.1)$$

When Reaction 3.4.1 is reversed and CO_2 is lost from the water, calcium carbonate deposits are formed. The concentration of CO_2 in water determines the extent of dissolution of calcium carbonate. The carbon dioxide that water may gain by equilibration with the atmosphere is not sufficient to account for the levels of calcium dissolved in natural waters, especially groundwaters. Rather, the respiration of microorganisms degrading organic matter in water, sediments, and soil,

$$\{CH_2O\} + O_2 \rightarrow CO_2 + H_2O \qquad (3.4.2)$$

accounts for the very high levels of CO_2 and HCO_3^- observed in water. This is an extremely important factor in aquatic chemical processes and geochemical transformations.

The equilibrium between dissolved carbon dioxide and calcium carbonate minerals is important in determining several natural water chemistry parameters such as alkalinity, pH, and dissolved calcium concentration (Figure 3.2). For fresh water, the typical figures quoted for the concentrations of both HCO_3^- and Ca^{2+} are 1.00×10^{-3} M. It may be shown that these are reasonable values when the water is in

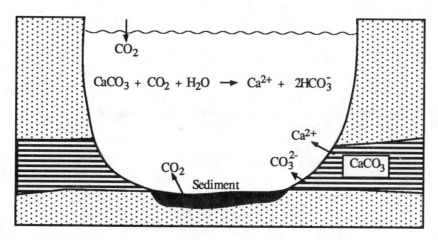

Figure 3.2. Carbon dioxide-calcium carbonate equilibria.

equilibrium with limestone, $CaCO_3$, and with atmospheric CO_2. The concentration of CO_2 in water in equilibrium with air has already been calculated as 1.146×10^{-5} M. The other constants needed to calculate $[HCO_3^-]$ and $[Ca^{2+}]$ are the acid dissociation constant for CO_2:

$$K_{a1} = \frac{[H^+][HCO_3^-]}{[CO_2]} = 4.45 \times 10^{-7} \tag{3.4.3}$$

the acid dissociation constant of HCO_3^-:

$$K_{a2} = \frac{[H^+][CO_3^{2-}]}{[HCO_3^-]} = 4.69 \times 10^{-11} \tag{3.4.4}$$

and the solubility product of calcium carbonate (calcite):

$$K_{sp} = [Ca^{2+}][CO_3^{2-}] = 4.47 \times 10^{-9} \tag{3.4.5}$$

The reaction between calcium carbonate and dissolved CO_2 is

$$CaCO_3(s) + CO_2(aq) + H_2O \xleftarrow{\longrightarrow} Ca^{2+} + 2HCO_3^- \tag{3.4.6}$$

for which the equilibrium expression is the following:

$$K' = \frac{[Ca^{2+}][HCO_3^{2-}]^2}{[CO_2]} = \frac{K_{sp}K_1}{K_{a2}} = 4.24 \times 10^{-5} \tag{3.4.7}$$

The stoichiometry of Reaction 3.4.6 gives a bicarbonate ion concentration that is twice that of calcium. Substitution of the value of CO_2 concentration into the expression for K' yields values of 4.99×10^{-4} M for $[Ca^{2+}]$ and 9.98×10^{-4} for $[HCO_3^-]$. Substitution into the expression for K_{sp} yields 8.96×10^{-6} M for $[CO_3^{2-}]$. When known concentrations are substituted into the product $K_{a1}K_{a2}$,

$$K_{a1}K_{a2} = \frac{[H^+]^2[CO_3^{2-}]}{[CO_2]} = 2.09 \times 10^{-17} \tag{3.4.8}$$

a value of 5.17×10^{-9} M is obtained for $[H^+]$ (pH 8.29). The alkalinity is essentially equal to the bicarbonate ion concentration, which is much higher than that of CO_3^{2-} or OH^-.

To summarize, for water in equilibrium with solid calcium carbonate and atmospheric CO_2, the following concentrations are calculated:

$$[CO_2] = 1.146 \times 10^{-5} \text{ M} \qquad [Ca^{2+}] = 4.99 \times 10^{-4} \text{ M}$$

$$[HCO_3^-] = 9.98 \times 10^{-4} \text{ M} \qquad [H^+] = 5.17 \times 10^{-9} \text{ M}$$

$$[CO_3^{2-}] = 8.96 \times 10^{-6} \text{ M} \qquad \text{pH} = 8.29$$

Factors such as nonequilibrium conditions, high CO_2 concentrations in bottom regions, and increased pH due to algal uptake of CO_2 cause deviations from these values. Nevertheless, they are close to the values found in a large number of natural water bodies.

Metal Species in Water

In Section 2.11 it was noted that metal ions in water, commonly denoted M^{n+}, exist in numerous forms. In order to secure the highest stability of their outer electron shells, metal ions in water are bonded, or *coordinated*, to water molecules in forms such as the hydrated metal cation $M(H_2O)_x^{n+}$, or other stronger bases (electron-donor partners) that might be present. Metal ions in aqueous solution seek to reach a state of maximum stability through chemical reactions including acid-base,

$$Fe(H_2O)_6^{3+} \longleftrightarrow FeOH(H_2O)_5^{2+} + H^+ \tag{3.4.9}$$

precipitation,

$$Fe(H_2O)_6^{3+} \longleftrightarrow Fe(OH)_3(s)^{2+} + 3H_2O + 3H^+ \tag{3.4.10}$$

and oxidation-reduction reactions:

$$Fe(H_2O)_6^{2+} \longleftrightarrow Fe(OH)_3(s)^+ + 3H_2O + e^- + 3H^+ \tag{3.4.11}$$

These all provide means through which metal ions in water are transformed to more stable forms. Because of reactions such as these and the formation of dimeric species, such as $Fe_2(OH)_2^{4+}$, the concentration of simple hydrated $Fe(H_2O)_6^{3+}$ ion in water is vanishingly small; the same holds true for many other ionic species dissolved in water.

The properties of metals dissolved in water depend largely upon the nature of metal species dissolved in the water. Therefore, **speciation** of metals plays a crucial role in their environmental chemistry in natural waters and wastewaters. In addition to the hydrated metal ions, for example, $Fe(H_2O)_6^{3+}$ and hydroxo species such as $FeOH(H_2O)_5^{2+}$ discussed above, metals may exist in water reversibly bound to inorganic anions or to organic compounds as **metal complexes**, or they may be present as **organometallic** compounds containing carbon-to-metal bonds. The solubilities, transport properties, and biological effects of such species are often vastly different from those of the metal ions themselves. Subsequent sections of this chapter consider metal species with an emphasis upon metal complexes. Special attention is given to *chelation*, in which particularly strong metal complexes are formed.

3.5. COMPLEXATION AND CHELATION

A metal ion in water may combine with an electron donor (Lewis base) to form a **complex** or **coordination compound** (or ion). Thus, cadmium ion in water combines with a ligand, cyanide ion, to form a complex ion as shown below:

$$Cd^{2+} + CN^- \longleftrightarrow CdCN^+ \tag{3.5.1}$$

Additional cyanide ligands may be added to form the progressively weaker (more easily dissociated) complexes $Cd(CN)_2$, $Cd(CN)_3^-$, and $Cd(CN)_4^{2-}$.

In this example, the cyanide ion is a **unidentate ligand**, which means that it possesses only one site that bonds to the cadmium metal ion. Complexes of unidentate ligands are of relatively little importance in solution in natural waters. Of considerably more importance are complexes with **chelating agents.** A chelating agent has more than one atom that may be bonded to a central metal ion at one time to form a ring structure. Thus, pyrophosphate ion, $P_4O_7^{4-}$, bonds to two sites on a calcium ion to form a chelate:

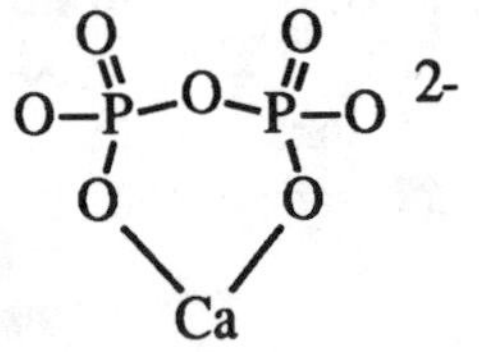

In general, since a chelating agent may bond to a metal ion in more than one place simultaneously, chelates are more stable than complexes involving unidentate ligands. Stability tends to increase with the number of chelating sites available on the ligand.

Structures of metal chelates take a number of different forms, all characterized by rings in various configurations. The structure of a tetrahedrally coordinated chelate of nitrilotriacetate ion is shown in Figure 3.3.

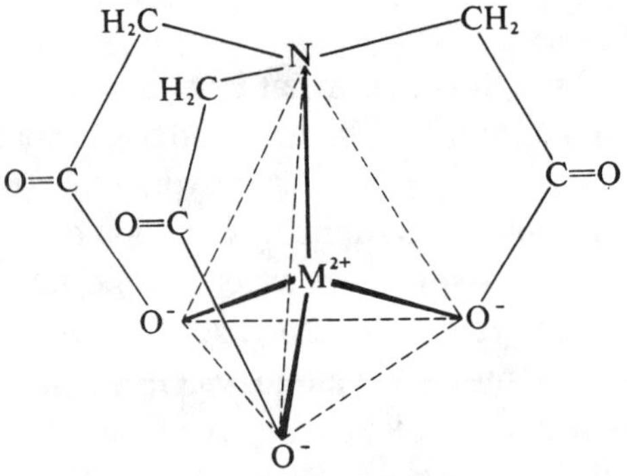

Figure 3.3. Nitrilotriacetate chelate of a divalent metal ion in a tetrahedral configuration.

The ligands found in natural waters and wastewaters contain a variety of functional groups which can donate the electrons required to bond the ligand to a metal ion.[1] Among the most common of these groups are:

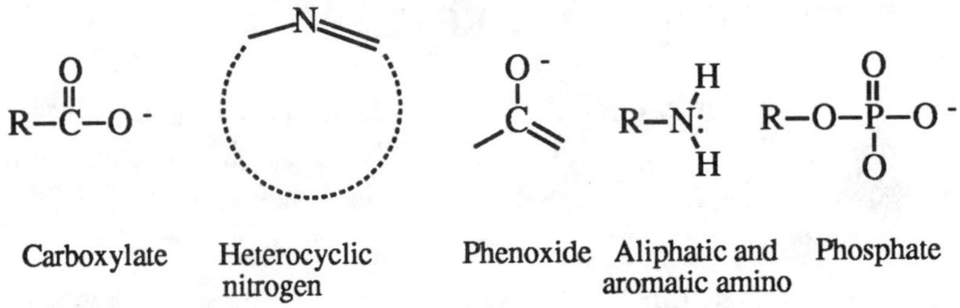

Carboxylate   Heterocyclic   Phenoxide Aliphatic and   Phosphate
               nitrogen   aromatic amino

These ligands complex most metal ions found in unpolluted waters and biological systems (Mg^{2+}, Ca^{2+}, Mn^{2+}, Fe^{2+}, Fe^{3+}, Cu^{2+}, Zn^{2+}, VO^{2+}). They also bind to contaminant metal ions such as Co^{2+}, Ni^{2+}, Sr^{2+}, Cd^{2+}, and Ba^{2+}.

Complexation may have a number of effects, including reactions of both ligands and metals. Among the ligand reactions are oxidation-reduction, decarboxylation, and hydrolysis. Complexation may cause changes in oxidation state of the metal and may result in a metal becoming solubilized from an insoluble compound. The formation of insoluble complex compounds removes metal ions from solution.

Complex compounds of metals such as iron (in hemoglobin) and magnesium (in chlorophyll) are vital to life processes. Naturally occurring chelating agents, such as humic substances and amino acids, are found in water and soil. The high concentration of chloride ion in seawater results in the formation of some chloro complexes. Synthetic chelating agents such as sodium tripolyphosphate, sodium ethylenediaminetetraacetate (EDTA), sodium nitrilotriacetate (NTA), and sodium citrate are produced in large quantities for use in metal-plating baths, industrial water treatment, detergent formulations, and food preparation. Small quantities of these compounds enter aquatic systems through waste discharges.

Occurrence and Importance of Chelating Agents in Water

Chelating agents are common potential water pollutants. These substances can occur in sewage effluent and industrial wastewater such as metal plating wastewater. Chelates formed by the strong chelating agent ethylenediaminetetraacetate (EDTA, structure illustrated at the beginning of Section 3.8) have been shown to greatly increase the migration rates of radioactive ^{60}CO from pits and trenches used by the Oak Ridge National Laboratory in Oak Ridge, Tennessee for disposal of intermediate-level radioactive waste[2]. EDTA was used as a cleaning and solubilizing agent for the decontamination of hot cells, equipment, and reactor components. Analysis of water from sample wells in the disposal pits showed EDTA concentrations of 3.4×10^{-7} M. The presence of EDTA 12-15 years after its burial attests to its low rate of biodegradation. In addition to cobalt, EDTA strongly chelates radioactive plutonium and radioisotopes of Am^{3+}, Cm^{3+}, and Th^{4+}. Such chelates with negative charges are much less strongly sorbed by mineral matter and are vastly more mobile than the unchelated metal ions.

Contrary to the above findings, only very low concentrations of chelatable radioactive plutonium were observed in groundwater near the Idaho Chemical Processing Plant's low-level waste disposal well.[3] No plutonium was observed in wells at any significant distance from the disposal well. The waste processing procedure used was designed to destroy any chelating agents in the waste prior to disposal, and no chelating agents were found in the water pumped from the test wells.

Complexing agents in wastewater are of concern primarily because of their ability to solubilize heavy metals from plumbing and from deposits containing heavy metals. Complexation may increase the leaching of heavy metals from waste disposal sites and reduce the efficiency with which heavy metals are removed with sludge in conventional biological waste treatment. Removal of chelated iron is difficult with conventional municipal water treatment processes. Iron(III) and perhaps several other essential micronutrient metal ions are kept in solution by chelation in algal cultures. The availability of chelating agents may be a factor in determining algal growth. The yellow-brown color of some natural waters is due to naturally-occurring chelates of iron.

3.6. BONDING AND STRUCTURE OF METAL COMPLEXES

This section discusses some of the fundamentals helpful in understanding complexation in water. A **complex** consists of a central atom to which ligands possessing electron-donor properties are bonded. The ligands may be negatively charged or neutral. The resulting complex may be neutral or may have a positive or negative charge. The ligands are said to be contained within the **coordination sphere** of the central metal atom. Depending upon the type of bonding involved, the ligands within the coordination sphere are held in a definite structural pattern. However, in solution, ligands of many complexes exchange rapidly between solution and the coordination sphere of the central metal ion.

The **coordination number** of a metal atom, or ion, is the number of ligand electron-donor groups that are bonded to it. The most common coordination numbers are 2, 4, and 6. Polynuclear complexes contain two or more metal atoms joined together through bridging ligands, frequently OH. An example of such a complex, the dinuclear complex of iron(III),

$$
\begin{array}{c}
\text{H} \\
\text{O} \\
\diagup \quad \diagdown \quad 4+ \\
\text{Fe} \qquad \text{Fe} \\
\diagdown \quad \diagup \\
\text{O} \\
\text{H}
\end{array}
$$

was mentioned in Section 2.11.

The nature and strength of bonds in metal complexes are crucial in determining their behavior and stability. In forming complexes, electron pairs are donated to the central metal ion by the ligand. These electron pairs fill empty orbitals on the metal ion, thus allowing the electron distribution of the metal ion to approach that of a noble gas.

Selectivity and Specificity in Chelation

Although chelating agents are never entirely specific for a particular metal ion, some complicated chelating agents of biological origin approach almost complete specificity for certain metal ions. One example of such a chelating agent is ferrichrome, synthesized by, and extracted from fungi, which forms extremely stable chelates with iron(III). It has been observed that blue-green algae of the *Anabaena* species secrete appreciable quantities of iron-selective hydroxamate chelating agents during periods of heavy algal bloom.[4] These blue-green algae readily take up iron chelated by hydroxamate-chelated iron, whereas some competing green algae, such as *Scenedesmus,* do not. Thus, the chelating agent serves a dual function of promoting the growth of certain blue-green algae while suppressing the growth of other species, allowing the former to predominate.

3.7. CALCULATIONS OF SPECIES CONCENTRATIONS

The stability of complex ions in solution is expressed in terms of **formation constants**. These can be **stepwise formation constants** (K expressions) representing the bonding of individual ligands to a metal ion or **overall formation constants** (β expressions) representing the binding of two or more ligands to a metal ion. These concepts are illustrated for complexes of zinc ion with ammonia by the following:

$$Zn^{2+} + NH_3 \longleftrightarrow ZnNH_3^{2+} \tag{3.7.1}$$

$$K_1 = \frac{[ZnNH_3^{2+}]}{[Zn^{2+}][NH_3]} = 3.9 \times 10^2 \text{ (Stepwise formation constant)} \tag{3.7.2}$$

$$ZnNH_3^{2+} + NH_3 \longleftrightarrow Zn(NH_3)_2^{2+} \tag{3.7.3}$$

$$K_2 = \frac{[Zn(NH_3)_2^{2+}]}{[ZnNH_3^{2+}][NH_3]} = 2.1 \times 10^2 \tag{3.7.4}$$

$$Zn^{2+} + 2NH_3 \longleftrightarrow Zn(NH_3)_2^{2+} \tag{3.7.5}$$

$$\beta_2 = \frac{[Zn(NH_3)_2^{2+}]}{[Zn^{2+}][NH_3]^2} = K_1 K_2 = 8.2 \times 10^4 \text{ (Overall formation constant)} \tag{3.7.6}$$

(For $Zn(NH_3)_3^{2+}$, $\beta_3 = K_1 K_2 K_3$ and for $Zn(NH_3)_4^{2+}$, $\beta_4 = K_1 K_2 K_3 K_4$.)

The following sections show some calculations involving chelated metal ions in aquatic systems. Because of their complexity, the details of these calculations may be beyond the needs of some readers, who may choose to simply consider the results. In addition to the complexation itself, consideration must be given to competition of H^+ for ligands, competition among metal ions for ligands, competition among different ligands for metal ions, and precipitation of metal ions by various precipitants. Not the least of the problems involved in such calculations is the lack of accurately known values of equilibrium constants to be used under the conditions being considered, a factor which can yield questionable results from even the most elegant computerized calculations. Furthermore, kinetic factors are often quite important. Such calculations can be quite useful to provide an overall view of aquatic systems in which complexation is important and as general guidelines to determine areas in which more data should be obtained.

3.8. COMPLEXATION BY DEPROTONATED LIGANDS

In most circumstances, metal ions and hydrogen ions compete for ligands, making the calculation of species concentrations more complicated. Before going into such calculations, however, it is instructive to look at an example in which the ligand has lost all ionizable hydrogen. At pH values of 11 or above, EDTA is essentially all in the completely ionized tetranegative form, Y^{4-}, illustrated below:

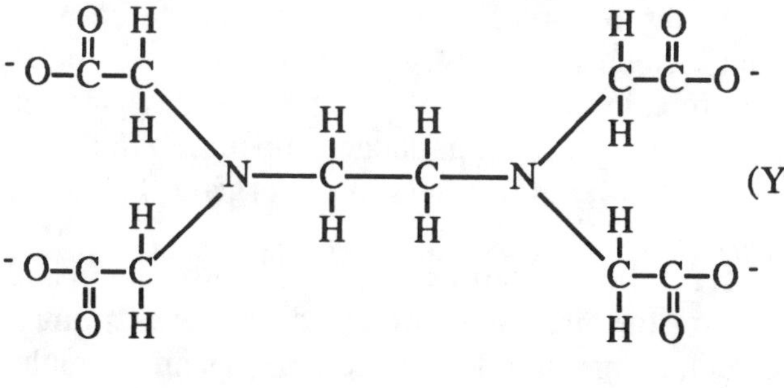

Consider a wastewater with an alkaline pH of 11 containing copper(II) at a total level of 5.0 mg/L and excess uncomplexed EDTA at a level of 200 mg/L (expressed as the disodium salt, $Na_2H_2C_{10}H_{12}O_8N_2 \cdot 2H_2O$, formula weight 372). At this pH uncomplexed EDTA is present as ionized Y^{4-}. The questions to be asked are: Will most of the copper be present as the EDTA complex? If so, what will be the equilibrium concentration of the hydrated copper(II) ion, Cu^{2+}? To answer the former question it is first necessary to calculate the molar concentration of uncomplexed excess EDTA, Y^{4-}. Since disodium EDTA with a formula weight of 372 is present at 200 mg/L (ppm), the total molar concentration of EDTA as Y^{4-} is 5.4×10^{-4} M. The formation constant K_1 of the copper-EDTA complex CuY^{2-} is given by

$$K_1 = \frac{[CuY^{2-}]}{[Cu^{2+}][Y^{4-}]} = 6.3 \times 10^{18} \qquad (3.8.1)$$

The ratio of complexed copper to uncomplexed copper is

$$\frac{[CuY^{2-}]}{[Cu^{2+}]} = [Y^{4-}]K_1 = 5.4 \times 10^{-4} \times 6.3 \times 10^{18} = 3.3 \times 10^{15} \qquad (3.8.2)$$

and therefore, essentially all of the copper is present as the complex ion. The molar concentration of total copper(II) in a solution containing 5.0 mg/L copper(II) is 7.9×10^{-5} M, which in this case is essentially all in the form of the EDTA complex. The very low concentration of uncomplexed, hydrated copper(II) ion is given by

$$[Cu^{2+}] = \frac{[CuY^{2-}]}{K_1[Y^{4-}]} = \frac{7.9 \times 10^{-5}}{6.3 \times 10^{18} \times 5.4 \times 10^{-4}} = 2.3 \times 10^{-20} \qquad (3.8.3)$$

It is seen that in the medium described, the concentration of hydrated copper(II) ion is extremely low compared to total copper(II) ion. Any phenomenon in solution that depends upon the concentration of the hydrated copper(II) ion (such as a physiological effect or an electrode response) would be very different in the medium described, as compared to the effect observed if all of the copper at a level of 5.0 mg/L were present as Cu^{2+} in a more acidic solution and in the absence of complexing agent. The phenomenon of reducing the concentration of hydrated metal ion to very low values through the action of strong chelating agents is one of the most important effects of complexation in natural aquatic systems.

3.9. COMPLEXATION BY PROTONATED LIGANDS

Generally, complexing agents, particularly chelating compounds, are conjugate bases of Brönsted acids. As examples, NH_3 is the conjugate base of the NH_4^+ acid cation, and glycinate anion, $H_2NCH_2CO_2^-$, is the conjugate base of glycine, $^+H_3NCH_2CO_2^-$. Therefore, in many cases hydrogen ion competes with metal ions for a ligand, so that the strength of chelation depends upon pH. In the nearly neutral pH range usually encountered in natural waters, most organic ligands are present in a conjugated acid form.

In order to understand the competition between hydrogen ion and metal ion for a ligand, it is useful to know the distribution of ligand species as a function of pH. Consider nitrilotriacetic acid, commonly designated H_3T, as an example. The trisodium salt of this compound, (NTA) is used as a detergent phosphate substitute and is a strong chelating agent. Biological processes are required for NTA degradation and under some conditions it persists for long times in water. Given the ability of NTA to solubilize and transport heavy metal ions, this material is of considerable environmental concern

Nitrilotriacetic acid, H_3T, loses hydrogen ion in three steps to form the nitrilotriacetate anion, T^{3-}, the structural formula of which is

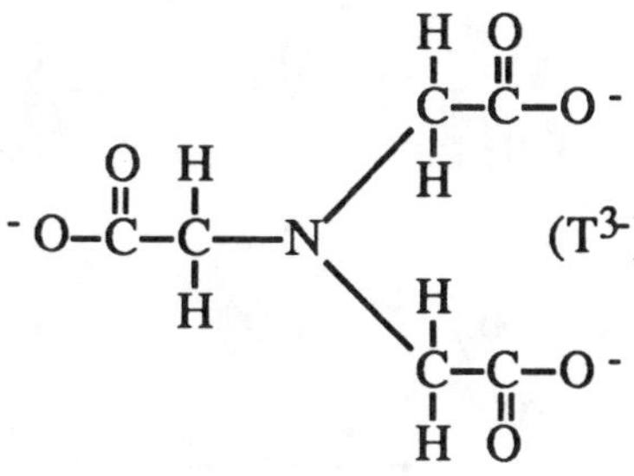

The T^{3-} species may coordinate through three $-CO_2^-$ groups and through the nitrogen atom, as shown in Figure 3.3. Note the similarity of the NTA structure to that of EDTA, discussed in Section 3.8. The stepwise ionization of H_3T is given by the following equilibria:

$$H_3T \;\longleftrightarrow\; H^+ + H_2T^- \tag{3.9.1}$$

$$K_{a1} = \frac{[H^+][H_2T^-]}{[H_3T]} = 2.18 \times 10^{-2} \qquad pK_{a1} = 1.66 \tag{3.9.2}$$

$$H_2T^- \;\longleftrightarrow\; H^+ + HT^{2-} \tag{3.9.3}$$

$$K_{a2} = \frac{[H^+][HT^{2-}]}{[H_2T^-]} = 1.12 \times 10^{-3} \qquad pK_{a2} = 2.95 \tag{3.9.4}$$

$$HT^{2-} \;\longleftrightarrow\; H^+ + T^{3-} \tag{3.9.5}$$

$$K_{a3} = \frac{[H^+][T^{3-}]}{[H_2T^-]} = 5.25 \times 10^{-11} \qquad pK_{a3} = 10.28 \tag{3.9.6}$$

These equilibrium expressions show that uncomplexed NTA may exist in solution as any one of the four species H_3T, H_2T^-, HT^{2-}, or T^{3-}, depending upon the pH of the solution. As was shown for the $CO_2/HCO_3^-/CO_3^{2-}$ system in Section 3.2 and Figure 3.1, fractions of NTA species can be illustrated graphically by a diagram of the distribution-of-species with pH as a master (independent) variable. The key points used to plot such a diagram for NTA are given in Table 3.1 and the plot of fractions of species (α values) as a function of pH is shown in Figure 3.4. Examination of the plot shows that the complexing anion T^{3-} is the predominant species only at relatively high pH values, much higher than usually would be encountered in natural waters. The HT^{2-} species has an extremely wide range of predominance, however, spanning the entire normal pH range of ordinary fresh waters.

Table 3.1. Fractions of NTA Species at Selected pH Values

pH value	α_{H_3T}	$\alpha_{H_2T^-}$	$\alpha_{HT^{2-}}$	$\alpha_{HT^{3-}}$
pH below 1.00	1.00	0.00	0.00	0.00
pH = pK_{a1}	0.49	0.49	0.02	0.00
pH = $^1/_2(pK_{a1} + pK_{a2})$	0.16	0.68	0.16	0.00
pH = pK_{a2}	0.02	0.49	0.49	0.00
pH = $^1/_2(pK_{a2} + pK_{a3})$	0.00	0.00	1.00	0.00
pH = pK_{a3}	0.00	0.00	0.50	0.50
pH above 12	0.00	0.00	0.00	1.00

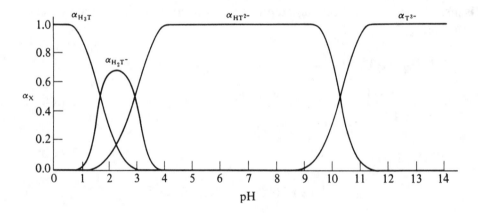

Figure 3.4. Plot of fraction of species α_x as a function of pH for NTA species in water.

3.10. SOLUBILIZATION OF LEAD ION FROM SOLIDS BY NTA

A major concern regarding the widespread introduction of strong chelating agents such as NTA into aquatic ecosystems from sources such as detergents or electroplating wastes is that of possible solubilization of toxic heavy metals from solids through the action of chelating agents. Experimentation is required to determine whether this may be a problem, but calculations are helpful in predicting probable effects. The extent of solubilization of heavy metals depends upon a number of factors, including the stability of the metal chelates, the concentration of the complexing agent in the water, pH, and the nature of the insoluble metal deposit. Several example calculations are given here.

Consider first the solubilization of lead from solid $Pb(OH)_2$ by NTA at pH 8.00. As illustrated in Figure 3.4, essentially all uncomplexed NTA is present as HT^{2-} ion at pH 8.00. Therefore, the solubilization reaction is

$$Pb(OH)_2(s) + HT^{2-} \longleftrightarrow PbT^- + OH^- + H_2O \qquad (3.10.1)$$

which may be obtained by adding the following reactions;

$$Pb(OH)_2(s) \longleftrightarrow Pb^{2+} + 2OH^- \qquad (3.10.2)$$

$$K_{sp} = [Pb^{2+}][OH^-]^2 = 1.61 \times 10^{-20} \qquad (3.10.3)$$

$$HT^{2-} \longleftrightarrow H^+ + T^{3-} \qquad (3.10.4)$$

$$K_{a3} = \frac{[H^+][T^{3-}]}{[HT^{2-}]} = 5.25 \times 10^{-11} \qquad (3.10.5)$$

$$Pb^{2+} + T^{3-} \longleftrightarrow PbT^- \qquad (3.10.6)$$

$$K_f = \frac{[PbT^-]}{[Pb^{2+}][T^{3-}]} = 2.45 \times 10^{11} \qquad (3.10.7)$$

$$H^+ + OH^- \longleftrightarrow H_2O \qquad (3.10.8)$$

$$\frac{1}{K_w} = \frac{1}{[H^+][OH^-]} = \frac{1}{1.00 \times 10^{-14}} \qquad (3.10.9)$$

$$Pb(OH)_2(s) + HT^{2-} \longleftrightarrow PbT^- + OH^- + H_2O \qquad (3.10.1)$$

$$K = \frac{[PbT^-][OH^-]}{[HT^{2-}]} = \frac{K_{sp}K_{a3}K_f}{K_w} = 2.07 \times 10^{-5} \qquad (3.10.10)$$

Assume that a sample of water contains 25 mg/L of $N(CH_2CO_2Na)_3$, the trisodium NTA salt, formula weight 257. The total concentration of both complexed and uncomplexed NTA is 9.7×10^{-5} mmol/mL. Assuming a system in which NTA at pH 8.00 is in equilibrium with solid $Pb(OH)_2$, the NTA may be primarily in the uncomplexed form, HT^{2-}, or in the lead complex, PbT^-. The predominant species may be determined by calculating the ratio of $[PbT^-]$ to $[HT^{2-}]$ from the expression for K, noting that at pH 8.00, $[OH^-] = 1.00 \times 10^{-6}$ M:

$$\frac{[PbT^-]}{[HT^{2-}]} = \frac{K}{[OH^-]} = \frac{2.07 \times 10^{-5}}{1.00 \times 10^{-6}} \qquad (3.10.11)$$

Since $[PbT^-]/[HT^{2-}]$ is approximately 20 to1, most of the NTA in solution is present as the lead chelate. The concentration of PbT^- on a molar basis is just slightly less than the 9.7×10^{-5} mmols/mL total NTA present. The atomic weight of lead is 207 so that the concentration of lead in solution is approximately 20 mg/L. This reaction is pH-dependent such that the fraction of NTA chelated decreases with increasing pH.

Reaction of NTA with Metal Carbonate

Carbonates are common forms of heavy metal ion solids. Solid lead carbonate, $PbCO_3$, is stable within the pH region and alkalinity conditions often found in natural waters and wastewaters. An example similar to one in the preceding section may be worked, assuming that equilibrium is established with $PbCO_3$ rather than with solid $Pb(OH)_2$. In this example it is assumed that 25 mg/L of trisodium NTA is in equilibrium with $PbCO_3$ at pH 7.00 and a calculation is made to determine whether the lead will be complexed appreciably by the NTA. The carbonate ion, CO_3^{2-}, reacts with H^+ to form HCO_3^-. As discussed in Section 3.2, the acid-base equilibrium reactions for the $CO_2/HCO_3^-/CO_3^{2-}$ system are

$$CO_2 + H_2O \leftrightarrow HCO_3^- + H^+ \tag{3.2.3}$$

$$K'_{a1} = \frac{[H^+][HCO_3^-]}{[CO_2]} = 4.45 \times 10^{-7} \quad pK'_{a1} = 6.35 \tag{3.2.4}$$

$$HCO_3^- + \leftrightarrow CO_3^{2-} + H^+ \tag{3.2.5}$$

$$K'_{a2} = \frac{[H^+][CO_3^{2-}]}{[HCO_3^-]} = 4.69 \times 10^{-11} \quad pK'_{a2} = 10.33 \tag{3.2.6}$$

where the acid dissociation constants of the carbonate species are designated as K'_a to distinguish them from the acid dissociation constants of NTA. Figure 3.1 shows that within a pH range of about 7 to 10 the predominant carbonic species is HCO_3^-; therefore, the CO_3^{2-} released by the reaction of NTA with $PbCO_3$ will go into solution as HCO_3^-:

$$PbCO_3(s) + HT^{2-} \leftrightarrow PbT^- + HCO_3^- \tag{3.10.12}$$

This reaction and its equilibrium constant are obtained as follows:

$$PbCO_3(s) \leftrightarrow Pb^{2+} + CO_3^{2-} \tag{3.10.13}$$

$$K_{sp} = [Pb^{2+}][CO_3^{2-}] = 1.48 \times 10^{-13} \tag{3.10.14}$$

$$Pb^{2+} + T^{3-} \leftrightarrow PbT^- \tag{3.10.6}$$

$$K_f = \frac{[PbT^-]}{[Pb^{2+}][T^{3-}]} = 2.45 \times 10^{11} \tag{3.10.7}$$

$$HT^{2-} \leftrightarrow H^+ + T^{3-} \tag{3.10.4}$$

$$K_{a3} = \frac{[H^+][T^{3-}]}{[HT^{2-}]} = 5.25 \times 10^{-11} \tag{3.10.5}$$

$$CO_3^{2-} + H^+ \leftrightarrow HCO_3^- \tag{3.10.15}$$

$$\frac{1}{K'_{a2}} = \frac{[HCO_3^-]}{[CO_3^{2-}][H^+]} = \frac{1}{4.69 \times 10^{-11}} \tag{3.10.16}$$

$$PbCO_3(s) + HT^{2-} \leftrightarrow PbT^- + HCO_3^- \tag{3.10.12}$$

$$K = \frac{[PbT^-][HCO_3^-]}{[HT^{2-}]} = \frac{K_{sp}K_{a3}K_f}{K'_{a2}} = 4.06 \times 10^{-2} \tag{3.10.17}$$

From the expression for K, Equation 3.10.17, it may be seen that the degree to which $PbCO_3$ is solubilized as PbT^- depends upon the concentration of HCO_3^-. Although this concentration will vary appreciably, the figure commonly used to describe natural waters is a bicarbonate ion concentration of 1.00×10^{-3}, as shown in Section 3.4. Using this value the following may be calculated:

$$\frac{[PbT^-]}{[HT^{2-}]} = \frac{K}{[HCO_3^-]} = \frac{4.06 \times 10^{-2}}{1.00 \times 10^{-3}} = 40.6 \tag{3.10.18}$$

Thus, under the given conditions, most of the NTA in equilibrium with solid $PbCO_3$ would be present as the lead complex. As in the previous example, at a trisodium NTA level of 25 mg/L the concentration of soluble lead(II) would be approximately 20 mg/L. At relatively higher concentrations of HCO_3^-, the tendency to solubilize lead would be diminished, whereas at lower concentrations of HCO_3^-, NTA would be more effective in solubilizing lead.

Effect of Calcium Ion upon the Reaction of Chelating Agents with Slightly Soluble Salts

Chelatable calcium ion, Ca^{2+}, which is generally present in natural waters and wastewaters, competes for the chelating agent with a metal in a slightly soluble salt, such as $PbCO_3$. At pH 7.00, the reaction between calcium ion and NTA is

$$Ca^{2+} + HT^{2-} \leftrightarrow CaT^- + H^+ \tag{3.10.19}$$

described by the following equilibrium expression:

$$K' = \frac{[CaT^-][H^+]}{[Ca^{2+}][HT^{2-}]} = 1.48 \times 10^8 \times 5.25 \times 10^{-11} = 7.75 \times 10^{-3} \tag{3.10.20}$$

The value of K' is the product of the formation constant of CaT^-, (1.48×10^8) and K_{a3} of NTA, 5.25×10^{-11}. The fraction of NTA bound as CaT^- depends upon the concentration of Ca^{2+} and the pH. Typically, $[Ca^{2+}]$ in water is 1.00×10^{-3} M. Assuming this value and pH 7.00, the ratio of NTA present in solution as the calcium complex to that present as HT^{2-} is:

$$\frac{[CaT^-]}{[HT^{2-}]} = \frac{[Ca^{2+}]}{[H^+]} K' = \frac{1.00 \times 10^{-3}}{1.00 \times 10^{-7}} \times 7.75 \times 10^{-3} \tag{3.10.21}$$

$$\frac{[CaT^-]}{[HT^{2-}]} = 77.5$$

Therefore, most of the NTA in equilibrium with 1.00×10^{-3} Ca^{2+} would be present as the calcium complex, CaT^-, which would react with lead carbonate as follows:

$$PbCO_3(s) + CaT^- + H^+ \longleftrightarrow Ca^{2+} + HCO_3^- + PbT^- \tag{3.10.22}$$

$$K'' = \frac{[Ca^{2+}][HCO_3^-][PbT^-]}{[CaT^-][H^+]} \tag{3.10.23}$$

Reaction 3.10.22 may be obtained by subtracting Reaction 3.10.19 from Reaction 3.10.12, and its equilibrium constant may be obtained by dividing the equilibrium constant of Reaction 3.10.19 by that of Reaction 3.10.12:

$$PbCO_3(s) + HT^{2-} \longleftrightarrow PbT^- + HCO_3^- \tag{3.10.12}$$

$$K = \frac{[PbT^-][HCO_3^-]}{[HT^{2-}]} = \frac{K_s K_{a3} K_f}{K'_{a2}} = 4.06 \times 10^{-2} \tag{3.10.17}$$

$$-(Ca^{2+} + HT^{2-} \longleftrightarrow CaT^- + H^+) \tag{3.10.19}$$

$$K' = \frac{[CaT^-][H^+]}{[Ca^{2+}][HT^{2-}]} = 7.75 \times 10^{-3} \tag{3.10.20}$$

$$PbCO_3(s) + CaT^- + H^+ \longleftrightarrow Ca^{2+} + HCO_3^- + PbT^- \tag{3.10.22}$$

$$K'' = \frac{K}{K'} = \frac{4.06 \times 10^{-2}}{7.75 \times 10^{-3}} = 5.24 \tag{3.10.24}$$

Having obtained the value of K", it is now possible to determine the distribution of NTA between PbT^- and CaT^-. Thus, for water containing NTA chelated to calcium at pH 7.00, a concentration of HCO_3^- of 1.00×10^{-3}, a concentration of Ca^{2+} of 1.00×10^{-3}, and in equilibrium with solid $PbCO_3$, the distribution of NTA between the lead complex and the calcium complex is:

$$\frac{[PbT^-]}{[CaT^-]} = \frac{[H^+]K''}{[Ca^{2+}][HCO_3^-]} = \frac{1.00 \times 10^{-7} \times 5.24}{1.00 \times 10^{-3} \times 1.00 \times 10^{-3}}$$

It may be seen that only about 1/3 of the NTA would be present as the lead chelate whereas under the identical conditions, but in the absence of Ca^{2+}, approximately all of the NTA in equilibrium with solid $PbCO_3$ was chelated to NTA. Since the fraction of NTA present as the lead chelate is directly proportional to the solubilization of $PbCO_3$, differences in calcium concentration will affect the degree to which NTA solubilizes lead from lead carbonate.

Competing Equilibria between Chloride and NTA Ligands in Seawater

The formation of chloro complexes is an additional consideration when chelation of metal ions is considered in seawater as shown in Figure 3.5 for the chelation of cadmium(II) with NTA. This scheme illustrates the complicated interactions involved in determining the degree of chelation of a heavy metal ion (cadmium) in seawater. It can be seen that the presence of excess chloride results in the formation of chloro complexes; higher concentrations of H^+ break up the cadmium-NTA chelate (CdT^-) by protonation of the ligand; and both Ca^{2+} and Mg^{2+} compete for the NTA ligands.

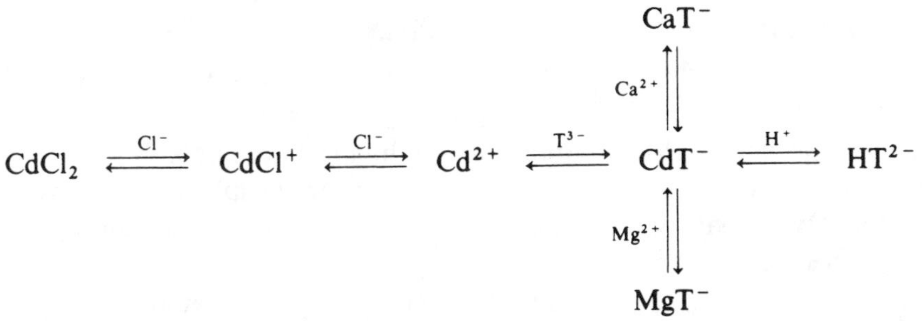

$$
\begin{array}{ccccccccc}
 & & & & & & CaT^- & & \\
 & & & & & & \Big\updownarrow {\scriptstyle Ca^{2+}} & & \\
CdCl_2 & \underset{\longleftarrow}{\overset{Cl^-}{\longrightarrow}} & CdCl^+ & \underset{\longleftarrow}{\overset{Cl^-}{\longrightarrow}} & Cd^{2+} & \underset{\longleftarrow}{\overset{T^{3-}}{\longrightarrow}} & CdT^- & \underset{\longleftarrow}{\overset{H^+}{\longrightarrow}} & HT^{2-} \\
 & & & & & & \Big\updownarrow {\scriptstyle Mg^{2+}} & & \\
 & & & & & & MgT^- & & \\
\end{array}
$$

Figure 3.5. Chelation scheme for cadmium in seawater.

3.11. POLYPHOSPHATES IN WATER

Phosphorus occurs as many oxoanions, anionic forms in combination with oxygen. Some of these are strong complexing agents. Since about 1930, salts of polymeric phosphorus oxoanions have been used increasingly for water treatment, for water softening, and as detergent builders. When used for water treatment, polyphosphates "sequester" calcium ion in a soluble or suspended form. The effect is to reduce the equilibrium concentration of calcium ion and prevent the precipitation of calcium carbonate in installations such as water pipes and boilers. Furthermore, when water is softened properly with polyphosphates, calcium does not form precipitates with soaps or interact detrimentally with detergents.

The simplest form of phosphate is orthophosphate, PO_4^{3-}:

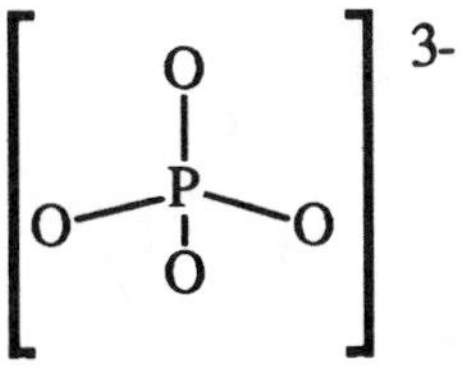

$$
\left[\begin{array}{c} O \\ \| \\ O-\!P\!-\!O \\ | \\ O \end{array} \right]^{3-}
$$

The orthophosphate ion possesses three sites for attachment of H^+. Orthophosphoric acid, H_3PO_4, has a pK_{a1} of 2.17, a pK_{a2} of 7.31, and a pK_{a3} of 12.36. Much of the orthophosphate in natural waters originates from the hydrolysis of polymeric phosphate species.

Pyrophosphate ion, $P_2O_7^{4-}$, is the first of a series of unbranched chain polyphosphates produced by the condensation of orthophosphate:

$$2PO_4^{3-} + H_2O \longleftrightarrow P_2O_7^{4-} + 2OH^- \tag{3.11.1}$$

A long series of linear polyphosphates may be formed, the second of which is triphosphate ion, $P_3O_{10}^{5-}$. These species consist of PO_4 tetrahedra with adjacent tetrahedra sharing a common oxygen atom at one corner. The structural formulas of the acidic forms, $H_4P_2O_7$ and $H_5P_3O_{10}$, are:

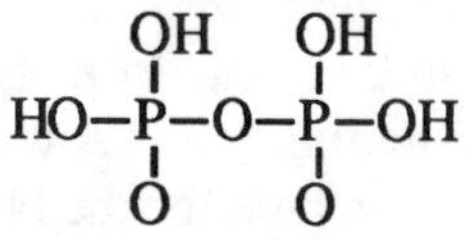

Pyrophosphoric
(diphosphoric) acid

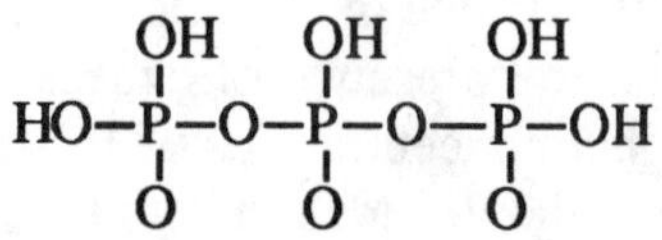

Triphosphoric acid

It is easy to visualize the longer chains composing the higher linear polyphosphates. **Vitreous sodium phosphates** are mixtures consisting of linear phosphate chains with from 4 to approximately 18 phosphorus atoms each. Those with intermediate chain lengths comprise the majority of the species present.

The acid-base behavior of the linear-chain polyphosphoric acids may be explained in terms of their structure by comparing them to orthophosphoric acid. Pyrophosphoric acid, $H_4P_2O_7$, has four ionizable hydrogens. The value of pK_{a1} is quite small (relatively strong acid), whereas pK_{a2} is 2.64, pK_{a3} is 6.76, and pK_{a4} is 9.42. In the case of triphosphoric acid, $H_5P_3O_{10}$, the first two pK_a values are small, pK_{a3} is 2.30, pK_{a4} is 6.50, and pK_{a5} is 9.24. When linear polyphosphoric acids are titrated with base, the titration curve has an inflection at a pH of approximately 4.5 and another inflection at a pH close to 9.5. To understand these phenomena consider the ionization of triphosphoric acid below:

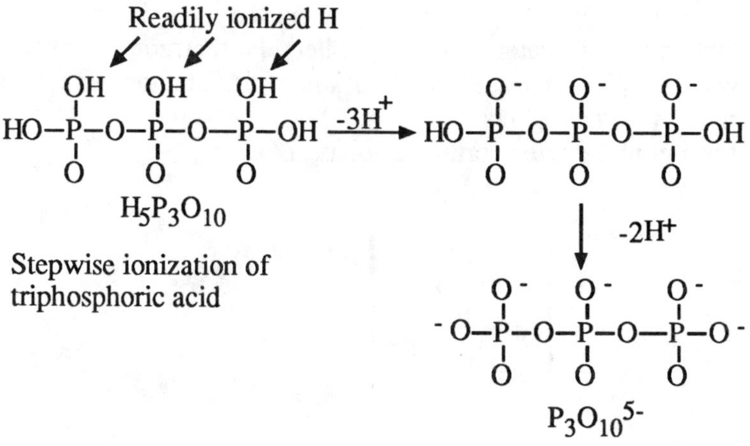

$P_3O_{10}^{5-}$ (3.11.2)

Each P atom in the polyphosphate chain is attached to an -OH group that has one readily ionizable hydrogen that is readily removed in titrating to the first equivalence point. The end phosphorus atoms have two OH groups each. One of the OH groups on an end phosphorus atom has a readily ionizable hydrogen, whereas the other loses its hydrogen much less readily. Therefore, one mole of triphosphoric acid, $H_5P_3O_{10}$, loses three moles of hydrogen ion at a relatively low pH (below 4.5), leaving the $H_2P_3O_{10}^{3-}$ species with two ionizable hydrogens. At intermediate pH values (below

9.5), an additional two moles of "end hydrogens" are lost to form the $P_3O_{10}^{5-}$ species. Titration of a linear-chain polyphosphoric acid up to pH 4.5 yields the number of moles of phosphorus atoms per mole of acid and titration from pH 4.5 to pH 9.5 yields the number of end phosphorus atoms. Orthophosphoric acid, H_3PO_4, differs from the linear chain polyphosphoric acids in that it has a third ionizable hydrogen, which is removed in only extremely basic media.

Ring polyphosphates constitute another type of polyphosphate in which the PO_4 tetrahedron is the basic structural member. The simplest member of this group is trimetaphosphoric acid, $H_3P_3O_{10}$, which has a six-member ring. The next higher member of the series is tetrametaphosphoric acid, $H_4P_4O_{12}$, possessing an eight-member ring. The structural formulas of these species are:

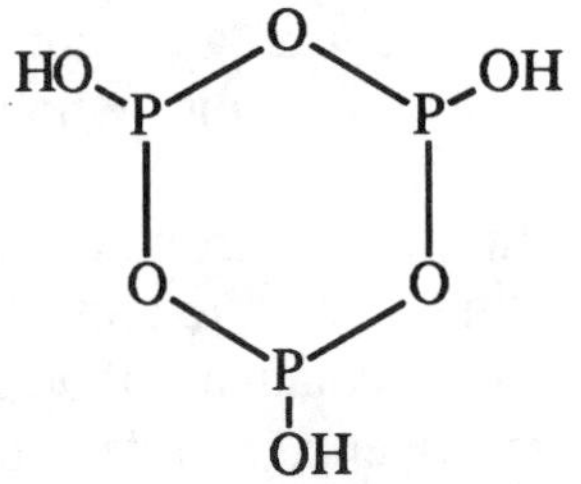

Trimetaphosphoric acid 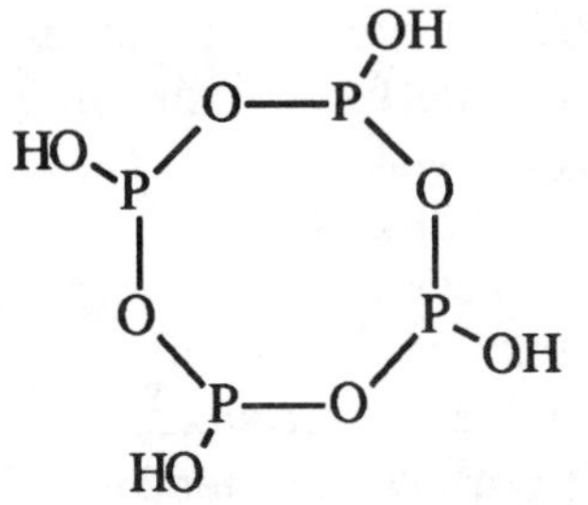   Tetrametaphosphoric acid

The cyclic phosphoric acids do not show two breaks in their titration curves because there are no end hydrogens.

Hydrolysis of Polyphosphates

All of the polymeric phosphates hydrolyze to simpler products in water. The rate of hydrolysis depends upon a number of factors, including pH, and the ultimate product is always some form of orthophosphate. The simplest hydrolytic reaction of a polyphosphate is that of pyrophosphoric acid to orthophosphoric acid:

$$H_4P_2O_7 + H_2O \rightarrow 2H_3PO_4 \tag{3.11.3}$$

Researchers have found evidence that algae and other microorganisms catalyze the hydrolysis of polyphosphates. Even in the absence of biological activity, polyphosphates hydrolyze chemically at a reasonable rate in water so that there is much less concern about the possibility of their transporting heavy metal ions than is the case with organic chelating agents such as NTA or EDTA, which must depend upon microbial degradation for their decomposition.

Complexation by Polyphosphates

In general, chain phosphates are good complexing agents and even form complexes with alkali-metal ions. Ring phosphates form much weaker complexes than do chain species. The different chelating abilities of chain and ring phosphates are due to structural hindrance of bonding by the ring polyphosphates.

3.12. COMPLEXATION BY HUMIC SUBSTANCES

The most important class of complexing agents that occur naturally are the **humic substances**.[5] These are degradation-resistant materials formed during the decomposition of vegetation that occur as deposits in soil, marsh sediments, peat, coal, lignite, or in almost any location where large quantities of vegetation have decayed. They are commonly classified on the basis of solubility. If a material containing humic substances is extracted with strong base, and the resulting solution is acidified, the products are (a) a nonextractable plant residue called **humin**; (b) a material that precipitates from the acidified extract, called **humic acid**; and (c) an organic material that remains in the acidified solution, called **fulvic acid**. Because of their acid-base, sorptive, and complexing properties, both the soluble and insoluble humic substances have a strong effect upon the properties of water. In general, fulvic acid dissolves in water and exerts its effects as the soluble species. Humin and humic acid remain insoluble and affect water quality through exchange of species, such as cations or organic materials, with water.

Humic substances are high-molecular-weight, polyelectrolytic macromolecules. Molecular weights range from a few hundred for fulvic acid to tens of thousands for the humic acid and humin fractions. These substances contain a carbon skeleton with a high degree of aromatic character and with a large percentage of the molecular weight incorporated in functional groups, most of which contain oxygen. The elementary composition of most humic substances is within the following ranges: C, 45-55%; O, 30-45%, H, 3-6%; N, 1-5%; and S, 0-1%. The terms *humin, humic acid,* and *fulvic acid* do not refer to single compounds but to a wide range of compounds of generally similar origin, with many properties in common. Humic substances have been known since before 1800, but their structural and chemical characteristics are still being explained.

Some feeling for the nature of humic substances may be obtained by considering the structure of a hypothetical molecule of fulvic acid below:

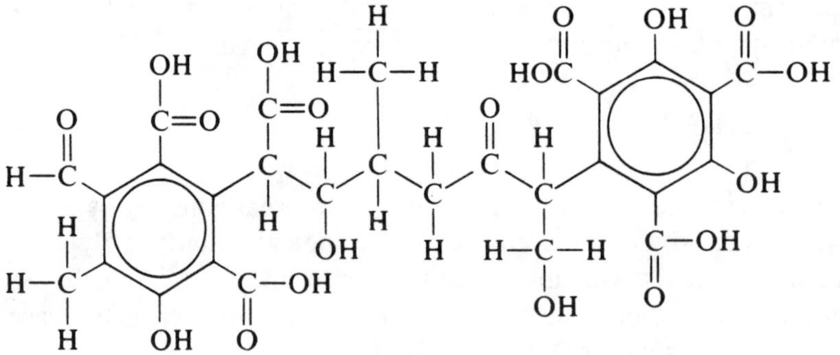

This structure is typical of the type of compound composing fulvic acid. The compound has a formula weight of 666, and its chemical formula may be represented by $C_{20}H_{15}(CO_2H)_6(OH)_5(CO)_2$. As shown in the hypothetical compound, the functional groups that may be present in fulvic acid are carboxyl, phenolic hydroxyl, alcoholic hydroxyl, and carbonyl. The functional groups vary with the particular acid sample. Approximate ranges in units of milliequivalents per gram of acid are: total acidity, 12-14; carboxyl, 8-9; phenolic hydroxyl, 3-6; alcoholic hydroxyl, 3-5; and carbonyl,

1-3. In addition, some methoxyl groups, $-OCH_3$, may be encountered at low levels.

The binding of metal ions by humic substances is one of the most important environmental qualities of humic substances. This binding can occur as chelation between a carboxyl group and a phenolic hydroxyl group, as chelation between two carboxyl groups, or as complexation with a carboxyl group (see below):

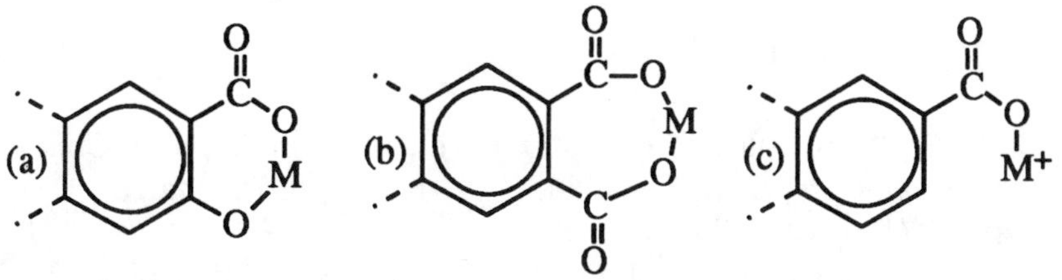

Figure 3.6. Binding of a metal ion, M^{2+}, by humic substances (a) by chelation between carboxyl and phenolic hydroxyl, (b) by chelation between two carboxyl groups, and (c) by complexation with a carboxyl group.

Iron and aluminum are very strongly bound to humic substances, whereas magnesium is rather weakly bound. Other common ions, such as Ni^{2+}, Pb^{2+}, Ca^{2+}, and Zn^{2+}, are intermediate in their binding to humic substances.

The role played by soluble fulvic-acid complexes of metals in natural waters is not well known. They probably keep some of the biologically important transition-metal ions in solution and are particularly involved in iron solubilization and transport. Fulvic acid-type compounds are associated with color in water. These yellow materials, called **Gelbstoffe**, frequently are encountered along with soluble iron.

Insoluble humic substances, the humins and humic acids, effectively exchange cations with water and may accumulate large quantities of metals. Lignite coal, which is largely a humic-acid material, tends to remove some metal ions from water.

Special attention has been given to humic substances since about 1970, following the discovery of **trihalomethanes** (THMs, such as chloroform and dibromochloromethane) in water supplies. It is now generally believed that these suspected carcinogens can be formed in the presence of humic substances during the disinfection of raw municipal drinking water by chlorination (see Chapter 8). The humic substances produce THMs by reaction with chlorine. The formation of THMs can be reduced by removing as much of the humic material as possible prior to chlorination.

3.13. COMPLEXATION AND REDOX PROCESSES

Complexation may have a strong effect upon equilibria by shifting reactions, such as that for the oxidation of lead,

$$Pb \leftarrow\rightarrow Pb^{2+} + 2e^- \qquad (3.13.1)$$

strongly to the right by binding to the product ion, thus cutting its concentration down to very low levels. Of perhaps more importance is the fact that upon oxidation,

$$M + \tfrac{1}{2}O_2 \leftarrow\rightarrow MO \qquad (3.13.2)$$

many metals form self-protective coatings of oxides, carbonates, or other insoluble species, which prevent further chemical reaction. Copper and aluminum roofing and structural iron are examples of materials which are thus self-protecting. A chelating agent in contact with such metals can result in continual dissolution of the protective coating so that the exposed metal corrodes readily. For example, chelating agents in wastewater may increase the corrosion of metal plumbing, thus adding heavy metals to effluents. Solutions of chelating agents employed to clean metal surfaces in metal plating operations have a similar effect.

LITERATURE CITED

1. Martell, A. E., "Principles of Complex Formation," in *Organic Compounds in Aquatic Environments*, S. D. Faust and J. V. Hunter, Eds., Marcell Dekker, Inc., New York, 1971, pp. 262–392.
2. Means, J. L., D. A. Crerar, and J. O. Duguid, "Migration of Radioactive Wastes: Radionuclide Mobilization by Complexing Agents," *Science*, **200**, 1978, pp. 1477-81.
3. Cleveland, KJ. M. and T. F. Rees, "Characterization of Plutonium in Ground Water near the Idaho Chemical Processing Plant," *Environmental Science and Technology*, **16**, 1982, pp. 437–439.
4. Murphy, T. P., D. R. S. Lean, and C. Nalewajko, "Blue–Green Algae: Their Excretion of Iron–Selective Chelators Enables them to Dominate other Algae," *Science*, **192**, 1976, pp. 900-2,
5. Manahan, Stanley E., "Humic Substances and the Fates of Hazardous Waste Chemicals," Chapter 6 in *Influence of Aquatic Humic Substances on Fate and Treatment of Pollutants*, Advances in Chemistry Series **219**, American Chemical Society, Washington, DC, 1989, pp. 83-92.

SUPPLEMENTARY REFERENCES

Suffet, I. H., and Patrick MacCarthy, Eds., *Aquatic Humic Substances: Influence on Fate and Treatment of Pollutants*, Advances in Chemistry Series **219**, American Chemical Society, Washington, DC, 1989.

Trace Inorganics in Water, ACS Advances in Chemistry Series No. **73**, American Chemical Society, Washington, DC, 1968.

Butler, J. N., *Ionic Equilibrium — a Mathematical Approach*, Addison-Wesley Publishing Co, Reading, Mass., 1964.

Hem, J. D., *Study and Interpretation of the Chemical Characteristics of Natural Water*, 2nd ed., U.S. Geological Survey Paper **1473**, Washington, DC, 1970.

Singer, Philip C., *Trace Metals and Metal-Organic Interactions in Natural Waters*, Ann Arbor Science Publishers, Inc., Ann Arbor, Mich., 1973.

Faust, Samuel D., and Hunter, Joseph V., *Organic Compounds in Aquatic Environments*, Marcel Dekker, Inc., 1971.

Stumm, Werner, "Metal Ions in Aqueous Solutions," in *Principles and Applications of Water Chemistry*, Samuel D. Faust and Joseph V. Hunter, Eds., John Wiley and Sons, Inc., New York, 1971, pp. 529-60

QUESTIONS AND PROBLEMS

1. Alkalinity is determined by titration with standard acid. The alkalinity is often expressed as mg/L of $CaCO_3$. If V_p mL of acid of normality N are required to titrate V_s mL of sample to the phenolphthalein endpoint, what is the formula for the phenolphthalein alkalinity as mg/L of $CaCO_3$?

2. Exactly 100 pounds of cane sugar (dextrose), $C_{12}H_{22}O_{11}$, were accidentally discharged into a small stream saturated with oxygen from the air at 25°C. How many liters of this water could be contaminated to the extent of removing all the dissolved oxygen by biodegradation?

3. Water with an alkalinity of 2.00×10^{-3} equivalents/liter has a pH of 7.00. Calculate $[CO_2]$, $[HCO_3^-]$, $[CO_3^{2-}]$, and $[OH^-]$.

4. Through the photosynthetic activity of algae, the pH of the water in problem 3 was changed to 10.00. Calculate all the preceding concentrations and the weight of biomass, $\{CH_2O\}$, produced. Assume no input of atmospheric CO_2.

5. Calcium chloride is quite soluble, whereas the solubility product of calcium flouride, CaF_2, is only 3.9×10^{-11}. A waste stream of 1.00×10^{-3} M HCl is injected into a formation of limestone, $CaCO_3$, where it comes into equilibrium. Give the chemical reaction that occurs and calculate the hardness and alkalinity of the water at equilibrium. Do the same for a waste stream of 1.00×10^{-3} M HF.

6. For a solution having 1.00×10^{-3} equivalents/liter total alkalinity (contributions from HCO_3^-, CO_3^{2-}, and OH^-) at $[H^+] = 4.69 \times 10^{-11}$, what is the percentage contribution to alkalinity from CO_3^{2-}?

7. A wastewater disposal well for carrying various wastes at different times is drilled into a formation of limestone ($CaCO_3$), and the wastewater has time to come to complete equilibrium with the calcium carbonate before leaving the formation through an underground aquifer. Of the following components in the wastewater, the one that would not cause an increase in alkalinity due either to the component, itself, or to its reaction with limestone, is (a) NaOH, (b) CO_2, (c) HF, (d) HCl, (e) all of the preceding would cause an increase in alkalinity.

8. Calculate the ratio $[PbT^-]/[HT^{2-}]$ for NTA in equilibrium with $PbCO_3$ in a medium having $[HCO_3^-] = 3.00 \times 10^{-3}$ M.

9. If the medium in problem 8 contained excess calcium such that the concentration of uncomplexed calcium, $[Ca^{2+}]$, were 5.00×10^{-3} M, what would be the ratio $[PbT^-]/[CaT^-]$ at pH 7?

10. A wastewater stream containing 1.00×10^{-3} M disodium NTA, Na_2HT, as the only solute is injected into a limestone ($CaCO_3$) formation through a waste disposal well. After going through this aquifer for some distance and reaching equilibrium, the water is sampled through a sampling well. What is the reaction between NTA species and $CaCO_3$? What is the equilibrium constant for the reaction? What are the equilibrium concentrations of CaT^-, HCO_3^-, and HT^{2-}? (The appropriate constants may be looked up in this chapter 3.)

11. If the wastewater stream in problem 10 were 0.100 M in NTA and contained other solutes that exerted a buffering action such that the final pH were 9.00, what would be the equilibrium value of HT^{2-} concentration in moles/liter?

12. Exactly 1.00×10^{-3} mole of $CaCl_2$, 0.100 mole of NaOH, and 0.100 mole of Na_3T were mixed and diluted to 1.00 liter. What was the concentration of Ca^{2+} in the resulting mixture?

13. How does chelation influence corrosion?

14. The following ligand has more than one site for binding to a metal ion. How many such sites does it have?

15. If a solution containing initially 25 mg/L trisodium NTA is allowed to come to equilibrium with solid $PbCO_3$ at pH 8.50 in a medium that contains 1.76×10^{-3} M HCO_3^- at equilibrium, what is the value of the ratio $[PbT^-]/[HT^{2-}]$?

16. After a low concentration of NTA has equilibrated with $PbCO_3$ at pH 7.00, in a medium having $[HCO_3^-] = 7.50 \times 10^{-4}$ M, what is the ratio $[PbT^-]/[HT^{2-}]$?

17. What detrimental effect may dissolved chelating agents have upon conventional biological waste treatment?

18. Why is chelating agent usually added to artificial algal growth media?

19. What common complex compound of magnesium is essential to certain life processes?

20. What is always the ultimate product of polyphosphate hydrolysis?

21. A solution containing initially 1.00×10^{-5} M CaT^- is brought to equilibrium with solid $PbCO_3$. At equilibrium, pH = 7.00, $[Ca^{2+}] = 1.50 \times 10^{-3}$ M, and $[HCO_3^-] = 1.10 \times 10^{-3}$ M. At equilibrium, what is the fraction of total NTA in solution as PbT^-?

22. What is the fraction of NTA present after HT^{2-} has been brought to equilibrium with solid $PbCO_3$ at pH 7.00, in a medium in which $[HCO_3^-] = 1.25 \times 10^{-3}$ M?

Oxidation-Reduction

4.1. THE SIGNIFICANCE OF OXIDATION-REDUCTION PHENOMENA

Oxidation-reduction (redox) reactions are those involving changes of oxidation states of reactants. Such reactions are easiest to visualize as the transfer of electrons from one species to another. For example, soluble cadmium ion, Cd^{2+}, is removed from wastewater by reaction with metallic iron. The overall reaction is

$$Cd^{2+} + Fe \rightarrow Cd + Fe^{2+} \tag{4.1.1}$$

This reaction is the sum of two **half-reactions**, a reduction half-reaction in which cadmium ion accepts two electrons and is reduced,

$$Cd^{2+} + 2e^- \rightarrow Cd \tag{4.1.2}$$

and an oxidation half-reaction in which elemental iron is oxidized:

$$Fe \rightarrow Fe^{2+} + 2e^- \tag{4.1.3}$$

When these two half-reactions are added algebraically, the electrons cancel on both sides and the result is the overall reaction given in Equation 4.1.1.

Oxidation-reduction phenomena are highly significant in the environmental chemistry of natural waters and wastewaters. The reduction of oxygen by organic matter in a lake,

$$\{CH_2O\}_{(oxidized)} + O_{2(reduced)} \rightarrow CO_2 + H_2O \tag{4.1.4}$$

results in oxygen depletion which can be fatal to fish. The rate at which sewage is oxidized is crucial to the operation of a waste-treatment plant. Reduction of insoluble iron(III) to soluble iron(II),

$$Fe(OH)_3(s) + 3H^+ + e^- \rightarrow Fe^{2+} + 3H_2O \tag{4.1.5}$$

in a reservoir contaminates the water with iron, which is hard to remove in the water-treatment plant. Oxidation of NH_4^+ to NO_3^- in water,

$$NH_4^+ + 2O_2 \rightarrow NO_3^- + 2H^+ + H_2O \tag{4.1.6}$$

is essential for getting the ammonium nitrogen into a form assimilable by algae in the water. Many other examples can be cited of the ways in which the types, rates, and equilibria of redox reactions largely determine the nature of important solute species in water.

This chapter discusses redox processes and equilibria in water. In so doing it emphasizes the concept of pE, analogous to pH and defined as the negative log of electron activity. Low pE values are indicative of reducing conditions and high pE values reflect oxidizing conditions.

Two important points should be stressed regarding redox reactions in natural waters and wastewaters. First, as is discussed in Chapter 6, "Aquatic Biochemistry," many of the most important redox reactions are catalyzed by microorganisms. Bacteria are the catalysts by which molecular oxygen reacts with organic matter, iron(III) is reduced to iron(II), and ammonia is oxidized to nitrate ion.

The second important point regarding redox reactions in the hydrosphere is their close relationship to acid-base reactions. Whereas the activity of the hydrogen ion, H^+, is used to express the extent to which water is acidic or basic, the activity of the electron, e^-, is used to express the degree to which an aquatic medium is oxidizing or reducing. Water with a high hydrogen ion activity, such as runoff from "acid rain", is *acidic*. By analogy, water with a high *electron* activity, such as that in the anaerobic digester of a sewage-treatment plant, is said to be *reducing*. Water with a low H^+ ion activity (high concentration of OH^-) — such as landfill leachate contaminated with waste sodium hydroxide — is *basic*, whereas water with a low electron activity — highly chlorinated water, for example — is said to be *oxidizing*. Actually, neither free electrons nor free H^+ ions as such are found dissolved in aquatic solution; they are always strongly associated with solvent or solute species. However, the concept of electron activity, like that of hydrogen ion activity, remains a very useful one to the aquatic chemist.

Many species in water undergo exchange of both electrons and H^+ ions. For example, acid mine water contains the hydrated iron(III) ion, $Fe(H_2O)_6^{3+}$, which readily loses H^+ ion

$$Fe(H_2O)_6^{3+} \longleftrightarrow Fe(H_2O)_5OH^{2+} + H^+ \qquad (4.1.7)$$

to contribute acidity to the medium. The same ion accepts an electron

$$Fe(H_2O)_6^{3+} + e^- \longleftrightarrow Fe(H_2O)_6^{2+} \qquad (4.1.8)$$

to give iron(II)

Generally, the transfer of electrons in a redox reaction is accompanied by H^+ ion transfer, and there is a close relationship between redox and acid-base processes. For example, if iron(II) loses an electron at pH 7, three hydrogen ions are also lost to form highly insoluble ferric hydroxide,

$$Fe(H_2O)_6^{2+} \longleftrightarrow e^- + Fe(OH)_3(s) + 3H_2O + 3H^+ \qquad (4.1.9)$$

an insoluble gelatinous solid.

The stratified body of water shown in Figure 4.1 can be used to illustrate redox phenomena and relationships in an aquatic system. The anaerobic sediment layer is so reducing that carbon can be produced in the -IV oxidation state as CH_4. If the lake becomes anaerobic, the hypolimnion may contain elements in their reduced states:

NH_4^+ for nitrogen, H_2S for sulfur, and soluble $Fe(H_2O)_6^{2+}$ for iron. Saturated with atmospheric oxygen, the surface layer may be a relatively oxidizing medium. If allowed to reach thermodynamic equilibrium, it is characterized by the more oxidized forms of the elements present: CO_2 for carbon, NO_3^- for nitrogen, iron as insoluble $Fe(OH)_3$, and sulfur as SO_4^{2-}. Substantial changes in the distribution of chemical species in water resulting from redox reactions are vitally important to aquatic organisms and have tremendous influence on water quality.

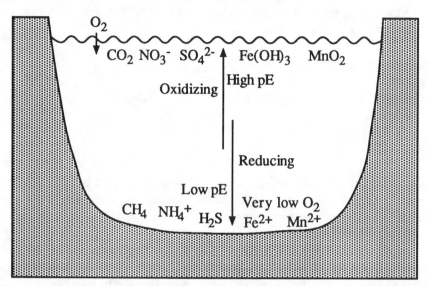

Figure 4.1. Predominance of various chemical species in a stratified body of water that has a high oxygen concentration (oxidizing, high pE) near the surface and a low oxygen concentration (reducing, low pE) near the bottom.

It should be pointed out that the systems presented in this chapter are assumed to be at equilibrium, a state almost never achieved in any real natural water or wastewater system. Most real aquatic systems are dynamic systems that may approach a steady-state, rather than true equilibrium. Nevertheless, the picture of a system at equilibrium is very useful in visualizing trends in natural water and wastewater systems, yet the model is still simple enough to comprehend. It is important to realize the limitations of such a model, however, especially in trying to make measurements of the redox status of water.

4.2. THE ELECTRON AND REDOX REACTIONS

In order to explain redox processes in natural waters it is necessary to have an understanding of redox reactions. In a formal sense such reactions can be viewed as the transfer of electrons between species. This section considers such reactions in a simple system. All redox reactions involve changes in the oxidation states of some of the species that take part in the reaction. Consider, for example, a solution containing iron(II) and iron(III) that is sufficiently acidic to prevent precipitation of solid $Fe(OH)_3$; such a medium might be acid mine water or a steel pickling liquor waste. Suppose that the solution is treated with elemental hydrogen gas over a suitable catalyst to bring about the reduction of iron(III) to iron(II). The overall reaction can be represented as

$$2Fe^{3+} + H_2 \leftarrow\rightarrow 2Fe^{2+} + 2H^+ \qquad (4.2.1)$$

The reaction is written with a double arrow, indicating that it is *reversible* and could proceed in either direction; for normal concentrations of reaction participants, this reaction goes to the right. As the reaction goes to the right, the hydrogen changes from an *oxidation state* (number) of 0 in elemental H_2 to a higher oxidation number of +1 in H^+ and is said to be *oxidized*. The oxidation state of iron goes from +3 in Fe^{3+} to +2 in Fe^{2+}; the oxidation number of iron decreases, which means that it is *reduced*.

All redox reactions such as this one can be broken down into a reduction *half-reaction*, in this case

$$2Fe^{3+} + 2e^- \leftarrow\rightarrow 2Fe^{2+} \qquad (4.2.2)$$

(for one electron, $Fe^{3+} + e^- \leftarrow\rightarrow Fe^{3+}$) and an oxidation half-reaction, in this case

$$H_2 \leftarrow\rightarrow 2H^+ + 2e^- \qquad (4.2.3)$$

Note that adding these two half-reactions together gives the overall reaction. *The addition of an oxidation half-reaction and a reduction half-reaction, each expressed for the same number of electrons so that the electrons cancel on both sides of the arrows, gives a whole redox reaction.*

The equilibrium of a redox reaction, that is, the degree to which the reaction as written tends to lie to the right or left, can be deduced from information about its constituent half-reactions. To visualize this, assume that the two half-reactions can be separated into two half-cells of an electrochemical cell as shown for Reaction 4.2.1 in Figure 4.2.

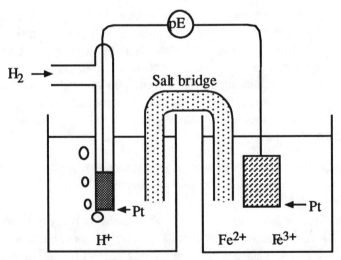

Figure 4.2. Electrochemical cell in which the reaction $2Fe^{3+} + H_2 \leftarrow\rightarrow 2Fe^{2+} + 2H^+$ can be carried out in two half-cells.

If the initial activities of H^+, Fe^{2+}, and Fe^{3+} were of the order of 1 (concentrations of 1 M) and if the pressure of H_2 were 1 atm, H_2 would be oxidized to H^+ in the left half-cell, Fe^{3+} would be reduced to Fe^{2+} in the right half-cell, and ions would migrate through the salt bridge to maintain electroneutrality in both half-cells. The net reaction occurring would be Reaction 4.2.1.

If a voltmeter were inserted in the circuit between the two electrodes, no significant current could flow and the two half-reactions could not take place. However, the voltage registered by the voltmeter would be a measure of the relative tendencies of the two half-reactions to occur. In the left half-cell the oxidation half-reaction,

$$H_2 \xleftarrow{\quad} \rightarrow 2H^+ + 2e^- \tag{4.2.3}$$

will tend to go to the right, releasing electrons to the platinum electrode in the half-cell and giving that electrode a relatively negative (-) potential. In the right half-cell the reduction half-reaction,

$$Fe^{3+} + e^- \xleftarrow{\quad} \rightarrow Fe^{2+} \tag{4.2.2}$$

will tend to go to the right, taking electrons from the platinum electrode in the half-cell and giving that electrode a relatively positive (+) potential. The difference in these potentials is a measure of the "driving force" of the overall reaction. If each of the reaction participants were at unit activity, the potential difference would be 0.77 volts.

The left electrode shown in Figure 4.2 is the standard electrode against which all other electrode potentials are compared. It is called the **standard hydrogen electrode, SHE** and has been assigned a value of exactly 0 volts by convention, and its half-reaction is written as the following:

$$2H^+ + 2e^- \xleftarrow{\quad} \rightarrow H_2 \qquad E^0 = +0.00 \text{ volts} \tag{4.2.4}$$

The measured potential of the right-hand electrode in Figure 4.2 versus the standard hydrogen electrode is called the **electrode potential, E.** If the Fe^{2+} and Fe^{3+} ions in solution are both at unit activity, the potential is the **standard electrode potential** (according to IUPAC convention, the **standard reduction potential**), E^0. The standard electrode potential for the Fe^{3+}/Fe^{2+} couple is 0.77 volts expressed conventionally as follows:

$$Fe^{3+} + e^- \xleftarrow{\quad} \rightarrow Fe^{2+} \qquad E^0 = +0.77 \text{ volts} \tag{4.2.5}$$

4.3. ELECTRON ACTIVITY AND pE

In this book, for the most part, pE and pE^0 are used instead of E and E^0 to more clearly illustrate redox equilibria in aquatic systems over many orders of magnitude of electron activity in a manner analogous to pH. Numerically, pE and pE^0 are simply the following:

$$pE = \frac{E}{\frac{2.303RT}{F}} = \frac{E}{0.0591} \quad \text{(at 25°C)} \tag{4.3.1}$$

For the standard hydrogen electrode the following is written:

$$pE^0 = \frac{E^0}{\frac{2.303RT}{F}} = \frac{E^0}{0.0591} \quad \text{(at 25°C)} \tag{4.3.2}$$

where R is the molar gas constant, T is the absolute temperature, and F is the Faraday constant. The "pE concept" is explained below.

Just as pH is defined as

$$pH = -\log(a_{H^+}) \tag{4.3.3}$$

where a_{H^+} is the activity of hydrogen ion in solution, pE is defined as

$$pE = -\log(a_{e^-}) \tag{4.3.4}$$

where a_{e^-} is the activity of the electron in solution. Since hydrogen ion concentration may vary over many orders of magnitude, pH is a convenient way of expressing a_{H^+} in terms of manageable numbers. Similarly, electron activities in water may vary over more than 20 orders of magnitude so that it is convenient to express a_{e^-} as pE.

Values of pE are defined in terms of the following half-reaction for which pE^0 is defined as exactly zero:*

$$2H^+(aq) + 2e^- \xleftarrow{} H_2(g) \qquad E^0 = +0.00 \text{ volts } pE = 0.00 \tag{4.3.5}$$

Whereas it is relatively easy to visualize the activities of ions in terms of concentration, it is harder to visualize the activity of the electron, and therefore pE, in similar terms. For example, at 25°C in pure water, a medium of zero ionic strength, the hydrogen ion concentration is 1.0×10^{-7}, the hydrogen-ion *activity* is 1.0×10^{-7}, and the pH is 7.0. The electron activity, however, must be defined in terms of Equation 4.3.5. When $H^+(aq)$ at unit activity is in equilibrium with hydrogen gas at 1 atmosphere pressure (and likewise at unit activity), the activity of the electron in the medium is exactly 1.00 and the pE is 0.0. If the electron activity were increased by a factor of 10 (as would be the case if $H^+(aq)$ at an activity of 0.100 were in equilibrium with H_2 at an activity of 1.00), the electron activity would be 10 and the pE value would be -1.0.

4.4. THE NERNST EQUATION

The **Nernst equation** is used to account for the effect of different activities upon electrode potential. Referring to Figure 4.2, if the Fe^{3+} ion concentration is increased relative to the Fe^{2+} ion concentration, it is readily visualized that the potential and the pE of the right electrode will become more positive because the higher concentration of electron-deficient Fe^{3+} ions clustered around it tends to draw electrons from the electrode. Decreased Fe^{3+} ion or increased Fe^{2+} ion concentration has the opposite effect. Such concentration effects upon E and pE are expressed by the **Nernst equation**. As applied to the half-reaction

---

* Thermodynamically the free energy change for this reaction is defined as exactly zero when all reaction participants are at unit activity. For ionic solutes, activity — the effective concentration in a sense — approaches concentration at low concentrations and low ionic strengths. The activity of a gas is equal to its partial pressure. Furthermore, the free energy, G, decreases for spontaneous processes occurring at constant temperature and pressure. Processes for which the free energy change, ΔG, is zero have no tendency toward spontaneous change and are in a state of equilibrium. Reaction 4.2.4 is the one upon which free energies of formation of all ions in aqueous solution are based. It also forms the basis for defining free energy changes for oxidation-reduction processes in water.

$$Fe^{3+} + e^- \longleftrightarrow Fe^{2+} \qquad E^0 = +0.77 \text{ volts} \qquad pE^0 = 13.2 \qquad (4.4.1)$$

the Nernst equation is

$$E = E^0 + \frac{2.303RT}{nF} \log \frac{[Fe^{3+}]}{[Fe^{2+}]} = E^0 + \frac{0.0591}{n} \log \frac{[Fe^{3+}]}{[Fe^{2+}]} \qquad (4.4.2)$$

where n is the number of electrons involved in the half-reaction (1 in this case), and the activities of Fe^{3+} and Fe^{2+} ions have been taken as their concentrations (a simplification valid for more dilute solutions, which will be made throughout this chapter). Considering that

$$pE = \frac{E}{\frac{2.303RT}{F}} \quad \text{and} \quad pE^0 = \frac{E^0}{\frac{2.303RT}{F}}$$

the Nernst equation can be expressed in terms of pE and pE^0

$$pE = pE^0 + \frac{1}{n} \log \frac{[Fe^{3+}]}{[Fe^{2+}]} \quad \text{(In this case n = 1)} \qquad (4.4.3)$$

The Nernst equation in this form is quite simple and offers some advantages in calculating redox relationships.

If, for example, the value of $[Fe^{3+}]$ is 2.35×10^{-3} M and $[Fe^{2+}] = 7.85 \times 10^{-5}$ M, the value of pE is

$$pE = 13.2 + \log \frac{2.35 \times 10^{-3}}{7.85 \times 10^{-5}} = 14.7 \qquad (4.4.4)$$

4.5. Reaction Tendency: Whole Reaction from Half-Reactions

This section discusses how half-reactions can be combined to give whole reactions and how the pE^0 values of the half-reactions can be used to predict the directions in which reactions will go. The half-reactions discussed here are the following:

$$Hg^{2+} + 2e^- \longleftrightarrow Hg \qquad pE^0 = 13.35 \qquad (4.5.1)$$

$$Fe^{3+} + e^- \longleftrightarrow Fe^{2+} \qquad pE^0 = 13.2 \qquad (4.5.2)$$

$$Cu^{2+} + 2e^- \longleftrightarrow Cu \qquad pE^0 = 5.71 \qquad (4.5.3)$$

$$2H^+ + 2e^- \longleftrightarrow H_2 \qquad pE^0 = 0.00 \qquad (4.5.4)$$

$$Pb^{2+} + 2e^- \longleftrightarrow Pb \qquad pE^0 = -2.13 \qquad (4.5.5)$$

Such half-reactions and their pE^0 values can be used to explain observations such as the following: A solution of Cu^{2+} flows through a lead pipe and the lead acquires a layer of copper metal, through the reaction

$$Cu^{2+} + Pb \rightarrow Cu + Pb^{2+} \tag{4.5.6}$$

This reaction occurs because the copper(II) ion has a greater tendency to acquire electrons than the lead ion has to retain them. This reaction can be obtained by subtracting the lead half-reaction, Equation 4.5.5, from the copper half-reaction, Equation 4.5.3:

$$Cu^{2+} + 2e^- \leftarrow\rightarrow Cu \qquad\qquad pE^0 = 5.71$$

$$-(Pb^{2+} + 2e^- \leftarrow\rightarrow Pb \qquad\qquad pE^0 = -2.13)$$

$$\overline{Cu^{2+} + Pb \leftarrow\rightarrow Cu + Pb^{2+} \quad pE^0 = 7.84} \tag{4.5.7}$$

The positive values of pE indicates that the reaction tends to go to the right as written. This occurs when lead metal directly contacts a solution of copper(II) ion. Therefore, if a waste solution containing cupric ion, a relatively innocuous pollutant, comes into contact with lead in plumbing, toxic lead may go into solution.

In principle, half-reactions may be allowed to occur in separate electrochemical half-cells, as could occur in the cell shown in Figure 4.3 if the meter (pE) were bypassed by an electrical conductor, are therefore called **cell reactions.**

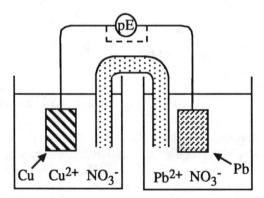

Figure 4.3. Cell for the measurement of pE between a lead half-cell and a copper half-cell. In this configuration "pE" has a very high resistance and current cannot flow.

If the activities of Cu^{2+} and Pb^{2+} are not unity, the direction of the reaction and value of pE are deduced from the Nernst equation. For Reaction 4.5.7 the Nernst equation is

$$pE = pE^0 + \frac{1}{n}\log\frac{[Cu^{2+}]}{[Pb^{2+}]} = 7.84 + \frac{1}{2}\log\frac{[Cu^{2+}]}{[Pb^{2+}]} \tag{4.5.8}$$

By combining the appropriate half-reactions it can be shown that copper metal will not cause hydrogen gas to be evolved from solutions of strong acid (hydrogen ion has less attraction for electrons than does cupric ion), whereas lead metal, in contrast, will displace hydrogen gas from acidic solutions.

4.6. THE NERNST EQUATION AND CHEMICAL EQUILIBRIUM

Refer again to Figure 4.3. Imagine that instead of the cell being set up to measure the potential between the copper and lead electrodes, the voltmeter, V, were removed and the electrodes directly connected with a wire so that the current might flow between them. The reaction

$$Cu^{2+} + Pb \longleftrightarrow Cu + Pb^{2+} \quad pE^0 = 7.84 \qquad (4.5.7)$$

will occur until the concentration of lead ion beomes so high, and that of copper ion so low, that the reaction stops. The system is at equilibrium and, since current no longer flows, pE is exactly zero. The equilibrium constant K for the reaction is given by the equation

$$K = \frac{[Pb^{2+}]}{[Cu^{2+}]} \qquad (4.6.1)$$

The equilibrium constant can be calculated from the Nernst Equation, noting that under equilibrium conditions pE is zero and $[Cu^{2+}]$ and $[Pb^{2+}]$ are at equilibrium concentrations:

$$pE = pE^0 + \frac{1}{n}\log \frac{[Cu^{2+}]}{[Pb^{2+}]} = pE^0 + \frac{1}{n}\log \frac{1}{K}$$

$$pE = 0.00 = 7.84 - \frac{1}{2}\log \frac{[Pb^{2+}]}{[Cu^{2+}]} = 7.84 - \frac{1}{2}\log K \qquad (4.6.2)$$

Note that the reaction products are placed over reactants in the log term, and a minus sign is placed in front to put the equilibrium constant in the correct form (a purely mathematical operation). The value of log K obtained from solving the above equation is 15.7.

The equilibrium constant for a redox reaction involving n electrons is given in terms of pE simply by

$$\log K = n(pE^0) \qquad (4.6.3)$$

4.7. THE RELATIONSHIP OF pE TO FREE ENERGY

Aquatic systems and the organisms that inhabit them — just like the steam engine or students hoping to pass physical chemistry — must obey the laws of thermodynamics. Bacteria, fungi, and human beings derive their energy from acting as mediators (catalysts) of chemical reactions and extracting a certain percentage of useful energy from them. In predicting or explaining the behavior of an aquatic system, it is helpful to be able to predict the useful energy that can be extracted from chemical reactions in the system, such as microbially mediated oxidation of organic matter to CO_2 and water, or the fermentation of organic matter to methane by anaerobic bacteria in the absence of oxygen. Such information may be obtained by knowing the free-energy change, ΔG, for the redox reaction; ΔG, in turn, may be obtained from pE for the reaction. The free-energy change for a redox reaction involving n electrons at an absolute temperature of T is given by

$$\Delta G = -2.303 nRT(pE) \tag{4.7.1}$$

where R is the gas constant. When all reaction participants are in their standard states (pure liquids, pure solids, solutes at an activity of 1.00) ΔG is the standard free energy change, ΔG^0, given by

$$\Delta G^0 = -2.303 nRT(pE^0) \tag{4.7.2}$$

4.8. REACTIONS IN TERMS OF ONE ELECTRON-MOLE

For comparing free energy changes between different redox reactions, it is most meaningful to consider the reactions in terms of the transfer of exactly 1 mole of electrons. This concept may be understood by considering two typical and important redox reactions that occur in aquatic systems — nitrification

$$NH_4^+ + 2O_2 \longleftrightarrow NO_3^- + 2H^+ + H_2O \qquad pE^0 = 5.85 \tag{4.8.1}$$

and oxidation of iron(II) to iron(III):

$$4Fe^{2+} + O_2 + 10H_2O \longleftrightarrow 4Fe(OH)_3(s) + 8H^+ \qquad pE^0 = 7.6 \tag{4.8.2}$$

What do reactions written in this way really mean? If any thermodynamic calculations are to be made involving the reactions, Reaction 4.8.1 means that one mole of ammonium ion reacts with two moles of oxygen to yield one mole of nitrate ion, two moles of hydrogen ion, and one mole of water. Reaction 4.8.2 is taken to mean that four moles of iron(II) ion react with one mole of oxygen and ten moles of water to produce four moles of $Fe(OH)_3$ and eight moles of hydrogen ions. The free-energy changes calculated for these quantities of reaction participants do not enable meaningful comparisons of their free energy changes. Such comparisons may be made, though, on the common basis of the transfer of one mole of electrons, writing each reaction in terms of one electron-mole. The advantage of this approach is illustrated by considering Reaction 4.8.1, which involves an eight-electron change, and Reaction 4.8.2, which involves a four-electron change. Rewriting Equation 4.8.1 for one electron-mole yields

$$\frac{1}{8}NH_4^+ + \frac{1}{4}O_2 \longleftrightarrow \frac{1}{8}NO_3^- + \frac{1}{4}H^+ + \frac{1}{8}H_2O \qquad pE^0 = 5.85 \tag{4.8.3}$$

whereas Reaction 4.8.2, when rewritten for one electron-mole rather than four, yields:

$$Fe^{2+} + \frac{1}{4}O_2 + \frac{5}{2}H_2O \longleftrightarrow Fe(OH)_3(s) + 2H^+ \qquad pE^0 = 7.6 \tag{4.8.4}$$

From Equation 4.7.2, the standard free energy change for a reaction is

$$\Delta G^0 = -2.303nRT(pE^0) \tag{4.7.2}$$

which, for a one electron-mole reaction is simply

$$\Delta G^0 = -2.303RT(pE^0) \tag{4.7.2}$$

Therefore, for reactions written in terms of 1 electron-mole, a comparison of pE^0 values provides a direct comparison of ΔG^0 values.

As shown in Equation 4.6.3 for a redox reaction involving n electrons, pE^0 is related to the equilibrium constant by

$$\log K = \frac{1}{n}(pE^0) \tag{4.8.5}$$

which for a one electron-mole reaction becomes simply

$$\log K = pE^0 \tag{4.8.6}$$

Reaction 4.8.3, the nitrification reaction written in terms of one electron-mole, has a pE^0 value of +5.85. The equilibrium-constant expression for this reaction is,

$$K = \frac{[NO_3^-]^{1/8}[H^+]^{1/4}}{[NH_4^+]^{1/8}P_{O_2}^{1/4}} \tag{4.8.7}$$

a computationally cumbersome form for which log K is simply

$$\log K = pE^0 = 5.85 \text{ or } K = 7.08 \times 10^5 \tag{4.8.8}$$

Table 4.1 is a compilation of pE^0 values for redox reactions that are especially important in aquatic systems. Most of these values are calculated from thermodynamic data rather than from direct potentiometric measurements in an electrochemical cell, as shown in Figure 4.2. Most electrode systems that might be devised do not give potential responses corresponding to the Nernst equation; that is, they do not behave *reversibly*. It is true that one may place a platinum electrode and a reference electrode in water and measure a potential. This potential, referred to the standard hydrogen electrode, is the so-called **E_H value**. Furthermore, the measured potential will be more positive (more oxidizing) in an oxidizing medium, such as the aerobic surface layers of a lake, than in a reducing medium, such as the anaerobic bottom regions of a body of water. However, attaching any quantitative significance to the E_H value measured directly with an electrode is a very dubious practice. Acid mine waters containing relatively high levels of sulfuric acid and dissoved iron give reasonably accurate E_H values by direct measurement, but most aquatic systems do not yield meaningful values of E_H.

Table 4.1. pE^0 Values of Redox Reactions Important in Natural Waters (at 25°C)

Reaction	pE^0	$pE^0(W)$[1]
(1) $\frac{1}{4}O_2(g) + H^+(W) + e \rightleftharpoons \frac{1}{2}H_2O$	+20.75	+13.75
(2) $\frac{1}{5}NO_3^- + \frac{6}{5}H^+(W) + e \rightleftharpoons \frac{1}{10}N_2(g) + \frac{3}{5}H_2O$	+21.05	+12.65
(3) $\frac{1}{2}MnO_2(s) + \frac{1}{2}HCO_3^-(10^{-3}) + \frac{3}{2}H^+(W) + e$ $\rightleftharpoons \frac{1}{2}MnCO_3(s) + \frac{3}{8}H_2O$	—	+8.5[2]
(4) $\frac{1}{2}NO_3^- + H^+(W) + e \rightleftharpoons \frac{1}{2}NO_2^- + \frac{1}{2}H_2O$	+14.15	+7.15
(5) $\frac{1}{8}NO_3^- + \frac{5}{4}H^+(W) + e \rightleftharpoons \frac{1}{8}NH_4^+ + \frac{3}{8}H_2O$	+14.90	+6.15
(6) $\frac{1}{6}NO_2^- + \frac{4}{3}H^+(W) + e \rightleftharpoons \frac{1}{6}NH_4^+ + \frac{1}{3}H_2O$	+15.14	+5.82
(7) $\frac{1}{2}CH_3OH + H^+(W) + e \rightleftharpoons \frac{1}{2}CH_4(g) + \frac{1}{2}H_2O$	+9.88	+2.88
(8) $\frac{1}{4}CH_2O + H^+(W) + e \rightleftharpoons \frac{1}{4}CH_4(g) + \frac{1}{4}H_2O$	+6.94	−0.06
(9) $FeOOH(s) + HCO_3^-(10^{-3}) + 2H^+(W) + e$ $\rightleftharpoons FeCO_3(s) + 2H_2O$	—	−1.67[2]
(10) $\frac{1}{2}CH_2O + H^+(W) + e \rightleftharpoons \frac{1}{2}CH_3OH$	+3.99	−3.01
(11) $\frac{1}{6}SO_4^{2-} + \frac{4}{3}H^+(W) + e \rightleftharpoons \frac{1}{6}S(s) + \frac{2}{3}H_2O$	+6.03	−3.30
(12) $\frac{1}{8}SO_4^{2-} + \frac{5}{4}H^+(W) + e \rightleftharpoons \frac{1}{8}H_2S(g) + \frac{1}{2}H_2O$	+5.75	−3.50
(13) $\frac{1}{8}SO_4^{2-} + \frac{9}{8}H^+ + e \rightleftharpoons \frac{1}{8}HS^- + \frac{1}{2}H_2O$	+4.13	−3.75
(14) $\frac{1}{2}S(s) + H^+(W) + e \rightleftharpoons \frac{1}{2}H_2S(g)$	+2.89	−4.11
(15) $\frac{1}{8}CO_2 + H^+ + e \rightleftharpoons \frac{1}{8}CH_4 + \frac{1}{4}H_2O$	+2.87	−4.13
(16) $\frac{1}{6}N_2(g) + \frac{4}{3}H^+(W) + e \rightleftharpoons \frac{1}{3}NH_4^+$	+4.68	−4.65
(17) $H^+(W) + e \rightleftharpoons \frac{1}{2}H_2(g)$	0.0	−7.00
(18) $\frac{1}{4}CO_2(g) + H^+(W) + e \rightleftharpoons \frac{1}{4}CH_2O + \frac{1}{4}H_2O$	−1.20	−8.20

[1] (W) indicates $a_{H^+} = 1.00 \times 10^{-7}M$ and $pE^0(W)$ is a pE^0 at $a_{H^+} = 1.00 \times 10^{-7}M$.

[2] These data correspond to $a_{HCO_3^-} = 1.00 \times 10^{-3}M$ rather than unity and so are not exactly $pE^0(W)$; they represent typical aquatic conditions more nearly than pE^0 values do.

Source: Werner Stumm and James J. Morgan, *Aquatic Chemistry* (New York: Wiley-Interscience, 1970), p. 318. Reproduced with permission of John Wiley and Sons, Inc.

4.9. THE LIMITS OF pE IN WATER

There are pH-dependent limits to the pE values at which water is thermodynamically stable. Water may be both oxidized

$$2H_2O \leftarrow\rightarrow O_2 + 4H^+ + 4e^- \tag{4.9.1}$$

or it may be reduced:

$$2H_2O + 2e^- \leftarrow\rightarrow H_2 + 2OH^- \tag{4.9.2}$$

These two reactions determine the limits of pE in water. On the oxidizing side (relatively more positive pE values), the pE value is limited by the oxidation of water, Half-reaction 4.9.1. The evolution of hydrogen, Half-reaction 4.9.2, limits the pE value on the reducing side.

The condition under which oxygen from the oxidation of water has a pressure of 1.00 atm can be regarded as the oxidizing limit of water whereas a hydrogen pressure of 1.00 atmosphere may be regarded as the reducing limit of water. These are **boundary conditions** that enable calculation of the stability boundaries of water. Writing the reverse of Reaction 4.9.1 for one electron and setting $P_{O_2} = 1.00$ yields:

$$\frac{1}{4}O_2 + H^+ + e^- \leftarrow\rightarrow \frac{1}{2}H_2O \qquad pE^0 = 20.75 \text{ (from Table 4.1)} \tag{4.9.3}$$

$$pE = pE^0 + \log (P_{O_2}^{1/4}[H^+]) \tag{4.9.4}$$

$$pE = 20.75 - pH \tag{4.9.5}$$

Thus, Equation 4.9.5 defines the oxidizing limit of water. At a specified pH, pE values more positive than the one given by Equation 4.9.5 cannot exist at equilibrium in water in contact with the atmosphere.

The pE-pH relationship for the reducing limit of water is given by the following derivation:

$$H^+ + e^- \leftarrow\rightarrow \frac{1}{2}H_2 \qquad pE^0 = 0.00 \tag{4.9.6}$$

$$pE = pE^0 + \log [H^+] \tag{4.9.7}$$

$$pE = -pH \tag{4.9.8}$$

For neutral water (pH = 7.00), substitution into Equations 4.9.8 and 4.9.5 yields 7.00 to 13.75 for the pE range of water. The pE-pH boundaries of stability for water are shown by the dashed lines in Figure 4.4 (Section 4.11).

The decomposition of water is very slow in the absence of a suitable catalyst. Therefore, water may have temporary nonequilibrium pE values more negative than the reducing limit or more positive than the oxidizing limit. An example of the latter is a solution of chlorine in water.

4.10. pE VALUES IN NATURAL WATER SYSTEMS

Although it is not generally possible to obtain accurate pE values by direct potentiometric measurements in natural aquatic systems, in principle, pE values may be calculated from the species present in water at equilibrium. An obviously significant pE value is that of neutral water in thermodynamic equilibrium with the atmosphere. In water under these conditions, $P_{O_2} = 0.21$ atm and $[H^+] = 1.00 \times 10^{-7}$ M. Substitution into Equation 4.9.4 yields:

$$pE = 20.75 + \log\{(0.21)^{1/4} \times 1.00 \times 10^{-7}\} = 13.8 \qquad (4.10.1)$$

According to this calculation, a pE value of around +13 is to be expected for water in equilibrium with the atmosphere, that is, an aerobic water. At the other extreme, consider anaerobic water in which methane and CO_2 are being produced by microorganisms. Assume $P_{CO_2} = P_{CH_4}$ and that pH = 7.00. The relevant half-reaction is

$$\frac{1}{8}CO_2 + H^+ + e^- \leftarrow\rightarrow \frac{1}{8}CH_4 + \frac{1}{4}H_2O \qquad (4.10.2)$$

for which the Nernst equation is

$$pE = 2.87 + \log\frac{P_{CO_2}^{1/8}[H^+]}{P_{CH_4}^{1/8}} = 2.87 + \log[H^+] = 2.87 - 7.00 = -4.13 \qquad (4.10.3)$$

Note that the pE value of -4.13 does not exceed the reducing limit of water at pH 7.00, which from Equation 4.9.8 is -7.00. It is of interest to calculate the pressure of oxygen in neutral water at this low pE value of -4.13. Substitution into Equation 4.9.4 yields

$$-4.13 = 20.75 + \log(P_{O_2}^{1/4} \times 1.00 \times 10^{-7})$$

from which the pressure of oxygen is calculated to be 3.0×10^{-72} atm. This incredibly low figure for the pressure of oxygen means that equilibrium with respect to oxygen partial pressure is not achieved under these conditions. Certainly, under any condition approaching equilibrium between comparable levels of CO_2 and CH_4, the partial pressure of oxygen must be extremely low.

4.11. pE-pH DIAGRAMS

The examples cited so far have shown the close relationships between pE and pH in water. This relationship may be expressed graphically in the form of a **pE-pH diagram.** Such diagrams show the regions of stability and the boundary lines for various species in water. Because of the numerous species that may be formed, such diagrams may become extremely complicated. For example, if a metal is being considered, several different oxidation states of the metal, hydroxy complexes, and different forms of the solid metal oxide or hydroxide may exist in different regions described by the pE-pH diagram. Most waters contain carbonate, and many contain sulfates and sulfides, so that various metal carbonates, sulfates, and sulfides may predominate in different regions of the diagram. In order to illustrate the principles involved, however, a simplified pE-pH diagram is considered here. The reader is referred to more advanced works on geochemistry and aquatic chemistry for more complicated (and more realistic) pE-pH diagrams[1,2].

A pE-pH diagram for iron may be constructed assuming a maximum concentration of iron in solution, in this case 1.0×10^{-5} M. The following equilibria will be considered:

$$Fe^{3+} + e^- \longleftrightarrow Fe^{2+} \qquad pE^0 = +13.2 \tag{4.11.1}$$

$$Fe(OH)_2(s) + 2H^+ \longleftrightarrow Fe^{2+} + 2H_2O \tag{4.11.2}$$

$$K_{sp} = \frac{[Fe^{2+}]}{[H^+]^2} = 8.0 \times 10^{12} \tag{4.11.3}$$

$$Fe(OH)_3(s) + 3H^+ \longleftrightarrow Fe^{3+} + 3H_2O \tag{4.11.4}$$

$$K_{sp}' = \frac{[Fe^{3+}]}{[H^+]^3} = 9.1 \times 10^3 \tag{4.11.5}$$

(The constants K_{sp} and K_{sp}' are derived from the solubility products of $Fe(OH)_2$ and $Fe(OH)_3$, respectively, and are expressed in terms of $[H^+]$ to facilitate the calculations.) Note that the formation of species such as $Fe(OH)^{2+}$, $Fe(OH)_2^+$, and solid $FeCO_3$ or FeS, all of which might be of significance in a natural water system, is not considered.

In constructing the pE-pH diagram, several boundaries must be considered. The first two of these are the oxidizing and reducing limits of water (see Section 4.9). At the high pE end, the stability limit of water is defined by Equation 4.9.5:

$$pE = 20.75 - pH \tag{4.9.5}$$

The low pE limit is defined by Equation 4.9.8:

$$pE = -pH \tag{4.9.8}$$

The pE-pH diagram constructed for the iron system must fall between the boundaries defined by these two equations.

Below pH 3, Fe^{3+} may exist in equilibrium with Fe^{2+}. The boundary line that separates these two species, where $[Fe^{3+}] = [Fe^{2+}]$, is given by the following calculation:

$$pE = 13.2 + \log \frac{[Fe^{3+}]}{[Fe^{2+}]} \tag{4.11.6}$$

$$[Fe^{3+}] = [Fe^{2+}] \tag{4.11.7}$$

$$pE = 13.2 \text{ (independent of pH)} \tag{4.11.8}$$

At pE exceeding 13.2, as the pH increases from very low values, $Fe(OH)_3$ precipitates from a solution of Fe^{3+}. The pH at which precipitation occurs depends of course upon the concentration of Fe^{3+}. In this example, a maximum soluble iron concentration of 1.00×10^{-5} M has been chosen so that at the $Fe^{3+}/Fe(OH)_3$ boundary, $[Fe^{3+}] = 1.00 \times 10^{-5}$ M. Substitution in Equation 4.11.5 yields:

$$[H^+]^3 = \frac{[Fe^{3+}]}{K_{sp}{}'} = \frac{1.00 \times 10^{-5}}{9.1 \times 10^3} \tag{4.11.9}$$

$$pH = 2.99 \tag{4.11.10}$$

In a similar manner, the boundary between Fe^{2+} and solid $Fe(OH)_2$ may be defined, assuming $[Fe^{2+}] = 1.00 \times 10^{-5}$ M (the maximum soluble iron concentration specified at the beginning of this exercise) at the boundary:

$$[H^+]^2 = \frac{[Fe^{2+}]}{K_{sp}} = \frac{1.00 \times 10^{-5}}{8.0 \times 10^{12}} \text{ (from Equation 4.11.3)} \tag{4.11.11}$$

$$pH = 8.95 \tag{4.11.12}$$

Throughout a wide pE-pH range, Fe^{2+} is the predominant soluble iron species in equilibrium with the solid hydrated iron(III) oxide, $Fe(OH)_3$. The boundary between these two species depends upon both pE and pH. Substituting Equation 4.11.5 into Equation 4.11.6 yields:

$$pE = 13.2 + \log\frac{K_{sp}{}'[H^+]^3}{[Fe^{2+}]} \tag{4.11.13}$$

$$pE = 13.2 + \log 9.1 \times 10^3 - \log 1.00 \times 10^{-5} + 3 \times \log [H^+]$$

$$pE = 22.2 - 3\,pH \tag{4.11.14}$$

The boundary between the solid phases $Fe(OH)_2$ and $Fe(OH)_3$ likewise depends upon both pE and pH, but it does not depend upon an assumed value for total soluble iron. The required relationship is derived from substituting both Equation 4.11.3 and Equation 4.11.5 into Equation 4.11.6:

$$pE = 13.2 + \log\frac{K_{sp}{}'[H^+]^3}{[Fe^{2+}]} \tag{4.11.13}$$

$$pE = 13.2 + \log\frac{K_{sp}{}'[H^+]^3}{K_{sp}[H^+]^2} \tag{4.11.15}$$

$$pE = 13.2 + \log \frac{9.1 \times 10^3}{8.0 \times 10^{12}} + \log [H^+]$$

$$pE = 4.3 - pH \tag{4.11.16}$$

All of the equations needed to prepare the pE-pH diagram for iron in water have now been derived. To summarize, the equations are (4.9.5), O_2–H_2O boundary; (4.9.8), H_2–H_2O boundary; (4.11.8), Fe^{3+}–Fe^{2+} boundary; (4.11.10), Fe^{3+}–$Fe(OH)_3$ boundary; (4.11.12), Fe^{2+}–$Fe(OH)_2$ boundary; (4.11.14), Fe^{2+}–$Fe(OH)_3$ boundary; and (4.11.16), $Fe(OH)_2$–$Fe(OH)_3$ boundary.

The pE-pH diagram for the iron system in water is shown in Figure 4.4. In this system, at a relatively high hydrogen ion activity and high electron activity (an acidic reducing medium), iron(II) ion, Fe^{2+}, is the predominant iron species; some groundwaters contain appreciable levels of iron(II) under these conditions. (In most natural water systems the solubility range of Fe^{2+} is very narrow because of the precipitation of FeS or $FeCO_3$.) At a very high hydrogen ion activity and low electron activity (an acidic oxidizing medium), Fe^{3+} ion predominates. In an oxidizing medium at lower acidity, solid, $Fe(OH)_3$ is the primary iron species present. Finally, in a basic reducing medium, with low hydrogen ion activity and high electron activity, solid $Fe(OH)_2$ is stable.

Note that within the pH regions normally encountered in a natural aquatic system (approximately pH 5 to 9) $Fe(OH)_3$ or Fe^{2+} are the predominant stable iron species. In fact, it is observed that in waters containing dissolved oxygen at any appreciable level, and therefore having a relatively high pE, hydrated iron(III) oxide ($Fe(OH)_3$) is essentially the only inorganic iron species found. Such waters contain a high level of suspended iron, but any truly soluble iron must be in the form of a complex (see Chapter 3).

In highly anaerobic, low pE water, appreciable levels of Fe^{3+} may be present. When such water is exposed to atmospheric oxygen, the pE rises and $Fe(OH)_3$ precipitates. The resulting deposits of hydrated iron(III) oxide can stain laundry and bathroom fixtures with a refractory red/brown stain. This phenomenon also explains why red iron oxide deposits are found near pumps and springs that bring deep, anaerobic water to the surface. In shallow wells, where the water may become aerobic, solid $Fe(OH)_3$ may precipitate on the well walls, clogging the aquifer outlet. This usually occurs through bacterially-mediated reactions which are discussed in Chapter 6.

One species not yet considered is elemental iron. For the half-reaction

$$Fe^{2+} + 2e^- \longleftrightarrow Fe \qquad pE^0 = -7.45 \qquad (4.11.17)$$

the Nernst equation gives pE as a function of $[Fe^{2+}]$

$$pE = -7.45 + \frac{1}{2}\log[Fe^{2+}] \qquad (4.11.18)$$

For iron metal in equilibirum with 1.00×10^{-5} M Fe^{2+}, the following pE value is obtained:

$$pE = -7.45 + -\frac{1}{2}\log 1.00 \times 10^{-5} = -9.95 \qquad (4.11.19)$$

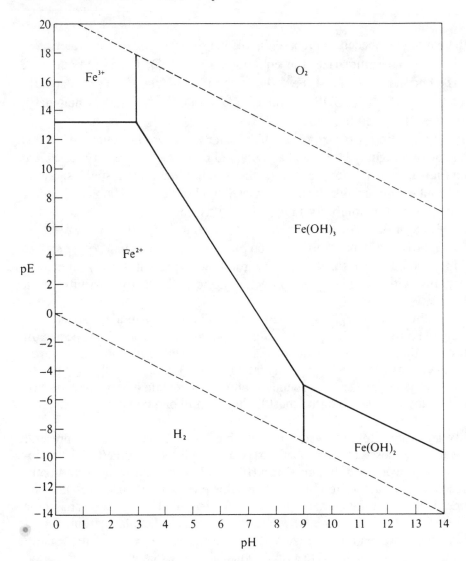

Figure 4.4. Simplifed pE-pH diagram for iron in water. The maximum soluble iron concentration is 1.00×10^{-5} M.

Examination of Figure 4.4 shows that the pE values for elemental iron in contact with Fe^{2+} is below the reducing limit of water. Iron metal in contact with water, therefore, is thermodynamically unstable, a factor that contributes to corrosion.

4.12. CORROSION

One of the most damaging redox phenomena is **corrosion**, defined as the destructive alteration of metal through interactions with its surroundings. In addition to its multi-billion dollar annual costs due to destruction of equipment and structures, corroson introduces metals into water systems and destroys pollution-control equipment and waste disposal pipes; it is aggravated by water and air pollutants and some kinds of hazardous wastes (see corrosive wastes in Section 17.6).

Thermodynamically all commonly-used metals are unstable relative to their environments. Elemental metals tend to undergo chemical changes to produce the more stable forms of ions, salts, oxides, and hydroxides. Fortunately, the rates of corrosion are normally slow, so that metals exposed to air and water may endure for long periods of time. However, protective measures are necessary. Sometimes these measures fail; for example, witness the gaping holes in automobile bodies exposed to salt used to control road ice.

Corrosion normally occurs when an electrochemical cell is set up on a metal surface. The area corroded is the anode, where the following oxidation reaction occurs, illustrated for the formation of a divalent metal ion from a metal, M:

$$M \rightarrow M^{2+} + 2e^- \tag{4.12.1}$$

Several cathodic reactions are possible. One of the most common of these is the reduction of H^+ ion:

$$2H^+ + 2e^- \rightarrow H_2 \tag{4.12.2}$$

Oxygen may also be involved in cathodic reactions, including reduction to hydroxide, reduction to water, and reduction to hydrogen peroxide:

$$O_2 + 2H_2O + 4e^- \rightarrow 4OH^- \tag{4.12.3}$$

$$O_2 + 4H^+ + 4e^- \rightarrow 2H_2O \tag{4.12.4}$$

$$O_2 + 2H_2O + 2e^- \rightarrow 2OH^- + 2H_2O_2 \tag{4.12.5}$$

Oxygen may either accelerate corrosion processes by participating in reactions such as these, or retard them by forming protective oxide films. As discussed in Chapter 6, bacteria are often involved with corrosion.

LITERATURE CITED

1. Stumm, Werner, and James J. Morgan, *Aquatic Chemistry*, 2nd ed., Wiley–Interscience, New York, 1981.

2. Garrels, R. M., and C. M. Christ, *Solutions, Minerals, and Equilibria*, Harper and Row, New York, 1965.

SUPPLEMENTARY REFERENCES

Brubaker, G. R., and P. Phipps, Eds., *Corrosion Chemistry*, American Chemical Society, 1979.

Bates, Roger G., "The Modern Meaning of pH," *Crit. Rev. Anal. Chem.*, **10**, 247-278 (1981).

Lingane, James J., *Electroanalytical Chemistry*, Wiley-Interscience, New York, 1958.

Baas Becking, L. G. M., I. R. Kaplan, and D. Moore, "Limits of the Natural Environment in Terms of pH and Oxidation-Reduction Potentials in Natural Waters," *J. Geol.*, **68**, 243-284 (1960).

Stumm, Werner, *Redox Potential as an Environmental Parameter: Conceptual Significance and Operational Limitation*, Third International Conference on Water Pollution Research, (Munich, Germany), Water Pollution Control Federation, Washington, DC, 1966.

Yu, T. R., *Physical Chemistry of Paddy Soils*, Springer-Verlag, Berlin, 1985.

West, J. M., *Basic Corrosion and Oxidation*, John Wiley and Sons, Inc., New York, 1980.

Dowdy, R. H., *Chemistry in the Soil Environment*, ASA Special Publication **40**, American Society of Agronomy, Madison, Wisconsin, 1981.

"Electrochemical Phenomena," Chapter 6 in *The Chemistry of Soils*, Garrison Sposito, Oxford University Press, New York, 1989, pp. 106-126.

Rowell, D. L., "Oxidation and Reduction," in *The Chemistry of Soil Processes*, D. J. Greenland and M. H. B. Hayes, Eds., Wiley, Chichester, U.K., 1981.

Kotrly, S., and L. Sucha, *Handbook of of Chemical Equilibria in Analytical Chemistry*, John Wiley and Sons, Inc., New York, 1985.

"Element Fixation in Soil," Chapter 3 in *The Soil Chemistry of Hazardous Materials*, James Dragun, Hazardous Materials Control Research Insitute, Silver Spring, MD, 1988, pp. 75-152.

QUESTIONS AND PROBLEMS

1. The acid-base reaction for the dissociation of acetic acid is

$$HOAc + H_2O \rightarrow H_3O^+ + OAc^-$$

with $K_a = 1.75 \times 10^{-5}$. Break this reaction down into two half-reactions involving H^+ ion. Break down the redox reaction

$$Fe^{2+} + H^+ \rightarrow Fe^{3+} + 1/2 H_2$$

into two half-reactions involving the electron. Discuss the analogies between the acid-base and redox processes.

2. Assuming a bicarbonate ion concentration $[HCO_3^-]$ of $1.00 \times 10^{-3} M$ and a value of 3.5×10^{-11} for the solubility product of $FeCO_3$, what would you expect to be the stable iron species at pH 9.5 and pE -8.0, as shown in Figure 4.4?

3. Assuming that the partial pressure of oxygen in water is that of atmospheric O_2, 0.21 atm, rather than the 1.00 atm assumed in deriving Equation 4.9.5, derive an equation describing the oxidizing pE limit of water as a function of pH.

4. Plot log P_{O_2} as a function of pE at pH 7.00.

5. Calculate the pressure of oxygen for a system in equilibrium in which $[NH_4^+]$ = $[NO_3^-]$ at pH 7.00.

6. Calculate the values of $[Fe^{3+}]$, pE, and pH at the point in Figure 4.4 where Fe^{2+} at a concentration of 1.00×10^{-5} M, $Fe(OH)_2$, and $Fe(OH)_3$ are all in equilibrium.

7. What is the pE value in a solution in equilibrium with air (21% O_2 by volume) at pH 6.00?

8. What is the pE value at the point on the Fe^{2+}–$Fe(OH)_2$ boundary line (see Figure 4.4) in a solution with a soluble iron concentration of 1.00×10^{-4} M at pH 6.00?

9. What is the pE value in an acid mine water sample having $[Fe^{3+}] = 7.03 \times 10^{-3}$ M and $[Fe^{2+}] = 3.71 \times 10^{-4}$ M?

10. At pH 6.00 and pE 2.58, what is the concentration of Fe^{2+} in equilibrium with $Fe(OH)_2$?

11. What is the calculated value of the partial pressure of O_2 in acid mine water of pH 2.00, in which $[Fe^{3+}] = [Fe^{2+}]$?

12. What is the major advantage of expressing redox reactions and half-reaction in terms of exactly one electron-mole?

13. Why are pE values that are determined by reading the potential of a platinum electrode versus a reference electrode generally not very meaningful?

14. What determines the oxidizing and reducing limits, respectively, for the thermodynamic stability of water?

15. How would you expect pE to vary with depth in a stratified lake?

16. Upon what half-reaction is the rigorous definition of pE based?

Phase Interactions

5.1. CHEMICAL INTERACTIONS INVOLVING SOLIDS, GASES, AND WATER

Homogeneous chemical reactions occurring entirely in aqueous solution are rather rare in natural waters and wastewaters. Instead, most significant chemical and biochemical phenomena in water involve interactions between species in water and another phase. Some of these important interactions are illustrated in Figure 5.1. Several examples of phase interactions in water illustrated by the figure are the following: Production of solid biomass through the photosynthetic activity of algae occurs

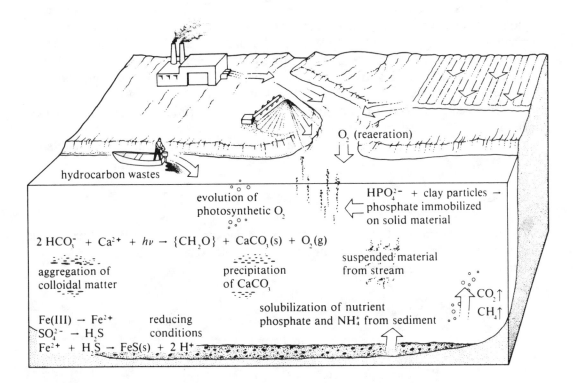

Figure 5.1. Most important environmental chemical processes in water involve interactions between water and another phase.

within a suspended algal cell and involves exchange of dissolved solids and gases between the surrounding water and the cell. Similar exchanges occur when bacteria degrade organic matter (often in the form of small particles) in water. Chemical reactions occur that produce solids or gases in water. Iron and many important trace-level elements are transported through aquatic systems as colloidal chemical compounds or are sorbed to solid particles. Pollutant hydrocarbons and some pesticides may be present on the water surface as an immiscible liquid film. Sediment can be washed physically into a body of water.

This chapter discusses the importance of interactions among different phases in aquatic chemical processes. In a general sense, in addition to water, these phases may be divided between *sediments* (bulk solids) and *suspended colloidal material*. The ways in which sediments are formed and the significance of sediments as repositories and sources of aquatic solutes are discussed. Mentioned in earlier chapters, solubilities of solids and gases (Henry's law) are covered here in some detail.

Much of this chapter deals with the behavior of colloidal material, which consists of very fine particles of solids, gases, or immiscible liquids suspended in water. Colloidal material is involved with many significant aquatic chemical phenomena. It is very reactive because of its high surface area to volume ratio.

5.2. FORMATION OF SEDIMENTS

Sediments, which typically consist of mixtures of clay, silt, sand, organic matter, and various minerals, may vary in composition from pure mineral matter to predominantly organic matter. Physical, chemical, and biological processes may all result in the deposition of sediments in the bottom regions of bodies of water. Sedimentary material may be simply carried into a body of water by erosion or through sloughing (caving in) of the shore. Thus, clay, sand, organic matter, and other materials may be washed into a lake and settle out as layers of sediment.

Sediments may be formed by simple precipitation reactions, several of which are discussed below. When a phosphate-rich wastewater enters a body of water containing a high concentration of calcium ion, the following reaction occurs to produce solid hydroxyapatite:

$$5Ca^{2+} + H_2O + 3HPO_4^{2-} \rightarrow Ca_5OH(PO_4)_3(s) + 4H^+ \qquad (5.2.1)$$

Calcium carbonate sediment may form when water rich in carbon dioxide and containing a high level of calcium as temporary hardness (see Section 3.4) loses carbon dioxide to the atmosphere,

$$Ca^{2+} + 2HCO_3^- \rightarrow CaCO_3(s) + CO_2(g) + H_2O \qquad (5.2.2)$$

or when the pH is raised by a photosynthetic reaction:

$$Ca^{2+} + 2HCO_3^- + h\nu \rightarrow \{CH_2O\} + CaCO_3(s) + O_2(g) \qquad (5.2.3)$$

Oxidation of reduced forms of an element can result in its transformation to an insoluble species, such as occurs when iron(II) is oxidized to iron(III) to produce a precipitate of insoluble ferric hydroxide:

$$4Fe^{2+} + 10H_2O + O_2 \rightarrow 4Fe(OH)_3(s) + 8H^+ \tag{5.2.4}$$

A decrease in pH can result in the production of an insoluble humic acid sediment from base-soluble organic humic substances in solution (see Section 3.12).

Biological activity is responsible for the formation of some aquatic sediments. Some bacterial species produce large quantities of iron(III) oxide (see Section 6.12) as part of their energy-extracting mediation of the oxidation of iron(II) to iron(III). In anaerobic bottom regions of bodies of water, some bacteria use sulfate ion as an electron receptor,

$$SO_4^{2-} \rightarrow H_2S \tag{5.2.5}$$

whereas other bacteria reduce iron(III) to iron(II):

$$Fe(OH)_3(s) \rightarrow Fe^{2+} \tag{5.2.6}$$

The net result is a precipitation reaction producing a black layer of iron(II) sulfide sediment:

$$Fe^{2+} + H_2S \rightarrow FeS(s) + 2H^+ \tag{5.2.7}$$

This frequently occurs during the winter, alternating with the production of calcium carbonate byproduct from photosynthesis (Reaction 5.2.3) during the summer. Under such conditions a layered bottom sediment is produced composed of alternate layers of FeS and $CaCO_3$ as shown in Figure 5.2.

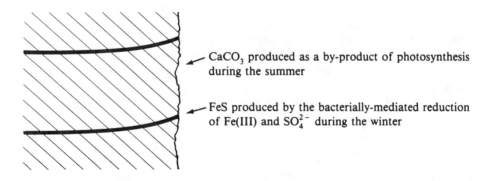

CaCO$_3$ produced as a by-product of photosynthesis during the summer

FeS produced by the bacterially-mediated reduction of Fe(III) and SO$_4^{2-}$ during the winter

Figure 5.2.   Alternate layers of FeS and $CaCO_3$ in a lake sediment. This phenomenon has been observed in Lake Zürich.

The preceding are only a few examples of reactions that result in the formation of bottom sediments in bodies of water. Eventually these sediments may become covered and form sedimentary minerals.

5.3. SOLUBILITIES

The formation and stabilities of non-aqueous phases in water are strongly dependent upon solubilities. Solubility calculations are addressed in this section with emphasis upon solids and gases.

Solubilities of Solids

Generally the solubility of a solid in water is of concern when the solid is slightly soluble, often having such a low solubility that it is called "insoluble." In section 3.10 the solubility of lead carbonate was considered. This salt can introduce toxic lead ion into water by reactions such as

$$PbCO_3(s) \longleftrightarrow Pb^{2+} + CO_3^{2-} \qquad (3.10.13)$$

A relatively straightforward calculation of the solubility of an ionic solid can be performed on barium sulfate,[1] which dissolves according to the reaction

$$BaSO_4(s) \longleftrightarrow Ba^{2+} + SO_4^{2-} \qquad (5.3.1)$$

for which the equilibrium constant is the following:

$$K_{sp} = [Ba^{2+}][SO_4^{2-}] = 1.23 \times 10^{-10} \qquad (5.3.2)$$

An equilibrium constant in this form that expresses the solubility of a solid that forms ions in water is a **solubility product** and is designated K_{sp}. In the simplest cases a solubility product can be used alone to calculate the solubility of a slightly soluble salt in water. The solubility (S, moles per liter) of barium sulfate is calculated as follows:

$$[Ba^{2+}] = [SO_4^{2-}] = S \qquad (5.3.3)$$

$$[Ba^{2+}][SO_4^{2-}] = S \times S = K_{sp} = 1.23 \times 10^{-10} \qquad (5.3.4)$$

$$S = (K_{sp})^{1/2} = (1.23 \times 10^{-10})^{1/2} = 1.11 \times 10^{-5} \qquad (5.3.5)$$

Even such a simple calculation may be complicated by variations in activity coefficients resulting from differences in ionic strength. **Intrinsic solubilities** account for the fact that a significant portion of the solubility of an ionic solid is due to the dissolution of the neutral form of the salt and must be added to the solubility calculated from K_{sp} to obtain the total solubility as illustrated below for the calculation of the solubility of calcium sulfate. When calcium sulfate dissolves in water the two major reactions are

$$CaSO_4(s) \longleftrightarrow CaSO_4(aq) \qquad (5.3.6)$$

$$[CaSO_4(aq)] = 5.0 \times 10^{-3} \text{ M} \quad (25°C) \qquad (5.3.7)$$
$$\text{(Intrinsic solubility of } CaSO_4)$$

$$CaSO_4(s) \longleftrightarrow Ca^{2+} + SO_4^{2-} \qquad (5.3.8)$$

$$[Ca^{2+}][SO_4^{2-}] = K_{sp} = 2.6 \times 10^{-5} \quad (25°C) \qquad (5.3.9)$$

and the total solubility of $CaSO_4$ is calculated as follows:

$$S = [Ca^{2+}] + [CaSO_4(aq)] \qquad (5.3.10)$$
$$\uparrow \qquad\qquad \uparrow$$

Contribution to solubility   Contribution to solubility
from solubility product   from intrinsic solubilty

$$S = (K_{sp})^{1/2} + [CaSO_4(aq)] = (2.6 \times 10^{-5})^{1/2} + 5.0 \times 10^{-3}$$
$$= 5.1 \times 10^{-3} + 5.0 \times 10^{-3} = 1.01 \times 10^{-2} \, M \tag{5.3.11}$$

It is seen that in this case the intrinsic solubility accounts for half of the solubility of the salt.

In Section 3.10 it was seen that solubilities of ionic solids can be very much affected by reactions of cations and anions. It was shown that the solubility of $PbCO_3$ is increased by the chelation of lead ion by NTA,

$$Pb^{2+} + T^{3-} \longleftrightarrow PbT^- \tag{3.10.6}$$

increased by reaction of carbonate ion with H^+,

$$H^+ + CO_3^{2-} \longleftrightarrow HCO_3^- \tag{5.3.12}$$

and decreased by the presence of carbonate ion from water alkalinity:

$$CO_3^{2-}(\text{from dissociation of } HCO_3^-) + Pb^{2+} \longleftrightarrow PbCO_3(s) \tag{5.3.13}$$

These examples illustrate that reactions of both cations and anions must often be considered in calculating the solubilities of ionic solids.

Solubilities of Gases

In Section 2.7 it was mentioned that the solubilities of gases in water are described by Henry's Law which states that *at constant temperature the solubility of a gas in a liquid is proportional to the partial pressure of the gas in contact with the liquid*. For a gas, "X," this law applies to equilibria of the type

$$X(g) \longleftrightarrow X(aq) \tag{5.3.14}$$

and does not account for additional reactions of the gas species in water such as,

$$NH_3 + H_2O \longleftrightarrow NH_4^+ + OH^- \tag{5.3.15}$$

$$SO_2 + HCO_3^- \ (\text{From water alkalinity}) \longleftrightarrow CO_2 + HSO_3^- \tag{5.3.16}$$

which may result in much higher solubilities than predicted by Henry's law alone.

Mathematically, Henry's Law is expressed as

$$[X(aq)] = K \, P_X \tag{5.3.17}$$

where $[X(aq)]$ is the aqueous concentration of the gas, P_X is the partial pressure of the gas, and K is the Henry's Law constant applicable to a particular gas at a specified temperature. For gas concentrations in units of moles per liter and gas pressures in atmospheres, the units of K are $mol \times L^{-1} \times atm^{-1}$. Some values of K for dissolved gases that are significant in water are given in Table 5.1.

Table 5.1. Henry's Law Constants for Some Gases in Water at 25°C.

Gas	K, mol x L^{-1} x atm^{-1}
O_2	1.28×10^{-3}
CO_2	3.38×10^{-2}
H_2	7.90×10^{-4}
CH_4	1.34×10^{-3}
N_2	6.48×10^{-4}
NO	2.0×10^{-4}

In calculating the solubility of a gas in water, a correction must be made for the partial pressure of water by subtracting it from the total pressure of the gas. At 25°C the partial pressure of water is 0.0313 atm; values at other temperatures are readily obtained from standard handbooks. The concentration of oxygen in water saturated with air at 1.00 atm and 25°C may be calculated as an example of a simple gas solubility calculation. Considering that dry air is 20.95% by volume oxygen and factoring in the partial pressure of water gives the following:

$$P_{O_2} = (1.0000 \text{ atm} - 0.0313 \text{ atm}) \times 0.2095 = 0.2029 \text{ atm} \tag{5.3.18}$$

$$[O_2(aq)] = K \times P_{O_2} = 1.28 \times 10^{-3} \text{ mol x } L^{-1} \text{ x atm}^{-1} \times 0.2029 \text{ atm}$$

$$= 2.60 \times 10^{-4} \text{ mol x } L^{-1} \tag{5.3.19}$$

Since the molecular weight of oxygen is 32, the concentration of dissolved oxygen in water in equilibrium with air under the conditions given above is 8.32 mg/L, or 8.32 parts per million (ppm).

The solubilities of gases decrease with increasing temperature. Account is taken of this factor with the **Clausius-Clapeyron** equation,

$$\log \frac{C_2}{C_1} = \frac{\Delta H}{2.303R} \left[\frac{1}{T_1} - \frac{1}{T_2} \right] \tag{5.3.20}$$

where C_1 and C_2 denote the gas concentration in water at absolute temperatures of T_1 and T_2, respectively, ΔH is the heat of solution, and R is the gas constant. The value of R is 1.987 cal x deg^{-1} x mol^{-1}, which gives ΔH in units of cal/mol.

5.4. THE NATURE OF COLLOIDAL PARTICLES

Many minerals, some organic pollutants, proteinaceous materials, some algae, and some bacteria are suspended in water as very small particles. Such particles, which have some characteristics of both species in solution and larger particles in suspension, which range in diameter from about 0.001 micrometer (μm) to about 1 μm, and which scatter white light as a light blue hue observed at right angles to the incident light, are classified as **colloidal particles**. The characteristic light scattering phenomenon of colloids results from their being the same order of size as the wavelength of light and is called the **Tyndall effect**. The unique properties and behavior of colloidal particles are strongly influenced by their physical-chemical characteristics, including high specific area, high interfacial energy, and high surface/charge density ratio.

Colloids may be classified as *hydrophilic colloids*, *hydrophobic colloids*, or *association colloids*. These three classes are briefly summarized below.

Hydrophilic colloids generally consist of macromolecules, such as proteins and synthetic polymers, that are characterized by strong interaction with water resulting in spontaneous formation of colloids when they are placed in water. In a sense, hydrophilic colloids are solutions of very large molecules or ions. Suspensions of hydrophilic colloids are less affected by the addition of salts to water than are suspensions of hydrophobic colloids.

Hydrophobic colloids interact to a lesser extent with water and are stable because of their positive or negative electrical charges as shown in Figure 5.3. The charged surface of the colloidal particle and the **counter-ions** that surround it compose an **electrical double layer**, which causes the particles to repel each other. Hydrophobic colloids are usually caused to settle from suspension by the addition of salts. Examples of hydrophobic colloids are clay particles, petroleum droplets, and very small gold particles.

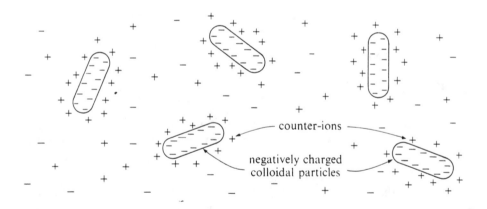

Figure 5.3. Representation of negatively charged hydrophopic colloidal particles surrounded in solution by positively charged counter-ions, forming an electrical double layer. (Colloidal particles suspended in water may have either a negative or positive charge.)

Association colloids consist of special aggregates of ions and molecules called **micelles**. To understand how this occurs, consider sodium stearate, a typical soap with the structural formula shown below:

$$H-\overset{H}{\underset{H}{C}}-\overset{H}{\underset{H}{C}}-\overset{H}{\underset{H}{C}}-\overset{H}{\underset{H}{C}}-\overset{H}{\underset{H}{C}}-\overset{H}{\underset{H}{C}}-\overset{H}{\underset{H}{C}}-\overset{H}{\underset{H}{C}}-\overset{H}{\underset{H}{C}}-\overset{H}{\underset{H}{C}}-\overset{H}{\underset{H}{C}}-\overset{H}{\underset{H}{C}}-\overset{H}{\underset{H}{C}}-\overset{H}{\underset{H}{C}}-\overset{H}{\underset{H}{C}}-\overset{H}{\underset{H}{C}}-\overset{H}{\underset{H}{C}}-\overset{O}{C}-O^-\ Na^+$$

Represented as ⌇⌇⌇⌇⌇⌇⌇⊖

The stearate ion has both a hydrophilic $-CO_2^-$ head and a long organophilic $CH_3(CH_2)_{16}^-$. As a result, stearate anions in water tend to form clusters consisting of as many as 100 anions clustered together with their hydrocarbon "tails" on the inside of a spherical colloidal particle and their ionic "heads" on the surface in contact with water and with Na^+ counterions. This results in the formation of **micelles** as illustrated in Figure 5.4.

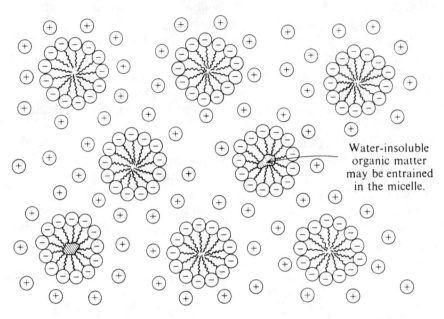

Water-insoluble organic matter may be entrained in the micelle.

Figure 5.4. Representation of colloidal soap micelle particles.

Colloid Stability

The stability of colloids is a prime consideration in determining their behavior. It is involved in important aquatic chemical phenomena including the formation of sediments, dispersion and agglomeration of bacterial cells, and dispersion and removal of pollutants (such as crude oil from an oil spill).

Discussed above, the two main phenomena contributing to the stabilization of colloids are **hydration** and **surface charge**. The layer of water on the surface of hydrated colloidal particles prevents contact, which would result in the formation of larger units. A surface charge on colloidal particles may prevent aggregation, since like-charged particles repel each other. The surface charge is frequently pH-dependent; around pH 7 most colloidal particles in natural waters are negatively charged. Negatively charged aquatic colloids include algal cells, bacterial cells, proteins, and colloidal petroleum droplets.

One of the three major ways in which a particle may acquire a surface charge is by **chemical reaction at the particle surface**. This phenomenon, which frequently involves hydrogen ion and is pH-dependent, is typical of hydroxides and oxides and is illustrated for manganese dioxide, MnO_2, in Figure 5.5.

As an illustration of pH-dependent charge on colloidal particle surfaces consider the effects of pH on the surface charge of hydrated manganese oxide, $MnO_2(H_2O)(s)$. In a relatively acidic medium, the reaction

$$MnO_2(H_2O)(s) + H^+ \rightarrow MnO_2(H_3O)^+(s) \qquad (5.4.1)$$

may occur on the surface giving the particle a net positive charge. In a more basic medium, hydrogen ion may be lost from the hydrated oxide surface to yield negatively charged particles:

$$MnO_2(H_2O)(s) \rightarrow MnO_2(OH)^-(s) + H^+ \tag{5.4.2}$$

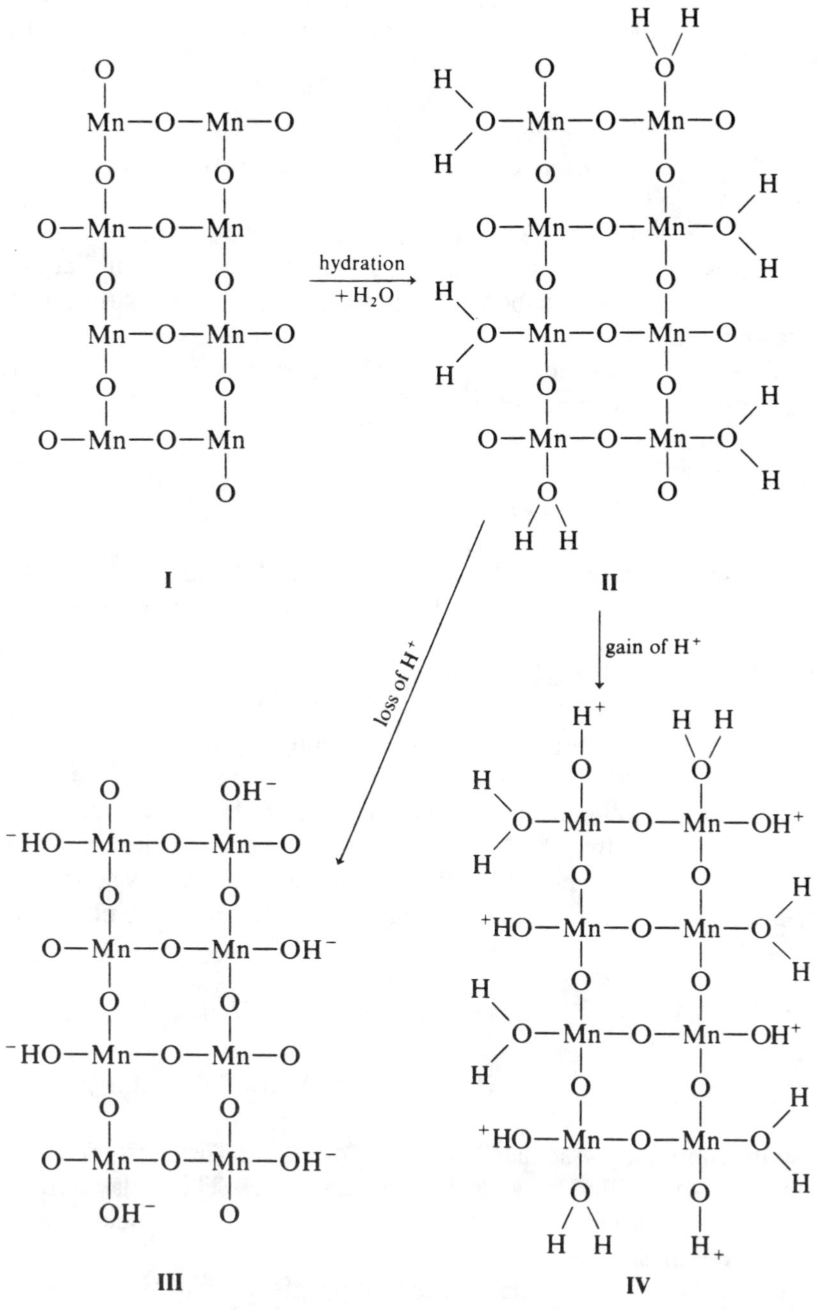

Figure 5.5. Acquisition of surface charge by colloidal MnO_2 in water. Anhydrous MnO_2 (I) has two O atoms per Mn atom. Suspended in water as a colloid, it binds to water molecules to form hydrated MnO_2 (II). Loss of H^+ from the bound H_2O yields a negatively charged colloidal particle (III). Gain of H^+ by surface O atoms yields a positively charged particle (IV). The former process (loss of H^+ ion) predominates for metal oxides.

At some intermediate pH value, called the **zero point of charge (ZPC)**, colloidal particles of a given hydroxide will have a net charge of zero, which favors aggregation of particles and precipitation of a bulk solid:

$$\text{Number of } MnO_2(H_3O)^+ \text{ sites } = \text{ Number of } MnO_2(OH)^- \text{ sites} \qquad (5.4.3)$$

Individual cells of microorganisms that behave as colloidal particles have a charge that is pH-dependent. The charge is acquired through the loss and gain of H^+ ion by carboxyl and amino groups on the cell surface:

$$^+H_3N(+ \text{ cell})CO_2H \quad ^+H_3N(\text{neutral cell})CO_2^- \quad H_2N(- \text{ cell})CO_2^-$$
$$\text{low pH} \qquad\qquad \text{intermediate pH} \qquad\qquad \text{high pH}$$

Ion absorption is a second way in which colloidal particles become charged. This phenomenon involves attachment of ions onto the colloidal particle surface by means other than conventional covalent bonding, including hydrogen bonding and London (Van der Waal) interactions.

Ion replacement is a third way in which a colloidal particle may gain a net charge. For example, replacement of some of the Si(IV) with Al(III) in the basic SiO_2 chemical unit in the crystalline lattice of some clay minerals,

$$[SiO_2] + Al(III) \rightarrow [AlO_2^-] + Si(IV) \qquad (5.4.4)$$

yields sites with a net negative charge. Similarly, replacement of Al(III) by a divalent metal ion such as Mg(II) in the clay crystalline lattice produces a net negative charge.

5.5. THE COLLOIDAL PROPERTIES OF CLAYS

Clays constitute the most important class of common minerals occurring as colloidal matter in water. The composition and properties of clays are discussed in some detail in Section 15.4 (as solid terrestrial minerals) and are briefly summarized here. **Clays** consist largely of hydrated aluminum and silicon oxides and are **secondary minerals**, which are formed by weathering and other processes acting on primary rocks (see Section 15.2). The general formulas of some common clays are given below:

- Kaolinite: $Al_2(OH)_4Si_2O_5$
- Montmorillonite: $Al_2(OH)_2Si_4O_{10}$

- Nontronite: $Fe_2(OH)_2Si_4O_{10}$
- Hydrous mica: $KAl_2(OH)_2(AlSi_3)O_{10}$

Iron and manganese are commonly associated with clay minerals. The most common clay minerals are illites, montmorillonites, chlorites, and kaolinites. These clay minerals are distinguished from each other by general chemical formula, structure, and chemical and physical properties.

Clays are characterized by layered structures consisting of sheets of silicon oxide alternating with sheets of aluminum oxide. Units of two or three sheets make up **unit layers**. Some clays, particularly the montmorillonites, may absorb large quantities of water between unit layers, a process accompanied by swelling of the clay.

As described in Section 5.4, clay minerals may attain a net negative charge by ion replacement, in which Si(IV) and Al(III) ions are replaced by metal ions of similar size but lesser charge. Compensation must be made for this negative charge by association of cations with the clay layer surfaces. Since these cations need not fit specific sites in the crystalline lattice of the clay, they may be relatively large ions, such as K^+, Na^+, or NH_4^+. These cations are called **exchangeable cations** and are exchangeable for other cations in water. The amount of exchangeable cations, expressed as milliequivalents (of monovalent cations) per 100 g of dry clay, is called the **cation-exchange capacity, CEC,** of the clay and is a very important characteristic of colloids and sediments that have cation-exchange capabilities.

Because of their structure and high surface area per unit weight, clays have a strong tendency to sorb chemical species from water. Thus, clays play a role in the transport and reactions of biological wastes, organic chemicals, gases, and other pollutant species in water. However, clay minerals also may effectively immobilize dissolved chemicals in water and so exert a purifying action. Some microbial processes occur at clay particle surfaces and, in some cases, sorption of organics by clay inhibits biodegradation. Thus, clay may play a role in the microbial degradation, or nondegradation, of organic wastes.

5.6. AGGREGATION OF PARTICLES

The processes by which particles aggregate and precipitate from colloidal suspension are quite important in the aquatic environment. For example, the settling of biomass during biological waste treatment depends upon the aggregation of bacterial cells. Other processes involving the aggregation of colloidal particles are the formation of bottom sediments and the clarification of turbid water for domestic or industrial use. Particle aggregation is complicated and may be divided into the two general classes of *coagulation* and *flocculation*. These are discussed below.

Colloidal particles are prevented from aggregating by the electrostatic repulsion between the electrical double layers (adsorbed-ion layer and counter-ion layer). **Coagulation** involves the reduction of this electrostatic repulsion, such that colloidal particles of identical materials may aggregate. **Flocculation** depends upon the presence of **bridging compounds,** which form chemically bonded links between colloidal particles and enmesh the particles in relatively large masses called **floc networks.**

Hydrophobic colloids often are readily coagulated by the addition of small quantities of salts that contribute ions to solution. Such colloids are stabilized by electrostatic repulsion. Therefore, the simple explanation of coagulation by ions in solution is that the ions reduce the electrostatic repulsion between particles to such an extent that the particles aggregate. Because of the double layer of electrical charge surrounding a charged particle, this aggregation mechanism is sometimes called **double-layer compression.** It is particularly noticeable in estuaries where sediment-laden fresh water flows into the sea, and is largely responsible for deltas formed where large rivers enter oceans.

The binding of positive ions to the surface of an initially negatively charged colloid can result in precipitation followed by colloid restabilization as shown in Figure 5.6. This kind of behavior is explained by an initial neutralization of the negative surface charge on the particles by sorption of positive ions, allowing coagulation to occur. As more of the source of positive ions is added, their sorption results in the formation of positive colloidal particles.

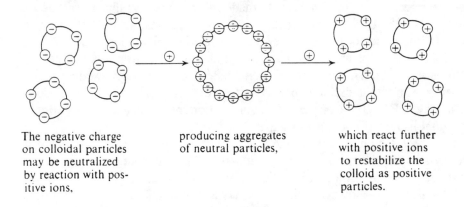

The negative charge on colloidal particles may be neutralized by reaction with positive ions,

producing aggregates of neutral particles,

which react further with positive ions to restabilize the colloid as positive particles.

Figure 5.6. Aggregation of negatively charged colloidal particles by reaction with positive ions, followed by restabilization as a positively charged colloid.

Flocculation of Colloids by Polyelectrolytes

Polyelectrolytes of both natural and synthetic origin are effective agents in flocculating colloids. Polyelectrolytes are polymers with a high formula weight that normally contain ionizable functional groups. Typical examples of synthetic polyelectrolytes are shown in Table 5.2.

It can be seen from Table 5.2 that anionic polyelectrolytes have negatively charged functional groups, such as $-SO_3^-$, and $-CO_2^-$. Cationic polyelectrolytes have positively charged functional groups, normally H^+ bonded to N. Nonionic polymers that serve as flocculants normally do not have charged functional groups.

Somewhat paradoxically, *anionic* polyelectrolytes may flocculate *negatively charged* colloidal particles. The mechanism by which this occurs involves bridging between the colloidal particles by way of the polyelectrolyte anions. Strong chemical bonding has to be involved, since both the particles and the polyelectrolytes are negatively charged. However, the process does occur and is particularly important in biological systems; for example, in the cohesion of tissue cells, clumping of bacterial cells, and antibody-antigen reactions.

The flocculation process induced by anionic polyelectrolytes is greatly facilitated by the presence of a low concentration of a metal ion capable of binding with the functional groups on the polyelectrolyte. The positively charged metal ion serves to form a bridge between the negatively charged anionic polyelectrolytes and negatively charged functional groups on the colloidal particle surface.

Flocculation of Bacteria by Polymeric Materials

The aggregation and settling of microorganism cells is a very important process in aquatic systems and is essential to the function of biological waste treatment systems. In biological waste treatment processes, such as the activated sludge process (Chapter 6), microorganisms utilize carbonaceous solutes in the water to produce biomass. The primary objective of biological waste treatment is the removal of carbonaceous material and, consequently, its oxygen demand. Part of the carbon is evolved from the water as CO_2, produced by the energy-yielding metabolic processes of the bacteria. However, a significant fraction of the carbon is removed as **bacterial floc**, consisting of aggregated bacterial cells that have settled from the water. The formation of this floc is obviously an important phenomenon in biological waste treatment. Polymeric substances, including polyelectrolytes, that are formed by the bacteria induce bacterial flocculation.

Table 5.2. Synthetic Polyelectrolytes and Neutral Polymers Used as Flocculants

Anionic polyelectrolytes   *Cationic polyelectrolytes*

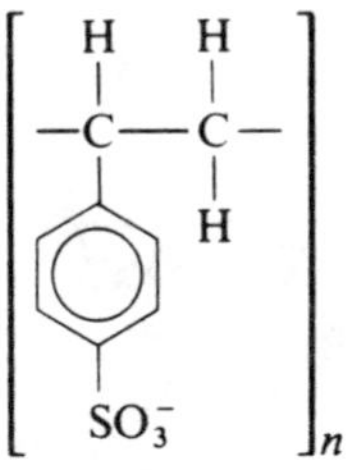

polystyrene
sulfonate

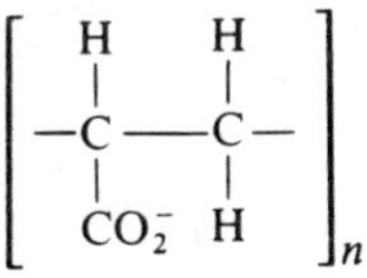

polyacrylate

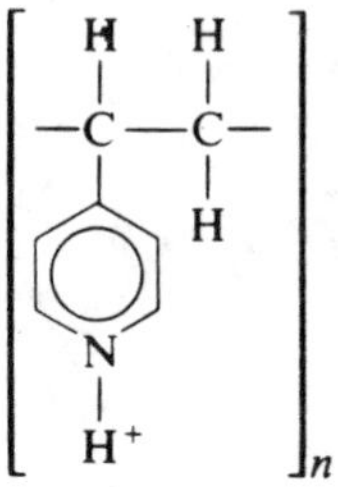

polyvinyl
pyridinium

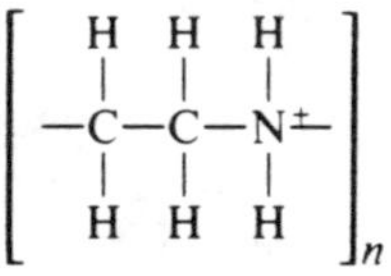

polyethylene
imine

Nonionic polymers

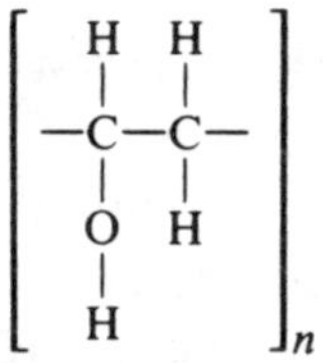

polyvinyl
alcohol

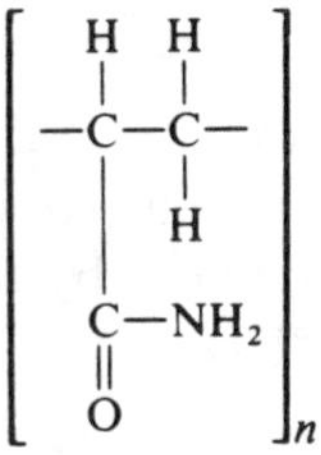

polyacrylamide

Within the pH range of normal natural waters (pH 5-9), bacterial cells are negatively charged. The ZPC of most bacteria is within the pH range 2-3. However, even at the ZPC, stable bacterial suspensions may exist. Therefore, surface charge is not necessarily required to maintain bacterial cells in suspension in water, and it is likely that bacterial cells remain in suspension because of the hydrophilic character of their surfaces. As a consequence, some sort of chemical interaction involving bridging species must be involved in bacterial flocculation.

5.7. SURFACE SORPTION BY SOLIDS

Many of the properties and effects of solids in contact with water have to do with the sorption of solutes by solid surfaces. Surfaces in finely divided solids tend to have excess surface energy because of an imbalance of chemical forces among surface atoms, ions, and molecules. Surface energy level may be lowered by a reduction in surface area. Normally this reduction is accomplished by aggregation of particles or by sorption of solute species.

Some kinds of surface interactions can be illustrated with metal oxide surfaces. (Such a surface, its reaction with water, and its subsequent acquisition of a charge by loss or gain of H^+ ion were shown in Figure 5.5 for MnO_2.) Such a surface may sorb metal ions, Mt^{z+}, by complexation to the surface,

$$M{-}OH + Mt^{z+} \longleftrightarrow M{-}OMt^{z-1} + H^+ \qquad (5.7.1)$$

or chelation with a metal ion:

$$
\begin{array}{c}
\text{M--OH} \\
\diagup\!\!\!\diagup \\
\diagup\!\!\!\diagup \quad + \ Mt^{z+} \\
\diagup\!\!\!\diagup \\
\text{M--OH}
\end{array}
\rightleftharpoons
\begin{array}{c}
\text{M--O} \\
\diagup\!\!\!\diagup \quad \diagdown \\
\diagup\!\!\!\diagup \quad\quad Mt^{z-2} + \ 2H^+ \\
\diagup\!\!\!\diagup \quad \diagup \\
\text{M--O}
\end{array}
\qquad (5.7.2)
$$

A metal ion complexed with a ligand, L, may bond by displacement of either H^+ or OH^-:

$$M{-}OH + MtL^{z+} \longleftrightarrow M{-}OMtL^{(z-1)} + H^+ \qquad (5.7.3)$$

$$M{-}OH + MtL^{z+} \longleftrightarrow M{-}(MtL)^{(z+1)} + OH^- \qquad (5.7.4)$$

Furthermore, in the presence of a ligand, dissociation of the complex and sorption of the metal complex and ligand must be considered as shown by the scheme below in which "(sorbed)" represents sorbed species and "(aq)" represents dissolved species:

$$Mt^{z+}(\text{sorbed}) \longleftrightarrow Mt^{z+}(\text{aq})$$

$$\uparrow\downarrow \qquad\qquad \uparrow\downarrow$$

$$MtL^{z+}(\text{sorbed}) \longleftrightarrow MtL^{z+}(\text{aq})$$

$$\uparrow\downarrow \qquad\qquad \uparrow\downarrow$$

$$L(\text{sorbed}) \longleftrightarrow L(\text{aq})$$

Some hydrated metal oxides, such as manganese(IV) oxide and iron(III) oxide, are especially effective in sorbing various species from aquatic solution. The sorption ability is especially pronounced for relatively fresh metal hydroxides or hydrated oxides such as colloidal MnO_2. This oxide usually is produced in natural waters by the oxidation of Mn(II) present in natural waters from the bacterially-mediated reduction of manganese oxides in anaerobic bottom sediments. Colloidal hydrated manganese(II) oxide can also be produced by the reduction of manganese(VII), which often is deliberately added to water as an oxidant in the form of permanganate salts to diminish taste and odor or to oxidize iron(II).

Freshly precipitated MnO_2 may have a surface area as large as several hundred square meters per gram. The hydrated oxide acquires a charge by loss and gain of H^+

ion and has a ZPC in an acidic pH range between 2.8 and 4.5. Since the pH of most normal natural waters exceeds 4.5, hydrous MnO_2 colloids are usually negatively charged.

The sorption of anions by solid surfaces is harder to explain than the sorption of cations. Phosphates may be sorbed on hydroxylated surfaces by displacement of hydroxides (ion exchange):

$$
\begin{array}{c}
\text{M—OH} \\
\\
\text{M—OH}
\end{array}
+ HPO_4^{2-} \rightleftharpoons
\begin{array}{c}
\text{M—O} \quad \text{OH} \\
\diagdown \diagup \\
\text{P} \\
\diagup \diagdown \\
\text{M—O} \quad \text{O}
\end{array}
+ 2OH^-
\qquad (5.7.5)
$$

The degree of anion sorption varies. As with phosphate, sulfate may be sorbed by chemical bonding, usually at a pH less than 7. Chloride and nitrate are sorbed by electrostatic attraction, such as occurs with positively charged colloidal particles in soil at a low pH. More specific bonding mechanisms may be involved in the sorption of fluoride, molybdate, selenate, selenite, arsenate, and arsenite anions.

5.8. ION EXCHANGE WITH BOTTOM SEDIMENTS

Bottom sediments are important sources of inorganic and organic matter in streams, fresh-water impoundments, estuaries, and oceans. It is incorrect to consider bottom sediments simply as wet soil. Normal soils are in contact with the atmosphere and are aerobic, whereas generally, the environment around bottom sediments is anaerobic, and they are subjected to reducing conditions. Bottom sediments undergo continuous leaching, whereas soils do not. The level of organic matter in sediments is generally higher than that in soils.

One of the most important characteristics of bottom sediments is their ability to exchange cations with the surrounding aquatic medium. **Cation-exchange capacity** (CEC) measures the capacity of a solid, such as a sediment, to sorb cations. It varies with pH and with salt concentration. Another parameter, **exchangeable cation status** (ECS), refers to the amounts of specific ions bonded to a given amount of sediment. Generally, both CEC and ECS are expressed as milliequivalents per 100 g of solid.

Because of the generally anaerobic nature of bottom sediments, special care must be exercised in their collection and treatment. Particularly, contact with atmospheric oxygen rapidly oxidizes exchangeable Fe^{2+} and Mn^{2+} to non-exchangeable oxides containing the metals in higher oxidation states as Fe_2O_3 and MnO_2. Therefore, sediment samples must be sealed and frozen as soon as possible after collection.

A common method for the determination of CEC consists of: (1) treating the sediment with a solution of an ammonium salt so that all exchangeable sites are occupied by NH_4^+ ion; (2) displacing the ammonium ion with a solution of NaCl and (3) determining the quantity of displaced ammonium ion. The CEC values may then be expressed as the number of milliequivalents of ammonium ion exchanged per 100 g of dried sample. Note that the sample must be dried *after* exchange.

The basic method for the determination of ECS consists of stripping all of the exchangeable metal cations from the sediment sample with ammonium acetate. Metal cations, including Fe^{2+}, Mn^{2+}, Zn^{2+}, Cu^{2+}, Ni^{2+}, Na^+, K^+, Ca^{2+}, and Mg^{2+}, are then determined in the leachate. Exchangeable hydrogen ion is very difficult to determine by direct methods. It is generally assumed that the total cation exchange capacity minus the sum of all exchangeable cations except hydrogen ion is equal to the exchangeable hydrogen ion.

Fresh water sediments typically have CEC values of 20-30 milliequivalents/100 g. The ECS values for individual cations typically range from less than 1 to 10-20 milliequivalents/100 g. Sediments are important repositories of metal ions that may be exchanged with surrounding waters. Furthermore, because of their capacity to sorb and release hydrogen ions, sediments have an important buffering effect in some waters.

Trace-Level Metals in Suspended Matter and Sediments

Sediments and suspended particles are important repositories for trace amounts of metals such as chromium, cadmium, copper, molybdenum, nickel, cobalt, and manganese. These metals may be present as discrete compounds, ions held by cation-exchanging clays, bound to hydrated oxides of iron or manganese, or chelated by insoluble humic substances. The form of the metals depends upon pE. Examples of specific trace-metal-containing compounds that may be stable in natural waters under oxidizing and reducing conditions are given in Table 5.3. Solubilization of metals from sedimentary or suspended matter is often a function of the complexing agents present. These include amino acids, such as histidine, tyrosine, or cysteine; citrate ion; and, in the presence of seawater, chloride ion. Suspended particles containing trace elements may be in the submicrometer size range. Although less available than metals in true solution, metals held by very small particles are more accessible than those in sediments. Among the factors involved in metal availability are the identity of the metal, its chemical form (type of binding, oxidation state), the nature of the suspended material, the type of organism, and the physical and chemical conditions in the water. The pattern of trace-metal occurrence in suspended matter in relatively unpolluted water tends to correlate well with that of the parent minerals from which the suspended solids originated; anomalies appear in polluted waters where industrial sources add to the metal content of the stream.

Table 5.3. Inorganic Trace Metal Compounds That May be Stable under Oxidizing and Reducing Conditions

| | Discrete compound that may be present | |
| | Oxidizing conditions | Reducing conditions |
Metal	(high pE)	(low pE, S(−II) present)
Cadmium	$CdCO_3$	CdS
Copper	$Cu_2(OH)_2CO_3$	CuS
Iron	$Fe_2O_3 \cdot x(H_2O)$	FeS, FeS_2
Mercury	HgO	HgS
Manganese	$MnO_2 \cdot x(H_2O)$	MnS, $MnCO_3$
Nickel	$Ni(OH)_2$, $NiCO_3$	NiS
Lead	$2\ PbCO_3 \cdot Pb(OH)_2$, $PbCO_3$	PbS
Zinc	$ZnCO_3$, $ZnSiO_3$	ZnS

Phosphorous Exchange with Bottom Sediments

Phosphorus is one of the key elements in aquatic chemistry and is thought to be the limiting nutrient in the growth of algae under many conditions. Exchange with sediments plays a role in making phosphorus available for algae and contributes, therefore, to eutrophication. Sedimentary phosphorus may be classified into the following types:

- **Phosphate minerals**, particularly hydroxyapatite, $Ca_5OH(PO_4)_3$

- **Nonoccluded phosphorus**, such as orthophosphate ion bound to the surface of SiO_2 or $CaCO_3$. Such phosphorous is generally more soluble and more available than occluded phosphorus (below).

- **Occluded phosphorus** consisting of orthophosphate ions contained within the matrix structures of amorphous hydrated oxides of iron and aluminum and amorphous aluminosilicates. Such phosphorus is not so readily available as nonoccluded phosphorus.

- **Organic phosphorus** incorporated within aquatic biomass, usually of algal or bacterial origin.

In some waters receiving heavy loads of domestic or industrial wastes, inorganic polyphosphates (from detergents, for example) may be present in sediments. Runoff from fields where liquid polyphosphate fertilizers have been used might possibly provide polyphosphates sorbed on sediments.

Organic Compounds on Sediments and Suspended Matter

Many organic compounds interact with suspended material and sediments in bodies of water. Settling of suspended material containing sorbed organic matter carries organic compounds into the sediment of a stream or lake. For example, this phenomenon is largely responsible for the presence of herbicides in sediments containing contaminated soil particles eroded from crop land. Some organics are carried into sediments by the remains of organisms or by fecal pellets from zooplankton that have accumulated organic contaminants.

Suspended particulate matter affects the mobility of organic compounds sorbed to particles. Furthermore, sorbed organic matter undergoes chemical degradation and biodegradation at different rates and by different pathways compared to organic matter in solution. There is, of course, a vast variety of organic compounds that get into water. As one would expect, they react with sediments in different ways, the type and strength of binding varying with the type of compound. The most common types of sediments considered for their organic binding abilities are clays, organic (humic) substances, and clay-humic substances complexes. Both clays and humic substances act as cation exchangers. Therefore, these materials sorb cationic organic compounds through ion exchange. This is a relatively strong sorption mechanism, greatly reducing the mobility and biological activity of the organic compound. When sorbed by clays, cationic organic compounds are generally held between the layers of the clay mineral structure where their biological activity is essentially zero.

Since most sediments lack strong anion exchange sites, negatively charged organics are not held strongly at all. Thus, these compounds are relatively mobile in water. Their biological activity (and biodegradability) remains high in water despite the presence of solids.

The degree of sorption of organic compounds is generally inversely proportional to their water solubility. The more water-insoluble compounds tend to be taken up strongly by lipophilic ("fat-loving") solid materials, such as humic substances (see Section 3.12). Compounds having a relatively high vapor pressure can be lost from water or solids by evaporation. When this happens, photochemical processes (see Chapter 9) can play an important role in their degradation.

The herbicide 2,4-D (2,4-dichlorophenoxyacetic acid) has been studied extensively in regard to sorption reactions. Most of these studies have dealt with pure clay minerals, however, whereas soils and sediments are likely to have a strong clay-fulvic acid complex component. The sorption of 2,4-D by such a complex can be described using an equation of the Freundlich isotherm type,

$$X = K C^n \tag{5.8.1}$$

where X is the amount sorbed per unit weight of solid, C is the concentration of 2,4-D at equilibrium, and n and K are constants. These values are determined by plotting log X versus log C. If a Freundlich-type equation is obeyed, the plot will be linear with a slope of n and an intercept of log K. In a study of the sorption of 2,4-D on an organoclay complex at 5°C,[2] n was found to be 0.76 and log K was 0.815. At 25°C, n was 0.83 and log K was 0.716.

Sorption of comparatively nonvolatile hydrocarbons by sediments removes these materials from contact with aquatic organisms but also greatly retards their biodegradation. Aquatic plants produce some of the hydrocarbons that are found in sediments. Photosynthetic organisms, for example, produce quantities of n-heptadecane. Pollutant hydrocarbons in sediments are indicated by a smooth chain-length distribution of n-alkanes and thus can be distinguished from hydrocarbons generated photosynthetically in the water. An analysis of sediments in Lake Zug, Switzerland, for example, has shown a predominance of pollutant petroleum hydrocarbons near densely populated areas.

The sorption of neutral species like petroleum obviously cannot be explained by ion-exchange processes. It probably involves phenomena such as Van der Waals forces (a term sometimes invoked when the true nature of an attractive force is not understood, but generally regarded as consisting of induced dipole-dipole interaction involving a neutral molecule) and hydrogen bonding.

Obviously, uptake of organic matter by suspended and sedimentary material in water is an important phenomenon. Were it not for this phenomenon, it is likely that pesticides in water would be much more toxic. Biodegradation is generally slowed down appreciably, however, by sorption by a solid. In certain intensively farmed areas, there is a very high accumulation of pesticides in the sediments of streams, lakes, and reservoirs. The sorption of pesticides by solids and the resulting influence on their biodegradation is an important consideration in the licensing of new pesticides.

The transfer of surface water to groundwater often results in sorption of some water contaminants by soil and mineral material.[3] To take advantage of this purification effect, some municipal water supplies are drawn from beneath the surface of natural or artificial river banks as a first step in water treatment. The movement of water from waste landfills to aquifers is also an important process (see Chapter 18) in which pollutants in the landfill leachate may be sorbed by solid material through which the water passes.

The sorption of dilute solutions of halogenated and aromatic hydrocarbons by soil and sand has been studied under simulated water infiltration conditions.[3] The relationship between the sorption equilibria observed may be expressed by the formula

$$S = K_pC \tag{5.8.2}$$

where S and C are the concentrations of hydrocarbons in the solid and liquid phases, respectively, and K_p is the partition coefficient. It was found that the two most important factors in estimating the sorption of nonpolar organic compounds were: (1) the fraction of organic carbon, f_{OC}, in the solid sorbents; and (2) the 1-octanol/water partition coefficient, K_{ow}, of the organic compound. (The K_{ow} value is a measure of the tendency of a solute to dissolve from water into immiscible 1-octanol. This long-chain alcohol mimics lipid (fat) tissue and K_{ow} is used to indicate a tendency toward bioaccumulation of solutes in water.) The K_p of individual compounds was determined using the following empirical relationship:

$$S = K_pC \tag{5.8.3}$$

The sorption was found to be reversible on the solids studied, which included natural aquifer material, river sediment, soil, sand, and sewage sludge. The organic compounds studied included methylbenzene compounds containing from 1 to 4 chlorine atoms, tetrachloroethylene, n-butylbenzene, benzene, acetophenone, tetrachloroethane, naphthalene, parathion, β-BHC, DDT (the latter three compounds are insecticides), pyrene, and tetracene.

5.9. Sorption of Gases–Gases in Interstitial Waters

Interstitial water consisting of water held by sediments is an important reservoir for gases in natural water systems. Generally, the gas concentrations in interstitial waters are different from those in the overlying water.

The results of the analyses of interstitial water in some sediments taken from Chesapeake Bay[4] are given in Table 5.4. Examination of this table shows that CH_4 could not be detected at the sediment surface, which is because the equilibrium

Table 5.4. Analysis of Interstitial Gases in Chesapeake Bay Sediment Samples

Gas	Depth	Gas concentration, mL/L
N_2	surface*	13.5
N_2	1.00 m	2.4
Ar	surface	0.35
Ar	1.00 m	0.12
CH_4	surface	0.00
CH_4	1.00 m	1.4×10^2

* Refers to interstitial water in the surface layer of sediment.

concentration of methane in air is very low, and it is biodegradable under aerobic conditions. However, of the gases analyzed, by far the highest concentration at a depth of 1 meter was that of methane. The methane is produced by the anaerobic fermentation of biodegradable organic matter, $\{CH_2O\}$, (see Section 6.6):

$$2\{CH_2O\} \;\rightarrow\; CH_4(g) \,+\, CO_2(g) \tag{5.9.1}$$

The concentrations of argon and nitrogen are much lower at a depth of 1 meter than they are at the sediment surface. This finding may be explained by the stripping action of the fermentation-produced methane rising to the sediment surface.

LITERATURE CITED

1. Manahan, Stanley E., *Quantitative Chemical Analysis*, Brooks/Cole Publishing Co., Pacific Grove, CA, 1986.

2. Means, J. C., and R. Wijayratne, "Role of Natural Colloids in the Transport of Hydrophobic Pollutants," *Science*, **215**, 968–970 (1982).

3. Schwarzenbach, R. P., and J. Westall, "Transport of Nonpolar Organic Compounds from Surface Water to Groundwater. Laboratory Sorption Studies," *Environmental Science and Technology* **15** 1360-7 (1982).

4. Reeburgh, W. S., "Determination of Gases in Sediments," *Environmental Science and Technology*, **2**, 140-1 (1968).

SUPPLEMENTARY REFERENCES

Hites, Ronald A., and S. J. Eisenreich, Eds., *Sources and Fates of Aquatic Pollutants*, Advances in Chemistry Series **216**, American Chemical Society, Washington, DC, 1987.

Rump, H. H., and H. Krist, *Laboratory Manual for the Examination of Water, Waste Water, and Soil*, VCH Publishers, New York, NY, 1989.

Jorgensen, S. E., and I. Johnsen, *Principles of Environmental Science and Technology*, second edition, Elsevier Science Publishers, Amsterdam, The Netherlands, 1989.

Southgate, D. A. T., I. T. Johnson, and G. R. Fenwick, Eds., *Nutrient Availability: Chemical and Biological Aspects*, Royal Society of Chemistry, Letchworth, England, 1989.

Baker, R. A., Ed., *Contaminants and Sediments*, 2 vols., Ann Arbor Science Publishers, Inc., Ann Arbor, Mich., 1980.

Drever, James L., *The Geochemistry of Natural Waters*, Prentice-Hall, Inc., Englewood Cliffs, New Jersey, 1982.

Faust, Samuel D., and Joseph V. Hunter, *Organic Compounds in Aquatic Environments*, Marcel Dekker, Inc., Marcel Dekker, Inc., New York, 1971.

Garrels, R. M., and C. L. Christ, *Solutions, Minerals and Equilibraia*, Harper and Row, Inc., New York, 1965.

Helmke, P. A., R. D. Koons, P. J. Schomberg, and I. K. Kiskander, "Determination of Trace Element Contamination of Sediments by Multielement Analysis of Clay-Size Fraction," *Environmental Science and Technology*, **10**, 984-989 (1977).

Hem., J. D., *Study and Interpretation of the Chemical Characteristics of Natural Water*, 2nd ed., U. S. Geological Survey Paper **1473**, U. S. Geological Survey, Washington, D. C., 1970.

Hiemenz, P. C., *Principles of Colloid and Surface Chemistry*, Marcel Dekker, Inc., New York, 1977.

Schindler, P. W., "Heterogeneous Equilibria Involving Oxides, Hydroxides, Carbonates, and Hydroxide Carbonates," Equilibrium Concepts in Natural Water Systems, Werner Stumm, Ed., ACS Advances in Chemistry Series No 67, American Chemical society, Washington, DC, 1967.

Shaw, D. J., *Introduction to Colloid and Surface Chemistry*, Butterworth Publishers, Inc., Woburn, Mass., 1980.

Stumm, Werner, and James J. Morgan, *Aquatic Chemistry*, 2nd ed., Wiley-Interscience, New York 1981.

van Olphen, H. *An Introduction to Clay Colloid Chemistry*, 2nd ed., John Wiley and Sons, Inc., New York, 1977.

QUESTIONS AND PROBLEMS

1. A sediment sample was taken from a lignite strip-mine pit containing highly alkaline (pH–10) water. Cations were displaced from the sediment by treatment with HCl. A total analysis of cations in the leachate yielded, on the basis of millimoles per 100 g of dry sediment, 150 millimole of Na^+, 5 millimoles of K^+, 20 millimoles of Mg^{2+}, and 75 millimoles of Ca^{2+}. What is the cation exchange capacity of the sediment in milliequivalents per 100 g of dry sediment?

2. What is the value of $[O_2(aq)]$ for water saturated with a mixture of 50% O_2, 50% N_2 by volume at 25°C and a total pressure of 1.00 atm?

3. Of the following, the least likely mode of transport of iron(III) in a normal stream is: (a) bound to suspended humic material, (b) bound to clay particles by cation exchange processes, (c) as suspended Fe_2O_3, (d) as soluble Fe^{3+} ion, (e) bound to colloidal clay-humic substance complexes.

4. How does freshly precipitated colloidal iron(III) hydroxide interact with many divalent metal ions in solution?

5. What stabilizes colloids made of bacterial cells in water?

6. The solubility of oxygen in water is 14.74 mg/L at 0°C and 7.03 mg/L at 35°C. Estimate the solubility at 50°C.

7. What is thought to be the mechanism by which bacterial cells aggregate?

8. What is a good method for the production of freshly precipitated MnO_2?

9. A sediment sample was equilibrated with a solution of NH_4^+ ion, and the NH_4^+ was later displaced by Na^+ for analysis. A total of 33.8 milliequivalents of NH_4^+ were bound to the sediment and later displaced by Na^+. After drying, the sediment weighed 87.2 g. What was its CEC in milliequivalents/100 g?

10. A sediment sample with a CEC of 67.4 milliequivalents/100 g was found to contain the following exchangeable cations in milliequivalents/100 g: Ca^{2+}, 21.3; Mg^{2+}, 5.2; Na^+, 4.4; K^+, 0.7. The quantity of hydrogen ion, H^+, was not measured directly. What was the ECS of H^+ in milliequivalents/100 g?

11. What is the meaning of *zero point of charge* as applied to colloids? Is the surface of a colloidal particle totally without charged groups at the ZPC?

12. The concentration of methane in an interstitial water sample was found to be 150 mL/L at STP. Assuming that the methane was produced by the fermentation of organic matter, $\{CH_2O\}$, what weight of organic matter was required to produce the methane in a liter of the interstitial water?

13. What is the difference between CEC and ECS?

14. Match the sedimentary mineral on the left with its conditions of formation on the right:
    (a) $FeS(s)$ (1) May be formed when anaerobic water is exposed to O_2.
    (b) $Ca_5OH(PO_4)_3$ (2) May be formed when aerobic water becomes anaerobic.
    (c) $Fe(OH)_3$ (3) Photosynthesis by-product.
    (d) $CaCO_3$ (4) May be formed when wastewater containing a particular kind of contaminant flows into a body of very hard water.

15. In terms of their potential for reactions with species in solution, how might metal atoms, M, on the surface of a metal oxide, MO, be described?

16. Air is 20.95% oxygen by volume. If air at 1.0000 atm pressure is bubbled through water at 25°C, what is the partial pressure of O_2 in the water?

17. The volume percentage of CO_2 in a mixture of that gas with N_2 was determined by bubbling the mixture at 1.00 atm and 25°C through a solution of 0.0100 M $NaHCO_3$ and measuring the pH. If the equilibrium pH was 6.50, what was the volume percentage of CO_2?

18. For what purpose is a polymer with the following general formula used?

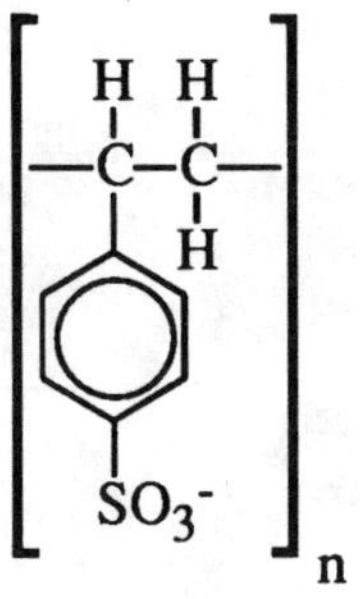

Aquatic Microbial Biochemistry

6.1. AQUATIC BIOCHEMICAL PROCESSES

Microorganisms — **bacteria, fungi, and algae** — are living catalysts that enable a vast number of chemical processes to occur in water and soil. A majority of the important chemical reactions occurring in water, particularly those involving organic matter and oxidation-reduction processes, occur through bacterial intermediaries. Algae are the primary producers of biological organic matter (biomass) in water. Microorganisms are responsible for the formation of many sediment and mineral deposits; they also play the dominant role in secondary waste treatment.

Pathogenic microorganisms must be eliminated from water purified for domestic use. In the past, major epidemics of typhoid, cholera, and other water-borne diseases resulted from pathogenic microorganisms in water supplies. Even today, constant vigilance is required to ensure that water for domestic use is free of pathogens.

Although this chapter considers primarily the role played by microorganisms in aquatic chemical transformations, special mention should be made of viruses in water. Viruses cannot grow by themselves, but reproduce in the cells of host organisms. They are only about 1/30-1/20 the size of bacterial cells, and they cause a number of diseases, such as polio, viral hepatitis, and perhaps cancer. It is thought that many of these diseases are waterborne.

Because of their small size (0.025-0.100 μm), and biological instability, viruses are difficult to isolate and culture. They often survive municipal water treatment, including chlorination. Thus, although viruses have no effect upon the overall environmental chemistry of water, they are an important consideration in the treatment and use of water.

Microorganisms important in aquatic chemistry may be divided among three categories: bacteria, fungi, and algae. Fungi and bacteria (with the exception of photosynthetic bacteria) are classified as reducers. Reducers break down chemical compounds to more simple species and thereby extract the energy needed for their growth and metabolism. Since reducers can utilize only chemical energy, and chemical transformation mediated by them must involve a net loss of free energy. However, compared to higher organisms, the energy utilization of bacteria and fungi is very efficient.

Algae are classified as producers, because they utilize light energy and store it as chemical energy. In the absence of sunlight, however, algae utilize chemical energy for their metabolic needs. In a sense, therefore, bacteria and fungi may be looked upon as environmental catalysts, whereas algae function as aquatic solar fuel cells. Some of the effects of microorganisms on the chemistry of water in nature are illustrated in Figure 6.1.

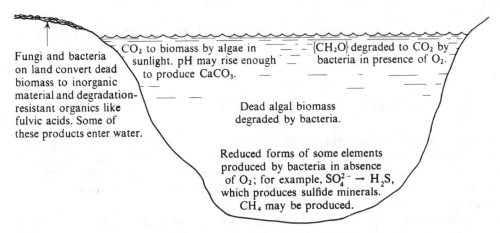

CO_2 to biomass by algae in sunlight. pH may rise enough to produce $CaCO_3$.

$\{CH_2O\}$ degraded to CO_2 by bacteria in presence of O_2.

Fungi and bacteria on land convert dead biomass to inorganic material and degradation-resistant organics like fulvic acids. Some of these products enter water.

Dead algal biomass degraded by bacteria.

Reduced forms of some elements produced by bacteria in absence of O_2; for example, $SO_4^{2-} \rightarrow H_2S$, which produces sulfide minerals. CH_4 may be produced.

Figure 6.1. Effects of microorganisms on the chemistry of water in nature.

6.2. ALGAE

The term algae is somewhat difficult to define. For the purposes of discussion here, algae may be considered as generally microscopic organisms that subsist on inorganic nutrients and produce organic matter from carbon dioxide by photosynthesis. The general nutrient requirements of algae are carbon (from CO_2 or HCO_3^-), nitrogen (generally as NO_3^-), phosphorus (as some form of orthophosphate), sulfur (as SO_4^{2-}), and trace elements including sodium, potassium, calcium, magnesium, iron, cobalt, and molybdenum.

In a highly simplified form, the production of organic matter by algal photosynthesis is described by the reaction

$$CO_2 + H_2O \xrightarrow{h\nu} \{CH_2O\} + O_2(g) \tag{6.2.1}$$

where $\{CH_2O\}$ represents a unit of carbohydrate and $h\nu$ stands for the energy of a quantum of light. Fogg[1] has represented the overall formula of the algae *Chlorella* as $C_{5.7}H_{9.8}O_{2.3}NP_{0.06}$. Using Fogg's formula for algal biomass exclusive of the phosphorus, the overall reaction for photosynthesis is:

$$5.7CO_2 + 3.4H_2O + NH_3 \xrightarrow{h\nu} C_{5.7}H_{9.8}O_{2.3}N + 6.25O_2(g) \tag{6.2.2}$$

In the absence of light, algae metabolize organic matter in the same manner as do nonphotosynthetic organisms. Thus, algae may satisfy their metabolic demands by utilizing chemical energy from the degradation of stored starches or oils, or from the consumption of algal protoplasm itself. In the absence of photosynthesis, the metabolic process consumes oxygen, so during the hours of darkness an aquatic system with a heavy growth of algae may become depleted in oxygen.

6.3. FUNGI

Fungi are nonphotosynthetic organisms. Most frequently they possess a filamentous structure. The morphology (structure) of fungi covers a wide range. Some fungi are as simple as the microscopic unicellular yeasts, whereas other fungi form large, intricate toadstools. The microscopic filamentous structures of fungi generally are much larger than bacteria and usually are 5-10 μm in width. Fungi are aerobic (oxygen-requiring) organisms and generally can thrive in more acidic media than can bacteria. They are also more tolerant of higher concentrations of heavy metal ions than are bacteria.

Perhaps the most important function of fungi in the environment is the breakdown of cellulose in wood and other plant materials. To accomplish this, fungal cells secrete a biological catalyst (enzyme, see Section 20.3) called cellulase. This enzyme breaks insoluble cellulose down to soluble carbohydrates that can be absorbed by the fungal cell. Because it acts outside the organism, it is called an extracellular enzyme or exoenzyme.

Although fungi do not grow well in water, they play an important role in determining the composition of natural waters and wastewaters because of the large amount of their decomposition products that enter water. An example of such a product is humic material, which interacts with hydrogen ions and metals (see Section 3.12).

6.4. BACTERIA

Bacteria may be shaped as rods, spheres, or spirals. They may occur individually or grow as groups ranging from two to millions of individual cells. Individual bacteria cells are very small and may be observed only through a microscope. Most bacteria fall into the size range of 0.5-3.0 microns. However, considering all species, a size range of 0.3-50 μm is observed. In general, it is assumed that a filter with 0.45-μm pores will remove all bacteria from water passing through it.

The metabolic activity of bacteria is greatly influenced by their small size. Their surface-to-volume ratio is extremely large, so that the inside of a bacterial cell is highly accessible to a chemical substance in the surrounding medium. Thus, for the same reason that a finely divided catalyst is more efficient than a more coarsely divided one, bacteria may bring about very rapid chemical reactions compared to those mediated by larger organisms. Bacteria excrete enzymes that can act outside the cell (exoenzymes) that break down solid food material to soluble components which can penetrate bacterial cell walls, where the digestion process is completed.

Bacterial cells have a number of separate components (Figure 6.2). Some cells are surrounded by a slime layer that is thought to protect the bacterial cells from attack by other microorganisms. The cell wall gives the cell form and rigidity. The layer immediately inside the cell wall is the cytoplasmic membrane, which controls the nature and quantity of materials transported into and out of the cell. The inside of the cell is filled with cytoplasm, the medium in which the cell's metabolic processes are carried out. The nuclear body (generally not considered to be true nucleus) controls metabolic processes and reproduction. In addition to these features, the cell may contain inclusions of reserve food material, consisting of fats, carbohydrates, and even elemental sulfur. Some bacteria possess moveable flagella, hair-like appendages that give the bacteria mobility. Like fungi and algae, bacteria produce spores, metabolically inactive bodies that form and survive under adverse conditions in a "resting" state until conditions favorable for growth occur.

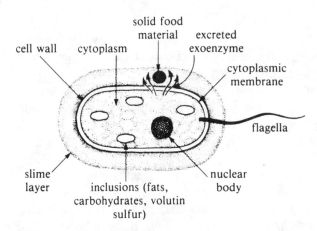

Figure 6.2. Schematic diagram of typical bacterial cell structure.

Although individual bacteria cells cannot be seen by the naked eye, bacterial colonies arising from individual cells are readily visible. A common method of counting individual bacterial cells in water consists of spreading a measured volume of water on a plate of agar gel containing bacterial nutrients. Wherever a viable bacterial cell adheres to the plate, a bacterial colony consisting of many cells will grow. These visible colonies are counted and related to the number of cells present initially. In most natural water samples it is necessary to dilute the sample with a sterile buffer solution to prevent formation of too many colonies.

Autotrophic and Heterotrophic Bacteria

Bacteria may be divided into two main categories, autotrophic and heterotrophic. **Autotrophic bacteria** are not dependent upon organic matter for growth and thrive in a completely inorganic medium; they use carbon dioxide or other carbonate species as a carbon source. A number of sources of energy may be used, depending upon the species of bacteria; however, a biologically mediated chemical reaction always supplies the energy.

An example of autotrophic bacteria is *Gallionella*. In the presence of oxygen, these bacteria are grown in a medium consisting of NH_4Cl, phosphates, mineral salts, CO_2 (as a carbon source), and solid FeS (as an energy source). It is believed that the following is the energy-yielding reaction for this species:

$$4FeS(s) + 9O_2 + 10H_2O \rightarrow 4Fe(OH)_3(s) + 4SO_4^{2-} + 8H^+ \qquad (6.4.1)$$

Starting with the simplest inorganic materials, autotrophic bacteria must synthesize all of the complicated proteins, enzymes, and other materials needed for life processes. It follows, therefore, that the biochemistry of autotrophic bacteria is quite complicated. Because of their consumption and production of a wide range of minerals, autotrophic bacteria are involved in many geochemical transformations.

Heterotrophic bacteria depend upon organic compounds, both for their energy and for the carbon required to build their biomass. They are much more common in occurrence than autotrophic bacteria. Heterotrophic bacteria are the microorganisms primarily responsible for the breakdown of pollutant organic matter in waters and of organic wastes in biological waste-treatment processes.

Algae are autotrophic organisms, using CO_2 as a carbon source and light as an energy source. Fungi are entirely heterotrophic, deriving carbon and energy from the degradation of organic matter.

Aerobic and Anaerobic Bacteria

Another classification system for bacteria depends upon their requirement for molecular oxygen. **Aerobic bacteria** require oxygen as an electron receptor:

$$O_2 + 4H^+ + 4e^- \rightarrow H_2O \tag{6.4.2}$$

Anaerobic bacteria function only in the complete absence of molecular oxygen. Frequently, molecular oxygen is quite toxic to anaerobic bacteria.

A third class of bacteria, facultative bacteria, utilize free oxygen when it is available and use other substances as electron receptors (oxidants) when molecular oxygen is not available. Common oxygen substitutes in water are nitrate ion (see Section 6.8) and sulfate ion (see Section 6.9).

Kinetics of Bacterial Growth

The population size of bacteria and unicellular algae as a function of time in a growth culture is illustrated by Figure 6.3, which shows a **population curve** for a bacterial culture. Such a culture is started by inoculating a rich nutrient medium with a small number of bacterial cells. The population curve consists of four regions. The first region is characterized by little bacterial reproduction and is called the *lag phase*. The lag phase occurs because the bacteria must become acclimated to the new medium. Following the lag phase comes a period of very rapid bacterial growth. This is the *log phase*, or exponential phase, during which the population doubles over a

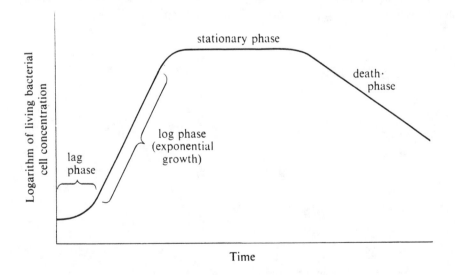

Figure 6.3. Population curve for a bacterial culture.

regular time interval called the generation time. This behavior can be described by a mathematical model in which growth rate is proportional to the number of individuals present and there are no limiting factors such as death or lack of food:

$$\frac{dn}{dt} = kN \tag{6.4.3}$$

This equation can be integrated to give

$$\ln \frac{N}{N_0} = kt \quad \text{or} \quad N = N_0 e^{kt} \tag{6.4.4}$$

where N is the population at time t and N_0 is the population at time t = 0. Thus, another way of describing population growth during the log phase is to say that the logarithm of bacterial population increases linearly with time. The generation time, or doubling time, is $(\ln 2)/k$, analogous to the half-life of radioactive decay. Fast growth during the log phase can cause very rapid microbial transformations of chemical species in water.

The log phase terminates and the *stationary phase* begins when a limiting factor is encountered. Typical factors limiting growth are depletion of an essential nutrient, build-up of toxic material, and exhaustion of oxygen. During the stationary phase, the number of viable cells remains virtually constant. Depending upon the bacterial species and other circumstances, the stationary phase may be either very long or very short in duration. After the stationary phase, the bacteria begin to die faster than they reproduce, and the population enters the *death phase*.

Temperature has a strong effect upon bacterial growth. However, a given temperature does not affect different kinds of bacteria in the same way, since they have different optimum temperatures for growth. **Psychrophilic bacteria** are bacteria having temperature optima below approximately 20°C. The temperature optima of **mesophilic bacteria** lie between 20°C and 45°C. Bacteria having temperature optima above 45°C are called **thermophilic bacteria**. The temperature range for optimum growth of bacteria is remarkably wide, with some bacteria being able to grow at 0°C, and some thermophilic bacteria existing as temperatures as high as 80°C.

Figure 6.4 shows the temperature dependence of the growth rate of a hypothetical species of bacteria with a temperature optimum of approximately 36°C. Typically,

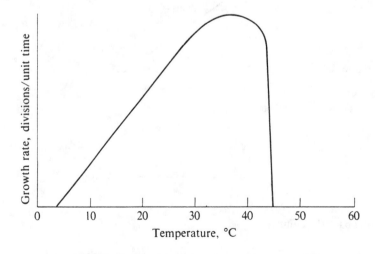

Figure 6.4. Bacterial growth rate as a function of temperature.

such curves have an optimum growth rate that is skewed toward the high temperature end of the curve and show an abrupt drop in growth rate beyond the temperature maximum. Presumably, vital enzymes are destroyed at temperatures not far above the optimum.

Over a relatively short range of temperature, a plot of bacterial growth rate as a function of the reciprocal of the absolute temperature, $1/K$, is linear. Students of physical chemistry will recognize such a relationship as an Arrhenius plot, which relates a kinetic rate constant to temperature.

As an interesting sidelight, some bacteria grow better, or even exclusively, at very high pressures in the oceans' depths, where pressures may be around 600 bars (a bar is 1.01 atm). Such bacteria are called barophilic bacteria.

6.5. MICROBIAL OXIDATION AND REDUCTION

Bacteria obtain the energy needed for their metabolic processes and reproduction by mediating redox reactions. Nature provides a large number of such reactions, and bacterial species have evolved that utilize many of these. The most environmentally important redox reactions occurring in water and soil through the action of bacteria are summarized in Table 6.1. As a consequence of their participation in such reactions, bacteria are involved in many biogeochemical processes in water and soil. Bacteria are essential participants in many important elemental cycles in nature, including those of nitrogen, carbon, and sulfur. They are responsible for the formation of many mineral deposits, including some of iron and manganese. On a smaller scale, some of these deposits form through bacterial action in natural water systems and even in pipes used to transport water.

Much of the remainder of this chapter is devoted to a discussion of important redox reactions mediated by bacteria, including those summarized in Table 6.1.

6.6. MICROBIAL TRANSFORMATIONS OF CARBON

Carbon is an essential life element and composes a high percentage of the dry weight of microorganisms. For most microorganisms, the bulk of net energy-yielding or energy-consuming metabolic processes involve changes in the oxidation state of carbon. These chemical transformations of carbon have important environmental implications. For example, when algae and other plants fix CO_2 as carbohydrate, represented as $\{CH_2O\}$,

$$CO_2 + H_2O \xrightarrow{h\nu} \{CH_2O\} + O_2(g) \tag{6.2.1}$$

carbon changes from the +4 to the 0 oxidation state. Energy from sunlight is stored as chemical energy in organic compounds. However, when the algae die, bacterial decomposition results in the reverse of the biochemical process represented by the above reaction, energy is released, and oxygen is consumed.

Table 6.1. Principal Microbially Mediated Oxidation and Reduction Reactions

TABLE 5.1 Principal Microbially Mediated Oxidation and Reduction Reactions

Oxidation	$pE^0(w)$[1]	Reduction	$pE^0(w)$[1]
(1) $\frac{1}{4}\{CH_2O\} + \frac{1}{4}H_2O \rightleftharpoons \frac{1}{4}CO_2(g) + H^+(w) + e$	-8.20	(A) $\frac{1}{4}O_2(g) + H^+(w) + e \rightleftharpoons \frac{1}{2}H_2O$	-13.75
(1a) $\frac{1}{2}HCOO^- \rightleftharpoons \frac{1}{2}CO_2(g) + \frac{1}{2}H^+(w) + e$	-8.73	(B) $\frac{1}{5}NO_3^- + \frac{6}{5}H^+(w) + e \rightleftharpoons \frac{1}{10}N_2(g) + \frac{3}{5}H_2O$	$+12.65$
(1b) $\frac{1}{2}\{CH_2O\} + \frac{1}{2}H_2O \rightleftharpoons \frac{1}{2}HCOO^- + \frac{3}{2}H^+(w) + e$	-7.68	(C) $\frac{1}{8}NO_3^- + \frac{5}{4}H^+(w) + e \rightleftharpoons \frac{1}{8}NH_4^+ + \frac{3}{8}H_2O$	$+6.15$
(1c) $\frac{1}{2}CH_3OH \rightleftharpoons \frac{1}{2}\{CH_2O\} + H^+(w) + e$	-3.01	(D) $\frac{1}{2}\{CH_2O\} + H^+(w) + e \rightleftharpoons \frac{1}{2}CH_3OH$	-3.01
(1d) $\frac{1}{2}CH_4(g) + \frac{1}{2}H_2O \rightleftharpoons \frac{1}{2}CH_3OH + H^+(w) + e$	$+2.88$	(E) $\frac{1}{8}SO_4^{2-} + \frac{9}{8}H^+(w) + e \rightleftharpoons \frac{1}{8}HS^- + \frac{1}{2}H_2O$	-3.75
(2) $\frac{1}{8}HS^- + \frac{1}{2}H_2O \rightleftharpoons \frac{1}{8}SO_4^{2-} + \frac{9}{8}H^+(w) + e$	-3.75	(F) $\frac{1}{8}CO_2(g) + H^+(w) + e \rightleftharpoons \frac{1}{8}CH_4(g) + \frac{1}{4}H_2O$	-4.13
(3) $\frac{1}{8}NH_4^+ + \frac{3}{8}H_2O \rightleftharpoons \frac{1}{8}NO_3^- + \frac{5}{4}H^+(w) + e$	$+6.16$	(G) $\frac{1}{6}N_2 + \frac{4}{3}H^+(w) + e \rightleftharpoons \frac{1}{3}NH_4^+$	-4.68
(4)[1] $FeCO_3(s) + 2\,H_2O \rightleftharpoons FeOOH(s) + HCO_3^-\ (10^{-3})$ $+ 2\,H^+(w) + e$	-1.67		
(5)[1] $\frac{1}{2}MnCO_3(s) + H_2O \rightleftharpoons \frac{1}{2}MnO_2(s) + \frac{1}{2}HCO_3^-\ (10^{-3})$ $+ \frac{3}{2}H^+(w) + e$	-8.5		

Sequence of Microbial Mediation

Model 1: Excess organic material (water initially contains O_2, NO_3^-, SO_4^{2-} and HCO_3^-). Examples: Hypolimniaan of a eutrophic lake, sediments, sewage treatment plant digester.

	Combination	$pE^0(w)^2$	$\Delta G^0(w)$, kcal
Aerobic respiration	(1) + (A)	21.95	−29.9
Denitrification	(1) + (B)	20.85	−28.4
Nitrate reduction	(1) + (C)	14.36	−19.6
Fermentation[3]	(1b) + (D)	4.67	−6.4
Sulfate reduction	(1) + (E)	4.45	−5.9
Methane fermentation	(1) + (F)	4.07	−5.6
N-fixation	(1) + (G)	3.52	−4.8

Model 2: Excess O_2 (water initially contains organic material, SH^-, NH_4^+, and possibly Fe(II) and Mn(II)). Examples: aerobic waste treatment, self-purification in streams, epilimnion of lake.

	Combination	$pE^0(w)^2$	$\Delta G^0(w)$, kcal
Aerobic respiration	(A) + (1)	21.95	−29.9
Sulfide oxidation	(A) + (2)	17.50	−23.8
Nitrification	(A) + (3)	7.59	−10.3
Ferrous oxidation[4]	(A) + (4)	15.42	−21.0
Mn(II) oxidation[4]	(A) + (5)	5.75	−7.2

[1] pE^0 at H^+ ion activity of 1.00×10^{-7}; $H^+(w)$ designates $[H^+] = 1.00 \times 10^{-7} M$. $pE^0(w)$ values in the left column are given for reduction, although the reaction is written as an oxidation.

[2] $pE^0(w) = \log K(w)$ for a reaction written for a one-electron transfer. The term $K(w)$ is the equilibrium constant for the reaction in which the activity of the hydrogen ion has been set at 1.00×10^{-7} and incorporated into the equilibrium constant.

[3] Fermentation is interpreted as an organic redox reaction where one organic substance is reduced by oxidizing another organic substance (for example, alcoholic fermentation; the products are metastable thermodynamically with respect to CO_2 and CH_4).

[4] The data for $pE^0(w)$ or $\Delta G^0(w)$ of these reactions correspond to $a_{HCO_3^-} = 1.00 \times 10^{-3} M$ rather than to unity.

Source: W. Stumm and J.J. Morgan, *Aquatic Chemistry*, New York: Wiley-Interscience, 1970, pp. 336–7. Reproduced with permission of John Wiley and Sons, Inc.

In the presence of oxygen, the principal energy-yielding reaction of bacteria is the oxidation of organic matter. Since it is generally more meaningful to compare reactions on the basis of the reaction of one electron-mole, the aerobic degradation of organic matter is conveniently written as

$$-\frac{1}{4}\{CH_2O\} + -\frac{1}{4}O_2(g) \rightarrow -\frac{1}{4}CO_2 + -\frac{1}{4}H_2O \qquad (6.6.1)$$

for which the free-energy change is -29.9 kcal (see aerobic respiration, Table 6.1). From this general type of reaction, bacteria and other microorganisms extract the energy needed to carry out their metabolic processes; to synthesize new cell material; for reproduction; and for locomotion.

Partial microbial decomposition of organic matter is a major step in the production of peat, lignite, coal, oil shale, and petroleum. Under reducing conditions, particularly below water, the oxygen content of the original plant material (approximate empirical formula, $\{CH_2O\}$) is lowered, leaving materials with relatively higher carbon contents.

Methane-Forming Bacteria

The production of methane in anoxic (oxygen-less) sediments is favored by high organic levels and low nitrate and sulfate levels. Methane production plays a key role in local and global carbon cycles as the final step in the anaerobic decomposition of organic matter. This process is the source of about 80% of the methane entering the atmosphere.

The carbon from microbially produced methane can come from either the reduction of CO_2 or the fermentation of organic matter, particularly acetate. The anoxic production of methane can be represented in the following simplified manner. When carbon dioxide acts as an electron receptor in the absence of oxygen, methane gas is produced:

$$-\frac{1}{8}CO_2 + H^+ + e^- \rightarrow -\frac{1}{8}CH_4 + -\frac{1}{4}H_2O \qquad (6.6.2)$$

This reaction is mediated by methane-forming bacteria. When organic matter is degraded microbially, the half-reaction for one electron-mole of $\{CH_2O\}$ is

$$-\frac{1}{4}\{CH_2O\} + -\frac{1}{4}H_2O \rightarrow -\frac{1}{4}CO_2 + H^+ + e^- \qquad (6.6.3)$$

Adding half-reactions 6.6.2 and 6.6.3 yields the overall reaction for the anaerobic degradation of organic matter by methane-forming bacteria, which involves a free-energy change of -5.55 kcal per electron-mole:

$$-\frac{1}{4}\{CH_2O\} \rightarrow -\frac{1}{8}CH_4 + -\frac{1}{8}CO_2 \qquad (6.6.4)$$

This reaction, in reality a series of complicated processes, is a **fermentation reaction**, defined as a redox process in which both the oxidizing agent and reducing agent are organic substances. It may be seen that only about one-fifth as much free energy is obtained from one electron-mole of methane formation as from a one electron-mole reaction involving complete oxidation of one electron-mole of the organic matter, Reaction 6.6.1.

There are four main categories of methane-producing bacteria. These bacteria, differentiated largely by morphology, are *Methanobacterium*, *Methanobacillus*, *Methanococcus*, and *Methanosarcina*. The methane-forming bacteria are *obligately anaerobic*; that is, they cannot tolerate the presence of molecular oxygen. The necessity of avoiding any exposure to oxygen makes the laboratory culture of these bacteria very difficult.

Methane formation is a valuable process responsible for the degradation of large quantities of organic wastes, both in biological waste-treatment processes (see Chapter 8) and in nature. Methane production is used in biological waste-treatment plants to further degrade excess sludge from the activated sludge process. In the bottom regions of natural waters, methane-forming bacteria degrade organic matter in the absence of oxygen. This eliminates organic matter which would otherwise require oxygen for its biodegradation. If this organic matter were transported to aerobic water containing dissolved O_2, it would exert a biological oxygen demand (BOD). Methane production is a very efficient means for the removal of BOD. The reaction,

$$CH_4 + 2O_2 \rightarrow CO_2 + 2H_2O \tag{6.6.5}$$

shows that 1 mole of methane requires 2 moles of oxygen for its oxidation to CO_2. Therefore, the production of 1 mole of methane and its subsequent evolution from water are equivalent to the removal of 2 moles of oxygen demand. In a sense, therefore, the removal of 16 grams (1 mole) of methane is equivalent to the addition of 64 grams (2 moles) of available oxygen to the water.

There is some potential for producing methane fuel from anaerobic digestion of organic wastes. Some installations use cattle feedlot wastes. Methane is routinely generated by the action of anaerobic bacteria and is used for heat and engine fuel at sewage treatment plants (see Chapter 8). Methane produced underground in old garbage dumps is being tapped by some municipalities; however, methane seeping into basements of buildings constructed on landfill containing garbage has caused serious explosions and fires.

Bacterial Utilization of Hydrocarbons

Methane is oxidized under aerobic conditions by a number of strains of bacteria. one of these, *Methanomonas*, is a highly specialized organism that cannot use any material other than methane as an energy source. Methanol, formaldehyde, and formic acid are intermediates in the microbial oxidation of methane to carbon dioxide. As discussed in Section 6.7, several types of bacteria can degrade higher hydrocarbons and use them as energy and carbon sources.

Microbial Utilization of Carbon Monoxide

Carbon monoxide is removed from the atmosphere by contact with soil. It has been found that carbon monoxide is removed rapidly from air in contact with soil. Since neither sterilized soil nor green plants grown under sterile conditions show any capacity to remove carbon monoxide from air, this ability must be due to microorganisms in the soil. Fungi capable of CO metabolism include some commonly-occurring strains of the ubiquitous *Penicillium* and *Aspergillus*. It is also possible that some bacteria are involved in CO removal. Whereas some microorganisms metabolize CO, other aquatic and terrestrial organisms produce this gas.

6.7. BIODEGRADATION OF ORGANIC MATTER

The biodegradation of organic matter in the aquatic and terrestrial environments is a crucial environmental process. Some organic pollutants are biocidal; for example, effective fungicides must be antimicrobial in action. Therefore, in addition to killing harmful fungi, fungicides frequently harm beneficial saprophytic fungi (fungi that decompose dead organic matter) and bacteria. Herbicides, which are designed for plant control, and insecticides, which are used to control insects, generally do not have any detrimental effect upon microorganisms.

The biodegradation of organic matter by microorganisms occurs by way of a number of stepwise, microbially catalyzed reactions. These reactions will be discussed individually with examples.

Oxidation

Oxidation occurs by the action of oxygenase enzymes (see Chapter 20 for a discussion of biochemical terms). The microbially catalyzed conversion of aldrin to dieldrin is an example of epoxide formation, a major step in many oxidation mechanisms. **Epoxidation** consists of adding an oxygen atom between two C atoms in an unsaturated system as shown below:

$$\text{(6.7.1)}$$

a particularly important means of metabolic attack upon aromatic rings that abound in many xenobiotic compounds.

Microbial Oxidation of Hydrocarbons

The degradation of hydrocarbons by microbial oxidation is an important environmental process because it is the primary means by which petroleum wastes are eliminated from water and soil. Bacteria capable of degrading hydrocarbons include *Micrococcus, Pseudomonas, Mycobacterium,* and *Nocardia.*

The most common initial step in the microbial oxidation of alkanes involves conversion of a terminal $-CH_3$ group to a $-CO_2$ group. More rarely, the initial enzymatic attack involves the addition of an oxygen atom to a nonterminal carbon, forming a ketone. After formation of a carboxylic acid from the alkane, further oxidation normally occurs by a process illustrated by the following reaction, a β–oxidation:

$$CH_3CH_2CH_2CH_2CH_2O_2H + 3O_2 \rightarrow CH_3CH_2CH_2O_2H + 2CO_2 + 2H_2O \quad \text{(6.7.2)}$$

Since 1904, it has been known that the oxidation of fatty acids involves oxidation of the β–carbon atom, followed by removal of two-carbon fragments. A complicated cycle with a number of steps is involved. The residue at the end of each cycle is an organic acid with two fewer carbon atoms than its precursor at the beginning of the cycle.

Hydrocarbon degradability varies and microorganisms show a strong preference for straight-chain hydrocarbons. A major reason for this preference is that branching inhibits β–oxidation at the site of the branch. The presence of a quaternary carbon (below) particularly inhibits alkane degradation.

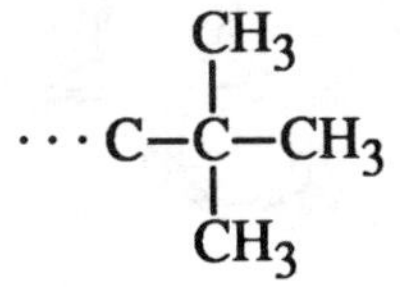

$$\cdots C-\underset{\underset{CH_3}{|}}{\overset{\overset{CH_3}{|}}{C}}-CH_3$$

Despite their chemical stability, aromatic rings are susceptible to microbial oxidation. The overall process leading to ring cleavage is

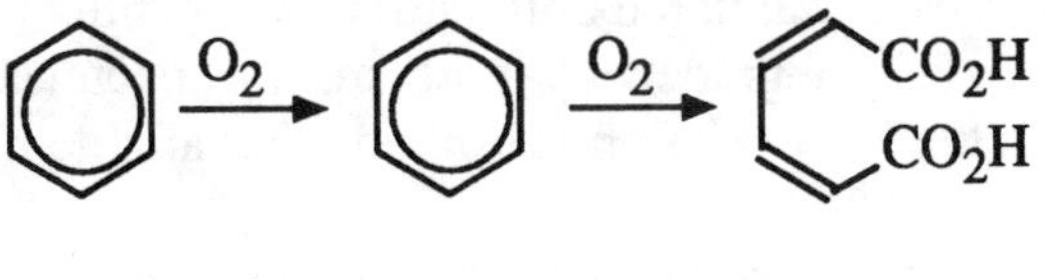

(6.7.2)

in which cleavage is preceded by addition of –OH to adjacent carbon atoms. Among the microorganisms that attack aromatic rings is the fungus *Cunninghamella elegans.*[2] It¢metabolizes a wide range of hydrocarbons including: C_3–C_{32} alkanes; alkenes; and aromatics, including toluene, naphthalene, anthracene, biphenyl, and phenanthrene. A study of the naphthalene by this organism led to the isolation of the following metabolites (the percentage yields are given in parentheses):

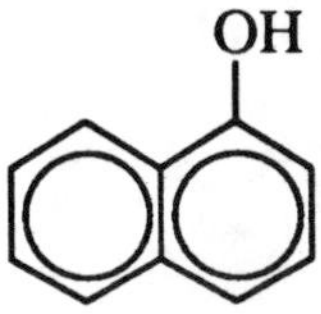

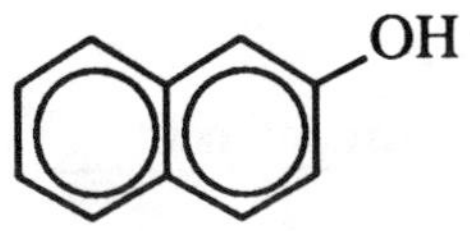

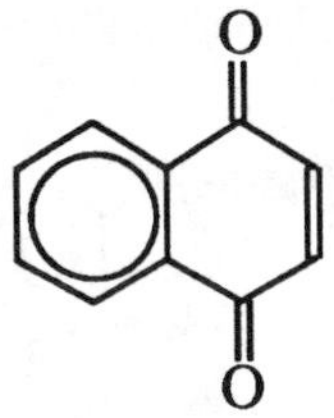

1-Naphthol (67.9%)   2-Naphthol (6.3%)   1,4-Naphthoquinone (2.8%)

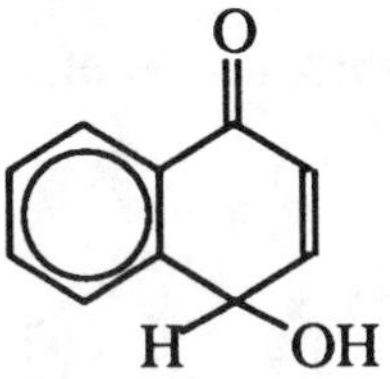

4-Hydroxy-1-tetralone (16.7%) 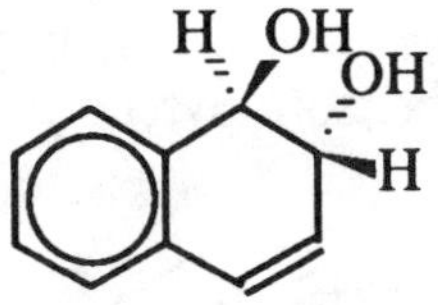   *Trans* -1,2-dihydroxy-
1,2-dihydronaphthalene (5.3%)

The initial attack of oxygen on naphthalene produces 1,2-naphthalene oxide (below), which reacts to form the other products shown above.

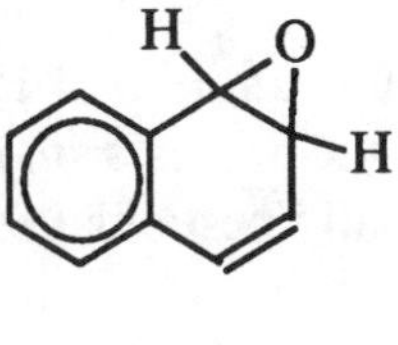

The biodegradation of an aromatic compound that is particularly important from the environmental viewpoint is that of the polynuclear aromatic hydrocarbon benzo-(a)pyrene:

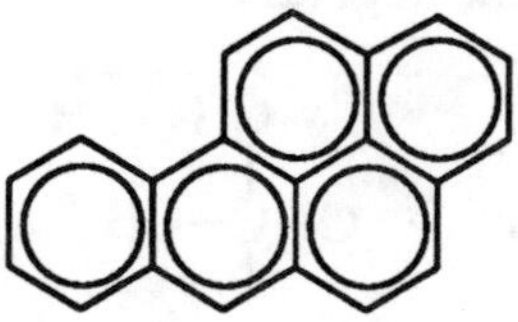

It proceeds by oxidation processes similar to those just described for aromatic rings and naphthalene. Metabolic products of benzo(a)pyrene bind to cellular DNA to cause cancer.

The biodegradation of petroleum is essential to the elimination of oil spills (about 5×10^6 metric tons per year). This oil is degraded by both marine bacteria and filamentous fungi. In some cases, the rate of degradation is limited by available nitrate and phosphate.

The physical form of crude oil makes a large difference in its degradability. Degradation in water occurs at the water-oil interface. Therefore, thick layers of crude oil prevent contact with bacterial enzymes and O_2. Apparently, bacteria synthesize an emulsifier that keeps the oil dispersed in the water as a fine colloid and therefore accessible to the bacterial cells.

Hydroxylation often accompanies microbial oxidation. It is the attachment of –OH groups to hydrocarbon chains or rings. It can follow epoxidation as shown by the following rearrangement reaction for benzene epoxide:

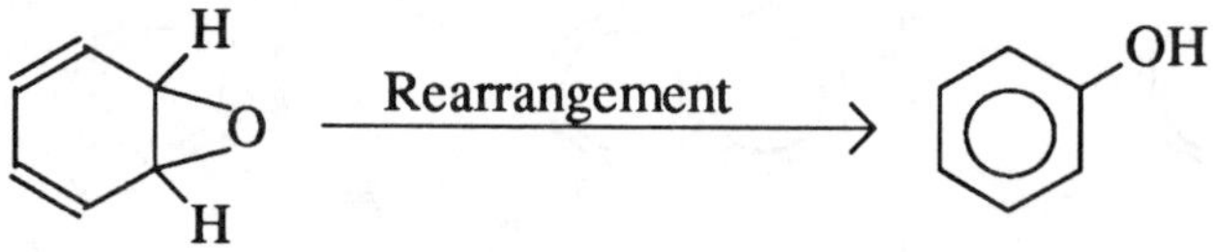

Hydroxylation can consist of the addition of more than one epoxide group. An example of epoxidation and hydroxylation is the metabolic production of the 7,8-diol-9,10-epoxide of benzo(a)pyrene as illustrated below:

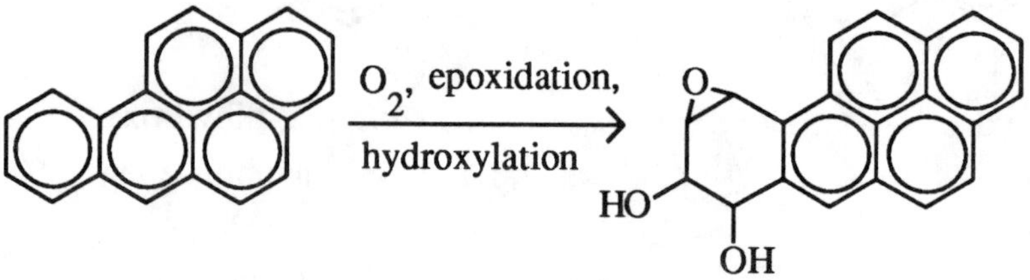

Other Biochemical Processes

Hydrolysis, which involves the addition of H_2O to a molecule accompanied by cleavage of the molecule into two species, is a major step in microbial degradation of many pollutant compounds, especially pesticidal esters, amides, and organophosphate esters. The types of enzymes that bring about hydrolysis are **hydrolase enzymes**, those that enable the hydrolysis of esters are called **esterases**, and those that hydrolyze amides are **amidases**. At least one species of *Pseudomonas* hydrolyzes malathion in a type of hydrolysis reaction typical of those by which pesticides are degraded:

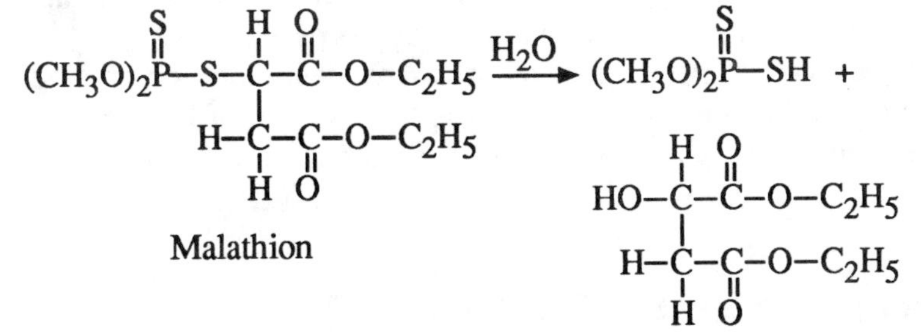

$$
\begin{array}{c}
\underset{\substack{\parallel \\ S}}{}\quad\underset{\substack{\mid \\ H}}{}\ \underset{\substack{\parallel \\ O}}{} \\
(CH_3O)_2P-S-C-C-O-C_2H_5 \xrightarrow{\ H_2O\ } (CH_3O)_2\overset{\displaystyle S}{\overset{\parallel}{P}}-SH\ + \\
H-C-C-O-C_2H_5 \\
\underset{H}{\mid}\ \underset{O}{\parallel} \\
\text{Malathion} \qquad\qquad\qquad\qquad\qquad
\end{array}
$$

(6.7.3)

HO–C–C–O–C₂H₅
H–C–C–O–C₂H₅

Reductions are carried out by **reductase enzymes**; for example, nitroreductase enzyme catalyzes the reduction of the nitro group. Table 6.2 gives the major kinds of functional groups reduced by microorganisms.

Table 6.2. Functional Groups that Undergo Microbial Reduction

Functional group	Process	Product
$R-\overset{\displaystyle O}{\overset{\parallel}{C}}-H$	Aldehyde reduction	$R-\overset{\displaystyle H}{\underset{\displaystyle H}{C}}-OH$
$R-\overset{\displaystyle O}{\overset{\parallel}{C}}-R'$	Ketone reduction	$R-\overset{\displaystyle H\ OH}{\underset{\displaystyle H}{C}}-R'$
$R-\overset{\displaystyle O}{\overset{\parallel}{S}}-R'$	Sulfoxide reduction	$R-S-R'$
$R-SS-R'$	Disulfide reduction	$R-SS-H$
$\overset{H}{\underset{H}{}}C{=}C\overset{H}{\underset{H}{}}$	Alkene reduction	$H-\overset{H}{\underset{H}{C}}-\overset{H}{\underset{H}{C}}-OH$
$R-NO_2$	Nitro reduction	$R-NO,\ \ R-N\overset{H}{\underset{H}{}},\ R-N\overset{H}{\underset{OH}{}}$

Dehalogenation reactions involving the bacterially-mediated replacement of a halogen atom with –OH are discussed in more detail in Section 6.10.

Ring cleavage is a crucial step in the ultimate degradation of organic compounds having aromatic rings. Normally, ring cleavage follows the addition of –OH groups (hydroxylation).

Many environmentally significant organic compounds contain alkyl groups, such as the methyl (–CH₃) group, attached to atoms of O, N, and S. An important step in the microbial metabolism of many of these compounds is **dealkylation**, replacement of alkyl groups by H as shown in Figure 6.5. Examples of these kinds of reactions include O-dealkylation of methoxychlor insecticides, N-dealkylation of carbaryl insecticide, and S-dealkylation of dimethyl mercaptan. Alkyl groups removed by dealkylation usually are attached to oxygen, sulfur, or nitrogen atoms; those attached to carbon are normally not removed directly by microbial processes.

An important step in the metabolism of the many xenobiotic compounds that contain covalently bound halogens (F, Cl, Br, I) is the removal of halogen atoms, a process called **dehalogenation**. This process is discussed further in Section 6.10.

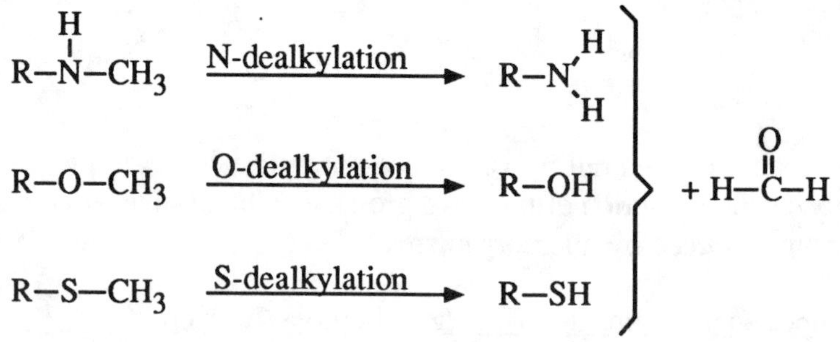

$$
\begin{array}{l}
\underset{H}{R-\overset{\displaystyle H}{N}-CH_3} \xrightarrow{\text{N-dealkylation}} R-N\overset{H}{\underset{H}{}} \\[2em]
R-O-CH_3 \xrightarrow{\text{O-dealkylation}} R-OH \\[2em]
R-S-CH_3 \xrightarrow{\text{S-dealkylation}} R-SH
\end{array}
\left.\right\} \; + \; H-\overset{\displaystyle O}{\underset{\displaystyle \|}{C}}-H
$$

Figure 6.5. Metabolic dealkylation reactions shown for the removal of CH_3 from N, O and S atoms in organic compounds.

6.8. MICROBIAL TRANSFORMATIONS OF NITROGEN

Some of the most important microorganism-mediated chemical reactions in aquatic and soil environments are those involving nitrogen compounds. They are summarized in the **nitrogen cycle** shown in Figure 6.6. This cycle describes the dynamic processes through which nitrogen is interchanged among the atmosphere, organic matter, and inorganic compounds. It is one of nature's most vital dynamic processes.

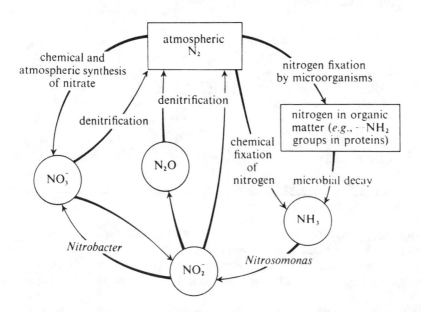

Figure 6.6. The nitrogen cycle.

Among the biochemical transformations in the nitrogen cycle are nitrogen fixation, whereby molecular nitrogen is fixed as organic nitrogen; nitrification, the process of oxidizing ammonia to nitrate; nitrate reduction, the process by which nitrogen in nitrate ion is reduced to form compounds having nitrogen in a lower oxidation state; and denitrification, the reduction of nitrate and nitrite to N_2, with a resultant net loss of nitrogen gas to the atmosphere. Each of these important chemical processes will be discussed separately.

Nitrogen Fixation

The overall microbial process for the **nitrogen fixation**, the binding of atmospheric nitrogen in a chemically combined form,

$$3\{CH_2O\} + 2N_2 + 2H_2O + 4H^+ \rightarrow 3\,CO_2 + 4NH_4^+ \tag{6.8.1}$$

is actually quite complicated and not completely understood. Biological nitrogen fixation is a key biochemical process in the environment and is essential for plant growth in the absence of synthetic fertilizers.

Only a few species of aquatic microorganisms have the ability to fix atmospheric nitrogen. Among the aquatic bacteria having this capability are photosynthetic bacteria, *Azotobacter*, and several species of *Clostridium*. Among the algae, blue-green algae can fix atmospheric nitrogen. In most natural fresh water systems, however, the fraction of nitrogen fixed by organisms in the water, relative to that originating from the decay of organic material, fertilizer runoff, and other external sources, is quite low.

The best-known and most important form of nitrogen-fixing bacteria is *Rhizobium*, which enjoys a symbiotic (mutually advantageous) relationship with leguminous plants such as clover or alfalfa. The *Rhizobium* bacteria are found in root nodules, special structures attached to the roots of legumes (see Figure 16.2). The nodules develop as a result of the bacteria "irritating" the root hairs of the developing legume plants. The nodules are connected directly to the vascular (circulatory) system of the plant, enabling the bacteria to derive photosynthetically-produced energy directly from the plant. Thus, the plant provides the energy required to break the strong triple bonds in the dinitrogen molecule, converting the nitrogen to a reduced form which is directly assimilated by the plant. When the legumes die and decay, NH_4^+ ion is released and is converted by microorganisms to nitrate ion which is assimilable by other plants. Some of the ammonium ion and nitrate released may be carried into natural water systems.

Some nonlegume angiosperms fix nitrogen through the action of actinomycetes bacteria contained in root nodules. Shrubs and trees in the nitrogen-fixing category are abundant in fields, forests, and wetlands throughout the world. Their rate of nitrogen fixation is comparable to that of legumes.

Free-living bacteria associated with some grasses are stimulated by the grasses to fix nitrogen. One such bacterium is *Spirillum lipoferum*. In tropical surroundings, the among of reduced nitrogen fixed by such bacteria can amount to the order of 100 kg per hectare per year.

Because of the cost of energy required to fix nitrogen synthetically, efforts are underway to increase the efficiency of natural means of nitrogen fixation. One approach uses recombinant DNA methodologies in attempts to transfer the nitrogen-fixing capabilities of nitrogen-fixing bacteria directly to plant cells. Though a fascinating possibility, this transfer has not yet been achieved on a practical basis. The other approach uses more conventional plant breeding and biological techniques in attempts to increase the range and effectiveness of the symbiotic relationship existing between some plants and nitrogen-fixing bacteria.

One matter of concern is that successful efforts to increase nitrogen fixation may upset the global nitrogen balance. Total global fixation of nitrogen was estimated at 237 million metric tons in 1974, which was up from an estimated 174 million metric tons in 1950 and 150 million metric tons in 1850. Potential accumulation of excess fixed nitrogen is the subject of some concern because of aquatic nitrate pollution and microbial production of N_2O gas. Some atmospheric scientists fear that excess N_2O gas may be involved in depletion of the protective atmospheric ozone layer (see Chapter 14).

Nitrification

Nitrification, the conversion of N(-III) to N(V), is a very common and extremely important process in water and in soil. As we have seen, aquatic nitrogen in thermodynamic equilibrium with air is in the +5 oxidation state as NO_2^-, whereas in most biological compounds, nitrogen is present as N(-III), such as $-NH2$ in amino acids. The equilibrium constant of the overall nitrification reaction, written for one electron-mole,

$$\frac{1}{4}O_2 + \frac{1}{8}NH_4^+ \rightarrow \frac{1}{8}NO_3^- + \frac{1}{4}H^+ + \frac{1}{8}H_2O \tag{6.8.2}$$

is $10^{7.59}$ (Table 6.1), so the reaction is highly favored from a thermodynamic viewpoint.

Nitrification is especially important in nature because nitrogen is absorbed by plants primarily as nitrate. When fertilizers are applied in the form of ammonium salts or anhydrous ammonia, a microbial transformation to nitrate enables maximum assimilation of nitrogen by the plants.

The nitrification conversion of ammoniacal nitrogen to nitrate ion takes place if extensive aeration is allowed to occur in the activated sludge sewage-treatment process (see Chapter 8). As the sewage sludge settles out in the settler, the bacteria in the sludge carry out while using this nitrate as an oxygen source (denitrification Reaction 6.8.8), producing N_2. The bubbles of nitrogen gas cause the sludge to rise, so that it does not settle properly. This can hinder the proper treatment of sewage through carryover of sludge into effluent water.

In nature, nitrification is catalyzed by two groups of bacteria, *Nitrosomonas* and *Nitrobacter*. *Nitrosomonas* bacteria bring about the transition of ammonia to nitrite,

$$NH_3 + \frac{3}{2}O_2 \rightarrow H^+ + NO_3^- + H_2O \tag{6.8.3}$$

whereas *Nitrobacter* mediate the oxidation of nitrite to nitrate:

$$NO_2^- + \frac{1}{2}O_2 \rightarrow NO_3^- \tag{6.8.4}$$

Both of these highly specialized types of bacteria are *obligate aerobes*; that is, they function only in the presence of molecular O_2. These bacteria are also *chemo-*

lithotrophic, meaning that they can utilize oxidizable inorganic materials as electron donors in oxidation reactions to yield needed energy for metabolic processes.

For the aerobic conversion of one electron-mole of ammoniacal nitrogen to nitrite ion at pH 7.00,

$$\frac{1}{4}O_2 + \frac{1}{6}NH_4^+ \rightarrow \frac{1}{6}NO_2^- + \frac{1}{3}H^+ + \frac{1}{6}H_2O \qquad (6.8.5)$$

the free-energy change is -10.8 kcal. The free-energy change for the aerobic oxidation of one electron-mole of nitrite ion to nitrate ion,

$$\frac{1}{4}O_2 + \frac{1}{2}NO_2^- \rightarrow \frac{1}{2}NO_3^- \qquad (6.8.6)$$

is -9.0 kcal. Both steps of the nitrification process involve an appreciable yield of free energy. It is interesting to note that the free-energy yield per electron-mole is approximately the same for the conversion of NH_4^+ to NO_2^- as it is for the conversion of NO_2^- to NO_3^-, about 10 kcal/electron-mole.

Nitrate Formation

As a general term, **nitrate reduction** refers to microbial processes by which nitrogen in chemical compounds is reduced to lower oxidation states. In the absence of free oxygen, nitrate may be used by some bacteria as an alternate electron receptor. The most complete possible reduction of nitrogen in nitrate ion involves the acceptance of 8 electrons by the nitrogen atom, with the consequent conversion of nitrate to ammonia (+V to -III oxidation state). Nitrogen is an essential component of protein, and any organism that utilizes nitrogen from nitrate for the synthesis of protein must first reduce the nitrogen to the -III oxidation state (ammoniacal form). However, incorporation of nitrogen into protein generally is a relatively minor use of the nitrate undergoing microbially mediated reactions and is more properly termed nitrate assimilation.

Generally, when nitrate ion functions as an electron receptor, the product is NO_2^-:

$$\frac{1}{2}NO_3^- + \frac{1}{4}\{CH_2O\} \rightarrow \frac{1}{2}NO_2^- + \frac{1}{4}H_2O + \frac{1}{4}H_2O \qquad (6.8.7)$$

The free-energy yield per electron-mole is only about 2/3 of the yield when oxygen is the oxidant; however, nitrate ion is a good electron receptor in the absence of O_2. One of the factors limiting the use of nitrate ion in this function is its relatively low concentration in most waters. Furthermore, nitrite, NO_2^-, is relatively toxic and tends to inhibit the growth of many bacteria after building up to a certain level. Sodium nitrate is sometimes used as a "first-aid" treatment in sewage lagoons that have become oxygen-deficient. It provides an emergency source of oxygen to reestablish normal bacterial growth.

Denitrification

An important special case of nitrate reduction is **denitrification**, in which the reduced nitrogen product is a nitrogen-containing gas. At pH 7.00, the free-energy change per electron-mole of reaction,

$$\frac{1}{5}NO_3^- + \frac{1}{4}\{CH_2O\} + \frac{1}{5}H^+ \rightarrow \frac{1}{10}N_2 + \frac{1}{4}CO_2 + \frac{7}{20}H_2O \qquad (6.8.8)$$

is -2.84 kcal. The free-energy yield per mole of nitrate reduced to N_2 (5 electron-moles) is lower than that for the reduction of the same quantity of nitrate to nitrite. More important, however, the reduction of a nitrate ion to N_2 gas consumes 5 electrons, compared to only 2 electrons for the reduction of NO_3^- to NO_2^-.

Denitrification is an important process in nature. It is the mechanism by which fixed nitrogen is returned to the atmosphere. Denitrification is also useful in advanced water treatment for the removal of nutrient nitrogen (see Chapter 8)., Because nitrogen gas is a nontoxic volatile substance that does not inhibit microbial growth, and since nitrate ion is a very efficient electron acceptor, denitrification allows the extensive growth of bacteria under anaerobic conditions.

Loss of nitrogen to the atmosphere may also occur through the formation of N_2O and NO by bacterial action on nitrate and nitrite catalyzed by the action of several types of bacteria. Production of N_2O relative to N_2 is enhanced during denitrification in soils by increased concentrations of NO_3^-, NO_2^-, and O_2

Competitive Oxidation of Organic Matter by Nitrate Ion and Other Oxidizing Agents

The successive oxidation of organic matter by dissolved O_2, NO_3^-, and SO_4^{2-} brings about an interesting sequence of nitrate-ion levels in sediments and hypolimnion waters initially containing O_2 but lacking a mechanism for reaeration[3]. This is shown in Figure 6.7, where concentrations of dissolved O_2, NO_3^-, and SO_4^{2-} are plotted as a function of total organic matter metabolized:

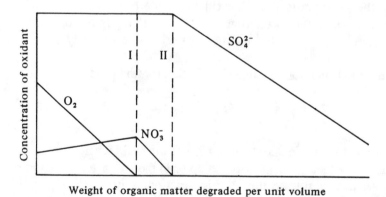

Figure 6.7. Oxidation of organic matter by O_2, NO_3^-, and SO_4^{2-}.

This behavior can be explained by the following sequence of biochemical processes:

$$O_2 + \text{organic matter} \rightarrow \text{products} \qquad (6.8.9)$$

$$NO_3^- + \text{organic matter} \rightarrow \text{products} \qquad (6.8.10)$$

$$SO_4^{2-} + \text{organic matter} \rightarrow \text{products} \qquad (6.8.11)$$

So long as some O_2 is present, some nitrate may be produced from organic matter. After exhaustion of molecular oxygen, nitrate is the favored oxidizing agent, and its concentration falls from a maximum value (I) to zero (II). Sulfate, which is usually present in a large excess over the other two oxidants, then becomes the favored electron receptor, enabling biodegradation of organic matter to continue.

6.9. MICROBIAL TRANSFORMATIONS OF PHOSPHORUS AND SULFUR

Phosphorus Compounds

Biodegradation of phosphorus compounds is important in the environment for two reasons. The first of these is that it provides a source of algal nutrient orthophosphate from the hydrolysis of polyphosphates (see Section 3.11). Secondly, biodegradation deactivates highly toxic organophosphate compounds, such as the organophosphate insecticides.

The organophosphorus compounds of greatest environmental concern tend to be sulfur-containing **phosphorothionate** and **phosphorodithioate** ester insecticides with the general formulas illustrated in Figure 6.8, where R represents a hydrocarbon or

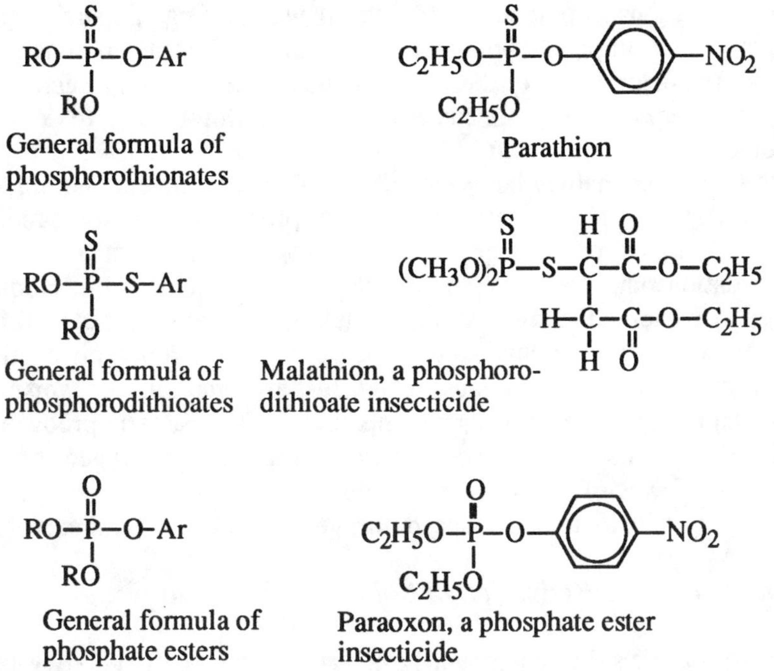

$$
\begin{array}{c}
S \\
\parallel \\
RO{-}P{-}O{-}Ar \\
\mid \\
RO
\end{array}
$$

General formula of
phosphorothionates

$$
\begin{array}{c}
S \\
\parallel \\
C_2H_5O{-}P{-}O{-}\bigcirc{-}NO_2 \\
\mid \\
C_2H_5O
\end{array}
$$

Parathion

$$
\begin{array}{c}
S \\
\parallel \\
RO{-}P{-}S{-}Ar \\
\mid \\
RO
\end{array}
$$

General formula of
phosphorodithioates

$$
\begin{array}{c}
S \quad\quad H\ \ O \\
\parallel \quad\quad \mid\ \ \parallel \\
(CH_3O)_2P{-}S{-}C{-}C{-}O{-}C_2H_5 \\
\mid \\
H{-}C{-}C{-}O{-}C_2H_5 \\
\mid\ \ \parallel \\
H\ \ O
\end{array}
$$

Malathion, a phosphorodithioate insecticide

$$
\begin{array}{c}
O \\
\parallel \\
RO{-}P{-}O{-}Ar \\
\mid \\
RO
\end{array}
$$

General formula of
phosphate esters

$$
\begin{array}{c}
O \\
\parallel \\
C_2H_5O{-}P{-}O{-}\bigcirc{-}NO_2 \\
\mid \\
C_2H_5O
\end{array}
$$

Paraoxon, a phosphate ester
insecticide

Figure 6.8. Phosphorothionate, phosphorodithioate, and phosphate ester insecticides.

substituted hydrocarbon moiety. These are used because they exhibit higher ratios of insect:mammal toxicity than do their non-sulfur analogs. The metabolic conversion of P=S to P=O (oxidative desulfuration, such as in the conversion of parathion to paraoxon) in organisms is responsible for the insecticidal activity and mammalian toxicity of phosphorothionate and phosphorodithioate insecticides. The biodegradation of these compounds is an important environmental chemical process. Fortunately, unlike the organohalide insecticides that they largely displaced, the organophosphates readily undergo biodegradation and do not bioaccumulate.

Hydrolysis is an important step in the biodegradation of phosphorothionate, phosphorodithioate, and phosphate ester insecticides as shown by the following general reactions where R is an alkyl group, Ar is a substituent group that is frequently aromatic, and X is either S or O:

$$\begin{array}{c} X \\ | \\ R-O-P-O-Ar \\ | \\ O \\ | \\ R \end{array} \quad \xrightarrow{\text{H}_2\text{O}} \quad \begin{array}{c} X \\ | \\ R-O-P-OH \\ | \\ O \\ | \\ R \end{array} + \text{HOAR} \qquad (6.9.1)$$

$$\begin{array}{c} X \\ | \\ R-O-P-O-Ar \\ | \\ O \\ | \\ R \end{array} \quad \xrightarrow{\text{H}_2\text{O}} \quad \begin{array}{c} X \\ | \\ R-O-P-OAr \\ | \\ O \\ | \\ R \end{array} + \text{HOR} \qquad (6.9.2)$$

Sulfur Compounds

Sulfur compounds are very common in water. Sulfate ion, SO_4^{2-}, is found in varying concentrations in practically all natural waters. Organic sulfur compounds, both those of natural origin and pollutant species, are very common in natural aquatic systems, and the degradation of these compounds is an important microbial process. Sometimes the degradation products, such as the odiferous and toxic H_2S, cause serious problems with water quality.

There is a strong analogy between sulfur in the environment and nitrogen in the environment. Sulfur in living material is present primarily in its most reduced state, for example, as the hydrosulfide group, -SH. Nitrogen in living material is present in the (-III) oxidation state, for example, as $-NH_2$. When organic sulfur compounds are decomposed by bacteria, the initial sulfur product is generally the reduced form, H_2S. When organic nitrogen compounds are decomposed by microorganisms, the reduced form of nitrogen, NH_3 or NH_3^+, is produced. Just as some microorganisms can produce elemental nitrogen from nitrogen compounds, some bacteria produce and store elemental sulfur from sulfur compounds. In the presence of oxygen, some bacteria convert reduced forms of sulfur to the oxidized form in SO_4^{2-} ion, whereas other bacteria catalyze the oxidation of reduced nitrogen compounds to nitrate ion.

Oxidation of H_2S and Reduction of Sulfate by Bacteria

Although organic sulfur compounds often are the source of H_2S in water, they are not required as the sulfur source for H_2S formation. The bacteria *Desulfovibrio* can reduce sulfate ion to H_2S. In so doing, they utilize sulfate as an electron acceptor in the oxidation of organic matter. The overall reaction is:

$$SO_4^{2-} + 2\{CH_2O\} + 2\,H^+ \rightarrow H_2S + 2CO_2 + 2H_2O \qquad (6.9.3)$$

Actually, other bacteria besides *Desulfovibrio* are required to oxidize organic matter completely to CO_2. The oxidation of organic matter by *Desulfovibrio* generally terminates with acetic acid, and accumulation of acetic acid is evident in bottom waters. Because of the high concentration of sulfate ion in seawater, bacterially-mediated formation of H_2S causes pollution problems in some coastal areas and is a major source of atmospheric sulfur. In waters where sulfide formation occurs, the sediment is often black in color due to the formation of FeS.

Bacterially-mediated reduction of sulfur in calcium sulfate deposits produces elemental sulfur interspersed in the pores of the limestone product. The highly generalized chemical reaction for this process is

$$2CaSO_4 + 3\{CH_2O\} \xrightarrow{\text{Bacteria}} 2CaCO_3 + 2S + CO_2 + 2H_2O \quad (6.9.4)$$

although the stoichiometric amount of free sulfur is never found in these deposits due to the formation of volatile H_2S, which escapes.

Whereas some bacteria can reduce sulfate ion to H_2S, others can oxidize hydrogen sulfide to higher oxidation states. The purple sulfur bacteria and green sulfur bacteria derive energy for their metabolic processes through the oxidation of H_2S. These bacteria utilize CO_2 as a carbon source and are strictly anaerobic. The aerobic colorless sulfur bacteria may use molecular oxygen to oxidize H_2S,

$$2H_2S + O_2 \rightarrow 2S + 2H_2O \quad (6.9.5)$$

elemental sulfur,

$$2S + 2H_2O + 3O_2 \rightarrow 4H^+ + 2SO_4^{2-} \quad (6.9.6)$$

or thiosulfate ion:

$$S_2O_3^{2-} + H_2O + 2O_2 \rightarrow 2H^+ + 2SO_4^{2-} \quad (6.9.7)$$

Oxidation of sulfur in a low oxidation state to sulfate ion produces sulfuric acid, a strong acid. One of the colorless sulfur bacteria, *Thiobacillus thiooxidans* is tolerant of 1 normal acid solutions, a remarkable acid tolerance. When elemental sulfur is added to excessively alkaline soils, the acidity is increased because of a microorganism-mediated reaction (6.9.6), which produces sulfuric acid. Elemental sulfur may be deposited as granules in the cells of purple sulfur bacteria and colorless sulfur bacteria. Such processes are important sources of elemental sulfur deposits.

Microorganism-Mediated Degradation of Organic Sulfur Compounds

Sulfur occurs in many types of biological compounds. As a consequence, organic sulfur compounds of natural and pollutant origin are very common in water. The degradation of these compounds is an important microbial process having a strong effect upon water quality.

Among some of the common sulfur-containing functional groups found in aquatic organic compounds are hydrosulfide (–SH), disulfide (–SS–), sulfide (–S–), sulfoxide ($-\overset{O}{\underset{\parallel}{S}}-$), sulfonic acid (–SO$_2$OH), thioketone ($-\overset{S}{\underset{\parallel}{C}}-$), and thiazole (a heterocyclic sulfur group). Protein contains some amino acids with sulfur functional groups — cysteine,

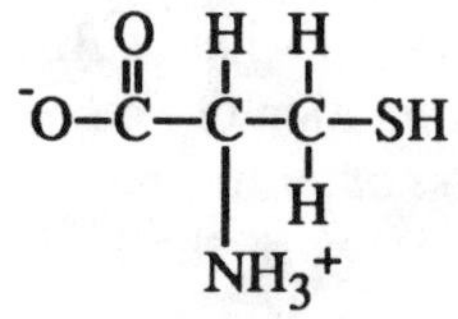

$$\overset{O}{\overset{\|}{}} \quad \overset{H}{\overset{|}{}} \quad \overset{H}{\overset{|}{}}$$
$$^-O-C-C-C-SH$$
$$\overset{|}{\underset{NH_3^+}{\overset{|}{H}}}$$

cystine, and methionine — whose breakdown is important in natural waters. The amino acids are readily broken down by bacteria and fungi.

The biodegradation of sulfur-containing amino acids can result in production of volatile organic sulfur compounds such as methyl thiol, CH_3SH, and dimethyl disulfide, CH_3SSCH_3. These compounds have strong, unpleasant odors. Their formation, in addition to that of H_2S, accounts for much of the odor associated with the biodegradation of sulfur-containing organic compounds.

Hydrogen sulfide is formed from a large variety of organic compounds through the action of a number of different kinds of microorganisms. A typical sulfur-cleavage reaction producing H_2S is the conversion of cysteine to pyruvic acid through the action of cysteine desulfhydrase enzyme in bacteria:

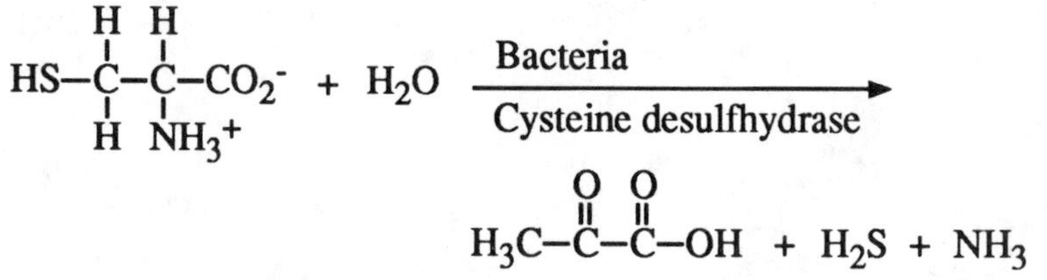

$$HS-\overset{\overset{\displaystyle H}{|}}{\underset{\underset{\displaystyle H}{|}}{C}}-\overset{\overset{\displaystyle H}{|}}{\underset{\underset{\displaystyle NH_3^+}{|}}{C}}-CO_2^- \ + \ H_2O \quad \xrightarrow[\text{Cysteine desulfhydrase}]{\text{Bacteria}}$$

$$H_3C-\overset{O}{\overset{\|}{C}}-\overset{O}{\overset{\|}{C}}-OH \ + \ H_2S \ + \ NH_3 \qquad (6.9.8)$$

Because of the numerous forms in which organic sulfur may exist, a variety of sulfur products and biochemical reaction paths must be associated with the biodegradation of organic sulfur compounds. Much remains to be learned in this area.

6.10. MICROBIAL TRANSFORMATIONS OF HALOGENS AND ORGANOHALIDES

Dehalogenation reactions involving the replacement of a halogen atom, for example,

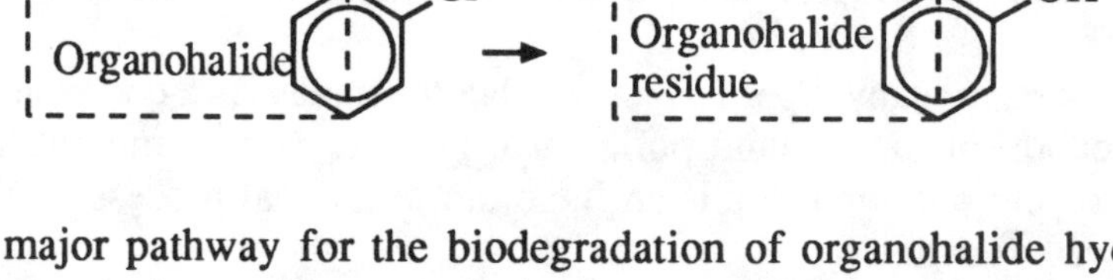

represent a major pathway for the biodegradation of organohalide hydrocarbons. Microorganisms need not utilize a particular organohalide compound as a sole carbon source in order to affect its degradation. This is due to the phenomenon of **cometabolism**, which results from a lack of specificity in the microbial degradation processes. Thus, bacterial degradation of small amounts of an organohalide compound may occur while the microorganism involved is metabolizing much larger quantities of another substance.

Bioconversion of DDT to replace Cl with H yields DDD:

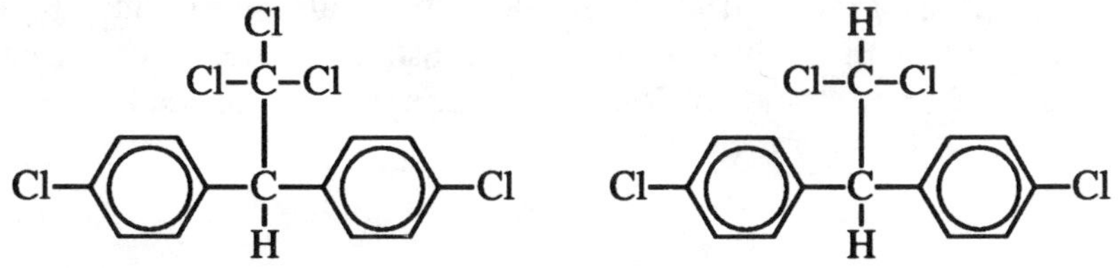

The latter compound is more toxic to some insects than DDT and is even manufactured as a pesticide. The same situation applies to microbially mediated conversion of aldrin to dieldrin:

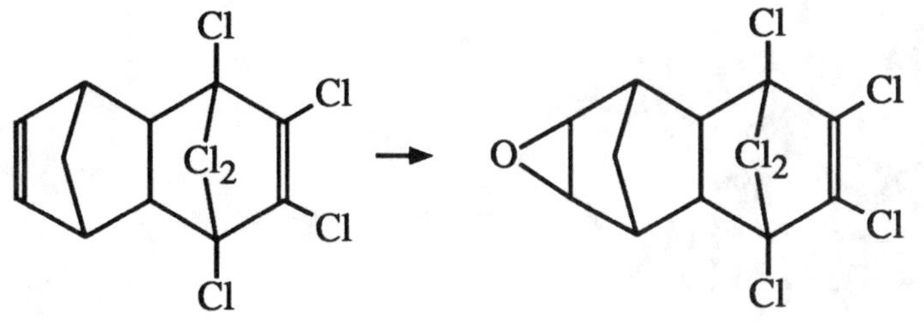

6.11. MICROBIAL TRANSFORMATIONS OF METALS AND METALLOIDS

Some bacteria, including *Ferrobacillus*, *Gallionella*, and some forms of *Sphaerotilus*, utilize iron compounds in obtaining energy for their metabolic needs. These bacteria catalyze the oxidation of iron(II) to iron(III) by molecular oxygen:

$$2Fe(II) + 4H^+ + O_2 \rightarrow 4Fe(II) + 2H_2O \tag{6.11.1}$$

The carbon source for some of these bacteria is CO_2. Since they do not require organic matter for carbon and because they derive energy from the oxidation of inorganic matter, these bacteria may thrive in environments where organic matter is absent.

The microorganism-mediated oxidation of iron(II) is not a particularly efficient means of obtaining energy for metabolic processes. For the reaction

$$FeCO_3(s) + \frac{1}{4}O_2 + \frac{3}{2}H_2O \rightarrow 4Fe(OH)_3(s) + CO_2 \tag{6.11.2}$$

the change in free energy is approximately 10 kcal/electron-mole. Approximately 220 g of iron(II) must be oxidized to produce 1.0 g of cell carbon. The calculation assumes CO_2 as a carbon source and a biological efficiency of 5%. The production of only 1.0 g of cell carbon would produce approximately 430 g of solid $Fe(OH)_3$. It follows that large deposits of hydrated iron(III) oxide form in areas where iron-oxidizing bacteria thrive.

Some of the iron bacteria, notably *Gallionella*, secrete large quantities of hydrated iron(III) oxide in the form of intricately branched structures. The bacterial cell grows at the end of a twisted stalk of the iron oxide. Individual cells of *Gallionella*, photographed through an electron microscope, have shown that the stalks consist of a number of strands of iron oxide secreted from one side of the cell (Figure 6.9).

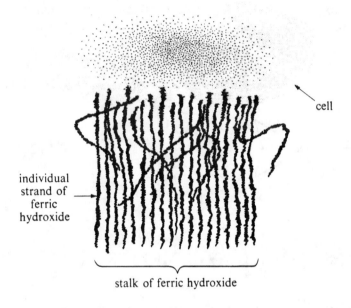

cell

individual
strand of
ferric
hydroxide

stalk of ferric hydroxide

Figure 6.9. Sketch of a cell of *Gallionella* showing iron(III) oxide secretion.

At nearly neutral pH values, bacteria deriving energy by mediating the air oxidation of iron(II) must compete with direct chemical oxidation of iron(II) by O_2. The latter process is relatively rapid at pH 7. As a consequence, these bacteria tend to grow in a narrow layer in the region between the oxygen source and the source of iron(II). Therefore, iron bacteria are sometimes called *gradient organisms*, and they grow at intermediate pE values.

Bacteria are strongly involved in the oceanic manganese cycle. Manganese nodules, a potentially important source of manganese, copper, nickel, and cobalt (see Section 21.2) yield different species of bacteria which enzymatically mediate both the oxidation and reduction of manganese.

Acid Mine Waters

One consequence of bacterial action on metal compounds is acid mine drainage, one of the most common and damaging problems in the aquatic environment. Many waters flowing from coal mines and draining from the "gob piles" left over from coal processing and washing are practically sterile due to high acidity.

Acid mine water results from the presence of sulfuric acid produced by the oxidation of pyrite, FeS_2. Microorganisms are closely involved in the overall process, which consists of several reactions. The first of these reactions is the oxidation of pyrite:

$$2FeS_2(s) + 2H_2O + 7O_2 \rightarrow 4H^+ + 4SO_4^{2-} + 2Fe^{2+} \qquad (6.11.3)$$

The next step is the oxidation of ferrous ion to ferric ion,

$$4Fe^{2+} + O_2 + 4H^+ \rightarrow 4Fe^{3+} + 2H_2O \qquad (6.11.4)$$

a process that occurs very slowly at the low pH values found in acid mine waters. Below pH 3.5, the iron oxidation is catalyzed by the iron bacterium *Thiobacillus ferrooxidans*, and in the pH range 3.5-4.5 it may be catalyzed by a variety of *Metallogenium*, a filamentous iron bacterium. Other bacteria that may be involved in acid mine water formation are *Thiobacillus thiooxidans* and *Ferrobacillus ferrooxidans*. The ferric ion further dissolves pyrite,

$$2FeS_2(s) + 14Fe^{3+} + 8H_2O \rightarrow 15Fe^{2+} + 2SO_4^{2-} + 16H^+ \qquad (6.11.5)$$

which in conjunction with Reaction 6.11.4 constitutes a cycle for the dissolution of pyrite. $Fe(H_2O)_6^{3+}$ is an acidic ion and at pH values much above 3, the iron(III) precipitates as the hydrated iron(III) oxide:

$$Fe^{3+} + 3H_2O \leftrightarrow Fe(OH)_3(s) + 3H^+ \qquad (6.11.6)$$

The beds of streams afflicted with acid mine drainage often are covered with "yellowboy," an unsightly deposit of amorphous, semigelatinous $Fe(OH)_3$. The most damaging component of acid mine water, however, is sulfuric acid. It is directly toxic and has other undesirable effects.

The prevention and cure of acid mine water is one of the major challenges facing the environmental chemist. One approach to eliminating excess acidity involves the use of carbonate rocks. When acid mine water is treated with limestone, the following reaction occurs:

$$CaCO_3(s) + 2H^+ + SO_4^{2-} \rightarrow Ca^{2+} + SO_4^{2-} + H_2O + CO_2(g) \qquad (6.11.7)$$

Unfortunately, because iron(III) is generally present, $Fe(OH)_3$ precipitates as the pH is raised (Reaction 6.11.6). The hydrated iron(III) oxide product covers the particles of carbonate rock with a relatively impermeable layer. This armoring effect prevents further neutralization of the acid.

Microbial Transitions of Selenium

Directly below sulfur in the periodic table, selenium is subject to bacterial oxidation and reduction. These transitions are important because selenium is a crucial element in nutrition, particularly of livestock. Diseases related to either selenium excesses or deficiency have been reported in at least half of the states of the U.S. and in 20 other countries, including the major livestock-producing countries. Livestock in New Zealand, in particular, suffer from selenium deficiency.

Microorganisms are closely involved with the selenium cycle, and microbial reduction of oxidized forms of selenium has been known for some time. A soil-dwelling strain of *Bacillus megaterium* has been found to be capable of oxidizing elemental selenium to selenite, SeO.[4]

6.12. MICROBIAL CORROSION

Corrosion is a redox phenomenon and was discussed in Section 4.12. Much corrosion is bacterial in nature. Bacteria involved with corrosion set up their own electrochemical cells in which a portion of the surface of the metal being corroded forms the anode of the cell and is oxidized. Structures called tubercles form in which bacteria pit and corrode metals as shown in Figure 6.10.

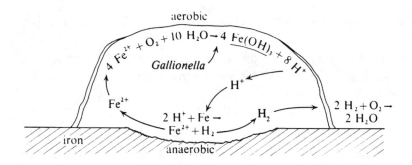

Figure 6.10. Tubercle in which the bacterially-mediated corrosion of iron occurs through the action of *Gallionella*.

It is beyond the scope of this book to discuss corrosion in detail. However, its significance and effects should be kept in mind by the environmental chemist.

LITERATURE CITED

1. Fogg, G. E., *The Metabolism of Algae*, John Wiley and Sons, Inc., New York, 1953.

2. Cerniglia, C. E., and D. T. Gibson, "Metabolism of Naphthalene by *Cunningham-ella elegans*," *Applied and Environmental Microbiology*, **34**, 363-70 (1977).

3. Bender, M. L., K. A. Fanning, P. H. Froehlich, and V. Maynard, "Interstitial Nitrate Profiles and Oxidation of Sedimentary Organic Matter in Eastern Equatorial Atlantic," *Science*, **198**, 605-8 (1977).

4. Sarathchandra, S. U., and J. H. Watkinson, "Oxidation of Elemental Selenium to Selenite by *Bacillus megaterium*," *Science* **211**, 600-601 (1981).

SUPPLEMENTARY REFERENCES

Scholze, R. J., *Biotechnology for Degradation of Toxic Chemicals in Hazardous Wastes*, Noyes Publications, Park Ridge, NJ, 1988.

Forster, C. F., *Biotechnology and Wastewater Treatment*, Cambridge University Press, New York, NY, 1985.

Broughton, W. J., Ed., *Nitrogen Fixation*, Vol 1, Oxford University Press, New York, 1981.

Dart, R. K., and R. J. Stretton, *Microbiological Aspects of Pollution Control*, Elsevier Scientific Publishing Co., Amsterdam, 1977.

Fenchel, T., and T. H. Blackburn, *Bacteria and Mineral Cycling*, Academic Press, New York, 1979.

Higgins, I. G., and R. G. Burns, *The Chemistry and Microbiology of Pollution*, Academic Press, New York, 1975.

Hill, I. R., and S. J. L. Wright, Eds., *Pesticide Microbiology*, Academic Press, New York, 1978.

Krumbein, W. E., Ed., *Environmental Biogeochemistry and Geomicrobiology*, Ann Arbor Science Publishers, Ann Arbor, Michigan, 1978.

Leisinger, T., R. Hutter, A. M. Cook, and J. Nüesch, Eds., *Microbial Degradation of Xenobiotics and Recalcitrant Compounds*, Academic Press, New York, 1981.

Maki, A. W., K. L. Dickson, and J. Cairns, Eds., *Biotransformation and Fate of Chemicals in the Aquatic Environment*, American Society for Microbiology, Washington, DC, 1980.

Matsumura, F., and C. R. Krishna Murti, Eds., *Biodegradation of Pesticides*, Plenum Publishing Co., New York, 1982.

McKinney, R. E., *Microbiology for Sanitary Engineers*, McGraw-Hill Book Company, New York, 1962.

Mitchell, R., *Water Pollution Microbiology*, Vol. 1, Wiley-Interscience, New York, 1970.

Mitchell, R., *Water Pollution Microbiology*, Vol. 2, Wiley-Interscience, New York, 1978.

Rheinheimer, G., *Aquatic Microbiology*, Wiley-Interscience, New York, 1981.

Round, F. E., *The Ecology of Algae*, Cambridge University Press, New York, 1981.

QUESTIONS AND PROBLEMS

1. As $CH_3CH_2CH_2CH_2CO_2H$ biodegrades in several steps to carbon dioxide and water, various chemical species are observed. What stable chemical species would be observed as a result of the first step of this degradation process?

2. Which of the following statements is true regarding the production of methane in water: (a) it occurs in the presence of oxygen, (b) it consumes oxygen, (c) It removes biological oxygen demand from the water, (d) it is accomplished by aerobic bacteria, (e) it produces more energy per electron-mole than does aerobic respiration.

3. At the time zero, the cell count of a bacterial species mediating aerobic respiration of wastes was 1×10^6 cells per liter. At 30 minutes it was 2×10^6; at 60 minutes it was 4×10^6; at 90 minutes, 7×10^6 ; at 120 minutes, 10×10^6; and at 150 minutes, 13×10^6. From these data, which of the following logical conclusions would you draw? (a) The culture was entering the log phase at the end of the 150-minute period, (b) the culture was in the log phase throughout the 150-minute period, (c) the culture was leaving the log phase at the end of the 150-minute period, (d) the culture was in the lag phase throughout the 150-minute period, (e) the culture was in the death phase throughout the 150-minute period.

4. Which of the following structures is that of a relatively biodegradable water pollutant?

(a)

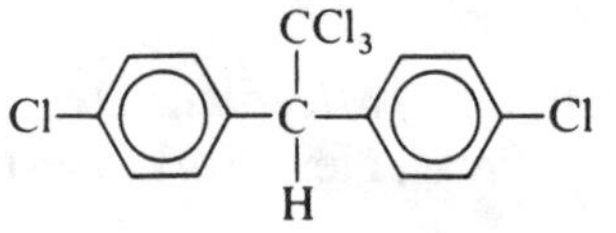

(b)

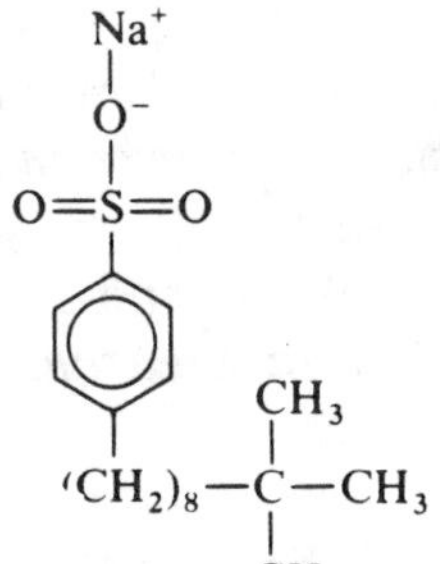

(c)

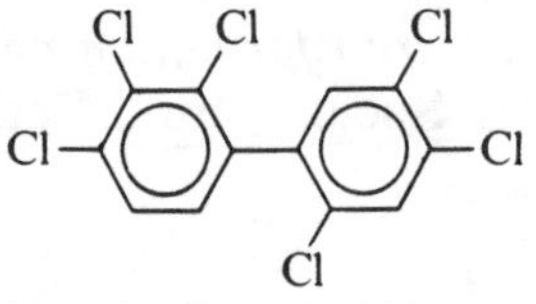

(d)

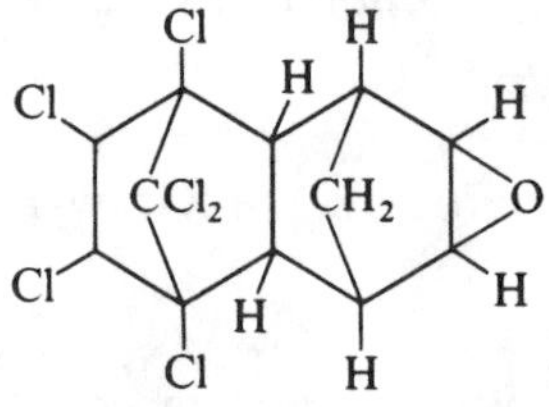

(e)

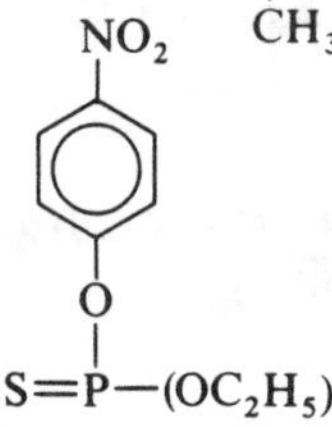

5. Suppose that the anaerobic fermentation of organic matter, $\{CH_2O\}$, in water yields 15.0 L of CH_4 (at standard temperature and pressure). How many grams of oxygen would be consumed by the aerobic respiration of the same quantity of $\{CH_2O\}$? (Recall the significance of 22.4 L in chemical reaction of gases.)

6. What weight of $FeCO_3(s)$, using Reaction 4 in Table 6.1, gives the same free energy yield as 1.00 g of organic matter, using Reaction 1, when oxidized by oxygen at pH 7.00?

7. How many bacteria would be produced after 10 hours by one bacterial cell, assuming exponential growth with a generation time of 20 minutes?

8. Referring to Reaction 6.8.2, calculate the concentration of ammonium ion in equilibrium with oxygen in the atmosphere and 1.00×10^{-5} M NO_3^- at pH 7.00.

9. When a bacterial nutrient medium is inoculated with bacteria grown in a markedly different medium, the lag phase (Fig. 6.3) often is quite long, even if the bacteria eventually grow well in the new medium. Can you explain this behavior?

10. Most plants assimilate nitrogen as nitrate ion. However, ammonia (NH_3) is a popular and economical fertilizer. What essential role do bacteria play when ammonia is used as a fertilizer? Do you think any problems might occur when using ammonia in a waterlogged soil lacking oxygen?

11. Why is the growth rate of bacteria as a function of temperature (Fig. 6.4) not a symmetrical curve?

12. Discuss the analogies between bacteria and a finely divided chemical catalyst.

13. Would you expect autotrophic bacteria to be more complex physiologically and biochemically than heterotrophic bacteria? Why?

14. Wastewater containing 8 mg/L O_2 (atomic weight O = 16), 1.00×10^{-3} M NO_3^-, and 1.00×10^{-2} M soluble organic matter, $\{CH_2O\}$, is stored isolated from the atmosphere in a container richly seeded with a variety of bacteria. Assume that denitrification is one of the processes which will occur during storage. After the bacteria have had a chance to do their work, which of the following statements will be true? (a) No $\{CH_2O\}$ will remain, (b) some O_2 will remain, (c) some NO_3^- will remain, (d) denitrification will have consumed more of the organic matter than aerobic respiration, (e) the composition of the water will remain unchanged.

15. Of the four classes of microorganisms — algae, fungi, bacteria, and virus — which has the least influence on water chemistry?

16. Figure 6.2 shows the main structural features of a bacterial cell. Which of these do you think might cause the most trouble in water-treatment processes such as filtration or ion exchange, where the maintenance of a clean, unfouled surface is critical? Explain.

17. A bacterium capable of degrading 2,4-D herbicide was found to have its maximum growth rate at 32 °C. Its growth rate at 12 °C was only 10% of the maximum. Do you think there is another temperature at which the growth rate would also be 10% of the maximum? If you believe this to be the case, of the following temperatures, choose the one at which it is most plausible for the bacterium to also have a growth rate of 10% of the maximum: 52 °C, 37 °C, 8 °C, 20 °C.

18. Addition of which two half-reactions in Table 6.1 is responsible for: (a) elimination of an algal nutrient in secondary sewage effluent using methanol as

a carbon source, (b) a process responsible for a bad-smelling pollutant when bacteria grow in the absence of oxygen, (c) A process that converts a common form of commercial fertilizer to a form that most crop plants can absorb, (d) a process responsible for the elimination of organic matter from wastewater in the aeration tank of an activated sludge sewage-treatment plant, (e) a characteristic process that occurs in the anaerobic digester of a sewage treatment plant.

19. The day after a heavy rain washed a great deal of cattle feedlot waste into a farm pond, the following counts of bacteria were obtained:

Time	Thousands of viable cells per mL
6:00 a.m.	0.10
7:00 a.m.	0.11
8:00 a.m.	0.13
9:00 a.m.	0.16
10:00 a.m.	0.20
11:00 a.m.	0.40
12:00 Noon	0.80
1:00 p.m.	1.60
2:00 p.m.	3.20

To which portion of the bacterial growth curve, Figure 6.3, does this time span correspond?

20. What is the surface area in square meters of 1.00 gram of spherical bacterial cells, 1.00 μm in diameter, having density of 1.00 g/cm^3?

21. What is the purpose of exoenzymes in bacteria?

22. Match each species of bacteria listed in the left column with its function on the right.

    (a) *Spirillum lipoferum*   (1) Reduces sulfate to H_2S
    (b) *Bacillus megaterium*   (2) Catalyzes oxidation of Fe^{2+} to Fe^{3+}
    (c) *Thiobacillus ferrooxidans* (3) Fixes nitrogen in grasses
    (d) *Desulfovibrio*   (4) Oxidizes elemental selenium to selenite

23. What factors favor the production of methane in anoxic surroundings?

WATER POLLUTION

7.1. NATURE AND TYPES OF WATER POLLUTANTS

Throughout history, the quality of drinking water has been a factor in determining human welfare. Fecal pollution of drinking water has frequently caused waterborne diseases that have decimated the populations of whole cities. Unwholesome water polluted by natural sources has caused great hardship for people forced to drink it or use it for irrigation.

Today there are still occasional epidemics of bacterial and viral diseases caused by infectious agents carried in drinking water — ominously, a major outbreak of deadly cholera in Peru in 1991, the first in the Western Hemisphere in this century. However, waterborne diseases have in general been well controlled, and drinking water in technologically advanced countries in the 1980s is remarkably free of the disease-causing agents that were very common water contaminants only a few decades earlier.

Currently, waterborne toxic chemicals pose the greatest threat to the safety of water supplies in industrialized nations. This is particularly true of groundwater in the U.S., which exceeds in volume the flow of all U.S. rivers, lakes, and streams. In some areas, the quality of groundwater is subject to a number of chemical threats. There are many possible sources of chemical contamination. These include wastes from industrial chemical production, metal plating operations, and pesticide runoff from agricultural lands. Some specific pollutants include industrial chemicals such as chlorinated hydrocarbons; heavy metals, including cadmium, lead, and mercury; saline water; bacteria, particularly coliforms; and general municipal and industrial wastes.

Since World War II there has been a tremendous growth in the manufacture and use of synthetic chemicals. Many of the chemicals have contaminated water supplies. Two examples are insecticide and herbicide runoff from agricultural land, and industrial discharge into surface waters. Most serious, though, is the threat to groundwater from waste chemical dumps and landfills, storage lagoons, treating ponds, and other facilities. These threats are discussed in more detail in Chapters 17-19.

It is clear that water pollution should be a concern of every citizen. Understanding the sources, interactions, and effects of water pollutants is essential for controlling pollutants in an environmentally safe and economically acceptable manner. Above all, an understanding of water pollution and its control depends upon a basic knowledge of aquatic environmental chemistry. That is why this text covers the basics of aquatic chemistry prior to discussing pollution. Water pollution may be studied much more effectively with a sound background in the basic properties of water, aquatic microbial reactions, sediment-water interactions, and other factors involved with the reactions, transport, and effects of these pollutants.

In considering water pollution, it is useful to keep in mind an overall picture of possible pollutant cycles as shown in Chapter 1, Figure 1. This figure illustrates the major routes of pollutant interchange among the biotic, terrestrial, atmospheric, and aquatic environments.

Water pollutants can be divided among some general classifications, as summarized in Table 7.1. Most of these categories of pollutants, and several subcategories, are discussed in this chapter. An enormous amount of material is published on this subject each year, and it is impossible to cover it all in one chapter. In order to be up to date on this subject the reader may want to survey journals and books dealing with water pollution, such as those listed in the Supplementary References section at the end of this chapter.

Table 7.1. General Types of Water Pollutants

Class of pollutant	Significance
Trace Elements	Health, aquatic biota
Metal-organic combinations	Metal transport
Inorganic pollutants	Toxicity, aquatic biota
Asbestos	Human health
Algal nutrients	Eutrophication
Radionuclides	Toxicity
Acidity, alkalinity, salinity (in excess)	Water quality, aquatic life
Sewage	Water quality, oxygen levels
Biochemical oxygen demand	Water quality, oxygen levels
Trace organic pollutants	Toxicity
Pesticides	Toxicity, aquatic biota, wildlife
Polychlorinated biphenyls	Possible biological effects
Chemical carcinogens	Incidence of cancer
Petroleum wastes	Effect on wildlife, esthetics
Pathogens	Health effects
Detergents	Eutrophication, wildlife, esthetics
Sediments	Water quality, aquatic biota, wildlife
Taste, odor, and color	Esthetics

7.2. ELEMENTAL POLLUTANTS

Trace element is a term that refers to those elements that occur at very low levels in a given system. The somewhat ambiguous term probably arose from the inadequacy of earlier analytical techniques — before modern methods such as atomic absorption, plasma emission, neutron-activation analysis, gas chromatography, and mass spectrometry extended the limits of detection to the very low levels currently attainable. In many early investigations, it was only possible to detect the presence of an element, as it was said to be present at a "trace level." A reasonable definition of a trace element is one that occurs at a level of a few parts per million or less. The term **trace substance** is a more general one applied to both elements and chemical compounds.

Table 7.2 summarizes the more important trace elements encountered in natural waters. Some of these are recognized as nutrients required for animal and plant life. Of these, many are essential at low levels but toxic at higher levels. This is typical behavior for many substances in the aquatic environment, a point that must be kept in mind in judging whether a particular element is beneficial or detrimental. Some of these elements, such as lead or mercury, have such toxicological and environmental significance that they are discussed in detail in separate sections.

Some of the **heavy metals** are among the most harmful of the elemental pollutants. These elements, which are in general the metals in the lower right-hand corner of the periodic table, include essential elements like iron as well as toxic metals like lead, cadmium, and mercury. Most of them have a tremendous affinity for sulfur and attack sulfur bonds in enzymes, thus immobilizing the enzymes. Protein carboxylic acid ($-CO_2H$) and amino ($-NH_2$) groups are also chemically bound by heavy metals. Cadmium, copper, lead and mercury ions bind to cell membranes, hindering transport processes through the cell wall. Heavy metals may also precipitate phosphate biocompounds or catalyze their decomposition. The biochemical effects of metals are discussed in Chapter 20.

Some of the **metalloids**, elements on the borderline between metals and nonmetals, are significant water pollutants. Arsenic, selenium, and antimony are of particular interest.

Inorganic chemicals manufacture has the potential to contaminate water with trace elements. Among the industries regulated for potential trace element pollution of water are those producing chlor-alkali, hydrofluoric acid, sodium dichromate (sulfate process and chloride ilmenite process), aluminum fluoride, chrome pigments, copper sulfate, nickel sulfate, sodium bisulfate, sodium hydrosulfate, sodium bisulfite, titanium dioxide, and hydrogen cyanide.

7.3. HEAVY METALS

Cadmium

Pollutant **cadmium** in water may arise from industrial discharges and mining wastes. Cadmium is widely used in metal plating. Chemically, cadmium is very similar to zinc, and these two metals frequently undergo geochemical processes together. Both metals are found in water in the +2 oxidation state.

The effects of acute cadmium poisoning in humans are very serious. Among them are high blood pressure, kidney damage, destruction of testicular tissue, and destruction of red blood cells. It is believed that much of the physiological action of cadmium arises from its chemical similarity to zinc. Specifically, cadmium may replace zinc in some enzymes, thereby altering the stereostructure of the enzyme and impairing its catalytic activity. Disease symptoms ultimately result.

Cadmium and zinc are common water and sediment pollutants in harbors surrounded with industrial installations. The estuarine migration and redistribution of zinc and cadmium from such sources has been the subject of a detailed study.[1] In the particular location studied, concentrations of cadmium as high as 130 ppm dry weight sediment were found in harbor sediments, and concentrations ranging up to 1.9 ppm were found in bay sediments outside the harbor. During the summer, when the harbor water was most stagnant, an interesting variation in cadmium concentration was observed with increasing depth. The aerobic surface layer contained a relatively high dissolved cadmium level, primarily as the soluble ion pair $CdCl^+$. The anaerobic bottom layer of the water was cadmium-poor because microbial reduction of sulfate had produced sulfide,

Table 7.2. Occurrence and Significance of Trace Elements in Natural Waters

Element	Sources	Effects and significance	U.S. Public Health Service Limit, mg/liter[1]	Occurrence: % of samples, highest and mean concentrations (μg/liter[2])
Arsenic	Mining by-product, pesticides, chemical waste	Toxic, possibly carcinogenic	0.05	5.5% (above 5 μg/L), 336, 64
Beryllium	Coal, nuclear power and space industries	Acute and chronic toxicity, possibly carcinogenic	Not given	Not given
Boron	Coal, detergent formulations, industrial wastes	Toxic to some plants	1.0	98% (above 1 μg/L), 5000, 101
Cadmium	Industrial discharge, mining waste, metal plating, water pipes	Replaces zinc biochemically, causes high blood pressure and kidney damage, destroys testicular tissue and red blood cells, toxic to aquatic biota.	0.01	2.5%, not given, 9.5
Chromium	Metal plating, cooling-tower water additive (chromate), normally found as Cr(VI) in polluted water	Essential trace element (glucose tolerance factor), possibly carcinogenic as Cr(VI)	0.05	24.5%, 112, 9.7
Copper	Metal plating, industrial and domestic wastes, mining, mineral leaching	Essential trace element, not very toxic to animals, toxic to plants and algae at moderate levels	1.0	74.4%, 280, 15
Fluorine (fluoride ion)	Natural geological sources, industrial waste, water additive	Prevents tooth decay at about 1 mg/L, causes mottled teeth and bone damage at around 5 mg/L in water	0.8–1.7 depending on temperature	Not given
Iodine (iodide)	Industrial waste, natural brines, seawater intrusion	Prevents goiter	Not given	Rare in fresh water

Element	Sources	Effects	Drinking water standard[1]	Occurrence[2]
Iron	Corroded metal, industrial wastes, acid mine drainage, low pE water in contact with iron minerals	Essential nutrient (component of hemoglobin), not very toxic, damages materials (bathroom fixtures and clothing)	0.05	75.6%, 4600, 52
Lead	Industry, mining, plumbing, coal, gasoline	Toxicity (anemia, kidney disease, nervous system), wildlife destruction	0.05	19.3% (above 2 µg/L), 140, 23
Manganese	Mining, industrial waste, acid mine drainage, microbial action on manganese minerals at low pE	Relatively nontoxic to animals, toxic to plants at higher levels, stains materials (bathroom fixtures and clothing)	0.05	51.4% (above 0.3 µg/L), 3230, 58
Mercury	Industrial waste, mining, pesticides, coal	Acute and chronic toxicity	Not given	Not given
Molybdenum	Industrial waste, natural sources, cooling-tower water additive	Possibly toxic to animals, essential for plants	Not given	32.7 (above 2 µg/L), 5400, 120
Selenium	Natural geological sources, sulfur, coal	Essential at low levels, toxic at higher levels, causes "alkali disease" and "blind staggers" in cattle, possibly carcinogenic	0.01	Not given
Silver	Natural geological sources, mining, electroplating, film-processing wastes, disinfection of water	Causes blue-grey discoloration of skin, mucous membranes, eyes	0.05	6.6% (above 0.1 µg/L), 38, 2.6
Zinc	Industrial waste, metal plating, plumbing	Essential element in many metallo-enzymes, aids wound healing, toxic to plants at higher levels; major component of sewage sludge, limiting land disposal of sludge	5.0	76.5% (above 2 µg/L), 1180, 64

[1] *Public Health Service Drinking Water Standards*, U.S. Public Health Service, 1962.

[2] J.F. Kopp and R.C. Kroner, *Trace Metals in Waters of the United States*, United States Environmental Protection Agency. The first figure is the percentage of samples showing the element; the second is the highest value found; the third is the mean value in positive samples (samples showing the presence of the metal at detectable levels).

$$2\{CH_2O\} + SO_4^{2-} + H^+ \rightarrow 2CO_2 + HS^- + 2H_2O \qquad (7.3.1)$$

which had precipitated cadmium as insoluble cadmium sulfide:

$$CdCl^+ + HS^- \rightarrow CdS(s) + H^+ + Cl^- \qquad (7.3.2)$$

Mixing of bay and harbor water by high winds during the winter resulted in desorption of cadmium from harbor sediments by aerobic bay water. This dissolved cadmium was carried out into the bay, where it reacted with suspended solid materials, which then became incorporated with the bay sediments. This is an example of the sort of complicated interaction of hydraulic, chemical solution-solid, and microbiological factors involved in the transport and distribution of a pollutant in an aquatic system.

Lead

Lead occurs in water in the +II oxidation state and arises from a number of industrial and mining sources. Lead from leaded gasoline used to be a major source of atmospheric and terrestrial lead, much of which eventually enters natural water systems. In addition to pollutant sources, lead-bearing limestone and galena (PbS) contribute lead to natural waters in some locations.

Despite greatly increased total use of lead by industry, evidence from hair samples and other sources indicates that body burdens of this toxic metal have decreased during recent decades. This may be the result of less lead used in plumbing and other products that come in contact with food or drink.

Acute lead poisoning in humans causes severe dysfunction in the kidneys, reproductive system, liver, and the brain and central nervous system. Sickness or death results. Lead poisoning from environmental exposure is thought to have caused mental retardation in many children. Mild lead poisoning causes anemia. The victim may have headaches and sore muscles and may feel generally fatigued and irritable.

Lead is probably not a major problem in drinking water, although the potential exists in cases where old lead pipe is still in use. Lead used to be a constituent of solder and some pipe-joint formulations, so that household water does have some contact with lead. Water that has stood in household plumbing for some time may accumulate spectacular levels of lead (along with zinc, cadmium, and copper) and should be let run for a while before use.

Mercury

Mercury generates the most concern of any of the heavy-metal pollutants. Mercury is found as a trace component of many minerals, with continental rocks containing an average of around 80 parts per billion, or slightly less, of this element. Cinnabar, red mercuric sulfide, is the chief commercial mercury ore. Fossil fuel coal and lignite contain mercury, often at levels of 100 parts per billion or even higher, a matter of some concern with increased use of these fuels for energy resources.

Metallic mercury is used, for example, in laboratory vacuum apparatus. The primary use of mercury metal is as an electrode in the electrolytic generation of chlorine gas. Large quantities of inorganic mercury(I) and mercury(II) compounds are used annually. Organic mercury compounds used to be widely applied as pesticides, particularly fungicides. These mercury compounds include aryl mercurials such as phenyl mercuric dimethyldithiocarbamate

$$\langle\!\!\!\bigcirc\!\!\!\rangle\!-\!Hg\!-\!S\!-\!\overset{\overset{\displaystyle S}{\|}}{C}\!-\!N\!\!\stackrel{\textstyle CH_3}{\underset{\textstyle CH_3}{}}$$

(used in paper mills as a slimicide and as a mold retardant for paper), and alkyl-mercurials such as ethylmercuric chloride, C_2H_5HgCl, used as a seed fungicide. The alkyl mercury compounds tend to resist degradation and are generally considered to be more of an environmental threat than either the aryl or inorganic compounds.

Mercury enters the environment from a large number of miscellaneous sources related to human use of the element. These include discarded laboratory chemicals, batteries, broken thermometers, lawn fungicides, amalgam tooth fillings, and pharmaceutical products. Taken individually, each of these sources may not contribute much of the toxic metal, but the total effect can be substantial. Sewage effluent sometimes contains up to 10 times the level of mercury found in typical natural waters.

The toxicity of mercury was tragically illustrated in the Minamata Bay area of Japan during the period 1953-1960. A total of 111 cases of mercury poisoning and 43 deaths were reported among people who had consumed seafood from the bay that had been contaminated with mercury waste from a chemical plant that drained into Minamata Bay. Congenital defects were observed in 19 babies whose mothers had consumed seafood contaminated with mercury. The level of metal in the contaminated seafood was 5-20 parts per million.

Among the toxicological effects of mercury are neurological damage, including irritability, paralysis, blindness, or insanity; chromosome breakage; and birth defects. The milder symptoms of mercury poisoning, such as depression and irritability, have a psychopathological character. Therefore, mild mercury poisoning may escape detection. Some forms of mercury are relatively nontoxic and have been used as medicines, in the treatment of syphilis, for example, for centuries. Other forms of mercury, particularly organic compounds, are highly toxic.

Because there are few major natural sources of mercury, and since most inorganic compounds of this element are relatively insoluble, it was assumed for some time that mercury was not a serious water pollutant in water. However, in 1970 alarming mercury levels were discovered in fish in Lake Saint Clair between Michigan and Ontario, Canada. A subsequent survey by the U.S. Federal Water Quality Administration revealed a number of other waters contaminated with mercury. It was found that several chemical plants, particularly caustic-chemical plants, were each releasing up to 14 or more kilograms of mercury in wastewaters each day.

The unexpectedly high concentrations of mercury found in water and in fish tissues result from the formation of soluble monomethylmercury ion, CH_3Hg^+, and volatile dimethylmercury, $(CH_3)_2Hg$, by anaerobic bacteria in sediments. Mercury from these compounds becomes concentrated in fish lipid (fat) tissue and the concentration factor from water to fish may exceed 10^3. The methylating agent by which inorganic mercury is converted to methylmercury compounds is methylcobalamin, a vitamin B_{12} analog:

$$HgCl_2 \xrightarrow{\text{Methylcobalamin}} CH_3HgCl + Cl^- \tag{7.3.3}$$

It is believed that the bacteria that synthesize methane produce methylcobalamin as an intermediate in the synthesis. Thus, waters and sediments in which anaerobic decay is occurring provide the conditions under which methylmercury production occurs. In neutral or alkaline waters, the formation of dimethyl mercury, $(CH_3)_2Hg$, is favored. This volatile compound can escape to the atmosphere.

7.4. METALLOIDS

The most significant water pollutant metalloid element is arsenic, a toxic element that has been the chemical villain of more than a few murder plots. Acute arsenic poisoning can result from the ingestion of more than about 100 mg of the element. Chronic poisoning occurs with the ingestion of small amounts of arsenic over a long period of time. There is some evidence that this element is also carcinogenic.

Arsenic occurs in the Earth's crust at an average level of 2-5 ppm. The combustion of fossil fuels, particularly coal, introduces large quantities of arsenic into the environment, much of it reaching natural waters. Arsenic occurs with phosphate minerals and enters into the environment along with some phosphorus compounds. Some formerly-used pesticides, particularly those from before World War II, contain highly toxic arsenic compounds. The most common of these are lead arsenate, $Pb_3(AsO_4)_2$; sodium arsenite, Na_3AsO_3; and Paris Green, $Cu_3(AsO_3)_2$. Another major source of arsenic is mine tailings. Arsenic produced as a byproduct of copper, gold, and lead refining greatly exceeds the commercial demand for arsenic, and it accumulates as waste material.

Like mercury, arsenic may be converted to more mobile and toxic methyl derivatives by bacteria, according to the following reactions:

$$H_3AsO_4 + 2H^+ + 2e^- \rightarrow H_3AsO_3 + H_2O \qquad (7.4.1)$$

$$H_3AsO_3 \xrightarrow{\text{Methylcobalamin}} \underset{\text{Methylarsinic acid}}{CH_3AsO(OH)_2} \qquad (7.4.2)$$

$$CH_3AsO(OH)_2 \xrightarrow{\text{Methylcobalamin}} \underset{\text{(Dimethylarsinic acid)}}{(CH_3)_2AsO(OH)} \qquad (7.4.3)$$

$$(CH_3)_2AsO(OH) + 4H^+ + 4e^- \rightarrow (CH_3)_2AsH + 2H_2O \qquad (7.4.4)$$

7.5. ORGANICALLY BOUND METALS AND METALLOIDS

An appreciation of the strong influence of complexation and chelation on heavy metals' behavior in natural waters and wastewaters may be gained by reading Sections 3.5-3.10, which deal with that subject. Methylmercury formation is discussed in Section 7.3. Both topics involve the combination of metals and organic entities in water. It must be stressed that the interaction of metals with organic compounds is of utmost importance in determining the role played by the metal in an aquatic system.

There are two major types of metal-organic interactions to be considered in an aquatic system. The first of these is complexation, usually chelation when organic ligands are involved. A reasonable definition of complexation by organics applicable to natural water and wastewater systems is a system in which a species is present that reversibly dissociates to a metal ion and an organic complexing species as a function of hydrogen ion concentration:

$$ML + 2H^+ \rightarrow M^{2+} + H_2L \qquad\qquad (7.5.1)$$

In this equation, M^{2+} is a metal ion and H_2L is the acidic form of a complexing — frequently chelating — ligand, L^{2-}.

Organometallic compounds, on the other hand, contain metals bound to organic entities by way of a carbon atom and do not dissociate reversibly at lower pH or greater dilution. Furthermore, the organic component, and sometimes the particular oxidation state of the metal involved, may not be stable apart from the organometallic compound. The simplest way to classify organometallic compounds for the purpose of discussing their toxicology is the following:

1. Those in which the organic group is an alkyl group such as ethyl in tetraethyllead, $Pb(C_2H_5)_4$:

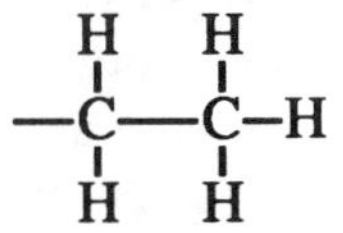

$$\begin{array}{cc} H & H \\ | & | \\ -C-\!\!-C-H \\ | & | \\ H & H \end{array}$$

2. **Carbonyls,** some of which are quite volatile and toxic, having carbon monoxide bonded to metals:

$$:C\equiv O:$$

(In the preceding Lewis formula of CO each dash, –, represents a pair of bonding electrons, and each pair of dots, :, represents an unshared pair of electrons.)

3. Those in which the organic group is a π electron donor, such as ethylene or benzene.

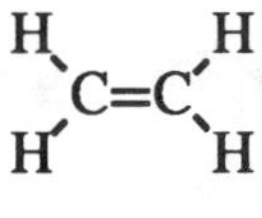

$$\begin{array}{c} H \\ {}^{\diagdown}\!C\!=\!C^{\diagup} \\ H^{\diagup} \qquad {}^{\diagdown}H \end{array}$$ Ethylene    Benzene

Combinations exist of the three general types of compounds outlined above, the most prominent of which are arene carbonyl species in which a metal atom is bonded to both an aromatic entity such as benzene and to several carbon monoxide molecules.

A large number of compounds exist that have at least one bond between the metal and a C atom on an organic group, as well as other covalent or ionic bonds between the metal and atoms other than carbon. Because they have at least one metal-carbon bond, as well as properties, uses, and toxicological effects typical of organometallic compounds, it is useful to consider such compounds along with organometallic compounds. Examples are monomethylmercury chloride, CH_3HgCl, in which the organometallic CH_3Hg^+ ion is ionically bonded to the chloride anion. Another example is phenyldichloroarsine, $C_6H_5AsCl_2$, in which a phenyl group is covalently bonded to arsenic through an As-C bond, and two Cl atoms are also covalently bonded to arsenic.

A number of compounds exist that consist of organic groups bonded to a metal atom through atoms other than carbon. Although they do not meet the strict definition thereof, such compounds can be classified as organometallics for the discussion of their toxicology and aspects of their chemistry. An example of such a compound is isopropyl titanate, $Ti(i-OC_3H_7)_4$, also called titanium isopropylate, a colorless liquid melting at 14.8°C and boiling at 104°C. Its behavior is more that of an organometallic compound than that of an inorganic compound, and by virtue of its titanium content it is not properly classified as an organic compound. The term "organometal" is sometimes applied to such a compound. For environmental considerations it may be regarded as an organometallic compound.

$$\begin{matrix} C_3H_7O & & OC_3H_7 \\ & Ti & \\ C_3H_7O & & OC_3H_7 \end{matrix}$$

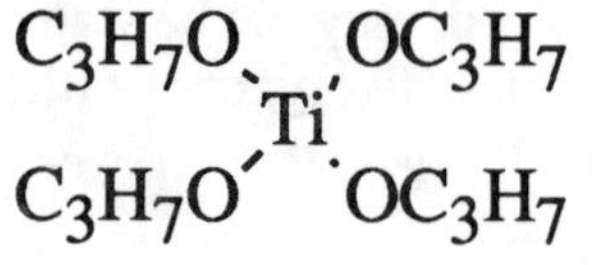

Some environmentally significant compounds have organometallic character, but also have formulas, structures, and properties of inorganic or organic compounds. These compounds could be called "mixed organometallics." However, so long as the differences are understood, compounds such as isopropyl titanate (see above) that do not meet all the criteria of organometallic compounds can be regarded as such for the discussion of their toxicities.

Organotin Compounds

Of all the metals, tin has the greatest number of organometallic compounds in commercial use, with global production on the order of 40,000 metric tons per year. In addition to synthetic organotin compounds, methylated tin species can be produced biologically in the environment. Figure 7.1 gives some examples of the many known organotin compounds.

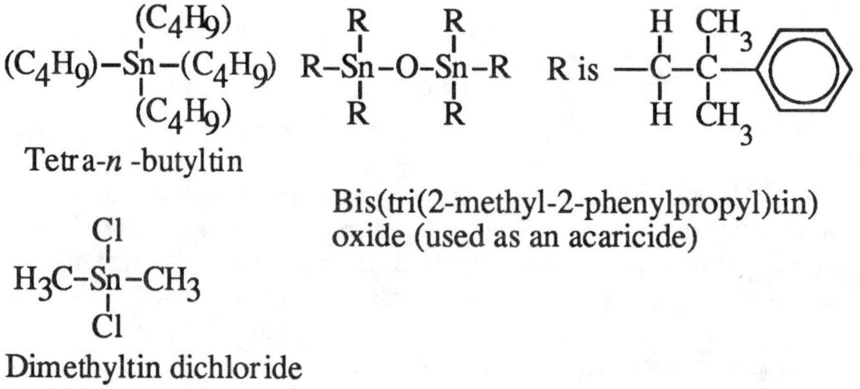

Figure 7.1. Examples of organotin compounds.

Major industrial uses of organotin compounds include applications of tin compounds in fungicides, acaricides, disinfectants, antifouling paints, stabilizers to lessen the effects of heat and light in PVC plastics, catalysts, and precursors for the formation of films of SnO_2 on glass. Tributyl tin chloride and related tributyl tin (TBT) compounds have bactericidal, fungicidal, and insecticidal properties and are of particular environmental significance because of growing use as industrial biocides. In addition to tributyl tin chloride, other tributyl tin compounds used as biocides include the hydroxide, the naphthenate, bis(tributyltin) oxide, and tris(tributylstannyl) phosphate. A major use of TBT is in boat and ship hull coatings to prevent the growth of fouling organisms. Other applications include preservation of wood, leather, paper, and textiles. Because of their antifungal activity TBT compounds are used as slimicides in cooling tower water.

Organotin compounds are readily absorbed through the skin, and skin rashes may result. Organotin compounds, especially those of the R_3SnX type, bind to proteins, probably through the sulfur on cysteine and histidine residues. Interference with mitochondrial function by several mechanisms appears to be the mode of biochemical action leading to toxic responses.

The interaction of trace metals with organic compounds in natural waters is too vast an area to cover in detail in this chapter; however, it may be noted that metal-organic interactions may involve organic species of both pollutant (such as EDTA) and natural (such as fulvic acids) origin. These interactions are influenced by, and sometimes play a role in, redox equilibria; formation and dissolution of precipitates; colloid formation and stability; acid-base reactions; and microorganism-mediated reactions in water. Metal-organic interactions may increase or decrease the toxicity of metals in aquatic ecosystems, and they have a strong influence on the growth of algae in water.

7.6. INORGANIC SPECIES

Many important inorganic water pollutants were considered in Section 7.2, as part of the discussion of pollutant trace elements. Inorganic pollutants that contribute acidity, alkalinity, or salinity to water are considered separately in this chapter. Still another class is that of algal nutrients. This leaves unclassified, however, some important inorganic pollutant species, of which cyanide ion, CN^-, is probably the most important. Others include ammonia, carbon dioxide, hydrogen sulfide, nitrite, and sulfite.

Cyanide

Cyanide, a deadly poisonous substance, exists in water as HCN, a weak acid with a K_a of 6×10^{-10}. The cyanide ion has a strong affinity for many metal ions, forming relatively less-toxic ferrocyanide, $Fe(CN)_6^{4-}$, with iron(II), for example. Volatile HCN is very toxic and has been used in gas chamber executions in the U.S.

Cyanide is widely used in industry, especially for metal cleaning and electroplating. It is also one of the main gas and coke scrubber effluent pollutants from gas works and coke ovens. Cyanide is widely used in certain mineral-processing operations. Until about 1970, a particular gold mine in the United States used almost 1-1/4 tons of cyanide per day to leach gold from a daily output of more than 5,000 tons of ore. Approximately 75 pounds of the cyanide were dumped into a creek each day. The same creek previously had received approximately 30 pounds of mercury daily, as a result of a leaching process that was replaced by cyanide leaching because of the polluting effect of the mercury. There was a distinct lack of aquatic life in the stream, and it was not recommended as a source of drinking water!

Ammonia and Other Inorganic Pollutants

Excessive levels of ammoniacal nitrogen cause water-quality problems. **Ammonia** is the initial product of the decay of nitrogenous organic wastes, and its presence frequently indicates the presence of such wastes. It is a normal constituent of low-pE groundwaters and is sometimes added to drinking water, where it reacts with chlorine to provide residual chlorine (see Section 8.11). Since the pK_a of ammonium ion, NH_4^+, is 9.26, most ammonia in water is present as the protonated form rather than as NH_3.

Hydrogen sulfide, H_2S, is a product of the anaerobic decay of organic matter containing sulfur. It is also produced in the anaerobic reduction of sulfate by microorganisms (see Chapter 6) and is evolved as a gaseous pollutant from geothermal waters. Wastes from chemical plants, paper mills, textile mills, and tanneries may also contain H_2S. Its presence is easily detected by its characteristic rotten-egg odor. In water, H_2S is a weak diprotic acid with pK_{a1} of 6.99 and pK_{a2} of 12.92; S^{2-} is not present in normal natural waters. The sulfide ion has tremendous affinity for many heavy metals, and precipitation of metallic sulfides often accompanies production of H_2S.

Free **carbon dioxide**, CO_2, is frequently present in water at high levels due to decay of organic matter. It is also added to softened water during water treatment as part of a recarbonation process (see Chapter 8). Excessive carbon dioxide levels may make water more corrosive and may be harmful to aquatic life.

Nitrite ion, NO_2^-, occurs in water as an intermediate oxidation state of nitrogen. Its pE range of stability is relatively narrow. Nitrite is added to some industrial process water to inhibit corrosion; it is rarely found in drinking water at levels over 0.1 mg/L.

Sulfite ion, SO_4^{2-}, is found in some industrial wastewaters. Sodium sulfite is commonly added to boiler feedwaters as an oxygen scavenger:

$$2SO_3^{2-} + O_2 \rightarrow 2SO_4^{2-} \tag{7.6.1}$$

Since pK_{a1} of sulfurous acid is 1.76 and pK_{a2} is 7.20, sulfite exists as either HSO_3^- or SO_3^{2-} in natural waters, depending upon pH. It may be noted that hydrazine, N_2H_4, also functions as an oxygen scavenger:

$$N_2H_4 + O_2 \rightarrow 2H_2O + N_2(g) \tag{7.6.2}$$

Asbestos in Water

The toxicity of inhaled asbestos is well established. The fibers scar lung tissue and cancer eventually develops, often 20 or 30 years after exposure. It is not known for sure whether asbestos is toxic in drinking water. This has been a matter of considerable concern because of the dumping of taconite (iron ore tailings) containing asbestos-like fibers into Lake Superior. The fibers have been found in drinking waters of cities around the lake. After having dumped the tailings into Lake Superior since 1952, the Reserve Mining Company at Silver Bay on Lake Superior solved the problem in 1980 by constructing a 6-square-mile containment basin inland from the lake. This $370-million facility keeps the taconite tailings covered with a 3-meter layer of water to prevent escape of fiber dust.

7.7. ALGAL NUTRIENTS AND EUTROPHICATION

The term **eutrophication**, derived from the Greek word meaning "well-nourished," describes a condition of lakes or reservoirs involving excess algal growth, which may eventually lead to severe deterioration of the body of water. The first step in eutrophication of a body of water is an input of plant nutrients (Table 7.3) from watershed runoff or sewage. The nutrient-rich body of water then produces a great deal of plant biomass by photosynthesis, along with a smaller amount of animal biomass. Dead biomass accumulates in the bottom of the lake, where it partially decays, recycling nutrient carbon dioxide, phosphorus, nitrogen, and potassium. If the lake is not too deep, bottom-rooted plants begin to grow, accelerating the accumulation of solid material in the basin. Eventually a marsh is formed, which finally fills in to produce a meadow or forest.

Table 7.3. Essential Plant Nutrients: Sources and Functions

Nutrient	Source	Function
Macronutrients		
Carbon (CO_2)	Atmosphere, decay	Biomass constituent
Hydrogen	Water	Biomass constituent
Oxygen	Water	Biomass constituent
Nitrogen (NO_3^-)	Decay, atmosphere (from nitrogen-fixing organisms), pollutants	Protein constituent
Phosphorus (phosphate)	Decay, minerals, pollutants	DNA/RNA constituent
Potassium	Minerals, pollutants	Metabolic function
Sulfur (sulfate)	Minerals	Proteins, enzymes
Magnesium	Minerals	Metabolic function
Calcium	Minerals	Metabolic function
Micronutrients		
B, Cl, Co, Cu, Fe, Mo, Mn, Na, Si V, Zn	Minerals, pollutants	Metabolic function and/or constituent of enzymes

Eutrophication is by no means a new phenomenon; for instance, it is basically responsible for the formation of huge deposits of coal and peat. However, human activity can greatly accelerate the process. To understand why this is so, refer to Table 7.3, which shows the chemical elements needed for plant growth. Most of these are present at a level more than sufficient to support plant life in the average lake or reservoir. Hydrogen and oxygen come from the water itself. Carbon is provided by CO_2 from the atmosphere or from decaying vegetation. Sulfate, magnesium, and calcium are normally present in abundance from mineral strata in contact with the water. The micronutrients are required at only very low levels (for example, approximately 40 ppb for copper). Therefore, the nutrients most likely to be limiting are the "fertilizer" elements: nitrogen, phosphorus, and potassium. These are all present in sewage and are, of course, found in runoff from heavily fertilized fields. They are also constituents of various kinds of industrial wastes. Each of these elements can also come from natural sources — phosphorus and potassium from mineral formations, and nitrogen fixed by bacteria, blue-green algae, or discharge of lightning in the atmosphere.

Generally, the single plant nutrient most likely to be limiting is phosphorus, and it is generally named as the culprit in excessive eutrophication. Household detergents are a common source of phosphate in wastewater, and eutrophication control has concentrated upon eliminating phosphates from detergents, removing phosphate at the sewage-treatment plant, and preventing phosphate-laden sewage effluents (treated or untreated) from entering bodies of water. (See Chapter 3 for additional details regarding phosphates and detergent phosphate substitutes in water.)

In some cases, nitrogen or even carbon may be limiting nutrients. This is particularly true of nitrogen in seawater.

The whole eutrophication picture is a complex one, and continued research is needed to solve the problem. It is indeed ironic that in a food-poor world, nutrient-rich wastes from over-fertilized fields or from sewage are causing excessive plant growth in many lakes and reservoirs. This perhaps illustrates a point that in general there is no such thing as a pollutant — there are only resources (in this case, plant nutrients) gone to waste.

7.8. ACIDITY, ALKALINITY, AND SALINITY

Aquatic biota are sensitive to extremes of pH. Largely because of osmotic effects, they cannot live in a medium having a salinity to which they are not adapted. Thus, a fresh-water fish soon succumbs in the ocean, and sea fish normally cannot live in fresh water. Excess salinity soon kills plants not adapted to it. There are, of course, ranges in salinity and pH in which organisms live. As shown in Figure 7.2, these ranges frequently may be represented by a reasonably symmetrical curve, along the fringes of which an organism may live without really thriving. These curves do not generally exhibit a sharp cutoff at one end or the other, as does the high-temperature end of the curve representing the growth of bacteria as a function of temperature (Figure 6.4).

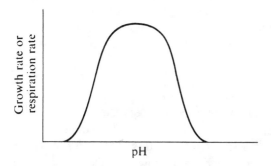

Figure 7.2. A generalized plot of the growth of an aquatic organism as a function of pH.

The most common source of **pollutant acid** in water is acid mine drainage. The sulfuric acid in such drainage arises from the microbial oxidation of pyrite or other sulfide minerals as described in Chapter 6. The values of pH encountered in acid-polluted water may fall below 3, a condition deadly to most forms of aquatic life except the culprit bacteria mediating the pyrite and iron(II) oxidation. Industrial wastes frequently contribute strong acid to water. Sulfuric acid produced by the air oxidation of pollutant sulfur dioxide (see Chapter 11) enters natural waters as acidic rainfall. In cases where the water does not have contact with a basic mineral, such as limestone, the water pH may become dangerously low. This condition occurs in some Canadian lakes, for example.

Excess **alkalinity**, and frequently accompanying high pH, generally are not introduced directly into water by human activity. However, in many geographic areas, the soil and mineral strata are alkaline and impart a high alkalinity to water. Human activity can aggravate the situation; for example, by exposure of alkaline overburden from strip mining to surface water or groundwater. Excess alkalinity in water is manifested by a characteristic fringe of white salts at the edges of a body of water or on the banks of a stream.

Water **salinity** may be increased by a number of human activities. Water passing through a municipal water system inevitably picks up salt from a number of processes; for example, recharging water softeners with sodium chloride. Salts can leach from spoil piles. One of the major environmental constraints on the production of shale oil, for example, is the high percentage of leachable sodium sulfate in piles of spent shale. Careful control of these wastes is necessary to prevent further saline pollution of water in areas where salinity is already a problem. Irrigation adds a great deal of salt to water, a phenomenon responsible for the Salton Sea in California, and is a source of conflict between the United States and Mexico over saline contamination of the Rio Grande and Colorado rivers. Irrigation and intensive agricultural production have caused saline seeps in some of the Western states. These occur when water seeps into a slight depression in tilled, sometimes irrigated, fertilized land, carrying salts (particularly sodium, magnesium, and calcium sulfates) along with it. The water evaporates in the dry summer heat, leaving a salt-laden area behind which no longer supports much plant growth. With time, these areas spread, removing productive crop land from production.

7.9. OXYGEN, OXIDANTS, AND REDUCTANTS

Oxygen is a vitally important species in water (see Chapter 2). In water, oxygen is consumed rapidly by the oxidation of organic matter, $\{CH_2O\}$:

$$\{CH_2O\} + O_2 \xrightarrow{\text{Microorganisms}} CO_2 + H_2O \tag{7.9.1}$$

Unless the water is reaerated efficiently, as by turbulent flow in a shallow stream, it rapidly becomes depleted in oxygen and will not support higher forms of aquatic life.

In addition to the microorganism-mediated oxidation of organic matter, oxygen in water may be consumed by the biooxidation of nitrogenous material,

$$NH_4^+ + 2O_2 \rightarrow 2H^+ + NO_3^- + H_2O \tag{7.9.2}$$

and by the chemical or biochemical oxidation of chemical reducing agents:

$$4Fe^{2+} + O_2 + 10H_2O \rightarrow 4Fe(OH)_3(s) + 8H^+ \tag{7.9.3}$$

$$2SO_3^{2-} + O_2 \rightarrow 2SO_4^{2-} \tag{7.9.4}$$

All these processes contribute to the deoxygenation of water.

The degree of oxygen consumption by microbially-mediated oxidation of contaminants in water is called the **biochemical oxygen demand** (or biological oxygen demand), **BOD**. This parameter is commonly measured by determining the quantity of oxygen utilized by suitable aquatic microorganisms during a five-day period. There is nothing particularly sacred about a five-day period for the BOD test; however, the test originated in England where the maximum stream flow is five days. It was assumed that any contaminant not decomposing in five days would reach the ocean. Despite its somewhat arbitrary nature, a five-day BOD test remains a respectable measure of the short-term oxygen demand exerted by a pollutant.

The addition of oxidizable pollutants to streams produces a typical **oxygen sag curve** as shown in Figure 7.3. Initially, a well-aerated, unpolluted stream is relatively free of oxidizable material; the oxygen level is high; and the bacterial population is relatively low. With the addition of oxidizable pollutant, the oxygen level drops because reaeration cannot keep up with oxygen consumption. In the decomposition zone, the bacterial population rises. The septic zone is characterized by a high bacterial population and very low oxygen levels. The septic zone terminates when the oxidizable pollutant is exhausted, and then the recovery zone begins. In the recovery zone, the bacterial population decreases and the dissolved oxygen level increases until the water regains its original condition.

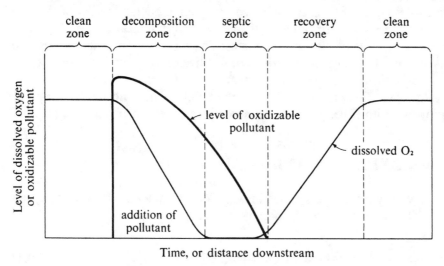

Figure 7.3.   Oxygen sag curve resulting from the addition of oxidizable pollutant material to a stream.

Although BOD is a reasonably realistic measure of water quality insofar as oxygen is concerned, the test for determining it is time-consuming and cumbersome to perform. Total organic carbon (TOC), is frequently measured by catalytically oxidizing carbon in the water and detecting CO_2. It is gaining in popularity because it is easily measured instrumentally.

7.10. ORGANIC POLLUTANTS

Sewage

As shown in Table 7.4, sewage from domestic, commercial, food-processing, and industrial sources contains a wide variety of pollutants, including organic pollutants. Some of these pollutants, particularly oxygen-demanding substances (see Section 7.15), oil, grease, and solids, are removed by primary and secondary sewage-treatment processes. Others, such as salts, heavy metals, and refractory (degradation-resistant) organics, are not efficiently removed.

Table 7.4. Some of the Primary Constituents of Sewage from a City Sewage System

Constituent	Potential Sources	Effects in Water
Oxygen-demanding substances	Mostly organic materials, particularly human feces	Consume dissolved oxygen
Refractory organics	Industrial wastes, household products	Toxic to aquatic life
Viruses	Human wastes	Cause disease (possibly cancer); major deterrent to sewage recycle through water systems
Detergents	Household detergents	Esthetics, prevent grease and oil removal, toxic to aquatic life
Phosphates	Detergents	Algal nutrients
Grease and oil	Cooking, food processing, industrial wastes	Esthetics, harmful to some aquatic life
Salts	Human wastes, water softeners, industrial wastes	Increase water salinity
Heavy metals	Industrial wastes, chemical laboratories	Toxicity
Chelating agents	Some detergents, industrial wastes	Heavy metal ion solubilization and transport
Solids	All sources	Esthetics, harmful to aquatic life

Disposal of inadequately treated sewage can cause severe problems. For example, offshore disposal of sewage results in the formation of beds of sewage residues. Municipal sewage typically contains about 0.1% solids, even after treatment, and these settle out in the ocean in a typical pattern, illustrated in Figure 7.4. The warm sewage water rises in the cold hypolimnion and is carried in one direction or another by tides or currents. It does not rise above the thermocline; instead, it spreads out as a cloud from which the solids rain down on the ocean floor. Aggregation of sewage colloids is aided by dissolved salts in seawater (see Chapter 5), thus promoting the formation of sediment.

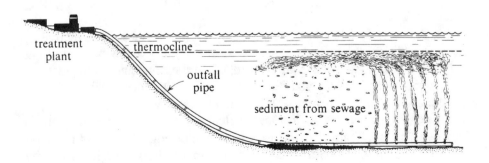

Figure 7.4. Settling of solids from an ocean-floor sewage effluent discharge.

Another major disposal problem with sewage is the sludge produced as a product of the sewage treatment process (see Chapter 8). This sludge contains organic material which continues to degrade slowly; refractory organics; and heavy metals. The amounts of sludge produced are truly staggering. For example, the city of Chicago produces about 3 million tons of sludge each year. A major consideration in the safe disposal of such amounts of sludge is the presence of potentially dangerous components such as heavy metals.

Better control of sewage sources is needed to minimize sewage pollution problems. Particularly, heavy metals and refractory organic compounds need to be controlled at the source to enable use of sewage, or treated sewage effluents, for irrigation, recycle to the water system, or groundwater recharge.

Soaps, Detergents, and Detergent Builders

Soaps, detergents, and associated chemicals are potential sources of organic pollutants. These pollutants are discussed briefly here.

Soaps are salts of higher fatty acids, such as sodium stearate, $C_{17}H_{33}COO^-Na^+$. The cleaning action of soap results largely from its emulsifying power. This concept may be understood by considering the dual nature of the soap anion. An examination of its structure shows that the stearate ion consists of an ionic carboxyl "head" and a long hydrocarbon "tail":

$$\underset{\text{(hydrocarbon chain)}}{\text{/\!}}\overset{\displaystyle O}{\underset{\displaystyle \parallel}{C}}-O^-\ Na^+$$

In the presence of oils, fats, and other water-insoluble organic materials, the tendency is for the "tail" of the anion to dissolve in the organic matter, whereas the "head" remains in aquatic solution. Thus, the soap emulsifies, or suspends, organic material in water. In the process, the anions form colloidal soap micelles, as shown in Figure 5.4.

Soap lowers the surface tension of water. At 25°C, the surface tension of pure water is 71.8 dynes/cm, whereas in the presence of dissolved soap the surface tension is lowered to around 25-30 dynes/cm.

The primary disadvantage of soap as a cleaning agent comes from its reaction with divalent cations to form insoluble salts of fatty acids:

$$C_{17}H_{33}COO^-Na^+ + Ca^{2+} \rightarrow Ca(C_{17}H_{33}CO_2)_2(s) + 2Na^+ \qquad (7.10.1)$$

These insoluble salts, usually salts of magnesium or calcium, are not at all effective as cleaning agents. In addition, the insoluble "curds" form unsightly deposits on clothing and in washing machines. If sufficient soap is used, all of the divalent cations may be removed by their reaction with soap and the water containing excess soap will have good cleaning qualities. This is the approach commonly used when soap is employed in the bathtub or wash basin, where the insoluble calcium and magnesium salts can be tolerated. However, in applications such as washing clothing, the water must be softened by the removal of calcium and magnesium or their complexation by substances such as polyphosphates (see Section 3.11).

Although the formation of insoluble calcium and magnesium salts has resulted in the essential elimination of soap as a cleaning agent for clothing, dishes, and most other materials, it has distinct advantages from the environmental standpoint. As soon as soap gets into sewage or an aquatic system, it generally precipitates as calcium and magnesium salts. Hence, any effects that soap might have in solution are eliminated. With eventual biodegradation, the soap is completely eliminated from the environment. Therefore, aside from the occasional formation of unsightly scum, soap does not cause any substantial pollution problems.

Synthetic **detergents** have good cleaning properties and do not form insoluble salts with "hardness ions" such as calcium and magnesium. Such synthetic detergents have the additional advantage of being the salts of relatively strong acids, and, therefore, they do not precipitate out of acidic waters as insoluble acids, an undesirable characteristic of soaps.

Synthetic detergents usually have a surface-active agent, or surfactant, added to them that lowers the surface tension of water to which the detergent is added, making the water "wetter." Until the early 1960s, the most common surfactant used was an alkyl benzene sulfonate, ABS, a sulfonation product of an alkyl derivative of benzene:

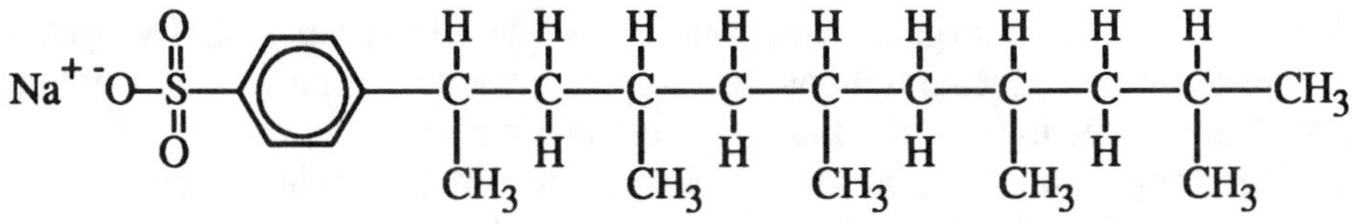

ABS suffered the distinct disadvantage of being only very slowly biodegradable because of its branched-chain structure (see Section 6.7). The most objectionable manifestation of the nonbiodegradable detergents, insofar as the average citizen was concerned, was the "head" of foam that began to appear in glasses of drinking water in areas where sewage was recycled through the domestic water supply. Sewage-plant operators were disturbed by spectacular beds of foam which appeared near sewage outflows and in sewage treatment plants. Occasionally, the entire aeration tank of an activated sludge plant would be smothered by a blanket of foam. Among the other undesirable effects of persistent detergents upon waste-treatment processes were lowered surface tension of water; deflocculation of colloids; flotation of solids; emulsification of grease and oil; and destruction of useful bacteria. Consequently, ABS was replaced by a biodegradable detergent known as LAS.

LAS, α–benzenesulfonate, has the general structure:

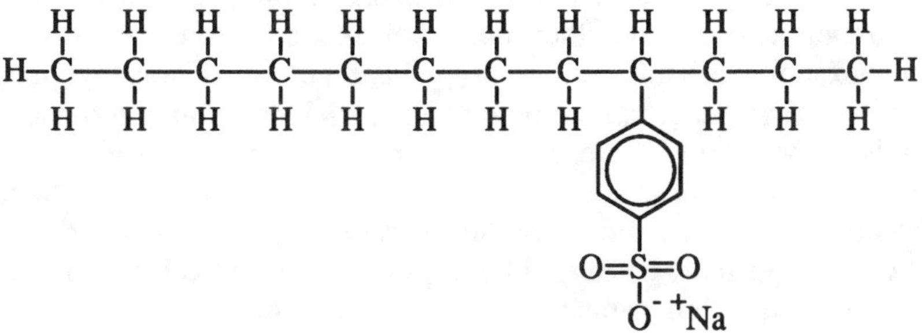

where the benzene ring may be attached at any point on the alkyl chain except at the ends. LAS is more biodegradable than ABS because the alkyl portion of LAS is not branched and does not contain the tertiary carbon which is so detrimental to biodegradability. Since LAS has replaced ABS in detergents, problems arising from the surface-active agent in the detergents (such as toxicity to fish fingerlings) have greatly diminished and the levels of surface-active agents found in water have decreased markedly.

Most of the environmental problems currently attributed to detergents do not arise from the surface-active agents, which basically improve the wetting qualities of water. The builders added to detergents continued to cause environmental problems for a longer time, however. Builders bind to hardness ions, making the detergent solution alkaline and greatly improving the action of the detergent surfactant. A commercial solid detergent contains only 10-30% surfactant. In addition, some detergents still contain polyphosphates added to complex calcium and to function as a builder. Other ingredients include anticorrosive sodium silicates, amide foam stabilizers, soil-suspending carboxymethylcellulose, diluent sodium sulfate, and water absorbed by other components. Of these materials, the polyphosphates have caused the most concern as environmental pollutants, although these problems have largely been resolved.

Biorefractory Organic Pollutants

Millions of tons of organic compounds are manufactured globally each year. Significant quantities of several thousand such compounds appear as water pollutants. Most of these compounds, particularly the less biodegradable ones, are substances to which living organisms have not been exposed until recent years. Frequently, their effects upon organisms are not known, particularly for long-term exposures at very low levels. The potential of synthetic organics for causing genetic damage, cancer, or other ill effects is uncomfortably high. On the positive side, organic pesticides enable a level of agricultural productivity without which millions would starve. Synthetic organic chemicals are increasingly taking the place of natural products in short supply. Thus it is that organic chemicals are essential to the operation of a modern society. Because of their potential danger, however, acquisition of knowledge about their environmental chemistry must have a high priority.

Biorefractory organics are the organic compounds of most concern in wastewater, particularly when they are found in sources of drinking water. These are nonbiodegradable, low-molecular-weight compounds of low volatility. Generally, biorefractory compounds are aromatic or chlorinated hydrocarbons, or both. Included in the list of refractory organic industrial wastes are acetone, benzene, bornyl alcohol, bromobenzene, bromochlorobenzene, butylbenzene, camphor chloroethyl ether, chloroform, chloromethylethyl ether, chloronitrobenzene, chloropyridine, dibromobenzene, dichlorobenzene, dichloroethyl ether, dinitrotoluene, ethylbenzene, ethylene dichloride, 2-ethylhexanol, isocyanic acid, isopropylbenzene, methylbiphenyl, methyl chloride, nitrobenzene, styrene, tetrachloroethylene, trichloroethane, toluene, and 1,2-dimethoxy benzene. Many of these compounds have been found in drinking water, and some are known to cause taste and odor problems in water. Biorefractory compounds are not completely removed by biological treatment, and water contaminated with these compounds must be treated by physical and chemical means, including air stripping, solvent extraction, ozonation, and carbon absorption.

Pesticides in Water

The introduction of DDT during World War II marked the beginning of a period of very rapid growth in pesticide use. Pesticides are employed for many different purposes. Chemicals used in the control of invertebrates include insecticides, molluscicides for the control of snails and slugs, and nematicides for the control of microscopic roundworms. Vertebrates are controlled by rodenticides which kill rodents, avicides used to repel birds, and piscicides used in fish control. Herbicides are used to kill plants. Plant growth regulators, defoliants, and plant desiccants are used for various purposes in the cultivation of plants. Fungicides are used against fungi, bactericides against bacteria, and algicides against algae. Annual U.S. pesticide production rose from about 330 million kilograms of active ingredients in 1962 to about 680 million kilograms of active ingredients in 1982. Although insecticide production has remained about level during the last two or three decades, herbicide production has increased greatly as chemicals have increasingly replaced cultivation of land in the control of weeds. Large quantities of pesticides enter water either directly, in applications such as mosquito control, or indirectly, primarily from drainage of agricultural lands. Major water pollutant pesticides are listed in Table 7.5. These generally belong to the classes of chlorinated hydrocarbons, organic phosphates, and carbamates, which are derived from carbamic acid,

$$\underset{\displaystyle H}{\overset{\displaystyle H}{\underset{|}{N}}}\ \ \overset{O}{\underset{||}{C}}$$

H–N–C–OH

The toxicities of pesticides vary widely. Parathion,

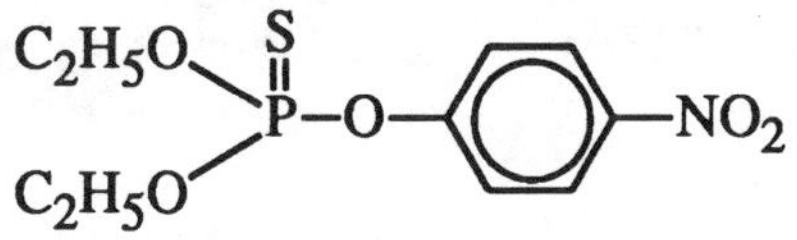

$$\begin{array}{c} C_2H_5O \\ \diagdown \\ C_2H_5O \end{array} \overset{S}{\underset{||}{P}}-O-\hspace{-2pt}\bigcirc\hspace{-2pt}-NO_2$$

has a toxicity rating of 6; as little as 120 mg of parathion has been known to kill an adult human and a dose of 2 mg has killed a child. Most accidental poisonings have occurred by absorption through the skin. Since its use began, several hundred people have been killed by parathion.

In contrast, **malathion** shows how differences in structural formula can cause pronounced differences in the properties of organophosphate pesticides. Malathion has two carboxyester linkages which are hydrolyzable by carboxylase enzymes to relatively non-toxic products as shown by the following reaction:

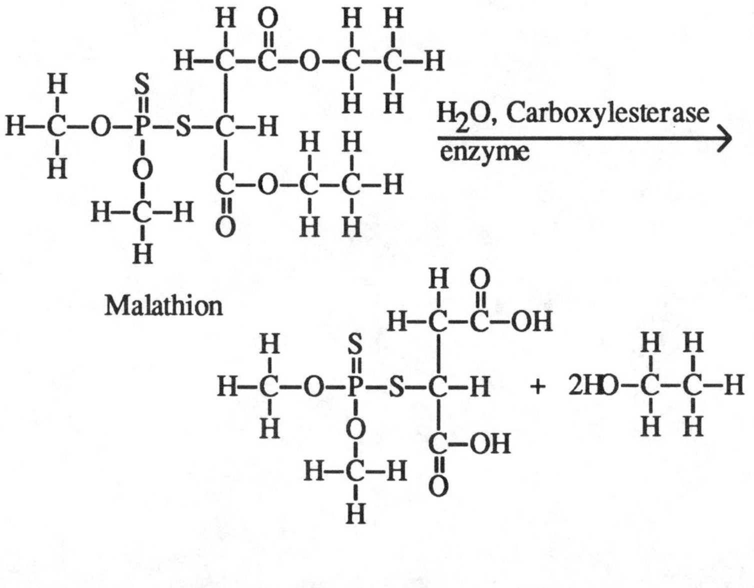

Malathion $\xrightarrow{\text{H}_2\text{O, Carboxylesterase enzyme}}$ (7.10.2)

Table 7.5. Pesticides Commonly Encountered in Water [1]

Pesticide	Formula	Fresh-water quality criteria [2]	Uses and characteristics
Aldicarb Temik®	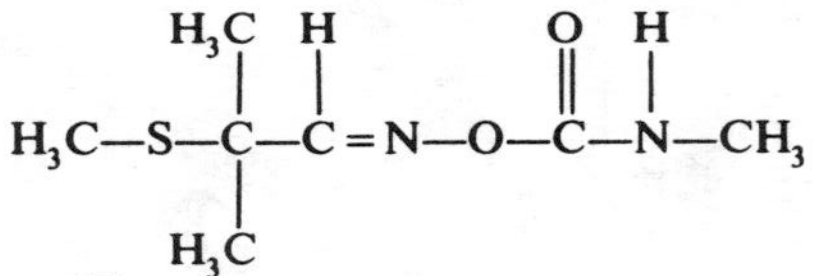 H_3C, $H_3C-S-C-C=N-O-C-N-CH_3$ (with H, H₃C, O, H, CH₃ groups)	—	Plant systemic insecticide taken up by roots or leaves; accused of contaminating well water on Long Island and banned in that area in 1982.
Aldrin–Dieldrin*	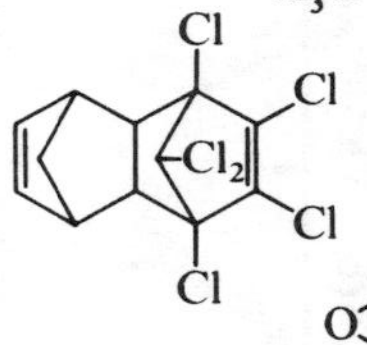 aldrin 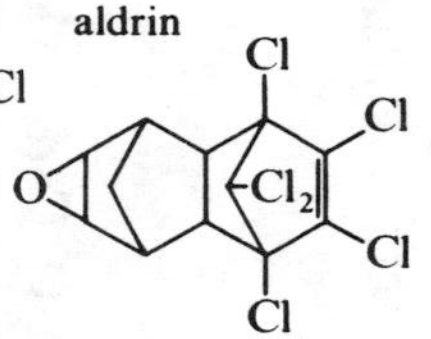 (Cl / Cl_2 structure); dieldrin (endo–exo) (Cl / Cl_2 / O structure)	0.003 µg/L	Persistent and stable in soil, effective against insects in soil. Organisms convert aldrin to dieldrin, known to be carcinogenic to mice. Banned in the U.S. for most uses in 1975.
Chlordane*	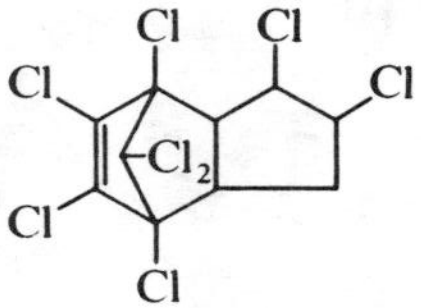 chlordane (Cl / Cl_2 structure)	0.01 µg/L	First of the cyclodiene[3] insecticides to be used, especially effective against termites. Potential carcinogen, use restricted in U.S. in 1978.
Chlorophenoxy herbicides*	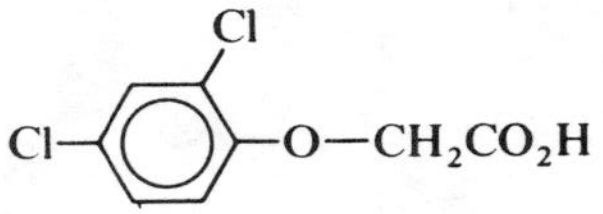 Cl—(ring)—$O-CH_2CO_2H$ 2,4-D; 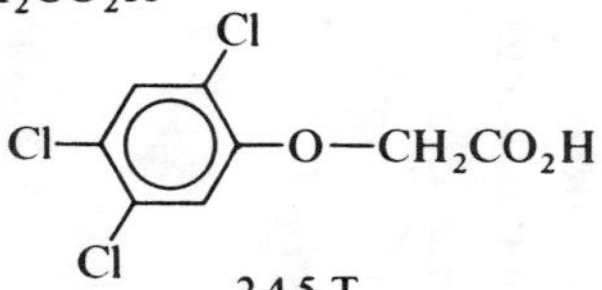 Cl—(ring)—$O-CH_2CO_2H$ 2,4,5-T	100 µg/L (domestic water supplies)	Highly selective for the control of broad-leaved weeds; relatively nontoxic to animals. May contain highly toxic tetrachlorodioxin (TCDD) as a manufacturing impurity.

Name	Structure	Concentration	Description
DDT*	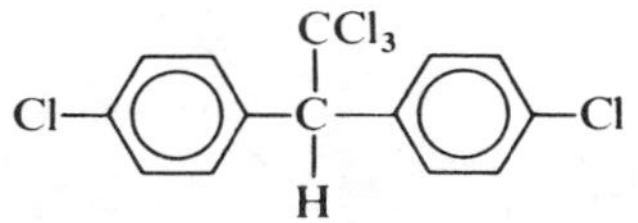	0.001 µg/L	Low acute toxicity to mammals; persistent; accumulates in food chain. Some evidence of carcinogenicity, use banned in the U.S. in 1972.
Demeton*	mixture of $\underset{\parallel}{O}$ $(C_2H_5O)_2P-S-CH_2CH_2-S-C_2H_5$ and $\underset{\parallel}{S}$ $(C_2H_5O)_2P-O-CH_2CH_2-S-C_2H_5$	0.1 µg/L	Lower acute toxicity than most organophosphate pesticides; used to control sucking insects. Action lasts 4–6 weeks under field conditions; systemic insecticide (taken up by roots, translocated to stems and leaves, toxic to insects sucking plant juices).
Endosulfan*	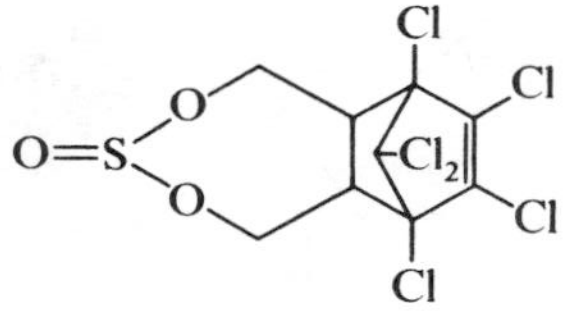	0.003 µg/L	Used on fruit and berries to control aphids, beetles, and caterpillars; lower chronic toxicity to mammals than most cyclodiene pesticides.
Endrin*	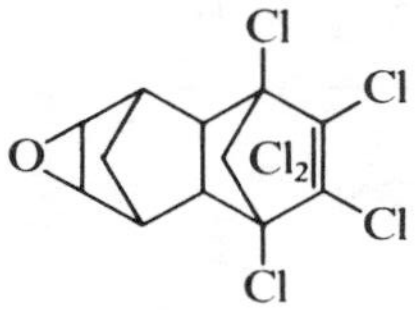 (endo–endo)	0.004 µg/L	Only pesticide effective against black currant mud mite; also used as a zoocide. Special precautions must be used to avoid skin contact during application; readily photolyzes to nontoxic ketone form.
Guthion (azinphosmethyl)*	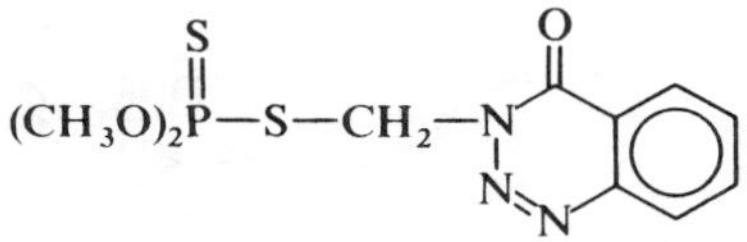	0.01 µg/L	Used for control of pests on fruit, cotton, other crops; effective acaricide (controls mites); relatively toxic to mammals.
Heptachlor*	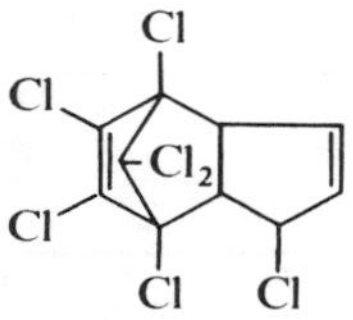	0.001 µg/L	Used to control pests in soil; insecticide in feed. Changes to the more toxic epoxide which persists for a long time in soil; use restricted in U.S. in 1978.

Table 7.5. (continued)

Pesticide	Formula	Fresh-water quality criteria[2]	Uses and characteristics
Lindane*	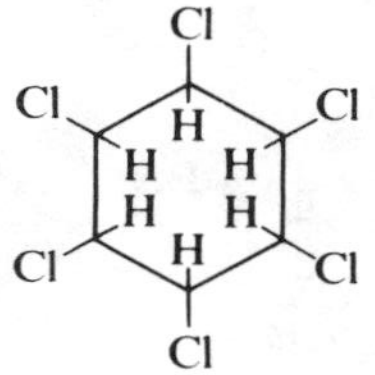 (γ-isomer)	0.01 µg/L	Used to control insects, plant pests, animal parasites; widely manufactured because of convenience, lack of odor, minimal residue.
Malathion*	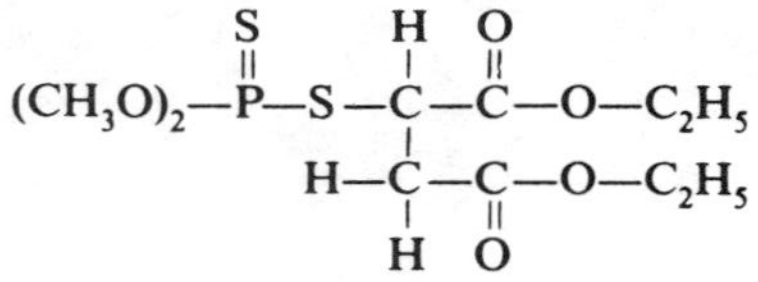	0.1 µg/L	Safer than most organophosphate pesticides, comparatively little hazard to mammals.
Methoxychlor*	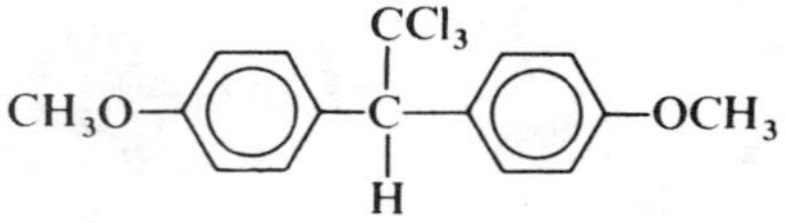	0.03 µg/L	Popular DDT substitute, reasonably biodegradable; low toxicity to mammals.
Mirex*	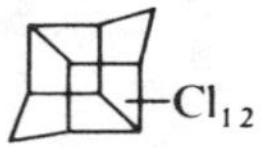	0.001 µg/L	Used almost exclusively to control the imported fire ant in the southeastern U.S.
Methyl parathion*	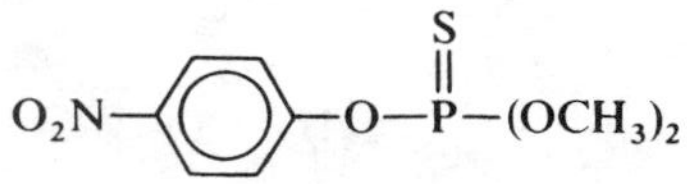	—	Used to control many plant pests; ranks second in U.S. pesticide consumption.
Toxaphene*	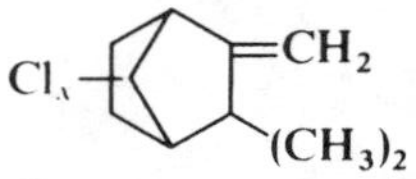 (x = average of 8)	5 µg/L	Pesticide most widely used in the U.S.; EPA has proposed banning it because of tests showing it is carcinogenic in mice and rats.

Name	Structure	Description
Carbaryl (Sevin®)	naphthalene–O–C(=O)–N(H)–CH₃	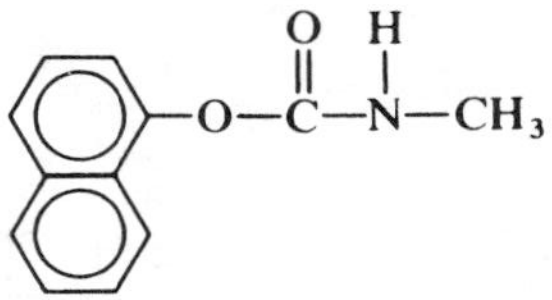 Carbamate pesticide widely used as a lawn and garden insecticide; low toxicity to mammals.
Diazinon	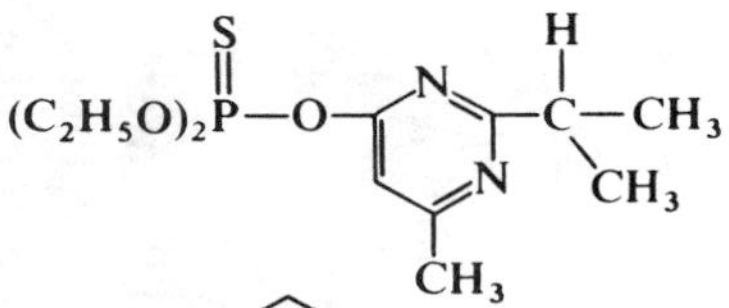 (C₂H₅O)₂P(=S)–O–[pyrimidine ring with CH₃ and CH(CH₃)CH₃ substituents]	Used to control plant pests and animal parasites; relatively high toxicity to mammals.
Carbofuran	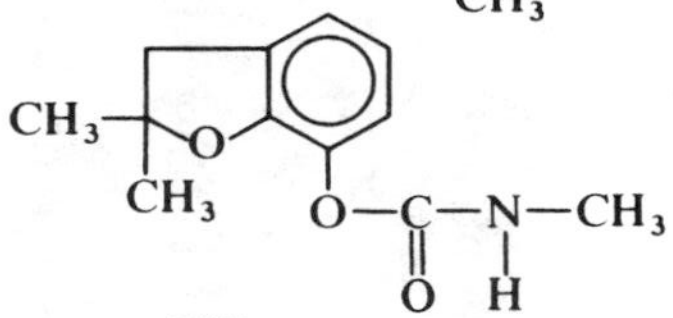 benzofuran ring (CH₃, CH₃)–O–C(=O)–N(H)–CH₃	Good plant systemic pesticide because of high water-solubility; readily taken up by roots and leaves; subject of litigation by exposed workers in Texas, resulting in a 1982 court order directing EPA to release health and safety information
Picloram	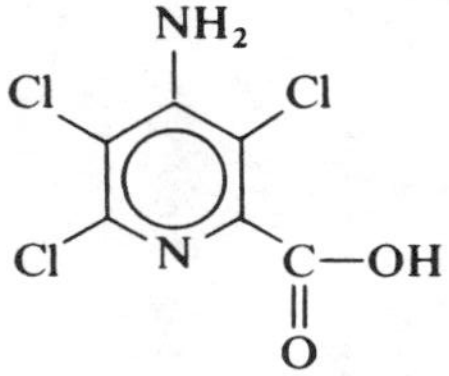 pyridine ring with NH₂, Cl, Cl, Cl, and C(=O)–OH substituents	Popular herbicide used against broad-leaved and woody plants; taken up by either roots or foliage; accused of causing cancer deaths in timbered areas of North Carolina.
Diquat	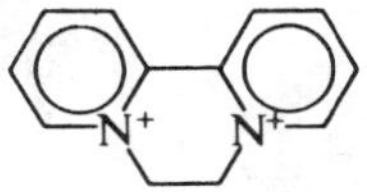 bipyridinium (two N⁺ pyridine rings joined by –CH₂CH₂–)	These are the only important pesticides of the bipyridylium (2 pyridine rings) type. They are contact herbicides applied directly to plant tissue, causing rapid cell membrane destruction and a frost-bitten appearance. Paraquat was used to spray marijuana in the southeastern U.S. in August, 1983, causing considerable controversy.
Paraquat 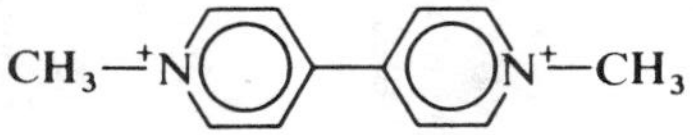	CH₃–⁺N(pyridine ring)–(pyridine ring)N⁺–CH₃	

---

[1] Pesticides designated by an asterisk are those discussed as particularly troublesome water pollutants in *Quality Criteria for Water*, U.S. Environmental Protection Agency, Washington, D.C., 1976. The others are noteworthy for large-scale use in the U.S.

[2] Except where noted, these are criteria for the protection of fresh-water marine life.

[3] "Cyclodiene," or "diene," pesticides are those whose synthesis is based upon hexachlorocyclopentadiene.

The enzymes that accomplish malathion hydrolysis are possessed by mammals, but not by insects, so that mammals can detoxify malathion, whereas insects cannot. The result is that malathion has selective insecticidal activity. For example, although malathion is a very effective insecticide, its LD_{50} (dose required to kill 50% of test subjects) for adult male rats is about 100 times that of parathion, reflecting the much lower mammalian toxicity of malathion compared to some of the more toxic organophosphate insecticides, such as parathion.

The chlorinated hydrocarbon DDT (dichlorodiphenyltrichloroethane or 1,1,1-trichloro-2,2-di-(4-chlorophenyl)-ethane) was used in massive quantities following World War II. Although its toxicity is generally low, its persistence and accumulation in food chains have led to a ban on DDT use in the United States; it is still employed in some countries.

A number of water pollution and health problems have been associated with the manufacture of pesticides. For example, degradation-resistant hexachlorobenzene,

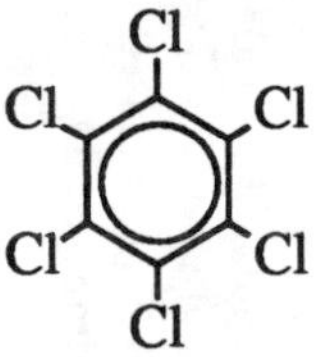

is used as a raw material for the synthesis of other pesticides and has often been found in water.

The most notorious byproducts of pesticide manufacture are **polychlorinated dibenzodioxins**, which have the same basic structure as that of TCDD (2,3,7,8 tetrachlorodibenzo-p-dioxin shown in Figure 7.5), but different numbers and

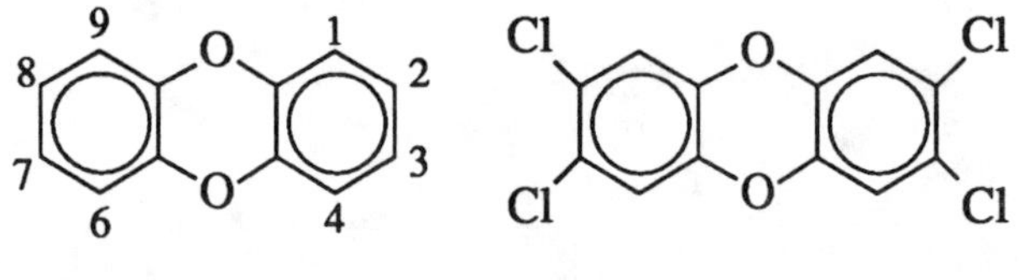

Dibenzo-p-dioxin   2,3,7,8-Tetrachlorodibenzo-
p-dioxin

Figure 7.5.   Dibenzo-p-dioxin and 2,3,7,8-tetrachlorodibenzo-p-dioxin (TCDD), often called simply "dioxin." In the structure of dibenzo-p-dioxin each number refers to a numbered carbon atom to which an H atom is bound and the names of derivatives are based upon the carbon atoms where another group has been substituted for the H atoms, as is seen by the structure and name of 2,3,7,8-tetrachlorodibenzo-p-dioxin.

arrangements of chlorine atoms on the ring structure. Commonly referred to as "dioxins," these species have a high environmental and toxicological significance.[2]

From 1 to 8 Cl atoms may be substituted for H atoms on dibenzo-p-dioxin, giving a total of 75 possible chlorinated derivatives. Of these, the most notable pollutant and hazardous waste compound is **2,3,7,8-tetrachlorodibenzo-p-dioxin (TCDD)**, often referred to simply as "**dioxin.**" This compound, which is one of the most toxic of all synthetic substances to some animals, was produced as a low-level contaminant in the manufacture of some aromatic, oxygen-containing organohalide compounds such as chlorophenoxy herbicides and hexachlorophene (Figure 7.6) manufactured by processes used until the 1960s.

The chlorophenoxy herbicides, including 2,4,5-trichlorophenoxyacetic acid (2,4,5-T) shown in Figure 7.6, were manufactured on a large scale for weed and brush control and as military defoliants. Fungicidal and bactericidal hexachlorophene was once widely applied to crops in the production of vegetables and cotton and was used as an antibacterial agent in personal care products, an application that has been discontinued because of toxic effects and possible TCDD contamination.

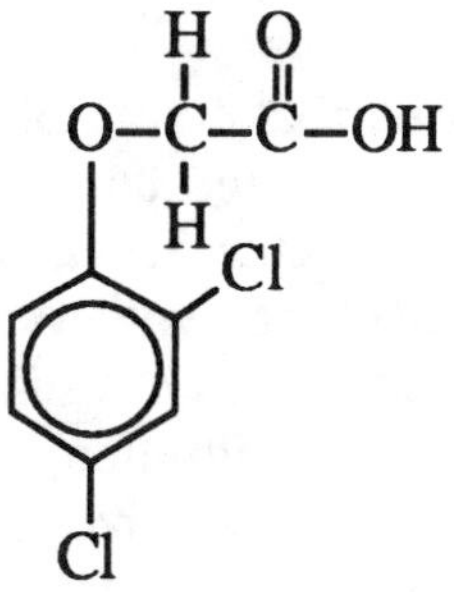

2,4–Dichlorophenoxy-
acetic acid (and esters)   Hexachlorophene

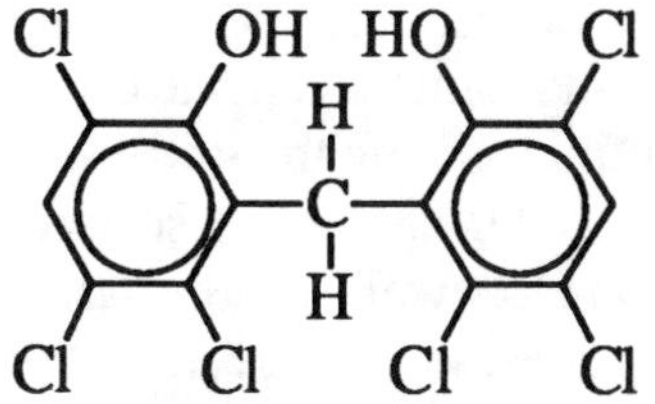

Figure 7.6. Two chemicals whose manufacture resulted in the production of byproduct TCDD contaminant.

TCDD has a very low vapor pressure of only 1.7×10^{-6} mm Hg at 25°C, a high melting point of 305°C, and a water solubility of only 0.2 µg/L. It is stable thermally up to about 700°C, has a high degree of chemical stability, and is poorly biodegradable. It is very toxic to some animals, with an LD_{50} of only about 0.6 µg/kg body mass in male guinea pigs. (The type and degree of its toxicity to humans is largely unknown; it is known to cause a severe skin condition called chloracne). Because of its properties, TCDD is a stable, persistent environmental pollutant and hazardous waste constituent of considerable concern. It has been identified in some municipal incineration emissions, and has been a widespread environmental pollutant from improper waste disposal.

The most notable case of TCDD contamination resulted from the spraying of waste oil mixed with TCDD on roads and horse arenas in Missouri in the early 1970s. The oil was used to try to keep dust down in these areas. The extent of contamination was revealed by studies conducted in late 1982 and early 1983. As a result, the U.S. EPA bought out the entire TCDD-contaminated town of Times Beach, Missouri, in March, 1983, at a cost of $33 million. On December 31, 1990 final permission was granted by the courts to incinerate the soil at Times Beach, as well as TCDD-contaminated soil from other areas at a total cost of $80 million.[3] TCDD has been released in a number of industrial accidents, the most massive of which exposed several tens of thousands of people to a cloud of chemical emissions spread over an approximately 3-square-mile area at the Givaudan-La Roche Icmesa manufacturing plant near Seveso, Italy, in 1976. On an encouraging note from a toxicological perspective, no abnormal occurrences of major malformations were found in a study of 15,291 children born in the area within 6 years after the release.[4]

One of the greater environmental disasters ever to result from pesticide manufacture involved the production of Kepone, structural formula

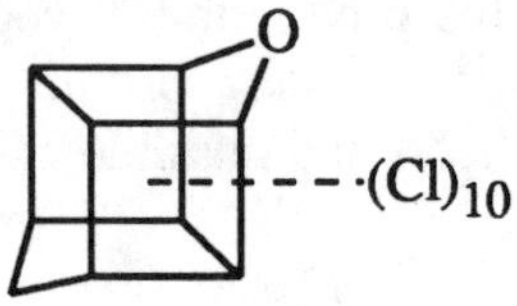

$$\cdots (Cl)_{10}$$

This pesticide has been used for the control of banana-root borer, tobacco wireworm, ants, and cockroaches. Kepone exhibits acute, delayed, and cumulative toxicity in birds, rodents, and humans, and it causes cancer in rodents. It was manufactured in Hopewell, Virginia, during the mid-1970s. During this time, workers were exposed to Kepone and are alleged to have suffered health problems as a result. The plant was connected to the Hopewell sewage system, and frequent infiltration of Kepone wastes caused the Hopewell sewage treatment plant to become inoperative at times. As much as 53,000 kg of Kepone may have been dumped into the sewage system during the years that the plant was operated. The sewage effluent was discharged to the James River, resulting in extensive environmental dispersion and toxicity to aquatic organisms. Decontamination of the river would have required dredging and detoxification of 135 million cubic meters of river sediment at a prohibitively high cost of several billion dollars.

Polychlorinated Biphenyls

First discovered as environmental pollutants in 1966, **polychlorinated byphenyls (PCB compounds)** have been found throughout the world in water, sediments, bird tissue, and fish tissue. These compounds constitute an important class of special wastes.[5] They are made by substituting from 1 to 10 Cl atoms onto the biphenyl aromatic structure as shown on the left in Figure 7.7. This substitution can produce 209 different compounds (congeners), of which one example is shown on the right in Figure 7.7.

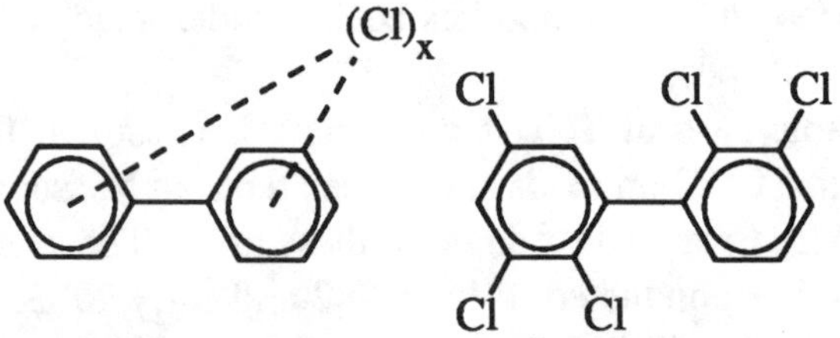

Figure 7.7. General formula of polychlorinated biphenyls (left, where X may range from 1 to 10) and a specific 5-chlorine congener (right).

Polychlorinated biphenyls have very high chemical, thermal, and biological stability; low vapor pressure; and high dielectric constants. These properties have led to the use of PCBs as coolant-insulation fluids in transformers and capacitors; for the impregnation of cotton and asbestos; as plasticizers; and as additives to some epoxy paints. The same properties that made extraordinarily stable PCBs so useful also contributed to their widespread dispersion and accumulation in the environment. By regulations issued in the U.S. under the authority of the Toxic Substances Control Act passed in 1976, the manufacture of PCBs was discontinued in the U.S. and their uses and disposal were strictly controlled.

Several chemical formulations have been developed to substitute for PCBs in electrical applications. Disposal of PCBs from discarded electrical equipment and other sources remains a problem, particularly since PCBs can survive ordinary incineration by escaping as vapors through the smokestack. However, they can be destroyed by special incineration processes.

PCBs are especially prominent pollutants in the sediments of the Hudson River[6] as a result of waste discharges from two capacitor manufacturing plants, which operated about 60 km upstream from the southernmost dam on the river from 1950 to 1976. The river sediments downstream from the plants exhibit PCB levels of about 10 ppm, 1-2 orders of magnitude higher than levels commonly encountered in river and estuary sediments.

In the fall of 1981 New York state hunters were warned to limit their consumption of wild ducks because of PCB contamination, and Montana hunters were given similar warnings because of contamination by Endrin.[7] Dissection of 63 ducks from the Hudson River and Lake Ontario regions showed contamination levels of 7.5 ppm PCBs, compared to limits of 3 ppm for chickens. As an example of the hardship that environmental contamination can cause, it was suggested that no more than two meals of duck be eaten per month and that the skin and fat should be carefully removed. It was further recommended that if the ducks were cooked with stuffing, the stuffing not be eaten!

Askarel

Askarel is the generic name of PCB-containing dielectric fluids in transformers. These fluids are 50-70 percent PCBs and may contain 30-50 percent trichlorobenzenes (TCBs). As of 1989 an estimated 100,000 askarel-containing transformers were still in use in the U.S.[8] Although the dielectric fluid may be replaced in these transformers enabling their continued use, PCBs tend to leach into the replacement fluid from the transformer core and other parts of the transformer over a several-month period.

Polybrominated Biphenyls

Polybrominated byphenyl (PBB) is a chemical fire retardant which was accidently mistaken for magnesium oxide and mixed with cattle feed distributed in Michigan in 1973[9]. As a result, over 30,000 cattle, approximately 6000 hogs, 1500 sheep, 1.5 million chickens, 18,000 pounds of cheese, 2700 pounds of butter, 34,000 pounds of dry milk products, and 5 million eggs had to be destroyed. Farm families eating PBB-contaminated foods had ingested the substance, and it was detected in the blood of many Michigan residents. Persons who had ingested PBB tended to have less disease resistance and increased incidence of rashes, liver ailments, and headaches. The economic cost of the Michigan PBB incident exceeded $100 million.

7.11. RADIONUCLIDES IN THE AQUATIC ENVIRONMENT

The massive production of **radionuclides** (radioactive isotopes) by weapons and nuclear reactors since World War II has been accompanied by increasing concern about the effects of radioactivity upon health and the environment. Radionuclides are produced as fission products of heavy nuclei of such elements as uranium or plutonium. They are also produced by the reaction of neutrons with stable nuclei. These phenomena are illustrated in Figure 7.8. Radionuclides are formed in large quantities as waste products in nuclear power generation. Their ultimate disposal is a problem that has caused much controversy regarding the widespread use of nuclear power. Artificially produced radionuclides are also widely used in industrial and medical applications, particularly as "tracers." With so many possible sources of radionuclides, it is impossible to entirely eliminate radioactive contamination of aquatic systems. Furthermore, radionuclides may enter aquatic systems from natural sources. Therefore, the transport, reactions, and biological concentration of radionuclides in aquatic ecosystems are of great importance to the environmental chemist.

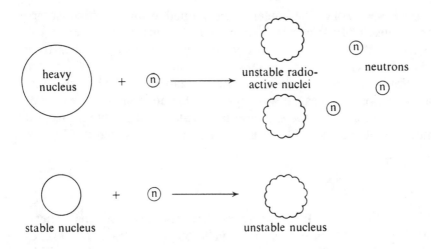

Figure 7.9. A heavy nucleus, such as that of ^{235}U, may absorb a neutron and break up (undergo fission), yielding lighter radioactive nuclei. A stable nucleus may absorb a neutron to produce a radioactive nucleus.

Radionuclides differ from other nuclei in that they emit **ionizing radiation** — alpha particles, beta particles, and gamma rays. The most massive of these emissions is the **alpha particle**, a helium nucleus of atomic mass 4, consisting of two neutrons and two protons. The symbol for an alpha particle is $^{4}_{2}\alpha$. An example of alpha production is found in the radioactive decay of uranium-238:

$$^{238}_{92}U \longrightarrow\ ^{234}_{90}Th\ +\ ^{4}_{2}\alpha\ + energy \tag{7.11.1}$$

This transformation consists of a uranium nucleus, atomic number 92 and atomic mass 238, losing an alpha particle, atomic number 2 and atomic mass 4 to yield a thorium nucleus, atomic number 90 and atomic mass 234.

Beta radiation consists of either highly energetic, negative electrons, which are designated $^{0}_{-1}\beta$, or positive electrons, called positrons and designated $^{0}_{1}\beta$. A typical beta emitter, chlorine-38, may be produced by irradiating chlorine with neutrons. The chlorine-37 nucleus, natural abundance 24.5%, absorbs a neutron to produce chlorine-38 and gamma radiation:

$$^{37}_{17}Cl + ^{1}_{0}n \longrightarrow ^{38}_{12}Cl + \gamma \qquad (7.11.2)$$

The chlorine-38 nucleus is radioactive and loses a negative beta particle to become an argon-38 nucleus:

$$^{38}_{17}Cl \longrightarrow ^{38}_{18}Ar + ^{0}_{-1}\beta \qquad (7.11.3)$$

Since the negative beta particle has essentially no mass and a -1 charge, the stable product isotope, argon-38, has the same mass and a charge 1 greater than chlorine-38.

Gamma rays are electromagnetic radiation similar to X-rays, though more energetic. Since the energy of gamma radiation is often a well-defined property of the emitting nucleus, it may be used in some cases for the qualitative and quantitative analysis of radionuclides.

The primary effect of alpha particles, beta particles, and gamma rays upon materials is the production of ions; therefore they are called **ionizing radiation**. Due to their large size, alpha particles do not penetrate matter deeply, but cause an enormous amount of ionization along their short path of penetration. Therefore, alpha particles present little hazard outside the body, but are very dangerous when ingested. Although beta particles are more penetrating than alpha particles, they produce much less ionization per unit path length. Gamma rays are much more penetrating than particulate radiation. Their degree of penetration is proportional to their energy.

The **decay** of a specific radionuclide follows first-order kinetics; that is, the number of nuclei disintegrating in a short time interval is directly proportional to the number of radioactive nuclei present. The rate of decay, $-dN/dt$, is given by the equation,

$$\text{Decay rate} = -\frac{dN}{dt} = \lambda N \qquad (7.11.4)$$

where N is the number of radioactive nuclei present and λ is the rate constant, which has units of reciprocal time. Since the exact number of disintegrations per second is difficult to determine in the laboratory, radioactive decay is often described in terms of the **activity**, A, which is proportional to the absolute rate of decay. The first-order decay equation may be expressed in terms of A,

$$A = A_0 e^{-\lambda t} \qquad (7.11.5)$$

where A is the activity at time t; A_0 is the activity when t is zero; and e is the natural logarithm base. The **half-life**, $t_{1/2}$, is generally used instead of λ to characterize a radionuclide:

$$t_{1/2} = \frac{0.693}{\lambda} \qquad (7.11.6)$$

As the term implies, a half-life is the period of time during which half of a given number of atoms of a specific kind of radionuclide decay. Ten half-lives are required for the loss of 99.9% of the activity of a radionuclide.

Radiation damages living organisms by initiating harmful chemical reactions in tissues. For example, bonds are broken in the macromolecules that carry out life processes. In cases of acute radiation poisoning, bone marrow which produces red blood cells is destroyed and the concentration of red blood cells is diminished. Radiation-induced genetic damage is of great concern. Such damage may not become apparent until many years after exposure. As humans have learned more about the effects of ionizing radiation, the dosage level considered to be safe has steadily diminished. For example, the United States Atomic Energy Commission has dropped the maximum permissible concentration of some radioisotopes to levels of less than one ten-thousandth of those considered safe in the early 1950s. Although it is possible that even the slightest exposure to ionizing radiation entails some damage, some radiation is unavoidably received from natural sources. For the majority of the population, exposure to natural radiation exceeds that from artificial sources.

The study of the ecological and health effects of radionuclides involves consideration of many factors. Among these are the type and energy of radiation emitter and the half-life of the source. In addition, the degree to which the particular element is absorbed by living species and the chemical interactions and transport of the element in aquatic ecosystems are important factors. Radionuclides having very short half-lives may be hazardous when produced but decay too rapidly to affect the environment into which they are introduced. Radionuclides with very long half-lives may be quite persistent in the environment but of such low activity that little environmental damage is caused. Therefore, in general, radionuclides with intermediate half-lives are the most dangerous. They persist long enough to enter living systems while still retaining a high activity. Because they may be incorporated within living tissue, radionuclides of "life elements" are particularly dangerous. Much concern has been expressed over strontium-90, a common waste product of nuclear testing. This element is interchangeable with calcium in bone. Strontium-90 fallout drops onto pasture and crop land and is ingested by cattle. Eventually, it enters the bodies of infants and children by way of cow's milk.

Some radionuclides found in water, primarily radium and potassium-40, originate from natural sources, particularly leaching from minerals. Others come from pollutant sources, primarily nuclear power plants and testing of nuclear weapons. The levels of radionuclides found in water typically are measured in units of picoCuries/liter, where a Curie is 3.7×10^{10} disintegrations per second, and a picoCurie is 1×10^{-10} that amount, or 3.7×10^{-2} disintegrations per second. (2.2 disintegrations per minute).

The radionuclide of most concern in drinking water is **radium**, Ra.[10] Areas in the United States where significant radium contamination of water has been observed include the uranium-producing regions of the western U.S., Iowa, Illinois, Wisconsin, Missouri, Minnesota, Florida, North Carolina, Virginia, and the New England states.

The maximum contaminant level (MCL) for total radium (^{226}Ra plus ^{228}Ra) in drinking water is specified by the U.S. Environmental Protection Agency as 5 pCi/L (picoCuries per liter). Perhaps as many as several hundred municipal water supplies in the U.S. exceed this level and require additional treatment to remove radium. Fortunately, conventional water softening processes, which are designed to take out excessive levels of calcium, are relatively efficient in removing radium from water.

As the use of nuclear power has increased, the possible contamination of water by fission-product radioisotopes has become more of a cause for concern. (If nations continue to refrain from testing nuclear weapons above ground, it is hoped that radioisotopes from this source will contribute only minor amounts of radioactivity to water.) Table 7.6 summarizes the major natural and artificial radionuclides likely to be encountered in water.

Table 7.6. Radionuclides in Water

Radionuclide	Half-life	Nuclear reaction, description, source
Naturally occurring and from cosmic reactions		
Carbon-14	5730 years	$^{14}N(n, p)$ ^{14}C,* thermal neutrons from cosmic or nuclear-weapon sources reacting with N_2
Silicon-32	~ 300 years	$^{40}Ar(p,x)$ ^{32}Si, nuclear spallation (splitting of the nucleus) of atmospheric argon by cosmic-ray protons
Potassium-40	~ 1.4×10^9 years	0.0119% of natural potassium
Naturally occurring from ^{238}U *series*		
Radium-226	1620 years	Diffusion from sediments, atmosphere
Lead-210	21 years	$^{226}Ra \rightarrow 6$ steps $\rightarrow ^{210}Pb$
Thorium-230	75,200 years	$^{238}U \rightarrow 3$ steps $\rightarrow ^{230}Th$ produced *in situ*
Thorium-234	24 days	$^{238}U \rightarrow ^{234}Th$ produced *in situ*
From reactor and weapons fission		
Strontium-90	28 years	These are the fission-product radioisotopes of greatest significance because of their high yields and biological activity.
Iodine-131	8 days	
Cesium-137	30 years	
Barium-140	13 days	The isotopes from barium-140 through krypton-85 are listed in generally decreasing order of fission yield.
Zirconium-95	65 days	
Cerium-141	33 days	
Strontium-89	51 days	
Ruthenium-103	40 days	
Krypton-85	10.3 years	
Cobalt-60	5.25 years	From nonfission neutron reactions in reactors
Manganese-54	310 days	From nonfission neutron reactions in reactors
Iron-55	2.7 years	$^{56}Fe(n, 2n)$ ^{55}Fe, from high-energy neutrons acting on iron in weapon hardware
Plutonium-239	24,300 years	$^{238}U(n, \gamma)$ ^{239}Pu, neutron capture by uranium

* This notation denotes the isotope nitrogen-14 reacting with a neutron, n, giving off a proton, p, and forming the isotope carbon-14; other nuclear reactions may be similarly deduced from the notation shown. (Note that x represents nuclear fragments from the spallation reaction.)

Transuranic elements are of growing concern in the oceanic environment. These alpha emitters are long-lived and highly toxic. As their production increases, so does the risk of environmental contamination. Included among these elements are various isotopes of neptunium, plutonium, americium, and curium. Specific isotopes, with half-lives in years given in parentheses, are: Np-237 (2.14×10^6); Pu-236 (2.85); Pu-238 (87.8); Pu-239 (2.44×10^4); Pu-240 (6.54×10^3); Pu-241 (15); Pu-242 (3.87×10^5); Am-241 (433); Am-243 (7.37×10^6); Cm-242 (0.22); and Cm-244 (17.9).

LITERATURE CITED

1. Holmes, C. W., E. A. Slade, and C. J. McLerran, "Migration and Redistribution of Zinc and Cadmium in a Marine Estuarine System," *Environmental Science and Technology*, **8**, 255-9 (1974).

2. Espositio, M. Pat, "Dioxin Wastes," Section 4.3 in *Standard Handbook of Hazardous Waste Treatment and Disposal*, Harry M. Freeman, Ed., McGraw Hill, New York, 1989, pp. 4.25-4.34.

3. "Judge Approves Plan to Incinerate Dioxin-Contaminated Waste in Missouri," *New York Times*, January 2, 1991, p. A9.

4. "Dioxin is Found Not to Increase Birth Defects," *New York Times*, March 18, 1988, p. 12.

5. McCoy, Drew E., "PCB Wastes," Section 4.2 in *Standard Handbook of Hazardous Waste Treatment and Disposal*, Harry M. Freeman, Ed., McGraw Hill, New York, 1989, pp. 4.13-4.23.

6. Bopp, R. F., H. J. Simpsom, C. R. Olsen, and N. Kostyk, "Polychlorinated Biphenyls in Sediments of the Tidal Hudson River," *Environmental Science and Technology*, **15**, 210-216 (1981).

7. Faber, H., "New York Hunters Warned on Eating Wild Ducks," *New York Times*, October 8, 1981.

8. Bishop, Jim, "Cleaning up Askarel," *Hazmat World*, June, 1989, p. 29.

9. Carter, L. J., "Michigan PBB Incident: Chemical Mix-up Leads to Disaster," *Science*, **192**, 240-3 (1976).

10. Valentine, Richard L., Roger C. Splinter, Timothy S. Mulholland, Jeffrey M. Baker, Thomas M. Nogaj and Jao-Jia Horng, *A Study of Possible Economical Ways of Removing Radium from Drinking Water*, EPA/600/S2-88/009, U. S. Environmental Protection Agency, Washington, DC, 1988.

SUPPLEMENTARY REFERENCES

Hassell, Kenneth A., *The Biochemistry and Uses of Pesticides*, 2nd ed., VCH Publishers, Inc., New York, 1991.

Thier, Hans-Peter, and Hans Zeumeer, Eds., *Manual of Pesticide Residue Analysis*, VCH Publishers, Inc., New York, 1988.

Grosse, Douglas, *Managing Hazardous Wastes Containing Heavy Metals*, SciTech Publishers, Matawan, NJ, 1990.

Lave, Lester B., and Arthur C. Upton, Eds., *Toxic Chemicals, Health, and the Environment*, The Johns Hopkins University Press, Baltimore, MD, 1987.

Wang, Rhoda G. M., Ed., *Biological Monitoring for Pesticide Exposure: Measurement, Estimation, and Risk Reduction*, ACS Symposium Series **382** American Chemical Society, Washington, D.C., 1989.

Halogenated-Organic-Containing Wastes: Treatment Technologies, Noyes Publications, Park Ridge, NJ, 1988.

Arenti, M., *Dioxin-Containing Wastes: Treatment Technologies*, Noyes Data Corp., Park Ridge, NJ, 1988.

Ram, Neil M., Edward J. Calabrese, and Russell F. Christman, Eds., *Organic Carcinogens in Drinking Water: Detection, Treatment, and Risk Assessment*, John Wiley & Sons, New York, NY, 1986.

Ware, George W., Ed., *Reviews of Environmental Contamination and Toxicology*, Vol. 99., Springer-Verlag, New York, NY, 1987.

Hites, Ronald A. and S. J. Eisenreich, Eds., *Sources and Fates of Aquatic Pollutants*, Advances in Chemistry Series 216. American Chemical Society, Washington, DC, 1987.

Milwidsky, B., and A. S. Davidsohn, *Synthetic Detergents*, 7th Ed., John Wiley and Sons, New York, NY, 1987.

Burnes, Michael E., Ed, *Low-Level Radioactive Waste Regulation: Science, Politics, and Fear*, Lewis Publishers, Chelsea, MI, 1988.

Tardiff, Robert G. and Joseph V. Rodricks, Eds., *Toxic Substances and Human Risk*, Plenum Press, New York, NY, 1987.

Worobec, Mary Devine, and Girard Ordway, *Toxic Substances Controls Guide*, BNA Books Distribution Center, Edison, NJ, 1989.

Howard, Philip H., *Handbook of Environmental Fate and Exposure Data for Organic Chemicals* (Three Volumes), Lewis Publishers, Chelsea, MI, 1989.

Brown, Lester R., *State of the World 1990*, Worldwatch Institute, Washington, DC, 1989.

Rump, H. H., and H. Krist, *Laboratory Manual for the Examination of Water, Waste Water, and Soil*. VCH Publishers, New York, NY, 1989.

Mance, Geoffrey, *Pollution Threat of Heavy Metals in Aquatic Environments*. Elsevier Applied Science Publishers, Essex, England, UK, 1987.

Paddock, Todd, *Dioxins and Furans: Questions and Answers*, Academy of Natural Sciences, Philadelphia, PA, 1989.

Cairns, John, Jr., and James R. Pratt, Eds. *Functional Testing of Aquatic Biota for Estimating Hazards of Chemicals*, ASTM, Philadelphia, PA, 1989.

Suffet, I. H., and Patrick MacCarthy, Eds. *Aquatic Humic Substances: Influence on Fate and Treatment of Pollution*, Advances in Chemistry Series No. 219, American Chemical Society, Washington, DC, 1989.

Drinking Water Health Advisory: Pesticides, United States Environmental Protection Agency, Lewis Publishers, Chelsea, MI, 1989.

Lewis, Timothy E., *Environmental Chemistry and Toxicology of Aluminum*, Lewis Publishers, Chelsea, MI, 1989.

Faust, S. D., and J. V. Hunter, *Organic Compounds in Aqueous Environments*, Marcell Dekker, Inc., New York, 1971.

Förstner, U., and G. T. W. Wittman, *Metal Pollution in the Aquatic Environment*, Springer-Verlag, New York, 1979.

Gehm, H. W., and J. I. Bregman, Eds., *Handbook of Water Resources and Pollution Control*, Van Nostrand Reinhold Company, New York, 1976.

Gould, R. F., Ed., *Fate of Organic Pesticides in the Aquatic Environment*, Advances in Chemistry Series III., American Chemical Society, Washington, D.C., 1972.

Laws, E. A., *Aquatic Pollution*, Wiley-Interscience, New York, 1981.

Novotny, V., and G. Chesters, *Handbook of Nonpoint Source Pollution*, Van Nostrand Reinhold Co., New York, 1981.

QUESTIONS AND PROBLEMS

1. Which of the following statements is true regarding chromium in water: (a) chromium(III) is suspected of being carcinogenic, (b) chromium(III) is less likely to be found in a soluble form than chromium(VI), (c) the toxicity of chromium(III) in electroplating wastewaters is decreased by oxidation to chromium(VI), (d) chromium is not an essential trace element, (e) chromium is known to form methylated species analogous to methylmercury compounds.

2. What do mercury and arsenic have in common in regard to their interactions with bacteria in sediments?

3. What are some characteristics of radionuclides that make them especially hazardous to humans?

4. To what class do pesticides containing the following group belong?

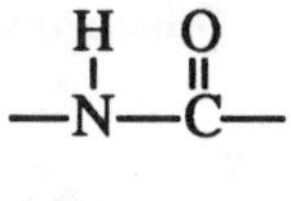

$$\begin{matrix} \text{H} & \text{O} \\ | & || \\ -\text{N} - \text{C} - \end{matrix}$$

5. Consider the following compound:

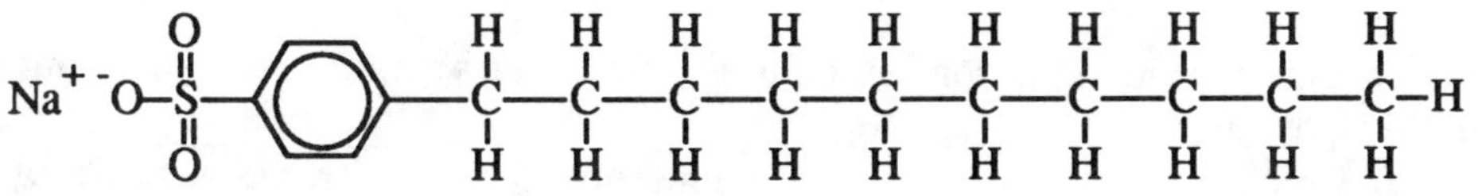

Which of the following characteristics is not possessed by the compound: (a) one end of the molecule is hydrophilic and the other end in hydrophobic, (b) surface-active qualities, (c) the ability to lower surface tension of water, (d) good bio-degradability, (e) tendency to cause foaming in sewage treatment plants.

6. A certain pesticide is fatal to fish fingerlings at a level of 0.50 parts per million in water. A leaking metal can containing 5.00 kg of the pesticide was dumped into a stream with a flow of 10.0 liters per second moving at 1 kilometer per hour. The container leaks pesticide at a constant rate of 5 mg/sec. For what distance (in km) downstream is the water contaminated by fatal levels of the pesticide by the time the container is empty?

7. What are two reasons that Na_3PO_4 is not used as a detergent builder instead of $Na_3P_3O_{10}$?

8. Of the compounds $CH_3(CH_2)10CO_2$, $(CH_3)_3C(CH_2)_2CO_2H$, $CH_3(CH_2)_{10}CH_3$, and ϕ-$(CH_2)_{10}CH_3$ (where ϕ represents a benzene ring), which is the most readily biodegradable?

9. A pesticide sprayer got stuck while trying to ford a stream flowing at a rate of 136 liters per second. Pesticide leaked into the stream for exactly 1 hour and at a rate that contaminated the stream at a uniform 0.25 ppm of methoxychlor. How much pesticide was lost from the sprayer during this time?

10. A sample of water contaminated by the accidental discharge of a radionuclide used for medicinal purposes showed an activity of 12,436 counts per second at the time of sampling and 8,966 cps exactly 30 days later. What is the half-life of the radionuclide?

11. What are the two reasons that soap is environmentally less harmful than ABS surfactant used in detergents?

12. What is the exact chemical formula of the specific compound designated as PCB?

13. A radioisotope has a nuclear half-life of 24 hours and a biological half-life of 16 hours (half of the element is eliminated from the body in 16 hours). A person accidentally swallowed sufficient quantities of this isotope to give an initial "whole body" count rate of 1000 counts per minute. What was the count rate after 16 hours?

14. What is the primary detrimental effect upon organisms of salinity in water arising from dissolved NaCl and Na_2SO_4?

15. Give a specific example of each of the following general classes of water pollutants: (a) trace elements, (b) metal-organic combinations, (c) pesticides

16. Match each compound in the left column with the description corresponding to it in the right column.

    (a) CdS

    (b) $(CH_3)_2AsH$

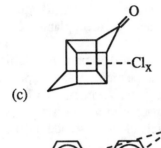

    (c)

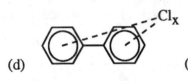

    (d)

(1) Pollutant released to a US stream by a poorly controlled manufacturing process.

(2) Insoluble form of a toxic trace element likely to be found in anaerobic sediments.

(3) Common environmental pollutant formerly used as a transformer coolant.

(4) Chemical species thought to be produced by bacterial action.

17. A polluted water sample is suspected of being contaminated with one of the following: soap, ABS surfactant, or LAS surfactant. The sample has a very low BOD relative to its TOC. Which is the contaminant?

8

Water Treatment

8.1. WATER TREATMENT AND WATER USE

The treatment of water may be divided into three major categories:

- Purification for domestic use

- Treatment for specialized industrial applications

- Treatment of wastewater to make it acceptable for release or reuse.

The type and degree of treatment are strongly dependent upon the source and intended use of the water. Water for domestic use must be thoroughly disinfected to eliminate disease-causing microorganisms, but may contain appreciable levels of dissolved calcium and magnesium (hardness). Water to be used in boilers may contain bacteria but must be quite soft to prevent scale formation. Wastewater being discharged into a large river may require less rigorous treatment than water to be reused in an arid region. As world demand for limited water resources grows, more sophisticated and extensive means will have to be employed to treat water.

Most physical and chemical processes used to treat water involve similar phenomena, regardless of their application to the three main categories of water treatment listed above. Therefore, after introductions to water treatment for municipal use, industrial use, and disposal, each major kind of treatment process is discussed as it applies to all of these applications.

8.2. MUNICIPAL WATER TREATMENT

The modern water treatment plant is often called upon to perform wonders with the water fed to it. The clear, safe, even tasteful water that comes from a faucet may have started as a murky liquid pumped from a polluted river laden with mud and swarming with bacteria. Or, its source may have been well water, much too hard for domestic use and containing high levels of stain-producing dissolved iron and manganese. The water treatment plant operator's job is to make sure that the water plant product presents no hazards to the consumer.

A schematic diagram of a typical municipal water treatment plant is shown in Figure 8.1. This particular facility treats water containing excessive hardness and a high level of iron. The raw water taken from wells first goes to an aerator. Contact of the water with air removes volatile solutes such as hydrogen sulfide, carbon dioxide, methane, and volatile odorous substances such as thiomethane (CH_3SH) and bacterial

183

metabolites. Contact with oxygen also aids iron removal by oxidizing soluble iron(II) to insoluble iron(III). The addition of lime as CaO or $Ca(OH)_2$ after aeration raises the pH and results in the formation of precipitates containing the hardness ions Ca^{2+} and Mg^{2+}. These precipitates settle from the water in a primary basin. Much of the solid material remains in suspension and requires the addition of coagulants (such as ferric and aluminum sulfates, which form gelatinous metal hydroxides) to settle the colloidal particles. Activated silica or synthetic polyelectrolytes may also be added to stimulate coagulation or flocculation. The settling occurs in a secondary basin after the addition of carbon dioxide to lower the pH. Sludge from both the primary and secondary basins is pumped to a sludge lagoon. The water is finally chlorinated, filtered, and pumped to the city water mains.

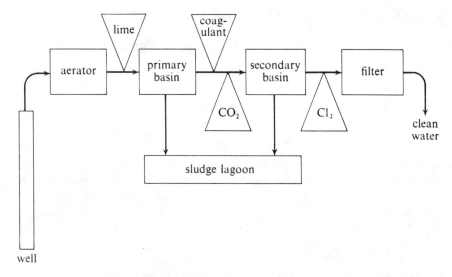

8.3. TREATMENT OF WATER FOR INDUSTRIAL USE

Water is widely used in various process applications in industry. Other major industrial uses are boiler feedwater and cooling water. The kind and degree of treatment of water in these applications depends upon the end use.[1] As examples, cooling water may require only minimal treatment, removal of corrosive substances and scale-forming solutes is essential for boiler feedwater, and water used in food processing must be free of pathogens and toxic substances. Improper treatment of water for industrial use can cause problems, such as corrosion, scale formation, reduced heat transfer in heat exchangers, reduced water flow, and product contamination. These effects may cause reduced equipment performance or equipment failure, increased energy costs due to inefficient heat utilization or cooling, increased costs for pumping water, and product deterioration. Obviously, the effective treatment of water at minimum cost for industrial use is a very important area of water treatment.

Numerous factors must be taken into consideration in designing and operating an industrial water treatment facility. These include the following:

- Water requirement
- Quantity and quality of available water sources
- Sequential use of water (successive uses for applications requiring progressively lower water quality)
- Water recycle
- Discharge standards

The various specific processes employed to treat water for industrial use are discussed in later sections of this chapter. **External treatment**, usually applied to the plant's entire water supply, uses processes such as aeration, filtration, and clarification to remove material from water that may cause problems. Such substances include suspended or dissolved solids, hardness, and dissolved gases. Following this basic treatment, the water may be divided into different streams, some to be used without further treatment and the rest to be treated for specific applications.

Internal treatment is designed to modify the properties of water for specific applications. Examples of internal treatment include the following:

- Reaction of dissolved oxygen with hydrazine or sulfite
- Addition of chelating agents to react with dissolved Ca^{2+} and prevent formation of calcium deposits
- Addition of precipitants, such as phosphate used for calcium removal
- Treatment with dispersants to inhibit scale
- Addition of inhibitors to prevent corrosion
- Adjustment of pH
- Disinfection for food processing uses or to prevent bacterial growth in cooling water

8.4. SEWAGE TREATMENT

Typical municipal sewage contains oxygen-demanding materials, sediments, grease, oil, scum, pathogenic bacteria, viruses, salts, algal nutrients, pesticides, refractory organic compounds, heavy metals, and an astonishing variety of flotsam ranging from children's socks to sponges. It is the job of the waste treatment plant to remove as much of this material as possible.

Several characteristics are used to describe sewage. These include turbidity (international turbidity units); suspended solids (ppm); total dissolved solids (ppm); acidity (H^+ ion concentration or pH); and dissolved oxygen (in ppm O_2). Biochemical oxygen demand is used as a measure of oxygen-demanding substances.

Current processes for the treatment of wastewater may be divided into three main categories of primary treatment, secondary treatment, and tertiary treatment, each of which is discussed separately. Also discussed are total wastewater treatment systems, based largely upon physical and chemical processes

Waste from a municipal water system is normally treated in a **publicly owned treatment works, POTW**. In the United States these systems are allowed to discharge only effluents that have attained a certain level of treatment, as mandated by Federal law.

Primary Waste Treatment

Primary treatment of wastewater consists of the removal of insoluble matter such as grit, grease, and scum from water. The first step in primary treatment normally is screening. Screening removes or reduces the size of trash and large solids that get into the sewage system. These solids are collected on screens and scraped off for subsequent disposal. Most screens are cleaned with power rakes. Comminuting devices shred and grind solids in the sewage. Particle size may be reduced to the extent that the particles can be returned to the sewage flow.

Grit in wastewater consists of such materials as sand and coffee grounds which do not biodegrade well and generally have a high settling velocity. **Grit removal** is practiced to prevent its accumulation in other parts of the treatment system, to reduce clogging of pipes and other parts, and to protect moving parts from abrasion and wear. Grit normally is allowed to settle in a tank under conditions of low flow velocity, and it is then scraped mechanically from the bottom of the tank.

Primary sedimentation removes both settleable and floatable solids. During primary sedimentation there is a tendency for flocculent particles to aggregate for better settling, a process that may be aided by the addition of chemicals. The material that floats in the primary settling basin is known collectively as grease. In addition to fatty substances, the grease consists of oils, waxes, free fatty acids, and insoluble soaps containing calcium and magnesium. Normally, some of the grease settles with the sludge and some floats to the surface, where it may be removed by a skimming device.

Secondary Waste Treatment by Biological Processes

The most obvious harmful effect of biodegradable organic matter in wastewater is BOD, consisting of a biochemical oxygen demand for dissolved oxygen by microorganism-mediated degradation of the organic matter. **Secondary wastewater treatment** is designed to remove BOD, usually by taking advantage of the same kind of biological processes that would otherwise consume oxygen in water receiving the wastewater. Secondary treatment by biological processes takes many forms but consists basically of the following: Microorganisms provided with added oxygen are allowed to degrade organic material in solution or in suspension until the BOD of the waste has been reduced to acceptable levels. The waste is oxidized biologically under conditions controlled for optimum bacterial growth and at a site where this growth does not influence the environment.

One of the simplest biological waste treatment processes is the **trickling filter** (Fig. 8.2) in which wastewater is sprayed over rocks or other solid support material covered with microorganisms. The structure of the trickling filter is such that contact of the wastewater with air is allowed and degradation of organic matter occurs by the action of the microorganisms.

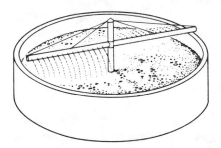

Figure 8.2. Trickling filter for secondary waste treatment.

Rotating biological reactors, another type of treatment system, consist of groups of large plastic discs mounted close together on a rotating shaft. The device is positioned such that at any particular instant half of each disc is immersed in wastewater and half exposed to air. The shaft rotates constantly, so that the submerged portion of the discs is always changing. The discs, usually made of high-density polyethylene or polystyrene, accumulate thin layers of attached biomass, which degrades organic matter in the sewage. Oxygen is absorbed by the biomass and by the layer of wastewater adhering to it during the time that the biomass is exposed to air.

Both trickling filters and rotating biological reactors are examples of fixed-film biological (FFB) processes. The greatest advantage of these processes is their low energy consumption. The energy consumption is minimal because it is not necessary to pump air or oxygen into the water, as is the case with the popular activated sludge process described below. The trickling filter has long been a standard means of wastewater treatment; for example, a 31-acre trickling filter installation in Baltimore which became operational in 1907 is still being used.

The **activated sludge process**, Figure 8.3, is probably the most versatile and effective of all waste treatment processes. Microorganisms in the aeration tank convert organic material in wastewater to microbial biomass and CO_2. Organic nitrogen is converted to ammonium ion or nitrate. Organic phosphorus is converted to orthophosphate. The microbial cell matter formed as part of the waste degradation processes is normally kept in the aeration tank until the microorganisms are past the log phase of growth (Section 6.3), at which point the cells flocculate relatively well to form settleable solids. These solids settle out in a settler and a fraction of them is discarded. Part of the solids, the return sludge, is recycled to the head of the aeration tank and comes into contact with fresh sewage. The combination of a high concentration of "hungry" cells in the return sludge and a rich food source in the influent sewage provides optimum conditions for the rapid degradation of organic matter.

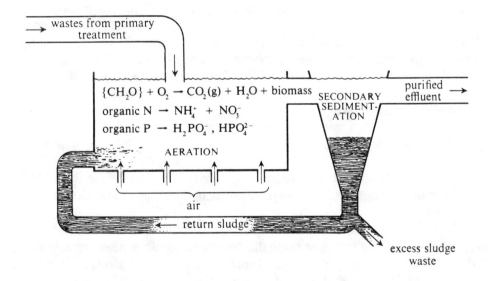

Figure 8.3. Activated sludge process.

The degradation of organic matter that occurs in an activated sludge facility also occurs in streams and other aquatic environments. However, in general, when a degradable waste is put into a stream, it encounters only a relatively small population of microorganisms capable of carrying out the degradation process. Thus, several days may be required for the buildup of a sufficient population of organisms to degrade the waste. In the activated sludge process, continual recycling of active organisms provides the optimum conditions for waste degradation, and a waste may be degraded within the very few hours that it is present in the aeration tank.

The activated sludge process provides two pathways for the removal of BOD, as illustrated schematically in Figure 8.4. BOD may be removed by (1) oxidation of

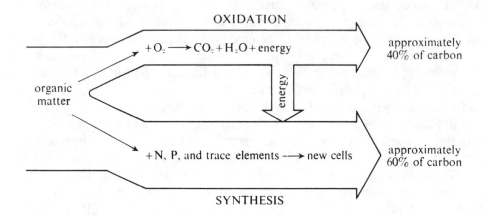

Figure 8.4. Pathways for the removal of BOD in biological wastewater treatment.

organic matter to provide energy for the metabolic processes of the microorganisms, and (2) synthesis, incorporation of the organic matter into cell mass. In the first pathway, carbon is removed in the gaseous form as CO_2. The second pathway provides for removal of carbon as a solid in biomass. That portion of the carbon converted to CO_2 is vented to the atmosphere and does not present a disposal problem. The disposal of waste sludge, however, is a problem, primarily because it is only about 1% solids and contains many undesirable components. Normally, partial water removal is accomplished by drying on sand filters, vacuum filtration, or centrifugation. The dewatered sludge may be incinerated or used as land fill. To a certain extent, sewage sludge may be digested in the absence of oxygen by methane-producing anaerobic bacteria to produce methane and carbon dioxide,

$$2\{CH_2O\} \rightarrow CH_4 + CO_2 \qquad (8.4.1)$$

a process that reduces both the volatile-matter content and the volume of the sludge by about 60%. A carefully designed plant may produce enough methane to provide for all of its power needs.

One of the most desirable means of sludge disposal is to use it to fertilize and condition soil. However, care has to be taken that excessive levels of heavy metals are not applied to the soil as sludge contaminants. Problems with various kinds of sludges resulting from water treatment are discussed further in Section 8.10.

Nitrification (the microbially mediated conversion of ammonium nitrogen to nitrate; see Section 6.8), is a significant process that occurs during biological waste treatment. Ammonium ion is normally the first inorganic nitrogen species produced in the biodegradation of nitrogenous organic compounds. It is oxidized, under the appropriate conditions, first to nitrite by *Nitrosomonas* bacteria,

$$2NH_4^+ + 3O_2 \rightarrow 4H^+ + 2NO_2^- + 2H_2O \qquad (8.4.2)$$

then to nitrate by *Nitrobacter*:

$$2NO_2^- + O_2 \rightarrow 2NO_3^-$$ (8.4.3)

These reactions occur in the aeration tank of the activated sludge plant and are favored in general by long retention times, low organic loadings, large amounts of suspended solids, and high temperatures. Nitrification can reduce sludge settling efficiency because the denitrification reaction

$$4NO_3^- + 5\{CH_2O\} + 4H^+ \rightarrow 2N_2(g) + 5CO_2(g) + 7H_2O$$ (8.4.4)

occurring in the oxygen-deficient settler causes bubbles to form on the sludge floc (aggregated sludge particles), making it so buoyant that it floats to the top. This prevents settling of the sludge and increases the organic load in the receiving waters. Under the appropriate conditions, however, advantage can be taken of this phenomenon to remove nutrient nitrogen from water (see Section 8.9).

Tertiary Waste Treatment

Unpleasant as the thought may be, many people drink used water — water that has been discharged from a municipal sewage treatment plant or from some industrial process. This raises serious questions about the presence of pathogenic organisms or toxic substances in such water. Because of high population density and rapid industrial development, the problem is especially acute in Europe. It is estimated[2] that 32 percent of the drinking water supplies in England and Wales contain "used" water; in Paris and surrounding areas, the extent of reused water ranges from 50 to 70 percent; and in West Germany, the Ruhr River has at times consisted of as much as 40 percent treated wastewater. Obviously, there is a great need to treat wastewater in a manner that makes it amenable to reuse. This requires treatment beyond the secondary processes.

Tertiary waste treatment (sometimes called **advanced waste treatment**) is a term used to describe a variety of processes performed on the effluent from secondary waste treatment. The contaminants removed by tertiary waste treatment fall into the general categories of (1) suspended solids; (2) dissolved organic compounds; and (3) dissolved inorganic materials, including the important class of algal nutrients. Each of these categories presents its own problems with regard to water quality. Suspended solids are primarily responsible for residual biological oxygen demand in secondary sewage effluent waters. The dissolved organics are the most hazardous from the standpoint of potential toxicity. The major problem with dissolved inorganic materials is that presented by algal nutrients, primarily nitrates and phosphates. In addition, potentially hazardous toxic metals may be found among the dissolved inorganics.

In addition to these chemical contaminants, secondary sewage effluent often contains a number of disease-causing microorganisms, requiring disinfection in cases where humans may later come into contract with the water. Among the bacteria that may be found in secondary sewage effluent are organisms causing tuberculosis, dysenteric bacteria (*Bacillus dysenteriae*, *Shigella dysenteriae*, *Shigella paradysenteriae*, *Proteus vulgaris*), cholera bacteria (*Vibrio cholerae*), bacteria causing mud fever (*Leptospira icterohemorrhagiae*), and bacteria causing typhoid fever (*Salmonella typhosa*, *Salmonella paratyphi*). In addition, viruses causing diarrhea, eye infections, infectious hepatitis, and polio may be encountered. Ingestion of sewage still causes disease, even in more developed nations.

Physical-Chemical Treatment of Municipal Wastewater

Complete physical-chemical wastewater treatment systems offer both advantages and disadvantages relative to biological treatment systems. The capital costs of these facilities can be less than those of biological treatment facilities, and they usually require less land. They are better able to cope with toxic materials and overloads. However, they require careful operator control and consume relatively large amounts of energy.

Basically, a physical-chemical treatment process involves:

- Removal of scum and solid objects.

- Clarification, generally with addition of a coagulant, and frequently with the addition of other chemicals (such as lime for phosphorus removal).

- Filtration to remove filterable solids.

- Activated carbon adsorption.

- Disinfection.

The basic steps of a complete physical-chemical wastewater treatment facility are shown in Figure 8.5.

During the early 1970s, it appeared likely that physical-chemical treatment would largely replace biological treatment. However, dramatically higher chemical and energy costs since then have changed the picture appreciably.

8.5. INDUSTRIAL WASTEWATER TREATMENT

Wastewater to be treated must be characterized fully, particularly with a thorough chemical analysis of possible waste constituents and their chemical and metabolic products. The biodegradability of wastewater constituents should also be determined. The options available for the treatment of wastewater are summarized briefly in this section and discussed in greater detail in later sections.

One of two major ways of removing organic wastes is biological treatment by an activated sludge, or related process (see Section 8.3 and Figure 8.3). It may be necessary to acclimate microorganisms to the degradation of constituents that are not normally biodegradable. Consideration needs to be given to possible hazards of biotreatment sludges, such as those containing excessive levels of heavy metal ions. The other major process for the removal of organics from wastewater is sorption by activated carbon (see Section 8.8), usually in columns of granular activated carbon. Activated carbon and biological treatment can be combined with the use of powdered activated carbon in the activated sludge process. The powdered activated carbon sorbs some constituents that may be toxic to microorganisms and is collected with the sludge. A major consideration with the use of activated carbon on wastewater is the hazard that spent activated carbon may present from the wastes it retains. These hazards may include those of toxicity or reactivity, such as those posed by explosives manufacture wastes sorbed to activated carbon. Regeneration of the carbon is expensive and can be hazardous in some cases.

Wastewater can be treated by a variety of chemical processes, including acid/base neutralization, precipitation, and oxidation/reduction. In some cases these treatment steps must precede biological treatment; for example, wastewater exhibiting extremes of pH must be neutralized in order for microorganisms to thrive in it.

Cyanide in the wastewater may be oxidized with chlorine and organics with ozone, hydrogen peroxide promoted with ultraviolet radiation, or dissolved oxygen at high temperatures and pressures. Heavy metals may be precipitated with base, carbonate, or sulfide.

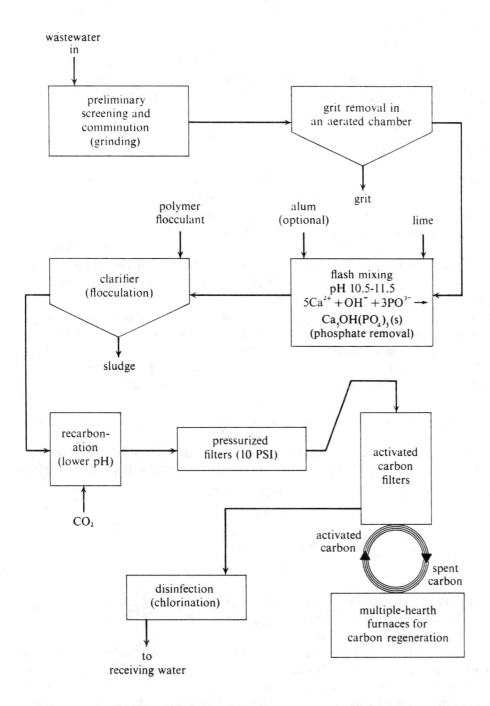

Figure 8.5. Major components of a complete physical-chemical treatment facility for municipal wastewater.

Wastewater can be treated by several physical processes. In some cases, simple density separation and sedimentation can be used to remove water-immiscible liquids and solids. Filtration is frequently required and flotation by gas bubbles generated on particle surfaces may be useful. Wastewater solutes can be concentrated by evaporation, distillation, and membrane processes, including reverse osmosis, hyper-filtration, and ultrafiltration. Organic constituents can be removed by solvent extrac-tion, air stripping, or steam stripping.

Synthetic resins are useful for removing some pollutant solutes from wastewater. Organophilic resins have proven useful for the removal of alcohols; aldehydes; ketones; hydrocarbons; chlorinated alkanes, alkenes, and aryl compounds; esters, including phthalate esters; and pesticides. Cation exchange resins are effective for the removal of heavy metals.

8.6. REMOVAL OF SOLIDS

Relatively large solid particles are removed from water by simple **settling** and **filtration**. A special type of filtration procedure known as **microstraining** is espec-ially effective in the removal of the very small particles. These filters are woven from stainless steel wire so fine that it is barely visible. This enables preparation of filters with openings only 60-70 μm across. These openings may be reduced to 5-15 μm by partial clogging with small particles, such as bacterial cells. The cost of this treatment is likely to be substantially lower than the costs of competing processes. High flow rates at low back pressures are normally achieved.

The removal of colloidal solids from water usually requires **coagulation**. Salts of aluminum and iron are the coagulants most often used in water treatment. Of these, alum or filter alum is most commonly used. This substance is a hydrated aluminum sulfate, $Al_2(SO_4)_3 \cdot 18H_2O$. When this salt is added to water, the aluminum ion hydrolyzes by reactions that consume alkalinity in the water, such as:

$$Al(H_2O)_6^{3+} + 3HCO_3^- \rightarrow Al(OH)_3(s) + 3CO_2 + 6H_2O \qquad (8.6.1)$$

The gelatinous hydroxide thus formed carries suspended material with it as it settles. In addition, however, it is likely that positively charged hydroxyl-bridged dimers such as

$$(H_2O)_4Al \underset{\underset{H}{O}}{\overset{\overset{H}{O}}{\Big[}} Al(H_2O)_4^{4+}$$

and higher polymers are formed which interact specifically with colloidal particles, bringing about coagulation. Metal ions in coagulants also react with virus proteins and destroy up to 99% of the virus in water.

Anhydrous iron(III) sulfate added to water forms ferric hydroxide in a reaction analogous to Equation 8.6.1. An advantage of iron(III) sulfate is that it works over a wide pH range of approximately 4-11. Hydrated iron(II) sulfate, $FeSO_4 \cdot 7H_2O$, or copperas, is also commonly used as a coagulant. It forms a gelatinous precipitate of hydrated iron(III) oxide; in order to function, it must be oxidized to iron(III) by dissolved oxygen in the water at a pH higher than 8.5, or by chlorine, which can oxidize iron(II) at lower pH values.

Sodium silicate partially neutralized by acid aids coagulation, particularly when used with alum. The chemical mechanism by which this activated silica operates is still not known with certainty.

Natural and synthetic polyelectrolytes are used in flocculating particles. Among the natural compounds so used are starch and cellulose derivatives, proteinaceous materials, and gums composed of polysaccharides. More recently, selected synthetic polymers that are effective flocculants have come into use. Neutral polymers and both anionic and cationic polyelectrolytes have been used succesfully as flocculants in various applications.

Coagulation-filtration is a much more effective procedure than filtration alone for the removal of suspended material from water. As the term implies, the process consists of the addition of coagulants that aggregate the particles into larger size particles, followed by filtration. Either alum or lime, often with added polyelectrolytes, is most commonly employed for coagulation, which proceeds by the mechanisms described in Section 8.4.

The filtration step of coagulation-filtration is usually performed on a medium such as sand or anthracite coal. Often, to reduce clogging, several media with progressively smaller interstitial spaces are used. One example is the **rapid sand filter**, which consists of a layer of sand supported by layers of gravel particles, the particles becoming progressively larger with increasing depth. The substance that actually filters the water is coagulated material that collects in the sand. As more material is removed, the buildup of coagulated material eventually clogs the filter and must be removed by back-flushing.

An important class of solids that must be removed from wastewater consists of suspended solids in secondary sewage effluent that arise primarily from sludge that was not removed in the settling process. These solids account for a large part of the BOD in the effluent and may interfere with other aspects of tertiary waste treatment. For example, these solids may clog membranes in reverse osmosis water treatment processes. The quantity of material involved may be rather high. Processes designed to remove suspended solids often will remove 10-20 mg/L of organic material from secondary sewage effluent. In addition, a small amount of the inorganic material is removed as well.

8.7. REMOVAL OF CALCIUM AND OTHER METALS

Calcium and magnesium salts, which generally are present in water as bicarbonates or sulfates, cause water hardness. One of the most common manifestations of water hardness is the insoluble "curd" formed by the reaction of soap with calcium or magnesium ions. The formation of these insoluble soap salts is discussed in Section 7.10. Although ions that cause water hardness do not form insoluble products with detergents, they do adversely affect detergent performance. Therefore, calcium and magnesium must be complexed or removed from water for detergents to function properly.

Another problem caused by hard water is the formation of mineral deposits. For example, when water containing calcium and bicarbonate ions is heated, insoluble calcium carbonate is formed:

$$Ca^{2+} + 2HCO_3^- \rightarrow CaCO_3(s) + CO_2(g) + H_2O \tag{8.7.1}$$

This product coats the surfaces of hot water systems, clogging pipes and reducing heating efficiency. Dissolved salts such as calcium and magnesium bicarbonates and sulfates can be especially damaging in boiler feedwater. Clearly, the removal of water hardness is essential for many uses of water.

Several processes are used for softening water. On a large scale, such as in community water-softening operations, the lime-soda process is used. This process involves the treatment of water with lime, $Ca(OH)_2$, and soda ash, Na_2CO_3. Calcium is precipitated as $CaCO_3$ and magnesium as $Mg(OH)_2$. When the calcium is present primarily as "bicarbonate hardness," it can be removed by the addition of $Ca(OH)_2$ alone:

$$Ca^{2+} + 2HCO_3^- + Ca(OH)_2 \rightarrow 2CaCO_3(s) + 2H_2O \qquad (8.7.2)$$

When bicarbonate ion is not present at substantial levels, a source of CO_3^{2-} must be provided at a high enough pH to prevent conversion of most of the carbonate to bicarbonate. These conditions are obtained by the addition of Na_2CO_3. For example, calcium present as the chloride can be removed from water by the addition of soda ash:

$$Ca^{2+} + 2Cl^- + 2Na^+ + CO_3^{2-} \rightarrow CaCO_3(s) + 2Cl^- + 2Na^+ \qquad (8.7.3)$$

Note that the removal of bicarbonate hardness results in a net removal of soluble salts from solution, whereas removal of nonbicarbonate hardness involves the addition of at least as many equivalents of ionic material as are removed.

The precipitation of magnesium as the hydroxide requires a higher pH than the precipitation of calcium as the carbonate:

$$Mg^{2+} + 2OH^- \rightarrow Mg(OH)_2(s) \qquad (8.7.4)$$

The high pH required may be provided by the basic carbonate ion from soda ash:

$$CO_3^{2-} + H_2O \rightarrow HCO_3^- + OH^- \qquad (8.7.5)$$

Some large-scale, lime-soda softening plants make use of the precipitated calcium carbonate product as a source of additional lime. The calcium carbonate is first heated to at least 825°C to produce quicklime, CaO:

$$CaCO_3 + heat \rightarrow CaO + CO_2(g) \qquad (8.7.6)$$

The quicklime is then slaked with water to produce calcium hydroxide:

$$CaO + H_2O \rightarrow Ca(OH)_2 \qquad (8.7.7)$$

The water softened by lime-soda softening plants usually suffers from two defects. First, because of super-saturation effects, some $CaCO_3$ and $Mg(OH)_2$ usually remain in solution. If not removed, these compounds will precipitate at a later time and cause harmful deposits or undesirable cloudiness in water. The second problem results from the use of highly basic sodium carbonate, which gives the product water an excessively high pH, up to pH 11. To overcome these problems, the water is recarbonated by bubbling CO_2 into it. The carbon dioxide converts the slightly soluble calcium carbonate and magnesium hydroxide to their soluble bicarbonate forms:

$$CaCO_3(s) + CO_2 + H_2O \rightarrow Ca^{2+} + 2HCO_3^- \qquad (8.7.8)$$

$$Mg(OH)_2(s) + 2CO_2 \rightarrow Mg^{2+} + 2HCO_3^- \tag{8.7.9}$$

The CO_2 also neutralizes excess hydroxide ion:

$$OH^- + CO_2 \rightarrow HCO_3^- \tag{8.7.10}$$

The pH generally is brought within the range 7.5-8.5 by recarbonation. The source of CO_2 used in the recarbonation process may be from the combustion of carbonaceous fuel. Scrubbed stack gas from a power plant frequently is utilized. Water adjusted to a pH, alkalinity, and Ca^{2+} concentration very close to $CaCO_3$ saturation is labeled *chemically stabilized*. It neither precipitates $CaCO_3$ in water mains, which can clog the pipes, nor dissolves protective $CaCO_3$ coatings from the pipe surfaces. Water with Ca^{2+} concentration much below $CaCO_3$ saturation is called an *aggressive* water.

Calcium may be removed from water very efficiently by the addition of orthophosphate:

$$5Ca^{2+} + 3PO_4^{2-} + OH^- \rightarrow Ca_5OH(PO_4)_3(s) \tag{8.7.11}$$

It should be pointed out that the chemical formation of a slightly soluble product for the removal of undesired solutes such as hardness ions, phosphate, iron, and manganese must be followed by sedimentation in a suitable apparatus. Frequently, coagulants must be added, and filtration employed for complete removal of these sediments.

Water may be purified by ion exchange, the reversible transfer of ions between aquatic solution and a solid material capable of bonding ions. The removal of NaCl from solution by two ion exchange reactions is a good illustration of this process. First the water is passed over a solid cation exchanger in the hydrogen form, represented by $H^{+-}\{Cat(s)\}$:

$$H^{+-}\{Cat(s)\} + Na^+ + Cl^- \rightarrow Na^{+-}\{Cat(s)\} + H^+ + Cl^- \tag{8.7.12}$$

Next the water is passed over an anion exchanger in the hydroxide ion form, represented by $OH^{-+}\{An(s)\}$:

$$OH^{-+}\{An(s)\} + H^+ + Cl^- \rightarrow Cl^{-+}\{An(s)\} + H_2O \tag{8.7.13}$$

Thus, the cations in solution are replaced by hydrogen ion and the anions by hydroxide ion, yielding water as the product.

The softening of water by ion exchange does not require the removal of all ionic solutes, just those cations responsible for water hardness. Generally, therefore, only a cation exchanger is necessary. Furthermore, the sodium rather than the hydrogen form of the cation exchanger is used, and the divalent cations are replaced by sodium ion. Sodium ion at low concentrations is harmless in water to be used for most purposes, and sodium chloride is a cheap and convenient substance with which to recharge the cation exchangers.

A number of materials have ion-exchanging properties. Among the minerals especially noted for their ion exchange properties are the aluminum silicate minerals, or **zeolites**. An example of a zeolite which has been used commercially in water softening is glauconite, $K_2(MgFe)_2Al_6(Si_4O_{10})_3OH_{12}$. Synthetic zeolites have been prepared by drying and crushing the white gel produced by mixing solutions of sodium silicate and sodium aluminate.

The discovery in the mid-1930s of synthetic ion exchange resins composed of organic polymers with attached functional groups marked the beginning of modern ion exchange technology. Structures of typical synthetic ion exchangers are shown in Figures 8.6 and 8.7. The cation exchanger shown in Figure 8.2 is called a **strongly acidic cation exchanger** because the parent $-SO_3^- H^+$ group is a strong acid. When the functional group binding the cation is the $-CO_2^-$ group, the exchange resin is called a **weakly acidic cation exchanger**, because the $-CO_2H$ group is a weak acid. Figure 8.7 shows a **strongly basic anion exchanger** in which the functional group is a quaternary ammonium group, $-N^+(CH_3)_3$. In the hydroxide form, $-N^+(CH_3)_3OH^-$, the hydroxide ion is readily released; hence the exchanger is classified as **strongly basic**.

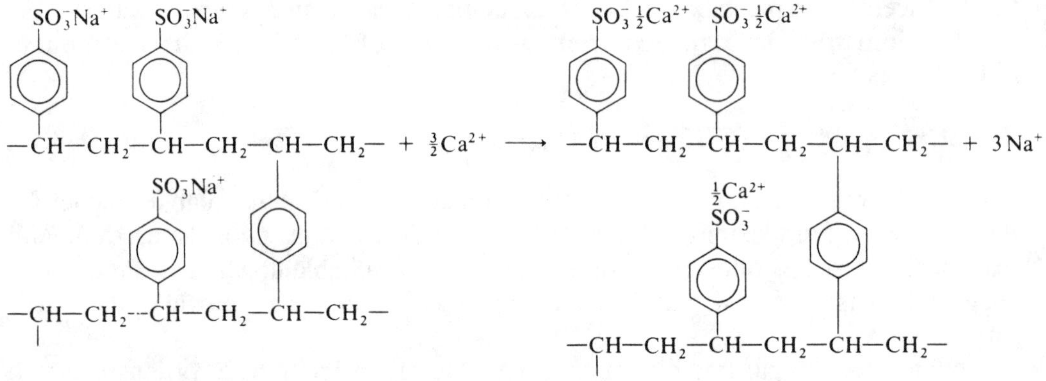

Figure 8.6. Strongly acidic cation exchanger. Sodium exchange for calcium in water is shown.

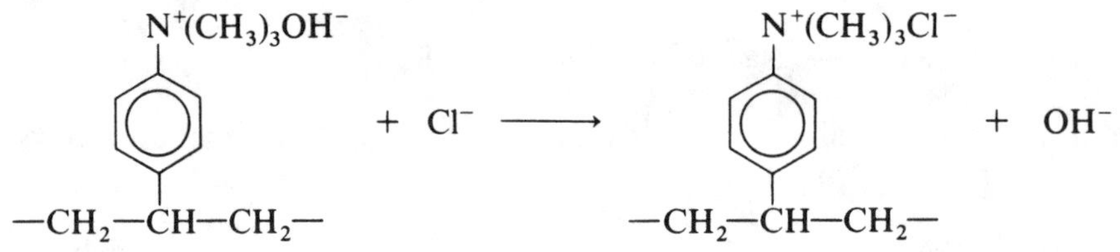

Figure 8.7. Strongly basic anion exchanger. Chloride exchange for hydroxide ion is shown.

The water-softening capability of a cation exchanger is shown in Figure 8.6, where sodium ion on the exchanger is exchanged for calcium ion in solution. The same reaction occurs with magnesium ion. Water softening by cation exchange is now a widely used, effective, and economical process. In many areas having a low water flow, however, it is not likely that home water softening by ion exchange may be used universally without some deterioration of water quality arising from the contamination of wastewater by sodium chloride. Such contamination results from the periodic need to regenerate a water softener with sodium chloride, in order to displace calcium and magnesium ions from the resin and replace these hardness ions with sodium ions:

$$Ca^{2+}\{Cat(s)\}_2 + 2Na^+ + 2Cl^- \rightarrow 2Na^+\{Cat(s)\} + Ca^{2+} + 2Cl^- \qquad (8.7.14)$$

During the regeneration process, a large excess of sodium chloride must be used — several pounds for a home water softener. Appreciable amounts of dissolved sodium chloride can be introduced into sewage by this route.

Strongly acidic cation exchangers are used for the removal of water hardness. Weakly acidic cation exchangers having the $-CO_2H$ group as a functional group are useful for removing alkalinity. Alkalinity generally is manifested by bicarbonate ion. This species is a sufficiently strong base to neutralize the acid of a weak acid cation exchanger:

$$2R-CO_2H + Ca^{2+} + 2HCO_3^- \rightarrow [R-CO_2^-]_2Ca^{2+} + 2H_2O + 2CO_2 \qquad (8.7.15)$$

However, weak bases such as sulfate ion or chloride ion are not strong enough to remove hydrogen ion from the carboxylic acid exchanger. An additional advantage of these exchangers is that they may be regenerated almost stoichiometrically with dilute strong acids, thus avoiding the potential pollution problem caused by the use of excess sodium chloride to regenerate strongly acidic cation changers.

Chelation or, as it is sometimes known, *sequestration*, is an effective method of softening water without actually having to remove calcium and magnesium from solution. A complexing agent is added which greatly reduces the concentrations of free hydrated cations, as shown by some of the example calculations in Chapter 3. For example, chelating calcium ion with excess EDTA anion (Y^{4-}),

$$Ca^{2+} + Y^{4-} \rightarrow CaY^{2-} \qquad (8.7.16)$$

reduces the concentration of hydrated calcium ion, preventing the precipitation of calcium carbonate:

$$Ca^{2+} + CO_3^{2-} \rightarrow CaCO_3(s) \qquad (8.7.17)$$

Polyphosphate salts, EDTA, and NTA (see Chapter 3) are chelating agents commonly used for water softening. Polysilicates are used to complex iron.

Removal of Iron and Manganese

Soluble iron and manganese are found in many groundwaters because of reducing conditions which favor the soluble +2 oxidation state of these metals (see Chapter 4). Iron is the more commonly encountered of the two metals. In groundwater, the level of iron seldom exceeds 10 mg/L, and that of manganese is rarely higher than 2 mg/L. The basic method for removing both of these metals depends upon oxidation to higher insoluble oxidation states. The oxidation is generally accomplished by aeration. The rate of oxidation is pH-dependent in both cases, with a high pH favoring more rapid oxidation. The oxidation of soluble Mn(II) to insoluble MnO_2 is a complicated process. It appears to be catalyzed by solid MnO_2, which is known to adsorb Mn(II). This adsorbed Mn(II) is slowly oxidized on the MnO_2 surface.

Chlorine and potassium permanganate are sometimes employed as oxidizing agents for iron and manganese. There is some evidence that organic chelating agents with reducing properties hold iron(II) in a soluble form in water. In such cases, chlorine is effective because it destroys the organic compounds and enables the oxidation of iron(II).

In water with a high level of carbonate, $FeCO_3$ and $MnCO_3$ may be precipitated directly by raising the pH above 8.5 by the addition of sodium carbonate or lime. This approach is less popular than oxidation, however.

Relatively high levels of insoluble iron(III) and manganese(IV) frequently are found in water as colloidal material which is difficult to remove. These metals may be associated with humic colloids or "peptizing" organic material that binds to colloidal metal oxides, stabilizing the colloid.

Heavy metals such as copper, cadmium, mercury, and lead are found in wastewaters from a number of industrial processes. Because of the toxicity of many heavy metals, their concentrations must be reduced to very low levels prior to release of the wastewater. A number of approaches are used in heavy metals removal.

Lime treatment (Section 8.4) removes heavy metals as insoluble hydroxides, basic salts, or coprecipitated with calcium carbonate or ferric hydroxide. This process does not completely remove mercury, cadmium, or lead, so their removal is aided by addition of sulfide (most heavy metals are sulfide-seekers):

$$Cd^{2+} + S^{2-} \rightarrow CaS(s) \qquad (8.7.18)$$

Heavy chlorination is frequently necessary to break down metal-solubilizing ligands (see Chapter 3). Lime precipitation does not normally permit recovery of metals and is sometimes undesirable from the economic viewpoint.

Electrodeposition (reduction of metal ions to metal by electrons at an electyrode), *reverse osmosis* (see Section 8.9), and *ion exchange* are frequently employed for metal removal. Solvent extraction using organic-soluble chelating substances is also effective in removing many metals. **Cementation**, a process by which a metal deposits by reaction of its ion with a more readily oxidized metal, may be employed:

$$Cu^{2+} + Fe \text{ (iron scrap)} \rightarrow Fe^{2+} + Cu \qquad (8.7.19)$$

Activated carbon adsorption effectively removes some metals from water at the part per million level. Sometimes a chelating agent is sorbed to the charcoal to increase metal removal.

The removal of trace metals from wastewater is summarized in Table 8.1. The second column in Table 8.1 shows the ranges of trace metal levels found in a survey of wastewater from industrial plants in Michigan. These data provide an idea of values to be encountered in industrial wastewaters in other areas.

Even when not specifically designed for the removal of heavy metals, most waste treatment processes remove appreciable quantities of the more troublesome heavy metals encountered in wastewater. The heavy metal removal resulting from biological waste treatment is shown in the third column of Table 8.1. These metals accumulate in the sludge from biological treatment, so sludge disposal must be given careful consideration. Average metal contents of biological waste treatment sludges from 33 biological treatment plants are given in the fourth column of Table 8.1.

Various physical-chemical treatment processes effectively remove heavy metals from wastewaters. One such treatment is lime precipitation followed by activated-carbon filtration. Activated-carbon filtration may also be preceded by treatment with ferric chloride to form a ferric hydroxide floc, which is an effective heavy metals scavenger. Similarly, alum, which forms aluminum hydroxide, may be added prior to activated-carbon filtration.

Table 8.1. Removal of Trace Metals by Wastewater Treatment Processes. [1]

Metal	Concentration range,[2] mg/L	% Removal by biological treatment	Concentration in digested sludge,[3] mg/kg	% Removal by lime precipitation–activated carbon	% Removal by ferric chloride–activated carbon	% Removal by alum–activated carbon
cadmium	<0.008–0.142	20–45	31	99.6	98.6	55.2
chromium(III)	<0.020–0.700	40–80	1100	98.2	99.3	99.3
copper	<0.020–3.36	0–70	1230	90	96	98.3
mercury	<0.0002–0.044	20–75	7	91	99	98.3
nickel	<0.0020–8.80	15–40	410	99.5	37	37
lead	<0.050–1.27	50–90	830	99.4	99.1	96.6
zinc	<0.030–8.31	35–80	2780	76	94	28

[1] Data from Reference [18]

[2] In wastewater from plants in Michigan

[3] Average of sludge from 33 biological waste treatment plants, mg metal/kg dry sludge

It should be noted that the form of the heavy metal has a strong effect upon the efficiency of metal removal. For instance, chromium(VI) is normally more difficult to remove than chromium(III). Chelation may prevent metal removal by solubilizing metals (see Chapter 3).

In the past, removal of heavy metals has been largely a fringe benefit of wastewater treatment processes. Currently, however, more consideration is being given to design and operating parameters that specifically enhance heavy-metals removal as part of wastewater treatment.

8.8. REMOVAL OF DISSOLVED ORGANICS

Very low levels of exotic organic compounds in drinking water are suspected of contributing to cancer and other maladies. Some of these are chlorinated organic compounds produced by chlorination of organics in water, especially humic substances. Removal of organics to very low levels prior to chlorination has been found to be effective in preventing trihalomethane formation. In addition, many organic compounds survive, or are produced by, secondary wastewater treatment. Almost half of these are humic substances (see Section 3.12) with a molecular-weight range of 1000-5000. Among the remainder are found ether-extractable materials, carbohydrates, proteins, detergents, tannins, and lignins. The humic compounds, because of their high molecular weight and anionic character, influence some of the physical and chemical aspects of waste treatment. The ether-extractables contain many of the compounds that are resistant to biodegradation and are of particular concern regarding potential toxicity, carcinogenicity, and mutagenicity. In the ether extract are found many fatty acids, hydrocarbons of the n-alkane class, naphthalene, diphenylmethane, diphenyl, methylnaphthalene, isopropyl benzene, dodecyl benzene, phenol, dioctyl-phthalate, and triethylphosphate.

The standard method for the removal of dissolved organic material is adsorption on activated carbon, a product that is produced from a variety of carbonaceous materials, including wood, pulp-mill char, peat, and lignite. The carbon is produced by charring the raw material anaerobically below 600°C, followed by an activation step consisting of partial oxidation. Carbon dioxide may be employed as an oxidizing agent at 600-700°C.

$$CO_2 + C \rightarrow 2CO \tag{8.8.1}$$

or the carbon may be oxidized by water at 800-900°C:

$$H_2O + C \rightarrow H_2 + CO \tag{8.8.2}$$

These processes develop porosity, increase the surface area, and leave the C atoms in arrangements that have affinities for organic compounds.

Activated carbon comes in two general types: granulated activated carbon, consisting of particles 0.1-1 mm in diameter, and powdered activated carbon, in which most of the particles are 50-100 μm in diameter.

The exact mechanism by which activated carbon holds organic materials is not known. However, one reason for the effectiveness of this material as an adsorbent is its tremendous surface area. A solid cubic foot of carbon particles may have a combined pore and surface area of approximately 10 square miles!

Although interest is increasing in the use of powdered activated carbon for water treatment, currently granular carbon is more widely used. It may be employed in a fixed bed, through which water flows downward. Accumulation of particulate matter requires periodic backwashing. An expanded bed in which particles are kept slightly separated by water flowing upward may be used with less chance of clogging.

Economics require regeneration of the carbon, which is accomplished by heating it to 950°C in a steam-air atmosphere. This process oxidizes adsorbed organics and regenerates the carbon surface, with an approximately 10% loss of carbon.

Removal of organics may also be accomplished by adsorbent synthetic polymers. Such polymers as Amberlite XAD-4 have hydrophobic surfaces and strongly attract relatively insoluble organic compounds, such as chlorinated pesticides. The porosity of these polymers is up to 50% by volume, and the surface area may be as high as 850 m^2/g. They are readily regenerated by solvents such as isopropanol and acetone. Under appropriate operating conditions, these polymers remove virtually all nonionic organic solutes; for example, phenol at 250 mg/L is reduced to less than 0.1 mg/L by appropriate treatment with Amberlite XAD-4.

Oxidation of dissolved organics holds some promise for their removal. Ozone, hydrogen peroxide, molecular oxygen (with or without catalysts), chlorine and its derivatives, permanganate, or ferrate can be used. Electrochemical oxidation may be possible in some cases.

8.9. REMOVAL OF DISSOLVED INORGANICS

In order for complete water recycling to be feasible, inorganic-solute removal is essential. The effluent from secondary waste treatment generally contains 300-400 mg/L more dissolved inorganic material than does the municipal water supply. It is obvious, therefore, that 100% water recycle without removal of inorganics would cause the accumulation of an intolerable level of dissolved material. Even when water is not destined for immediate reuse, the removal of the inorganic nutrients phosphorus and nitrogen is highly desirable to reduce eutrophication downstream. In some cases, the removal of toxic trace metals is needed.

One of the most obvious methods for removing inorganics from water is distillation. Unfortunately, the energy required for distillation is generally too high for the process to be economically feasible. Furthermore, volatile materials such as ammonia and odorous compounds are carried over to a large extent in the distillation process, unless special preventative measures are taken. Freezing produces a very pure water, but is considered uneconomical with present technology. Membrane processes considered most promising for bulk removal of inorganics from water are electro-dialysis, ion exchange, and reverse osmosis. (Other membrane processes used in water purification are nanofiltration, ultrafiltration, microfiltration, and dialysis.[3])

Electrodialysis

Electrodialysis consists of applying a direct current across a body of water separated into vertical layers by membranes alternately permeable to cations and anions. Cations migrate toward the cathode and anions toward the anode. Cations and anions both enter one layer of water, and both leave the adjacent layer. Thus, layers of water enriched in salts alternate with those from which salts have been removed. The water in the brine-enriched layers is recirculated to a certain extent to prevent excessive accumulation of brine. The principles involved in electrodialysis treatment are shown in Figure 8.8.

Although the relatively small ions constituting the salts dissolved in wastewater readily pass through the membranes, large organic ions (proteins, for example) and charged colloids migrate to the membrane surfaces, often fouling or plugging the membranes and reducing efficiency. In addition, growth of microorganisms on the membranes can cause fouling.

Experience with pilot plants indicates that electrodialysis has the potential to be a practical and economical method of removing up to 50% of the dissolved inorganics from secondary sewage effluent, once the effluent has been carefully pretreated to eliminate fouling substances. Such a level of efficiency would permit repeated recycle of water without dissolved inorganic materials reaching unacceptably high levels.

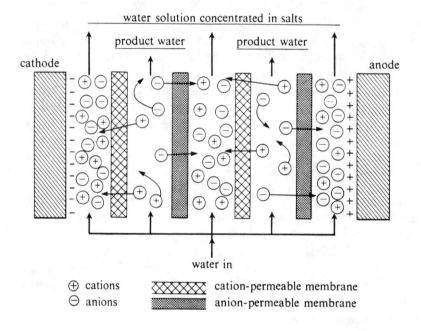

Figure 8.8. Electrodialysis apparatus for the removal of ionic material from water.

Ion Exchange

The ion exchange method for softening water is described in detail in Section 8.7. The ion exchange process used for removal of inorganics consists of passing the water successively over a solid cation exchanger and a solid anion exchanger, which replace cations and anions by hydrogen ion and hydroxide ion, respectively. The net result is that each equivalent of salt is replaced by a mole of water. For the hypothetical ionic salt MX, the reactions are:

$$H^{+-}\{Cat(s)\} + M^+ + X^- \rightarrow M^{+-}\{Cat(s)\} + H^+ + X^- \tag{8.9.1}$$

$$OH^{-+}\{An(s)\} + H^+ + X^- \rightarrow X^{-+}\{An(s)\} + H_2O \tag{8.9.2}$$

where $^-\{Cat(s)\}$ represents the solid cation exchanger and $OH^{-+}\{An(s)\}$ represents the solid anion exchanger. The cation exchanger is regenerated with strong acid and the anion exchanger with strong base.

Demineralization by ion exchange generally produces water of a very high quality. Unfortunately, some organic compounds in wastewater foul ion exchangers, and microbial growth on the exchangers can diminish their efficiency. In addition, regeneration of the resins is expensive, and the concentrated wastes from regeneration require disposal in a manner that will not damage the environment.

Reverse Osmosis

Reverse osmosis is a very useful technique for the purification of water. Basically, reverse osmosis consists of forcing pure water through a semipermeable membrane that allows the passage of water but not of other material. This process depends on the preferential sorption of water on the surface of the membrane, which is composed of porous cellulose acetate or polyamide. Pure water from the sorbed layer is forced through pores in the membrane under pressure. If the thickness of the sorbed water layer is d, the pore diameter for optimum separation should be 2d. The optimum pore diameter depends upon the thickness of the sorbed pure water layer and may be several times the diameters of the solute and solvent molecules. Therefore, reverse osmosis is not a simple sieve separation or ultrafiltration process. The principle of reverse osmosis is illustrated in Figure 8.9.

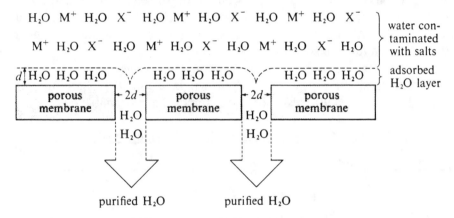

Figure 8.9. Solute removal from water by reverse osmosis.

Phosphorus Removal

Advanced waste treatment normally requires removal of phosphorus to reduce algal growth. Algae may grow at PO_4^{3-} levels as low as 0.05 mg/L. Growth inhibition requires levels well below 0.5 mg/L. Since municipal wastes typically contain approximately 25 mg/L of phosphate (as orthophosphates, polyphosphates, and insoluble phosphates), the efficiency of phosphate removal must be quite high to prevent algal growth. This removal may occur in the sewage treatment process (1) in the primary settler; (2) in the aeration chamber of the activated sludge unit; or (3) after secondary waste treatment.

Normally, the activated sludge process removes about 20% of the phosphorus from sewage. Thus, an appreciable fraction of largely biological phosphorus is removed with the sludge. Detergents and other sources contribute significant amounts of phosphorus to domestic sewage and considerable phosphate ion remains in the effluent. However, some wastes, such as carbohydrate wastes from sugar refineries, are so deficient in phosphorus that supplementation of the waste with inorganic phosphorus is required for proper growth of the microorganisms degrading the wastes.

Under some sewageplant operating conditions, much greater than normal phosphorus removal has been observed. In such plants, characterized by high dissolved oxygen and high pH levels in the aeration tank, removal of 60-90% of the phosphorus has been attained, yielding two or three times the normal level of phosphorus in the sludge. In a conventionally operated aeration tank of an activated sludge plant, the CO_2 level is relatively high because of release of the gas by the degradation of organic material. A high CO_2 level results in a relatively low pH, due to the presence of carbonic acid. The aeration rate generally is not very high because oxygen transfer to water is more efficient at lower dissolved oxygen concentrations. Therefore, the aeration rate normally is not high enough to sweep out sufficient dissolved carbon dioxide to bring its concentration down to low levels. Thus, the pH generally is low enough that phosphate is maintained primarily in the form of the $H_2PO_4^-$ ion. However, at a higher rate of aeration in a relatively hard water, the CO_2 is swept out, the pH rises, and reactions such as the following occur:

$$5Ca^{2+} + 3HPO_4^{2-} + H_2O \rightarrow Ca_5OH(PO_4)_3(s) + 4H^+ \tag{8.9.3}$$

The precipitated hydroxyapatite or other form of calcium phosphate is incorporated in the sludge floc. Reaction 8.9.3 is strongly hydrogen ion-dependent, and an increase in the hydrogen ion concentration drives the equilibrium back to the left. Thus, under anaerobic conditions when the sludge medium becomes more acidic due to higher CO_2 levels, the calcium returns to solution.

Chemically, phosphate is most commonly removed by precipitation. Some common precipitants and their products are shown in Table 8.2. Precipitation processes are capable of at least 90-95% phosphorus removal at reasonable cost.

Table 8.2. Chemical Precipitants for Phosphate and Their Products

Precipitant(s)	Products
$Ca(OH)_2$	$Ca_5OH(PO_4)_3$ (hydroxyapatite)
$Ca(OH)_2 + NaF$	$Ca_5F(PO_4)_3$ (fluorapatite)
$Al_2(SO_4)_3 + NaF$	$Al(PO_4)_3$
$FeCl_3$	$FePO_4$
MSO_4	$MgNH_4PO_4$

Lime, $Ca(OH)_2$, is the chemical most commonly used for phosphorus removal:

$$5Ca(OH)_2 + 3HPO_4^{2-} \rightarrow Ca_5OH(PO_4)_3(s) + 3H_2O + 6OH^- \tag{8.9.4}$$

Lime has the advantages of low cost and ease of regeneration. The efficiency with which phosphorus is removed by lime is not as high as would be predicted by the low solubility of hydroxyapatite, $Ca_5OH(PO_4)_3$. Some of the possible reasons for this are slow precipitation of $Ca_5OH(PO_4)_3$; formation of nonsettling colloids; precipitation of calcium as $CaCO_3$ in certain pH ranges; and the fact that phosphate may be present as condensed phosphates (polyphosphates) which form soluble complexes with calcium ion.

Phosphate can be removed from solution by adsorption on some solids, particularly activated alumina, Al_2O_3. Removals of up to 99.9% of orthophosphate have been achieved with this method.

Nitrogen Removal

Next to phosphorus, nitrogen is the algal nutrient most commonly removed as part of advanced wastewater treatment. The techniques most often used for nitrogen removal are summarized in Table 8.3 (see page 207). Nitrogen in municipal wastewater generally is present as organic nitrogen or ammonia. Ammonia is the primary nitrogen product produced by most biological waste treatment processes. This is because it is expensive to aerate sewage sufficiently to oxidize the ammonia to nitrate through the action of nitrifying bacteria. If the activated sludge process is operated under conditions such that the nitrogen is maintained in the form of ammonia, the latter may be stripped in the form of NH_3 gas from the water by air. For ammonia stripping to work, the ammoniacal nitrogen must be converted to volatile NH_3 gas, which requires a pH substantially higher than the pK_a of the NH_4^+ ion. In practice, the pH is raised to approximately 11.5 by the addition of lime (which also serves to remove phosphate). The ammonia is stripped from the water by air.

Nitrification followed by denitrification is a promising technique for the removal of nitrogen from wastewater. The first step is an essentially complete conversion of ammonia and organic nitrogen to nitrate under strongly aerobic conditions, achieved by more extensive than normal aeration of the sewage:

$$NH_4^+ + 2O_2 \text{ (Nitrifying bacteria)} \rightarrow NO_3^- + 2H^+ + H_2O \qquad (8.9.5)$$

The second step is the reduction of nitrate to nitrogen gas. This reaction is also bacterially catalyzed and requires a carbon source and a reducing agent such as methanol, CH_3OH.

$$6NO_3^- + 5CH_3OH + 6H^+ \text{ (Denitrifying bacteria)} \rightarrow$$
$$3N_2(g) + 5CO_2 + 13H_2O \qquad (8.9.6)$$

The denitrification process may be carried out either in a tank or on a carbon column. In pilot plant operation, conversions of 95% of the ammonia to nitrate and 86% of the nitrate to nitrogen have been achieved.

8.10. SLUDGE

Perhaps the most pressing water treatment problem at this time has to do with sludge collected or produced during water treatment. Finding a safe place to put the sludge or a use for it has proven troublesome, and the problem is aggravated by the growing numbers of water treatment systems.

Improper disposal of wastes continues to be a subject of public and governmental concern. One of the more recent problems to be addressed by legislative action in the U.S. is ocean dumping of sewage sludge. For many years sludge from New York and New Jersey has been disposed in the Atlantic Ocean's 106-Mile Deepwater Municipal Sewage Sludge Disposal Site.[4] In response to this kind of activity, the U.S. Congress passed the Ocean Dumping Ban Act of 1988. Among the act's requirements are an immediate prohibition against new ocean dumping of sewage sludge and stiff penalties for ocean disposal by communities already engaged in the practice starting January 1, 1992.

Table 8.2. Common Processes for the Removal of Nitrogen from Wastewater[1]

Process	Principles and conditions
Air stripping ammonia	Ammonium ion is the initial product of biodegradation of nitrogenous waste. It is removed by raising the pH to approximately 11 with lime and stripping ammonia gas from the water by air in a stripping tower. Scaling, icing, and air pollution are major disadvantages.
Ammonium ion exchange	This is an attractive alternative to air stripping, made possible by the development of clinoptilolite, a natural zeolite selective for ammonia: Na^+(clinoptilolite) + NH_4^+ → Na^+ + NH_4^+ (clinoptilolite). Regenerated with sodium or calcium salts.
Biosynthesis	The production of biomass in the sewage treatment system and its subsequent removal from the sewage effluent result in a net loss of nitrogen from the system.
Nitrification-denitrification	Several schemes are based on the conversion of ammonium nitrogen to nitrate under aerobic conditions, $$2\ NH_4^+ + 3\ O_2 \xrightarrow{Nitrosomonas} 4\ H^+ + 2\ NO_2^- + 2\ H_2O$$ $$2\ NO_2^- + O_2 \xrightarrow{Nitrobacter} 2\ NO_3^-$$ followed by production of elemental nitrogen (denitrification): $$4\ NO_3^- + 5\ \{CH_2O\} + 4\ H^+ \xrightarrow[\text{bacteria}]{\text{denitrifying}}$$ $$2\ N_2(g) + 5\ CO_2(g) + 7\ H_2O$$ Denitrification may be accomplished in an anaerobic activated sludge system or in an anaerobic column. Sometimes additional organic matter (methanol) is added.
Chlorination	Reaction of ammonium ion and hypochlorite (from chlorine) results in denitrification by chemical reactions: $$NH_4^+ + HOCl \rightarrow NH_2Cl + H_2O + H^+$$ $$2\ NH_2Cl + HOCl \rightarrow N_2(g) + 3\ H^+ + 3\ Cl^- + H_2O$$

[1]For details, see C. E. Adams, Jr., 1974, Removing nitrogen from waste water, *Environmental Science and Technology* 8: 696–701.

Some sludge is present in wastewater and may be collected from it. Such sludge includes human wastes, garbage grindings, organic wastes and inorganic silt and grit from storm water runoff, and organic and inorganic wastes from commercial and industrial sources. There are two major kinds of sludge generated in a waste treatment plant. The first of these is organic sludge from activated sludge, trickling filter, or rotating biological reactors. The second is inorganic sludge from the addition of chemicals, such as in phosphorus removal (see Section 8.9).

Most commonly, sewage sludge is subjected to anaerobic digestion in a digester designed to allow bacterial action to occur in the absence of air. This reduces the mass and volume of sludge and ideally results in the formation of a stabilized humus. Disease agents are also destroyed in the process.

Following digestion, sludge is generally conditioned and thickened to concentrate and stabilize it and make it more dewaterable. Relatively inexpensive processes, such as gravity thickening, may be employed to get the moisture content down to about 95%. Sludge may be further conditioned chemically by the addition of iron or aluminum salts, lime, or polymers.

Sludge dewatering is employed to convert the sludge from an essentially liquid material to a damp solid containing not more than about 85% water. This may be accomplished on sludge drying beds consisting of layers of sand and gravel. Mechanical devices may also be employed, including vacuum filtration, centrifugation, and filter presses. Heat may be used to aid the drying process.

Some of the alternatives for the ultimate disposal of sludge include land spreading, ocean dumping, and incineration. Each of these choices has disadvantages, such as the presence of toxic substances in sludge spread on land, or the high fuel cost of incineration.

Some of the undesirable components found in sewage sludge are shown in Table 8.3. This table refers to sludge from a primary settler, although many of the same components are found in secondary settler sludges.

Table 8.3. Undesirable Components Typically Found in Sewage Sludge [1]

Component	Level, ppm by dry weight unless otherwise stated
Organics	
PCB	0–105
DDT	0–1 (found much less frequently now)
DDD	0–0.5 (found much less frequently now)
dieldrin	0–2
aldrin	0–16
phenol	sometimes encountered
Heavy metals	
cadmium	0–100
lead	up to 400
mercury	3–15
chromium	up to 700
copper	80–1000
nickel	25–400
zinc	300–2000 (common deterrent to use of sludge as a soil conditioner due to its toxicity to plants)
Pathogenic microorganisms	
human viruses	generally present
salmonella (in raw sludge)	500 viable cells/100mL
salmonela (in digested sludge)	30 viable cells/100mL
fecal coliforms (raw sludge)	1×10^7 viable cells/100 mL
fecal coliforms (digested sludge)	4×10^5 viable cells/100mL

[1] Estimates based in part on 1974, Containing the flow of sewage sludge, *Environmental Science and Technology* 8: 702-3.

Rich in nutrients, waste sewage sludge contains around 5% N, 3% P, and 0.5% K on a dry-weight basis and can be used to fertilize and condition soil. The humic material in the sludge improves the physical properties and cation-exchange capacity of the soil. Among the factors limiting this application of sludge are excess nitrogen pollution of runoff water and groundwater, survival of pathogens, and the presence of heavy metals in the sludge.

Possible accumulation of heavy metals is of the greatest concern insofar as the use of sludge on cropland is concerned. Sewage sludge is an efficient heavy metals scavenger. On a dry basis, sludge samples from industrial cities have shown levels of up to 9,000 ppm zinc, 6,000 ppm copper, 600 ppm nickel, and up to 800 ppm cadmium! These and other metals tend to remain immobilized in soil by chelation with organic matter, adsorption on clay minerals, and precipitation as insoluble compounds, such as oxides or carbonates. However, increased application of sludge on cropland has caused distinctly elevated levels of zinc and cadmium in both leaves and grain of corn. Therefore, caution has been advised in heavy or prolonged application of sewage sludge to soil. The problem of heavy metals in sewage sludge is one of the many reasons for not allowing mixture of wastes to occur prior to treatment. Sludge does, however, contain nutrients which should not be wasted, given the possibility of eventual fertilizer shortages. Prior control of heavy metal contamination from industrial sources should greatly reduce the heavy metal content of sludge and enable it to be used more extensively on soil.

An increasing problem in sewage treatment arises from sludge sidestreams. These consist of water removed from sludge by various treatment processes[5]. Sewage treatment processes can be divided into mainstream treatment processes (primary clarification, trickling filter, activated sludge, and rotating biological reactor) and sidestream processes. During sidestream treatment sludge is dewatered, degraded, and disinfected by a variety of processes, including gravity thickening, dissolved air flotation, anaerobic digestion, aerobic digestion, vacuum filtration, centrifugation, belt-filter press filtration, sand-drying-bed treatment, sludge-lagoon settling, wet air oxidation, pressure filtration, and Purifax treatment. Each of these produces a liquid by-product sidestream which is circulated back to the mainstream. These add to the biochemical oxygen demand and suspended solids of the mainstream.

A variety of chemical sludges are produced by various water treatment and industrial processes. Among the most abundant of such sludges is alum sludge produced by the hydrolysis of Al(III) salts used in the treatment of water, which creates gelatinous aluminum hydroxide:

$$Al^{3+} + 3OH^- (aq) \rightarrow Al(OH)_3 (aq) \tag{8.10.1}$$

Alum sludges normally are 98% or more water and are very difficult to dewater.

Both iron(II) and iron(III) compounds are used for the precipitation of impurities from wastewater via the precipitation of $Fe(OH)_3$. The sludge contains $Fe(OH)_3$ in the form of soft, fluffy precipitates that are difficult to dewater beyond 10 or 12% solids.

The addition of either lime, $Ca(OH)_2$, or quicklime, CaO, to water is used to raise the pH to about 11.5 and cause the precipitation of $CaCO_3$, along with metal hydroxides and phosphates. Calcium carbonate is readily recovered from lime sludges and can be recalcined to produce CaO, which can be recycled through the system.

Metal hydroxide sludges are produced in the removal of metals such as lead, chromium, nickel, and zinc from wastewater by raising the pH to such a level that the corresponding hydroxides or hydrated metal oxides are precipitated. The disposal of these sludges is a substantial problem because of their toxic heavy metal content. Reclamation of the metals is an attractive alternative for these sludges.

Pathogenic (disease-causing) microorganisms may persist in the sludge left from the treatment of sewage. Many of these organisms present potential health hazards, and there is risk of public exposure when the sludge is applied to soil. Therefore, it is necessary both to be aware of pathogenic microorganisms in municipal wastewater treatment sludge and to find a means of reducing the hazards caused by their presence.

The most significant organisms in municipal sewage sludge include the following: (1) indicators, including fecal and total coliform: (2) pathogenic bacteria, including *Salmonellae* and *Shigellae*; (3) enteric (intestinal) viruses, including enterovirus and poliovirus; and (4) parasites, such as *Entamoeba histolytica* and *Ascaris lumbricoides*.

Several ways are recommended to significantly reduce levels of pathogens in sewage sludge. Aerobic digestion involves aerobic agitation of the sludge for periods of 40 to 60 days (longer times are employed with low sludge temperatures). Air drying involves draining and/or drying of the liquid sludge for at least three months in a layer 20-25 cm thick. This operation may be performed on underdrained sand beds or in basins. Anaerobic digestion involves maintenance of the sludge in an anaerobic state for periods of time ranging from 60 days at 20°C to 15 days at temperatures exceeding 35°C. Composting involves mixing dewatered sludge cake with bulking agents subject to decay, such as wood chips or shredded municipal refuse, and allowing the action of bacteria to promote decay at temperatures ranging up to 45-65°C. The higher temperatures tend to kill pathogenic bacteria. Finally, pathogenic organisms may be destroyed by lime stabilization in which sufficient lime is added to raise the pH of the sludge to 12 or higher.

8.11. WATER DISINFECTION

Chlorine is the most commonly used disinfectant employed for killing bacteria in water. When chlorine is added to water, it rapidly hydrolyzes according to the reaction

$$Cl_2 + H_2O \rightarrow H^+ + Cl^- + HOCl \tag{8.11.1}$$

which has the following equilibrium constant:

$$K = \frac{[H^+][Cl^-][HOCl]}{[Cl_2]} = 4.5 \times 10^{-4} \tag{8.11.2}$$

Hypochlorous acid, HOCl, is a weak acid that dissociates according to the reaction,

$$HOCl \leftarrow\rightarrow H^+ + OCl^- \tag{8.11.3}$$

with an ionization constant of 2.7×10^{-8}. From the above it can be calculated that the concentration of elemental Cl_2 is negligible at equilibrium above pH 3 when chlorine is added to water at levels below 1.0 g/L.

Sometimes, hypochlorite salts are substituted for chlorine gas as a disinfectant. Calcium hypochlorite, $Ca(OCl)_2$, is commonly used. The hypochlorites are safer to handle than gaseous chlorine.

The two chemical species formed by chlorine in water, HOCl and OCl⁻, are known as **free available chlorine**. Free available chlorine is very effective in killing bacteria. In the presence of ammonia, monochloramine, dichloramine, and trichloramine are formed:

$$NH_4^+ + HOCl \rightarrow NH_2Cl \text{ (monochloramine)} + H_2O + H^+ \tag{8.11.4}$$

$$NH_2Cl + HOCl \rightarrow NHCl_2 \text{ (dichloramine)} + H_2O \tag{8.11.5}$$

$$NHCl_2 + HOCl \rightarrow NCl_3 \text{ (trichloramine)} + H_2O \tag{8.11.6}$$

The chloramines are called **combined available chlorine**. Chlorination practice frequently provides for formation of combined available chlorine which, although a weaker disinfectant than free available chlorine, is more readily retained as a disinfectant throughout the water distriburtion system. Too much ammonia in water is considered undesirable because it exerts excess demand for chlorine.

At sufficiently high Cl:N molar ratios in water containing ammonia, some HOCl and OCL$^-$ remain unreacted in solution, and a small quantity of NCl_3 is formed. The ratio at which this occurs is called the **breakpoint**. Chlorination beyond the breakpoint ensures disinfection. It has the additional advantage of destroying the more common materials that cause odor and taste in water.

At moderate levels of NH_3–N (approximately 20 mg/L), when the pH is between 5.0 and 8.0, chlorination with a minimum 8:1 weight ratio of Cl to NH_3–nitrogen produces efficient denitrification:

$$NH_4^+ + HOCl \rightarrow NH_2Cl + H_2O + H^+ \tag{8.11.4}$$

$$2NH_2Cl + HOCl \rightarrow N_2(g) + 3H^+ + 3Cl^- + H_2O \tag{8.11.7}$$

This reaction is used to remove pollutant ammonia from wastewater. However, problems can arise from chlorination of organic wastes. Typical of such by-products is chloroform, produced by the chlorination of humic substances in water.

Chlorine is used to treat water other than drinking water. It is employed to disinfect effluent from sewage treatment plants, as an additive to the water in electric power plant cooling towers, and to control microorganisms in food processing.

Chlorine Dioxide

Chlorine dioxide, ClO_2, is an effective water disinfectant that is of particular interest because, in the absence of impurity Cl_2, it does not produce impurity trihalomethanes in water treatment.[6] In acidic and neutral water, respectively, the two half-reactions for ClO_2 acting as an oxidant are the following:

$$ClO_2 + 4H^+ + 5e^- \leftarrow\rightarrow Cl^- + 2H_2O \tag{8.11.8}$$

$$ClO_2 + e^- \leftarrow\rightarrow ClO^- \tag{8.11.9}$$

In the neutral pH range, chlorine dioxide in water remains largely as molecular ClO_2 until it contacts a reducing agent with which to react. Chlorine dioxide is a gas that is violently reactive with organic matter and explosive when exposed to light. For these reasons, ClO_2 is not shipped, but is generated on-site by processes such as the reaction of chlorine gas with solid sodium hypochlorite:

$$2NaClO_2(s) + Cl_2(g) \;\longleftrightarrow\; ClO_2(g) + NaCl(s) \qquad (8.11.10)$$

A high content of elemental chlorine in the product may require its purification to prevent unwanted side-reactions from Cl_2.

As a water disinfectant, chlorine dioxide does not chlorinate or oxidize ammonia or other nitrogen-containing compounds. Some concern has been raised over possible health effects of its main degradation byproducts, ClO_2^- and ClO_3^-.

Ozone

Ozone is sometimes used as a disinfectant in place of chlorine, particularly in Europe. Figure 8.10 shows the basic components of an ozone water treatment system. Basically, air is filtered, cooled, dried, and pressurized, then subjected to an electrical

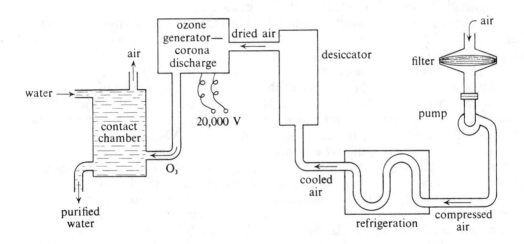

Figure 8.10. A schematic diagram of a typical ozone water-treatment system.

discharge of approximately 20,000 volts. The ozone produced is then pumped into a contact chamber where water contacts the ozone for 10-15 minutes. Concern over possible production of toxic organochlorine compounds by water chlorination processes has increased interest in ozonation. Furthermore, ozone is more destructive to viruses than is chlorine. Unfortunately, the solubility of ozone in water is relatively low, which limits its disinfective power.

A major consideration with ozone is the rate at which it decomposes spontaneously in water, according to the overall reaction,

$$2O_3 \rightarrow 3O_2(g) \qquad (8.11.11)$$

This rate of decomposition has been found to follow the empirical equation below:[7]

$$-\frac{d[O_3]}{dt} = k_0[OH^-]^{0.55}[O_3]^2 \tag{8.11.12}$$

Because of the decomposition of ozone in water, some chlorine must be added to maintain disinfectant throughout the water distribution system.

Iron(VI) in the form of ferrate ion, FeO_4^{2-}, is a strong oxidizing agent with excellent disinfectant properties. It has the additional advantage of removing heavy metals, viruses, and phosphate. It may well find limited application for disinfection in the future.

8.12. NATURAL WATER PURIFICATION PROCESSES

Virtually all of the materials that waste treatment processes are designed to eliminate may be absorbed by soil or degraded in soil. In fact, most of these materials are essential for soil fertility. Wastewater may provide the water that is essential to plant growth, in addition to the nutrients — phosphorus, nitrogen and potassium — usually provided by fertilizers. Wastewater also contains essential trace elements and vitamins. Stretching the point a bit, the degradation of organic wastes provides the CO_2 essential for photosynthetic production of plant biomass.

Soil may be viewed as a natural filter for wastes. Most organic matter is readily degraded in soil and, in principle, soil constitutes an excellent treatment system — primary, secondary and tertiary — for water. Soil has physical, chemical, and biological characteristics that can enable wastewater detoxification, biodegradation, chemical decomposition, and physical and chemical fixation. A number of soil characteristics are important in determining its use for land treatment of wastes. These characteristics include physical form, ability to retain water, aeration, organic content, acid-base characteristics, and oxidation-reduction behavior.[8] Soil is a natural medium for a number of living organisms that may have an effect upon biodegradation of wastewaters, including those that contain industrial wastes. Of these, the most important are bacteria, including those from the genera *Agrobacterium*, *Arthrobacteri*, *Bacillus*, *Flavobacterium*, and *Pseudomonas*. Actinomycetes and fungi are important in decay of vegetable matter and may be involved in biodegradation of wastes. Other unicellular organisms that may be present in or on soil are protozoa and algae. Soil animals, such as earthworms, affect soil parameters such as soil texture. The growth of plants in soil may have an influence on its waste treatment potential in such aspects as uptake of soluble wastes and erosion control.

Early civilizations, such as the Chinese, used human organic wastes to increase soil fertility, and the practice continues today. The ability of soil to purify water was noted well over a century ago. In 1850 and 1852, J. Thomas Way, a consulting chemist to the Royal Agricultural Society in England, presented two papers to the Society entitled "Power of Soils to Absorb Manure." Mr. Way's experiments showed that soil is an ion exchanger. Much practical and theoretical information on the ion exchange process resulted from his work.

Experiments involving the direct application of wastewater to soil have shown appreciable increases in soil productivity. A process called *overland flow* has been described[7] in which wastewater containing pulverized solids is allowed to trickle over sloping soil. Suspended solids, BOD, and nutrients are largely removed. This method is most applicable to rural communities in relatively warm climates. Approximately one acre of land is required to handle the sewage from 200 persons.

If such systems are not properly designed and operated, odor can become an overpowering problem. The author of this book is reminded of driving into a small town, recalled from some years before as a very pleasant place, and being assaulted with a virtually intolerable odor. The disgruntled residents pointed to a large spray irrigation system on a field in the distance — unfortunately upwind — spraying liqui-fied pig manure as part of an experimental feedlot waste treatment operation. The experiment was not deemed a success and was discontinued by the investigators, presumably before they met with violence from the local residents.

Industrial Wastewater Treatment by Soil

Wastes that are amenable to land treatment are biodegradable organic substances, particularly those contained in municipal sewage and in wastewater from some industrial operations, such as food processing. However, through acclimation over a long period of time, soil bacterial cultures may develop that are effective in degrading normally recalcitrant compounds that occur in industrial wastewater. Acclimated microorganisms are found particularly at contaminated sites, such as those where soil has been exposed to crude oil for many years.

A variety of enzyme activities are exhibited by microorganisms in soil that enable them to degrade synthetic substances. Even sterile soil may show enzyme activity due to extracellular enzymes secreted by microorganisms in soil. Some of these enzymes are hydrolase enzymes (see Chapter 20), such as those that catalyze the hydrolysis of organophosphate compounds as shown by the reaction,

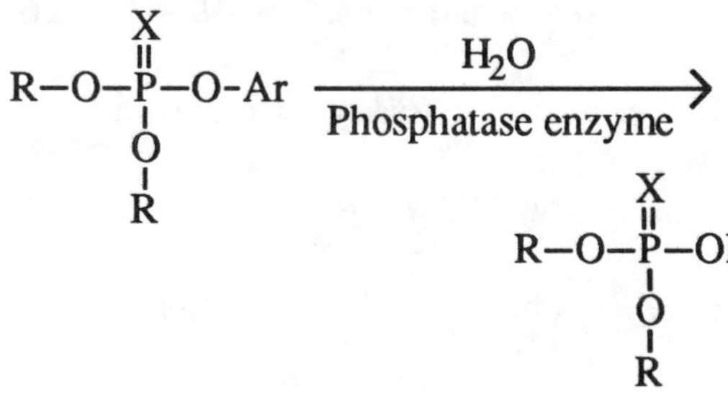

$$\underset{\substack{| \\ R}}{\overset{\substack{X \\ \|}}{R-O-P-O\text{-}Ar}} \xrightarrow[\text{Phosphatase enzyme}]{H_2O}$$

$$\underset{\substack{| \\ R}}{\overset{\substack{X \\ \|}}{R-O-P-OH}} + \text{HOAr} \qquad (8.12.1)$$

where R is an alkyl group, Ar is a substituent group that is frequently aromatic, and X is either S or O. Another example of a reaction catalyzed by soil enzymes is the oxidation of phenolic compounds by diphenol oxidase:

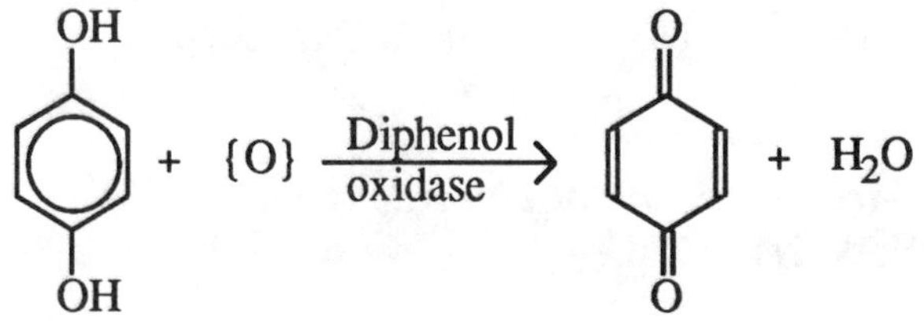

$$(8.12.2)$$

Land treatment is most used for petroleum refining wastes and is applicable to the treatment of fuels and wastes from leaking underground storage tanks. It can also be applied to biodegradable organic chemical wastes, including some organohalide compounds. Land treatment is not suitable for the treatment of wastes containing acids, bases, toxic inorganic compounds, salts, heavy metals, and organic compounds that are excessively soluble, volatile, or flammable.

LITERATURE CITED

1. Sussman, S., "Industrial Water Treatment," in *Kirk-Othmer Concise Encyclopedia of Chemical Technology*, David Eckroth, Ed., John Wiley and Sons, New York, 1985.

2. Miller, S., "Water Reuse," *Environmental Science and Technology*, **15**, 499-501 (1981).

3. Blanton, T. Clay, Dale Rohe, Joseph G. Jacangelo, and Benito J. Mariñas, "Emerging Membrane Processes for Drinking Water Treatment," *Waterworld News*, January/February, 1991, pp. 10-13.

4. "Ocean Dumping," *SFI Bulletin*, No. 403, Sport Fishing Institute, Washington, DC, April, 1989, pp. 1-3.

5. Ball, R., M. Harris, and K. Deeny, *Evaluation and Control of Sidestreams Generated in Publicly Owned Treatment Works*, EPA-600/S2-82-0216, Municipal Environmental Research Laboratory, U. S. Environmental Protection Agency, Cincinnati, Ohio, 1982.

6. "Disinfection," Chapter 12 in *Water Treatment Principles and Design*, James M. Montgomery, Consulting Engineers, John Wiley and Sons, Inc., 1985, pp. 262-300.

7. Thomas, R. E., K. Jackson, and L. Penrod, *Feasibility of Overland Flow for Treatment of Raw Domestic Wastewater*, EPA-660/2-74-087, U. S. Environmental Protection Agency, Washington, DC, 1974.

8. Sposito, Garrison, *The Chemistry of Soils*, Oxford University Press, New York, 1989.

SUPPLEMENTARY REFERENCES

Hammer, Donald A., Ed., *Constructed Wetlands for Wastewater Treatment*, Lewis Publishers, Inc., Chelsea, MI, 1989.

Perkins, Richard J., *Onsite Wastewater Disposal*, Lewis Publishers, Chelsea, MI, 1989.

Freeman, Harry M., Ed., *Standard Handbook of Hazardous Waste Treatment and Disposal*, McGraw-Hill, New York, NY, 1988.

Scholze, R. J., Ed., *Biotechnology for Degradation of Toxic Chemicals in Hazardous Wastes*, Noyes Publications, Park Ridge, NJ, 1988.

Forster, C. F., *Biotechnology and Wastewater Treatment*, Cambridge University Press, New York, NY, 1985.

Bauman, D. D., and D. M. Dworkin, *Planning for Water Reuse*, Maarouga Press, Chicago, 1978.

Cooper, W. J., Ed., *Chemistry in Water Reuse*, Vol. 1, Ann Arbor Science Publishers, Inc., Ann Arbor, Mich., 1981.

D'Itri, F. M., Ed., *Land Treatment of Municipal Wastewater*, Ann Arbor Science Publishers, Inc., Ann Arbor, Mich., 1982.

D'Itri, F. M., J. A. Martinez, and M. A. Lámbari, *Municipal Wastewater in Agriculture*, Ann Arbor Science Publishers, Inc., Ann Arbor, Mich., 1981.

Eikkum, A. S., and R. W. Seabloom, Eds., *Alternative Wastewater Treatment*, D. Reidel Publishing Co., Hingham, Mass., 1982.

Gillies, M. T., Ed., *Potable Water from Wastewater*, Noyes Data Corp., Park Ridge, New Jersey, 1981.

Hudson, H. E., Jr., *Water Clarification Processes*, Van Nostrand Reinhold Co., New York, 1981.

Mattox, G., Ed., *New Processes of Waste Water Treatment and Recovery*, John Wiley and Sons, Inc., New York, 1978.

Ramalho, R. S., *Introduction to Wastewater Treatment*, Academic Press, New York, 1977.

Vernick, A. S., and E. C. Walker, Eds., *Handbook of Wastewater Treatment Processes*, Marcel Dekker, Inc., 1981.

White, G. C., *Disinfection of Wastewater and Water for Reuse*, Van Nostrand Reinhold Co., New York, 1978.

Winkler, M. A., *Biological Treatment of Wastewater*, John Wiley and Sons, Inc., New York, 1981.

QUESTIONS AND PROBLEMS

1. Consider the equilibrium reactions and expressions covered in Chapter 3. How many moles of NTA should be added to 1000 liters of water having a pH of 9 and containing CO_3^{2-} at 1.00×10^{-4} M to prevent precipitation of $CaCO_3$? Assume a total calcium level of 40 mg/L.

2. What is the purpose of the return sludge step in the activated sludge process?

3. What are the two processes by which the activated sludge process removes soluble carbonaceous material from sewage?

4. Why might hard water be desirable as a medium if phosphorus is to be removed by an activated sludge plant operated under conditions of high aeration?

5. How does reverse osmosis differ from a simple sieve separation or ultrafiltration process?

6. How many liters of methanol would be required daily to remove the nitrogen from a 200,000-L/day sewage treatment plant producing an effluent containing 50 mg/L of nitrogen? Assume that the nitrogen has been converted to NO_3^- in the plant. The denitrifying reaction is Reaction 8.9.6.

7. Discuss some of the advantages of physical-chemical treatment of sewage as opposed to biological wastewater treatment. What are some disadvantages?

8. Why is recarbonation necessary when water is softened by the lime-soda process?

9. Assume that a waste contains 300 mg/L of biodegradable $\{CH_2O\}$ and is processed through a 200,000-L/day sewage-treatment plant which converts 40% of the waste to CO_2 and H_2O. Calculate the volume of air (at 25 °, 1 atm) required for this conversion. Assume that the O_2 is transferred to the water with 20% efficiency.

10. If all of the $\{CH_2O\}$ in the plant described in Question 9 could be converted to methane by anaerobic digestion, how many liters of methane (STP) could be produced daily?

11. Assuming that aeration of water does not result in the precipitation of calcium carbonate, which of the following would not be removed by aeration: hydrogen sulfide; carbon dioxide; volatile odorous bacterial metabolites; alkalinity; iron.

12. In which of the following water supplies would moderately high water hardness be most detrimental: municipal water; irrigation water; boiler feedwater; drinking water (in regard to potential toxicity).

13. An *increase* in which of the following *decreases* the rate of oxidation of iron(II) to iron(III) in water? [Fe(II)]; pH; $[H^+]$; $[O_2]$; $[OH^-]$.

14. A wastewater containing dissolved Cu^{2+} ion is to be treated to remove copper. Which of the following processes would *not* remove copper in an insoluble form; lime precipitation; cementation; treatment with NTA; ion exchange; reaction with metallic Fe.

15. Match each water contaminant in the left column with its preferred method of removal in the right column.

    (a) Mn^{2+} (1) Activated carbon
    (b) Ca^{2+} and HCO_3^- (2) Raise pH by addition of Na_2CO_3
    (c) Trihalomethane compounds   (3) Addition of lime
    (d) Mg^{2+} (4) Oxidation

16. A cementation reaction employs iron to remove Cd^{2+} present at a level of 350 mg/L from a wastewater stream. Given that the atomic weight of Cd is 112.4 and that of Fe is 55.8, how many kg of Fe are consumed in removing all the Cd from 4.50×10^6 liters of water?

17. Consider municipal drinking water from two different kinds of sources, one a flowing, well aerated stream with a heavy load of particulate matter and the other an anaerobic groundwater. Describe possible differences in the water treatment strategies for these two sources of water.

18. In treating water for industrial use, consideration is often given to "sequential use of the water." What is meant by this term? Give some plausible examples of sequential use of water.

19. Active biomass is used in the secondary treatment of municipal wastewater. Describe three ways of supporting a growth of the biomass, contacting it with wastewater, and exposing it to air.

20. Using appropriate chemical reactions for illustration, show how calcium present as the dissolved HCO_3^- salt in water is easier to remove than other forms of hardness, such as dissolved $CaCl_2$.

The Atmosphere and Atmospheric Chemistry

9.1. IMPORTANCE OF THE ATMOSPHERE

The atmosphere is a protective blanket which nurtures life on the Earth and protects it from the hostile environment of outer space. The atmosphere is the source of carbon dioxide for plant photosynthesis and of oxygen for respiration. It provides the nitrogen that nitrogen-fixing bacteria and ammonia-manufacturing plants use to produce chemically-bound nitrogen, an essential component of life molecules. As a basic part of the hydrologic cycle (Figure 2.1) the atmosphere transports water from the oceans to land, thus acting as the condenser in a vast solar-powered still. Unfortunately, the atmosphere also has been used as a dumping ground for many pollutant materials—ranging from sulfur dioxide to refrigerant Freon—a practice which causes damage to vegetation and materials, shortens human life, and alters the characteristics of the atmosphere itself.

The atmosphere serves a vital protective function. It absorbs most of the cosmic rays from outer space and protects organisms from their effects. It also absorbs most of the electromagnetic radiation from the sun, allowing transmission of significant amounts of radiation only in the regions of 300-2500 nm (near-ultraviolet, visible, and near-infrared radiation) and 0.01-40 m (radio waves). By absorbing electromagnetic radiation below 300 nm the atmosphere filters out damaging ultraviolet radiation that would otherwise be very harmful to living organisms. Furthermore, because it reabsorbs much of the infrared radiation by which absorbed solar energy is re-emitted to space, the atmosphere stabilizes the Earth's temperature, preventing the tremendous temperature extremes that occur on planets and moons lacking substantial atmospheres.

9.2. PHYSICAL CHARACTERISTICS OF THE ATMOSPHERE

Atmospheric science deals with the movement of air masses in the atmosphere, atmospheric heat balance, and atmospheric chemical composition and reactions. In order to understand atmospheric chemistry and air pollution, it is important to have an overall appreciation of the atmosphere, its composition, and physical characteristics as discussed in the first parts of this chapter.

Atmospheric Composition

Dry air within several kilometers of ground level consists of two **major components**

- Nitrogen, 78.08 %
  (by volume)

- Oxygen, 20.95 %

minor components

- Argon, 0.934 %

- Carbon dioxide, 0.035 %

noble gases

- Neon, 1.818×10^{-3} %
- Krypton, 1.14×10^{-4} %

- Helium, 5.24×10^{-4} %
- Xenon, 8.7×10^{-6} %

and **trace gases** as given in Table 9.1. Atmospheric air may contain 0.1–5% water by volume, with a normal range of 1–3%.

Variation of Pressure and Density with Altitude

As anyone who has exercised at high altitudes well knows, the density of the atmosphere decreases sharply with increasing altitude as a consequence of the gas laws and gravity. More than 99% of the total mass of the atmosphere is found within approximately 30 km (about 20 miles) of the Earth's surface. Such an altitude is miniscule compared to the Earth's diameter, so it is not an exaggeration to characterize the atmosphere as a "tissue-thin" protective layer. The total mass of the global atmosphere is approximately 5.14×10^{15} metric tons, a huge figure, but still only approximately one millionth of the Earth's total mass.

The fact that atmospheric pressure decreases as an approximately exponential function of altitude largely determines the characteristics of the atmosphere. Ideally, in the absence of mixing and at a constant absolute temperature, T, the pressure at any given height, P_h, is given in the exponential form,

$$P_h = P_0 e^{-mgh/RT} \tag{9.2.1}$$

where P_0 is the pressure at zero altitude (sea level); M is the average gram molecular mass of air (28.97 g/mole in the troposphere); g is the acceleration of gravity (981 cm $\times$ sec^{-1} at sea level); h is the altitude in cm; and R is the gas constant (8.314×10^7 erg $\times$ deg^{-1} $\times$ mole^{-1}). These units are given in the cgs (centimeter-gram-sec) system for consistency; altitude can be converted to meters or kilometers as appropriate.

The factor RT/Mg is defined as the **scale height**, which represents the increase in altitude by which the pressure drops by e^{-1}. At an average sea-level temperature of 288 K, the scale height is 8×10^5 cm or 8 km; at an altitude of 8 km, the pressure is only about 39% that at sea-level.

Table 9.1. Trace Gases in Dry Air, Volume Percent[1]

Gas or species	Volume percent[1]	Major sources	Process for removal from the atmosphere
CH_4	1.6×10^{-4}	Biogenic[2]	Photochemical[3]
CO	$\sim 1.2 \times 10^{-5}$	Photochemical, anthropogenic[4]	Photochemical
N_2O	3×10^{-5}	Biogenic	Photochemical
NO_x[5]	10^{-10}–10^{-6}	Photochemical, lightning, anthropogenic	Photochemical
HNO_3	10^{-9}–10^{-7}	Photochemical	Washed out by precipitation
NH_3	10^{-8}–10^{-7}	Biogenic	Photochemical, washed out by precipitation
H_2	5×10^{-5}	Biogenic, photochemical	Photochemical
H_2O_2	10^{-8}–10^{-6}	Photochemical	Washed out by precipitation
$HO\cdot$[6]	10^{-13}–10^{-10}	Photochemical	Photochemical
$HO_2\cdot$[6]	10^{-11}–10^{-9}	Photochemical	Photochemical
H_2CO	10^{-8}–10^{-7}	Photochemical	Photochemical
CS_2	10^{-9}–10^{-8}	Anthropogenic, biogenic	Photochemical
OCS	10^{-8}	Anthropogenic, biogenic, photochemical	Photochemical
SO_2	$\sim 2 \times 10^{-8}$	Anthropogenic, photochemical, volcanic	Photochemical
I_2	O–trace	—	—
CCl_2F_2[7]	2.8×10^{-5}	Anthropogenic	Photochemical
H_3CCCl_3[8]	$\sim 1 \times 10^{-8}$	Anthropogenic	Photochemical

[1] Levels in the absence of gross pollution.
[2] From biological sources.
[3] Reactions induced by the absorption of light energy as described later in this chapter.
[4] Sources arising from human activities.
[5] Sum of NO and NO_2.
[6] Reactive free radical species with one unpaired electron; described later in the chapter; these are transient species whose concentrations become much lower at night.
[7] A chlorofluorocarbon, Freon F-12.
[8] Methyl chloroform.

Conversion of Equation 9.2.1 to the logarithmic (base 10) form and expression of h in km yields

$$\text{Log } P_h = \text{Log } P_0 - \frac{Mgh \times 10^5}{2.303RT} \tag{9.2.2}$$

and taking the pressure at sea level to be exactly 1 atm gives the following expression:

$$\text{Log } P_h = - \frac{Mgh \times 10^5}{2.303\,RT} \tag{9.2.3}$$

Plots of P_h and temperature *versus* altitude are shown in Figure 9.1. The plot of P_h is nonlinear because of variations arising from temperature differences and the mixing of air masses. The plot reflects non-linear variations in temperature with altitude that are discussed later in this section.

The characteristics of the atmosphere vary widely with altitude, time (season), location (latitude), and even solar activity. Extremes of pressure and temperature are illustrated in Figure 9.1. At very high altitudes normally reactive species, such as atomic oxygen, O, persist for long periods of time. That occurs because the pressure is very low at these altitudes such that the distance travelled by a reactive species before it collides with a potential reactant — its **mean free path** — is quite high. A particle with a mean free path of 1×10^{-6} cm at sea level has a mean free path greater than 1×10^{6} cm at an altitude of 500 km, where the pressure is lower by many orders of magnitude.

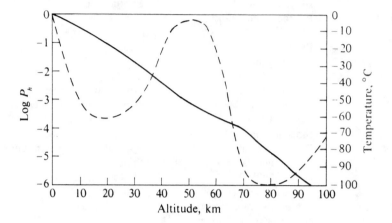

Figure 9.1. Variation of pressure (solid line) and temperature (dashed line) with altitude.

Stratification of the Atmosphere

As shown in Figure 9.2, the atmosphere is stratified on the basis of temperature/density relationships resulting from interrelationships between physical and photochemical (light-induced chemical phenomena) processes in air.

The lowest layer of the atmosphere extending from sea level to an altitude of 10-16 km is the **troposphere**, characterized by a generally homogeneous composition of major gases other than water and decreasing temperature with increasing altitude from the heat-radiating surface of the earth. The upper limit of the troposphere, which has a temperature minimum of about -56°C, varies in altitude by a kilometer or more with atmospheric temperature, underlying terrestrial surface, and time. The homogeneous composition of the troposphere results from constant mixing by circulating air masses. However, the water vapor content of the troposphere is extremely variable because of cloud formation, precipitation, and evaporation of water from terrestrial water bodies.

The very cold temperature of the **tropopause** layer at the top of the troposphere serves as a barrier that causes water vapor to condense to ice so that it cannot reach altitudes at which it would photodissociate through the action of intense high-energy ultraviolet radiation. If this happened, the hydrogen produced would escape the Earth's atmosphere and be lost. (Much of the hydrogen and helium originally present in the Earth's atmosphere was lost by this process.)

The atmospheric layer directly above the troposphere is the **stratosphere**, in which the temperature rises to a maximum of about -2°C with increasing altitude. This phenomenon is due to the presence of ozone, O_3, which may reach a level of around 10 ppm by volume in the mid-range of the stratosphere. The heating effect is caused by the absorption of ultraviolet radiation energy by ozone, a phenomenon discussed later in this chapter.

The absence of high levels of radiation-absorbing species in the **mesosphere** immediately above the stratosphere results in a further temperature decrease to about −92°C at an altitude around 85 km. The upper regions of the mesosphere and higher define a region called the exosphere from which molecules and ions can completely escape the atmosphere. Extending to the far outer reaches of the atmosphere is the **thermosphere,** in which the highly rarified gas reaches temperatures as high as 1200°C by the absorption of very energetic radiation of wavelengths less than approximately 200 nm by gas species in this region.

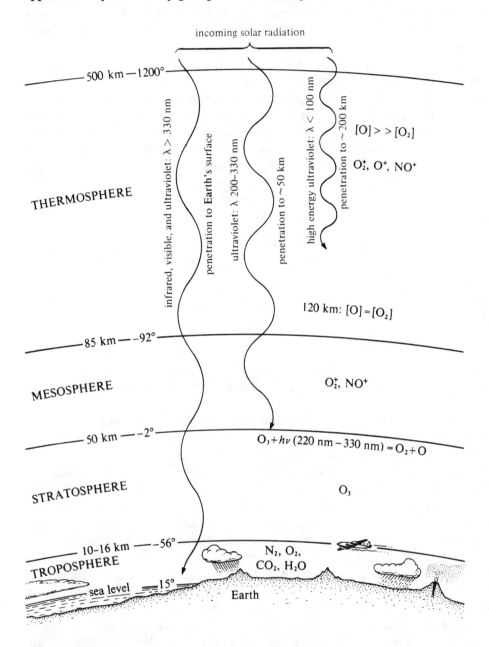

incoming solar radiation

500 km — 1200°

THERMOSPHERE

infrared, visible, and ultraviolet: λ > 330 nm

penetration to Earth's surface

ultraviolet: λ 200–330 nm

penetration to ~50 km

high energy ultraviolet: λ < 100 nm

penetration to ~200 km

$[O] >> [O_2]$

O_2^+, O^+, NO^+

120 km: $[O] = [O_2]$

85 km — −92°

MESOSPHERE

O_2^+, NO^+

50 km — −2°

$O_3 + h\nu$ (220 nm − 330 nm) $= O_2 + O$

STRATOSPHERE

O_3

10–16 km — −56°

TROPOSPHERE

$N_2, O_2,$
CO_2, H_2O

sea level — 15°

Earth

Figure 9.2. Major regions of the atmosphere (not to scale).

9.3. ENERGY AND MASS TRANSFER IN THE ATMOSPHERE

The physical and chemical characteristics of the atmosphere and the critical heat balance of the Earth are determined by energy and mass transfer processes in the atmosphere. These phenomena are addressed in this section.

Energy Transfer

The solar energy flux reaching the atmosphere is huge, amounting to 1.34×10^3 watts per square meter (19.2 kcal per minute per square meter) perpendicular to the line of solar flux at the top of the atmosphere, as illustrated in Figure 9.3. This value is the **solar constant**. If all this energy reached the Earth's surface and were retained, the planet would have vaporized long ago. As it is, the complex factors involved in maintaining the Earth's heat balance within very narrow limits are crucial to retaining conditions of climate that will support present levels of life on Earth. The great changes of climate that resulted in ice ages during some periods or tropical conditions during others were caused by variations of only a few degrees in average temperature. Marked climate changes within recorded history have been caused by much smaller average temperature changes. The mechanisms by which the Earth's average temperature is retained within its present narrow range are complex and not completely understood, but the main features are explained here.

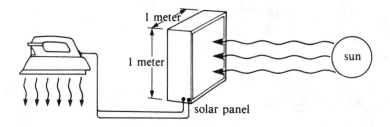

Figure 9.3. The solar flux at the distance of the Earth from the sun is 1.34×10^3 watts/m^2.

About half of the solar radiation entering the atmosphere reaches the Earth's surface either directly or after scattering by clouds, atmospheric gases, or particles. The remaining half of the radiation is either reflected directly back or absorbed in the atmosphere and its energy radiated back into space at a later time as infrared radiation. Most of the solar energy reaching the surface is absorbed and it must be returned to space in order to maintain heat balance. In addition, a very small amount of energy (less than 1% of that received from the sun) reaches the Earth's surface by convection and conduction processes from the Earth's hot mantle, and this, too, must be lost.

Energy transport, which is crucial to eventual reradiation of energy from the Earth is accomplished by three major mechanisms — conduction, convection, and radiation. **Conduction** of energy occurs through the interaction of adjacent atoms or molecules without the bulk movement of matter. **Convection** involves the movement of whole masses of air, which may be either relatively warm or cold. It is the mechanism by which abrupt temperature variations occur when large masses of air move across an area. As well as carrying **sensible heat** due to the kinetic energy of molecules, convection carries **latent heat** in the form of water vapor which releases heat as it condenses. An appreciable fraction of the Earth's surface heat is transported to clouds in the atmosphere by conduction and convection before being lost ultimately by radiation.

Radiation of energy occurs through electromagnetic radiation in the infrared region of the spectrum. As the only way in which energy is transmitted through a vacuum, radiation is the means by which all energy lost from the planet to maintain its heat balance is ultimately returned to space. The electromagnetic radiation that carries energy away from the Earth is of a much longer wavelength that the sunlight that brings energy to the Earth. This is a crucial factor in maintaining the Earth's heat balance and one susceptible to upset by human activities. The maximum intensity of incoming radiation occurs at 0.5 micrometers (500 nanometers) in the visible region, with essentially none outside the range of 0.2 μm to 3 μm. This range encompasses the whole visible region and small parts of the ultraviolet and infrared adjacent to it. Outgoing radiation is in the infrared region, with maximum intensity at about 10 μm, primarily between 2 μm and 40 μm. Thus the Earth loses energy by electromagnetic radiation of a much lower wavelength (lower energy per photon) than the radiation by which it receives energy.

Earth's Radiation Budget

The Earth's radiation budget is illustrated in Figure 9.4. The average surface temperature is maintained at a relatively comfortable 15°C because of an atmospheric "greenhouse effect" in which water vapor and, to a lesser extent carbon dioxide, reabsorb much of the outgoing radiation and reradiate about half of it back to the surface. Were this not the case, the surface temperature would average around -18°C. Most of the absorption of infrared radiation is done by water molecules in the atmosphere. Absorption is weak in the regions 7-8.5 μm and 11-14 μm and nonexistent between 8.5 μm and 11 μm, leaving a "hole" in the infrared absorption spectrum through which radiation may escape. Carbon dioxide, though present at a much lower concentration than water vapor, absorbs strongly between 12 μm and 16.3 μm, and plays a key role in maintaining the heat balance. There is concern that an increase in the carbon dioxide level in the atmosphere could prevent sufficient energy loss to cause a perceptible and damaging increase in the Earth's temperature. This phenomenon, discussed in more detail in Section 9.7 and Chapter 14, is popularly known as the **greenhouse effect** and may occur from elevated CO_2 levels caused by increased use of fossil fuels and the destruction of massive quantities of forests.

Mass Transfer and Meteorology

Meteorology is the science of atmospheric phenomena, encompassing the study of the movement of air masses as well as physical forces in the atmosphere such as heat, wind, and transitions of water, primarily liquid to vapor, or *vice versa*. Meteorological phenomena affect, and in turn are affected by, the chemical properties of the atmosphere. For example, meteorological phenomena determine whether or not power plant stack gas heavily laced with sulfur dioxide is dispersed high in the atmosphere, with little direct effect upon human health, or settles as a choking chemical blanket in the vicinity of the power plant. Los Angeles largely owes its susceptibility to smog to the meteorology of the Los Angeles basin, which holds hydrocarbons and nitrogen oxides long enough to cook up an unpleasant brew of damaging chemicals under the intense rays of the sun (see Chapter 13).

Short-term variations in the state of the atmosphere are described as **weather**. The weather is defined in terms of seven major factors: temperature, clouds, winds, humidity, horizontal visibility (as affected by fog, etc.), type and quantity of precipitation, and atmospheric pressure. All of these factors are closely interrelated. Cold air holds less water than warm air. Therefore, the cooling of warm moist air can

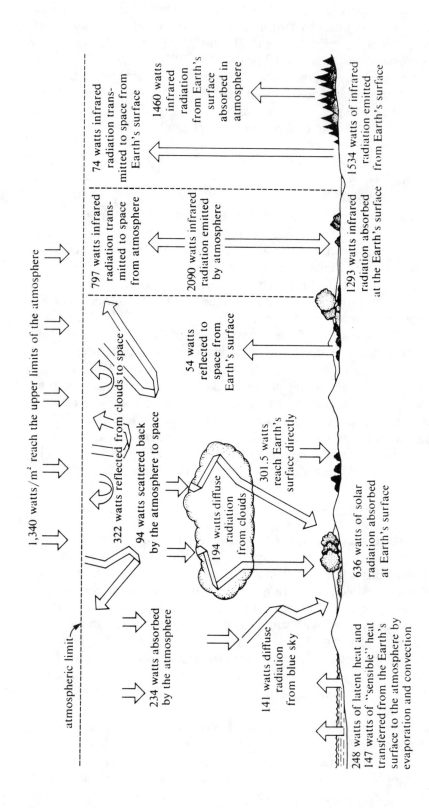

Figure 9.4. Earth's radiation budget expressed on the basis of portions of the 1340 watts/m² composing the solar flux

result in the formation of clouds, fog, and precipitation. Warm air tends to rise because of its lower density. Air flows from a region of high pressure to one of low pressure, creating winds.

In the absence of major weather changes, temperature varies predictably during the day. The Earth radiates heat during the night. The lowest temperatures occur within a few weeks of the winter solstice, and the highest temperatures occur right after the summer solstice. Near oceans, temperatures vary much less because of the heat-stabilizing effects of water (see Chapter 2). In the troposphere, temperature decreases with increasing altitude because of increasing distance from the heat source (Earth); lower concentration of water vapor with its high heat capacity; and cooling due to expansion of rising air.

Horizontally moving air is called wind, whereas vertically moving air is referred to as an air current. Wind and air currents are strongly involved with air pollution phenomena. Wind carries and disperses air pollutants. Prevailing wind direction is an important factor in determining the areas most affected by an air pollution source.

Atmospheric water can be present as vapor, liquid, or ice. The water content of air can be expressed as humidity. Relative humidity, expressed as a percentage, describes the amount of water vapor in the air as a ratio of the maximum amount that the air can hold at that temperature. Air with a given relative humidity can undergo any of several processes to reach the saturation point at which water vapor condenses. These processes are cooling by contact with a cold surface, adiabatic cooling by rising and expansion, mixing of cold and warm air masses, and radiation of heat by the air itself. The temperature below which moisture condensation occurs is called the dew point.

Clouds normally form when rising, adiabatically cooling air can no longer hold water in the vapor form, and the water forms very small aerosol droplets. Clouds may be classified in three major forms. Cirrus clouds occur at great altitudes and have a thin feathery appearance. Cumulus clouds are detached masses with a flat base and frequently a "bumpy" upper structure. Stratus clouds occur in large sheets and may cover all of the sky visible from a given point as overcast. Clouds are important absorbers and reflectors of radiation (heat). Their formation is affected by human activities, especially particulate matter pollution and emission of deliquescent gases, such as SO_2 and HCl.

Condensation of water vapor, which forms clouds, must occur prior to the formation of precipitation in the form of rain or snow. For this condensation to happen, air must be cooled below the dew point, and nuclei of condensation must be present. These nuclei are hydroscopic substances such as salts, sulfuric acid droplets, and some organic materials, including bacterial cells. Air pollution is now an important source of condensation nuclei.

The formation of precipitation is a complicated and important process. Cloud droplets normally take somewhat longer than a minute to form by condensation. They average about 0.04 mm across and do not exceed 0.2 mm in diameter. Raindrops range from 0.5–4 mm in diameter. Condensation processes do not form particles large enough to fall as precipitation (rain, snow, sleet, or hail). The small condensation droplets must collide and coalesce to form precipitation-size particles. When droplets reach a threshold diameter of about 0.04 mm, they grow more rapidly by coalescence with other particles than by condensation of water vapor.

Distinct air masses are a major feature of the troposphere. These air masses are uniform and are horizontally homogeneous. Their temperature and water-vapor content are particularly uniform. These characteristics are determined by the nature of the surface over which a large air mass forms. Polar continental air masses form over cold land regions; polar maritime air masses form over polar oceans. Air masses originating in the tropics may be similarly classified as tropical continental air masses or tropical maritime air masses. The movement of air masses and the conditions in them may have important effects upon pollutant reactions, effects, and dispersal.

The relatively sharp border areas between air masses are called **fronts**. A cold front is one in which warm air is replaced by cold air, whereas replacement of cold air by warm air produces a warm front. At the front between a warm air mass and a cold air mass, the cold air forms a wedge beneath the warm air as a consequence of their different densities. Warm, moist air rising along this wedge is cooled, often causing precipitation to occur.

A detailed discussion of the general circulation of air masses making up the global climate is beyond the scope of this book. One of the major factors involved is that warm air produced in the tropics around the equator tends to rise, cools by loss of heat through radiation as it migrates northward and southward from the equator, then returns to the tropical regions as cooled air. This circulating air is subject to a coriolis force, an inertial force which causes air moving back to the equator to lag behind the Earth's surface in its motion. As a result, wind moving toward the equator acquires a westward component of velocity.

The region of rising air around the equator comprises the intertropical convergence zone, an area with a great deal of cloudiness and rain. There is an absence of strong north–south winds across this zone, so that mixing of atmospheric constituents across the hemispheres requires a long time—one to two years. By comparison, mixing within a hemisphere requires one to two months.

The complicated movement of air across the Earth's surface is a crucial factor in the creation and dispersal of air pollution phenomena. When air movement ceases, air stagnation can occur with a resultant build-up of air pollutants in localized regions. Although the temperature of air relatively near the Earth's surface normally decreases with increasing altitude, certain atmospheric conditions can result in the opposite condition—increasing temperature with increasing altitude. Such conditions are characterized by high atmospheric stability and are known as **temperature inversions**. Typically, an inversion occurs by the collision of a warm air mass (warm front) with a cold air mass (cold front). The warm air mass overrides the cold air mass in the frontal area, producing the inversion. Because they limit the vertical circulation of air, temperature inversions result in air stagnation and the trapping of air pollutants in localized areas.

Air circulation on a relatively small scale is involved in thunderstorms, valley winds, and sea-land breezes. These fall in the category of mesometeorology or micrometeorology (as compared to macrometeorology, which involves the movement of very large masses of air).

Human activities have succeeded in changing the meteorology of whole cities, a mesometeorological effect. By paving over large areas of a city with nonreflecting asphalt, and by activities that generate heat, humans have created conditions under which the center of a city may be as much as 5°C warmer than the surrounding area. In such a case, the warmer air rises, bringing in a breeze from the surrounding area. Large cities have been described as "heat islands." Pollutants and carbon dioxide given off from cities may absorb emitted infrared radiation, causing a local greenhouse effect that probably is largely counterbalanced by reflection of incoming solar energy by particulate matter above cities.

Some human activities may even be changing the global climate. This possibility is discussed in Chapter 14.

In discussing atmospheric pollutants and their global cycles, it is convenient to define and use the following terms and concepts. A **reservoir** is a domain such as the atmosphere or biosphere where a pollutant may "reside" for a time. The amount of a specific pollutant in a reservoir is known as its **burden**. Burdens are commonly expressed in units of 10^{12} g (10^6 metric tons) called teragrams (Tg). The rate of transfer of a pollutant from one sphere or domain to another is called the flux. Flux frequently is expressed in units of teragrams per year.

It is useful to determine the **residence time** of a pollutant in a given reservoir. If a reservoir initially contains a pollutant of mass M_i, that leaves at a rate R_i governed by a first-order equation,

$$\frac{dm_i}{dt} = \beta_i M_i \tag{9.3.1}$$

the mean residence time, τ_i, is given by the following equation:

$$\tau_i = \frac{m_i}{R_i} = \frac{1}{\beta_i} \tag{9.3.2}$$

In this equation, β_i is a constant for a given system and is called the net transfer coefficient for removal. It is the sum of removal transfer coefficients involving all major removal processes. If the reservoir is in secular equilibrium, or at a steady state, the pollutant production rate P_i equals the removal rate:

$$\frac{dm_i}{dt} = P_i - R_i \tag{9.3.3}$$

9.4. CHEMICAL AND PHOTOCHEMICAL REACTIONS IN THE ATMOSPHERE

The study of atmospheric chemical reactions is difficult. One of the primary obstacles encountered in studying atmospheric chemistry is that the chemist generally must deal with incredibly low concentrations, so that the detection and analysis of reaction products is quite difficult. Simulating high-altitude conditions in the laboratory can be extremely hard because of interferences, such as those from species given off from container walls under conditions of very low pressure. Many chemical reactions that require a third body to absorb excess energy occur very slowly in the upper atmosphere, where there is a sparse concentration of third bodies, but occur readily in a container whose walls effectively absorb energy. Container walls may serve as catalysts for some important reactions, or they may absorb important species and react chemically with the more reactive ones.

Photochemical Processes

The absorption of light by chemical species can bring about reactions, called **photochemical reactions**, which do not otherwise occur under the conditions (particularly the temperature) of the medium in the absence of light. Thus, photochemical reactions, even in the absence of a chemical catalyst, occur at temperatures much lower than those which otherwise would be required. Photochemical reactions, which are induced by intense solar radiation, play a very important role in determining the nature and ultimate fate of a chemical species in the atmosphere.

Nitrogen dioxide, NO_2, is one of the most photochemically active species found in a polluted atmosphere and is an essential participant in the smog-formation process. A species such as NO_2 may absorb light of energy hv, producing an **electronically excited molecule,**

$$NO_2 + h\nu \rightarrow NO_2* \tag{9.4.1}$$

designated in the reaction above by an asterisk, *. The photochemistry of nitrogen dioxide is discussed in greater detail in Chapters 11 and 13.

Electronically excited molecules are one of the three relatively reactive and unstable species that are encountered in the atmosphere and are strongly involved with atmospheric chemical processes. The other two species are atoms or molecular fragments with unshared electrons, called **free radicals,** and **ions** consisting of ionized atoms or molecular fragments.

Electronically excited molecules are produced when stable molecules absorb energetic electromagnetic radiation in the ultraviolet or visible regions of the spectrum. A molecule may possess several possible excited states, but generally ultraviolet or visible radiation is energetic enough to excite molecules only to several of the lowest energy levels. The nature of the excited state may be understood by considering the disposition of electrons in a molecule. Most molecules have an even number of electrons. The electrons occupy orbitals, with a maximum of two electrons with opposite spin occupying the same orbital. The absorption of light may promote one of these electrons to a vacant orbital of higher energy. In some cases the electron thus promoted retains a spin opposite to that of its former partner, giving rise to an **excited singlet state**. In other cases the spin of the promoted electron is reversed, such that it has the same spin as its former partner; this gives rise to an **excited triplet state**.

| Ground state | Singlet state | Triplet state |

These excited states are relatively energized compared to the ground state and are chemically reactive species. Their participation in atmospheric chemical reactions, such as those involved in smog formation, will be discussed later in detail.

In order for a photochemical reaction to occur, light must be absorbed by the reacting species. If the absorbed light is in the visible region of the sun's spectrum, the absorbing species is colored. Colored NO_2 is a common example of such a species in the atmosphere. Normally, the first step in a photochemical process is the activation of the molecule by the absorption of a single unit of photochemical energy characteristic of the frequency of the light called a **quantum** of light. The energy of one quantum is equal to the product hν, where h is Planck's constant, 6.62×10^{-27} erg sec, and ν is the frequency of the absorbed light in sec^{-2} (inversely proportional to its wavelength, λ).

The reactions that occur following absorption of a photon of light to produce an electronically excited species are largely determined by the way in which the excited species loses its excess energy. This may occur by one of the following processes:

- Loss of energy to another molecule or atom (M) by **physical quenching**, followed by dissipation of the energy as heat

$$O_2^* + M \rightarrow O_2 + M(\text{higher translational energy}) \tag{9.4.2}$$

- **Dissociation** of the excited molecule (the process responsible for the predominance of atomic oxygen in the upper atmosphere)

$$O_2^* \rightarrow O + O \tag{9.4.3}$$

- **Direct reaction** with another species

$$O_2^* + O_2 \rightarrow 2O_2 + O \tag{9.4.4}$$

- **Luminescence** consisting of loss of energy by the emission of electromagnetic radiation

$$NO_2^* \rightarrow NO_2 + h\nu \tag{9.4.5}$$

If the re-emission of light is almost instantaneous, luminescence is called **fluorescence** and if it is significantly delayed, the phenomenon is **phosphorescence**. **Chemiluminescence** is said to occur when the excited species is formed by a chemical process.

$$O_2^* + M \rightarrow O_2 + M(\text{higher energy}) \tag{9.4.6}$$

- **Intermolecular energy transfer** in which an excited species transfers energy to another species

$$O_2^* + Na \rightarrow O_2 + Na^* \tag{9.4.7}$$

A subsequent reaction by the second species is called a **photosensitized** reaction.

- **Intramolecular transfer** in which energy is transferred within a molecule

$$XY^* \rightarrow XY^\dagger(\text{where } ^\dagger \text{ denotes another excited state of the same molecule}) \tag{9.4.8}$$

- **Spontaneous isomerization** as in the conversion of *o*-nitrobenzaldehyde to *o*-nitrosobenzoic acid, a reaction used in chemical actinometers to measure exposure to electromagnetic radiation:

$$\text{(structure: benzene ring with } \overset{\overset{O}{\|}}{C}-H \text{ group and } NO_2 \text{ group)} + h\nu \longrightarrow \text{(benzene ring with } \overset{\overset{O}{\|}}{C}-OH \text{ group and } NO \text{ group)} \qquad (9.4.9) \qquad (9.4.9)$$

- **Photoionization** through loss of an electron

$$N_2^* \rightarrow N_2^+ + e^- \qquad\qquad (9.4.10)$$

Electromagnetic radiation absorbed in the infrared region is not sufficiently energetic to break chemical bonds, but does cause the receptor molecules to gain vibrational and rotational energy. The energy absorbed as infrared radiation ultimately is dissipated as heat and raises the temperature of the whole atmosphere. As noted in Section 9.3, the absorption of infrared radiation is very important in the Earth's acquiring heat from the sun and in the retention of energy radiated from the Earth's surface.

Ions and Radicals in the Atmosphere

One of the characteristics of the upper atmosphere which is difficult to duplicate under laboratory conditions is the presence of significant levels of electrons and positive ions. Because of the rarefied conditions, these ions may exist in the upper atmosphere for long periods before recombining to form neutral species.

At altitudes of approximately 50 km and up, ions are so prevalent that the region is called the **ionosphere**. The presence of the ionosphere has been known since about 1901, when it was discovered that radio waves could be transmitted over long distances, where the curvature of the Earth makes line-of-sight transmission impossible. These radio waves bounce off the ionosphere.

Ultraviolet light is the primary producer of ions in the ionosphere. In darkness, the positive ions slowly recombine with free electrons. The process is especially rapid in the lower regions of the ionosphere, where the concentration of species is relatively high. Thus, the lower limit of the ionosphere lifts at night and makes possible the transmission of radio waves over much greater distances.

The Earth's magnetic field has a strong influence upon the ions in the upper atmosphere. Probably the best-known manifestation of this phenomenon is found in the Van Allen belts, discovered in 1958. These regions consist of two belts of ionized particles which circle the Earth. If they are visualized as two doughnuts, then the axis of the Earth's magnetic field extends through the holes in the doughnuts. In the inner belt, the highly energetic ionizing radiation consists of protons. In the outer belt, it consists of electrons. A schematic diagram of the Van Allen belts is shown in Figure 9.5.

Although ions are produced in the upper atmosphere primarily by the action of energetic electromagnetic radiation, they may also be produced in the troposphere by the shearing of water droplets during precipitation. The shearing may be caused by the compression of descending masses of cold air or by strong winds over hot, dry land masses. The last phenomenon is known as the foehn, sharav (in the Near East), or Santa Ana (in southern California). These hot, dry winds cause severe discomfort. The ions produced by them consist of electrons and positively charged molecular species.

Given sufficient humidity, these ions quickly become surrounded by clusters of one to eight water molecules forming what are called small air ions that may last for up to several minutes. Small air ions undergo further reactions, including incorporation of trace gases from the atmosphere. Eventually, they are neutralized by combination with ions of opposite charge or with uncharged condensation particles.

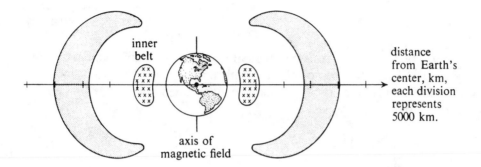

Figure 9.5. Cross section of the Van Allen belts encircling the Earth.

Free Radicals

In addition to forming ions by photoionization, energetic electromagnetic radiation in the atmosphere may produce atoms or groups of atoms with unpaired electrons called **free radicals**:

$$H_3C-\overset{\overset{O}{\|}}{C}-H + h\nu \;\longrightarrow\; H_3C\bullet \;+\; H\overset{\bullet}{C}O \qquad (9.4.11) \qquad (9.4.11)$$

Free radicals are involved with most significant atmospheric chemical phenomena and are of the utmost importance in the atmosphere. Because of their unpaired electrons and the strong pairing tendencies of electrons under most circumstances, free radicals are highly reactive. The upper atmosphere is so rarefied, however, that at very high altitudes radicals may have half-lives of several minutes, or even longer. Radicals can take part in chain reactions in which one of the products of each reaction is a radical. Eventually, through processes such as reaction with another radical, one of the radicals in a chain is destroyed and the chain ends:

$$H_3C\bullet + H_3C\bullet \rightarrow C_2H_6 \qquad\qquad (9.4.12)$$

This process is a **chain-terminating reaction**. Reactions involving free radicals are responsible for smog formation, discussed in Chapter 13.

Free radicals are quite reactive; therefore, they generally have short lifetimes. It is important to distinguish between high reactivity and instability. A totally isolated free radical or atom would be quite stable. Therefore, free radicals and single atoms from diatomic gases tend to persist under the rarefied conditions of very high altitudes because they can travel long distances before colliding with another reactive species. However, electronically excited species have a finite, generally very short, lifetime because they can lose energy through radiation without having to react with another species.

Hydroxyl and Hydroperoxyl Radicals in the Atmosphere

As illustrated in Figure 9.6, the hydroxyl radical, HO·, is the single most important reactive intermediate species in atmospheric chemical processes. It is formed by several mechanisms. At higher altitudes it is produced by photolysis of water:

$$H_2O + h\nu \rightarrow HO^· + H \tag{9.4.13}$$

In the presence of organic matter, hydroxyl radical is produced in abundant quantities as an intermediate in the formation of photochemical smog (see Chapter 13). To a certain extent in the atmosphere, and for laboratory experimentation, HO· is made by the photolysis of nitrous acid vapor:

$$HONO + h\nu \rightarrow HO^· + NO$$

In the relatively unpolluted troposphere, hydroxyl radical is produced as the result of the photolysis of ozone,

$$O_3 + h\nu(\lambda < 315\,nm) \rightarrow O^* + O_2 \tag{9.4.14}$$

followed by the reaction of a fraction of the excited oxygen atoms with water molecules:

$$O^* + H_2O \rightarrow 2HO^· \tag{9.4.15}$$

Involvement of the hydroxyl radical in chemical transformations of a number of trace species in the atmosphere is summarized in Figure 9.6, and some of the pathways illustrated are discussed in later chapters. Among the important atmospheric trace species that react with hydroxyl radical are carbon monoxide, sulfur dioxide, hydrogen sulfide, methane, and nitric oxide.

Hydroxyl radical is most frequently removed from the troposphere by reaction with methane or carbon monoxide:

$$CH_4 + HO^· \rightarrow H_3C^· + H_2O \tag{9.4.16}$$

$$CO + HO^· \rightarrow CO_2 + H \tag{9.4.17}$$

The highly reactive methyl radical, $H_3C^·$, reacts with O_2,

$$H_3C^· + O_2 \rightarrow H_3COO^· \tag{9.4.18}$$

to form **methylperoxyl radical**, $H_3COO^·$. (Further reactions of this species are discussed in Chapter 13.) The hydrogen atom produced in Reaction 9.4.17 reacts with O_2 to produce **hydroperoxyl radical**:

$$H + O_2 \rightarrow HOO^· \tag{9.4.19}$$

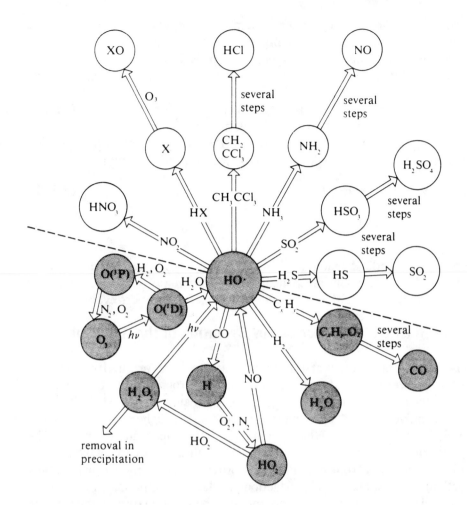

Figure 9.6. Control of trace gas concentrations by HO• radical in the troposphere. Processes below the dashed line are those largely involved in controlling the concentrations of HO• in the troposphere; those above the line control the concentrations of the associated reactants and products. Reservoirs of atmospheric species are shown in circles, reactions denoting conversion of one species to another are shown by arrows, and the reactants or photons needed to bring about a particular conversion are shown along the arrows. Hydrogen halides are denoted by HX and hydrocarbons by H_xY_y. (Source: D. D. Davis and W. L. Chameides, "Chemistry in the Troposphere," *Chemical and Engineering News*, October 4, 1982, pp. 39-52. Reprinted by permission of the American Chemical Society.)

The hydroperoxyl radical can undergo chain termination reactions, such as

$$HOO^\bullet + HO^\bullet \rightarrow H_2O + O_2 \tag{9.4.20}$$

$$HOO^\bullet + HOO^\bullet \rightarrow H_2O_2 + O_2 \tag{9.4.21}$$

or reactions that regenerate hydroxyl radical:

$$HOO^\bullet + NO \rightarrow NO_2 + HO^\bullet \tag{9.4.22}$$

$$HOO^\bullet + O_3 \rightarrow 2O_2 + HO^\bullet \tag{9.4.23}$$

The global concentration of hydroxyl radical, averaged diurnally and seasonally, is estimated to range from 2×10^5 to 1×10^6 radicals per cm^3 in the troposphere.

Because of the higher humidity and higher incident sunlight, which result in elevated O* levels, the concentration of HO• is higher in tropical regions. The southern hemisphere probably has about a 20% higher level of HO• than does the northern hemisphere because of greater production of anthropogenic HO•–consuming CO in the northern hemisphere.

The hydroperoxyl radical, HOO•, is an intermediate in some important chemical reactions. In addition to its production by the reactions discussed above, in polluted atmospheres, hydroperoxyl radical is made by the following two reactions, starting with photolytic dissociation of formaldehyde to produce a reactive formyl radical:

$$HCHO + hv \rightarrow H + H\overset{\bullet}{C}O \qquad (9.4.25)$$

$$H\overset{\bullet}{C}O + O_2 \rightarrow HOO^{\bullet} + CO \qquad (9.4.26)$$

The hydroperoxyl radical reacts more slowly with other species than does the hydroxyl radical. The kinetics and mechanisms of hydroperoxyl radical reactions are difficult to study because it is hard to retain these radicals free of hydroxyl radicals.

Chemical and Biochemical Processes in Evolution of the Atmosphere

It is now widely believed that the Earth's atmosphere originally was very different from its present state and that the changes were brought about by biological activity and accompanying chemical changes. Approximately 3.5 billion years ago, when the first primitive life molecules were formed, the atmosphere was chemically reducing, consisting primarily of methane, ammonia, water vapor, and hydrogen. The atmosphere was bombarded by intense, bond-breaking ultraviolet light, which, along with lightning and radiation from radionuclides, provided the energy to bring about chemical reactions that resulted in the production of relatively complicated molecules, including even amino acids and sugars. From the rich chemical mixture in the sea, life molecules evolved. Initially, these very primitive life forms derived their energy from fermentation of organic matter formed by chemical and photochemical processes, but eventually they gained the capability to produce organic matter, "$\{CH_2O\}$," by photosynthesis,

$$CO_2 + H_2O + hv \rightarrow \{CH_2O\} + O_2(g) \qquad (9.4.27)$$

and the stage was set for the massive biochemical transformation that resulted in the production of almost all the atmosphere's oxygen.

The oxygen initially produced by photosynthesis was probably quite toxic to primitive life forms. However, much of this oxygen was converted to iron oxides by reaction with soluble iron(II):

$$4Fe^{2+} + O_2 + 4H_2O \rightarrow 2Fe_2O_3 + 8H^+ \qquad (9.4.28)$$

This resulted in the formation of enormous deposits of iron oxides, the existence of which provides major evidence for the liberation of free oxygen in the primitive atmosphere.

Eventually, enzyme systems developed that enabled organisms to mediate the reaction of waste-product oxygen with oxidizable organic matter in the sea. Later, this mode of waste-product disposal was utilized by organisms to produce energy by respiration, which is now the mechanism by which nonphotosynthetic organisms obtain energy.

In time, O_2 accumulated in the atmosphere, providing an abundant source of oxygen for respiration. It had an additional benefit in that it enabled the formation of an ozone shield (see Section 9.5). The ozone shield absorbs bond-rupturing ultraviolet light. With the ozone shield protecting tissue from destruction by high-energy ultraviolet radiation, the Earth became a much more hospitable environment for life, and life forms were enabled to move from the sea to land.

9.5. REACTIONS OF ATMOSPHERIC OXYGEN

Some of the primary features of the exchange of oxygen among the atmosphere, lithosphere, hydrosphere, and biosphere are summarized in Figure 9.7. The oxygen cycle is critically important in atmospheric chemistry, geochemical transformations, and life processes.

Oxygen in the troposphere plays a strong role in processes that occur on the Earth's surface. Atmospheric oxygen takes part in energy-producing reactions, such as the burning of fossil fuels:

$$CH_4(\text{in natural gas}) + 2O_2 \rightarrow CO_2 + 2H_2O \tag{9.5.1}$$

Atmospheric oxygen is utilized by aerobic organisms in the degradation of organic material. Some oxidative weathering processes consume oxygen, such as

$$4FeO + O_2 \rightarrow 2Fe_2O_3 \tag{9.5.2}$$

Oxygen is returned to the atmosphere through plant photosynthesis:

$$CO_2 + H_2O + h\nu \rightarrow \{CH_2O\} + O_2 \tag{9.5.3}$$

All molecular oxygen now in the atmosphere is thought to have originated through the action of photosynthetic organisms, which shows the importance of photosynthesis in the oxygen balance of the atmosphere. It can be shown that most of the carbon fixed by these photosynthetic processes is dispersed in mineral formations as humic material (Section 3.11); only a very small fraction is deposited in fossil fuel beds. Therefore, although combustion of fossil fuels consumes large amounts of O_2, there is no danger of running out of atmospheric oxygen.

Molecular oxygen is somewhat unusual in that its ground state is a triplet state with two unpaired electrons, designated here as $^3O_2$, which can be excited to singlet molecular oxygen, designated here as $^1O_2$. The latter can be produced by several processes, including direct photochemical excitation, transfer of energy from other electronically excited molecules, ozone photolysis, and high-energy oxygen-producing reactions.[1]

Because of the extremely rarefied atmosphere and the effects of ionizing radiation, elemental oxygen in the upper atmosphere exists to a large extent in forms other than diatomic O_2. In addition to O_2, the upper atmosphere contains oxygen atoms, O; excited oxygen molecules, O_2^*, and ozone, O_3.

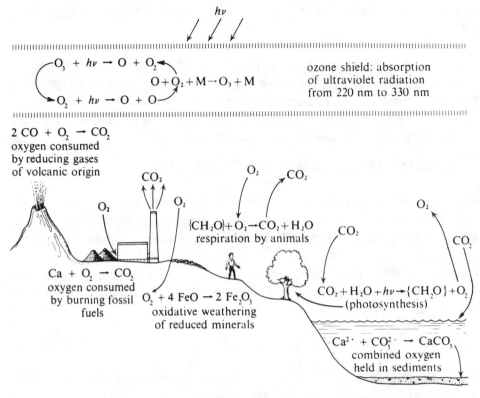

Figure 9.7. Oxygen exchange among the atmosphere, lithosphere, hydrosphere, and biosphere.

Atomic oxygen, O, is stable primarily in the thermosphere, where the atmosphere is so rarefied that the three-body collisions necessary for the chemical reaction of atomic oxygen seldom occur (the third body in this kind of three-body reaction absorbs energy to stabilize the products). Atomic oxygen is produced by a photochemical reaction:

$$O_2 + h\nu \rightarrow O + O \tag{9.5.4}$$

The oxygen-oxygen bond is strong (120 kcal/mole) and ultraviolet radiation in the wavelength regions 135-176 nm and 240-260 nm is most effective in causing dissociation of molecular oxygen. Because of photochemical dissociation, O_2 is virtually nonexistent at very high altitudes and less than 10% of the oxygen in the atmosphere at altitudes exceeding approximately 400 km is present in the molecular form. At altitudes exceeding about 80 km, the average molecular weight of air is lower than the 28.97 g/mole observed at sea level because of the high concentration of atomic oxygen. The resulting division of the atmosphere into a lower section with a uniform molecular weight and a higher region with a nonuniform molecular weight is the basis for classifying these two atmospheric regions as the **homosphere** and **heterosphere**, respectively.

Oxygen atoms in the atmosphere can exist in the ground state (O) and in excited states (O*). These are produced by the photolysis of ozone, which has a relatively weak bond energy of 26 kcal/mole, at wavelengths below 308 nm,

$$O_3 + h\nu(\lambda < 308 \, nm) \rightarrow O^* + O_2 \tag{9.5.5}$$

or by highly energetic chemical reactions such as

$$O + O + O \rightarrow O_2 + O^* \tag{9.5.6}$$

Excited atomic oxygen emits visible light at wavelengths of 636 nm, 630 nm, and 558 nm. This emitted light is partially responsible for **airglow**, a very faint electromagnetic radiation continuously emitted by the earth's atmosphere. Although its visible component is extremely weak, airglow is quite intense in the infrared region of the spectrum.

Oxygen ion, O^+, which may be produced by ultraviolet radiation acting upon oxygen atoms,

$$O + h\nu \rightarrow O^+ + e^- \tag{9.5.7}$$

is the predominant positive ion in some regions of the ionosphere. It may react with molecular oxygen or nitrogen,

$$O^+ + O_2 \rightarrow O_2^+ + O \tag{9.5.8}$$

$$O^+ + N_2 \rightarrow NO^+ + O \tag{9.5.9}$$

to form other positive ions.

In intermediate regions of the ionosphere, O_2^+ is produced by absorption of ultraviolet radiation at wavelengths of 17-103 nm. This diatomic oxygen ion can also be produced by the photochemical reaction of low-energy X-rays.

$$O_2 + h\nu \rightarrow O_2^+ + e^- \tag{9.5.10}$$

and by the following reaction:

$$N_2^+ + O_2 \rightarrow N_2 + O_2^+ \tag{9.5.11}$$

Ozone, O_3, has an essential protective function because it absorbs harmful ultraviolet radiation in the stratosphere and serves as a radiation shield, protecting living beings on the Earth from the effects of excessive amounts of such radiation. It is produced by a photochemical reaction,

$$O_2 + h\nu \rightarrow O + O \tag{9.5.12}$$

(where the wavelength of the exciting radiation must be less than 242.4 nm), followed by a three-body reaction,

$$O + O_2 + M \rightarrow O_3 + M(\text{increased energy}) \tag{9.5.13}$$

in which M is another species, such as a molecule of N_2 or O_2, which absorbs the excess energy given off by the reaction and enables the ozone molecule to stay together. The region of maximum ozone concentration is found within the range of 25-30 km high in the stratosphere where it may reach 10 ppm.

Ozone absorbs ultraviolet light very strongly in the region 220-330 nm. If this light were not absorbed by ozone, severe damage would result to exposed forms of life on the Earth. Absorption of electromagnetic radiation by ozone converts the radiation's energy to heat and is responsible for the temperature maximum encountered at the boundary between the stratosphere and the mesosphere at an altitude of approximately 50 km. The reason that the temperature maximum occurs at a higher altitude than that of the maximum ozone concentration arises from the fact that ozone is such an effective absorber of ultraviolet light, so that most of this radiation is absorbed in the upper stratosphere, where it generates heat, and only a small fraction reaches the lower altitudes, which remain relatively cool.

The overall reaction,

$$2O_3 \rightarrow 3O_2 \tag{9.5.14}$$

is favored thermodynamically so that ozone is inherently unstable. Its decomposition in the stratosphere is catalyzed by a number of natural and pollutant trace constituents, including NO, NO_2, H, $HO\cdot$, $HOO\cdot$, ClO, Cl, Br, and BrO. Ozone decomposition also occurs on solid surfaces, such as metal oxides and salts produced by rocket exhausts.

Although the mechanisms and rates for the photochemical production of ozone in the stratosphere are reasonably well known, the natural pathways for ozone removal are less well understood. The earliest known reaction for ozone removal is the reaction of ozone with atomic oxygen,

$$O_3 + O \rightarrow O_2 + O_2 \tag{9.5.15}$$

which obtains some of the atomic oxygen required from the following ozone-destroying reaction:

$$O_3 + h\nu \rightarrow O_2 + O \tag{9.5.16}$$

These reactions can account for only about 20% of the ozone removal. Another approximately 10% removal is accounted for by the reactive hydroxyl radical, $HO\cdot$, produced by photochemical reactions of H_2, O_2, and H_2O in the stratosphere. A plausible reaction sequence is

$$O_3 + HO\cdot \rightarrow O_2 + HOO\cdot \tag{9.5.17}$$

$$HOO\cdot + O \rightarrow HO\cdot + O_2 \tag{9.5.18}$$

Most stratospheric ozone is probably removed by the action of nitric oxide, which reacts as follows:

$$O_3 + NO \rightarrow NO_2 + O_2 \tag{9.5.19}$$

$$NO_2 + O \rightarrow NO + O_2 \tag{9.5.20}$$

Some NO in the stratosphere is produced by high-flying supersonic aircraft. At altitudes of approximately 30 km in the stratosphere, NO is largely generated by the reaction of photochemically-produced excited oxygen atoms with N_2O:

$$N_2O + O \rightarrow 2NO \tag{9.5.21}$$

Recall that N_2O is a natural component of the atmosphere and is a major product of the denitrification process by which fixed nitrogen is returned to the atmosphere in gaseous form. This is shown in the nitrogen cycle, Figure 6.6.

At altitudes exceeding 30 km, ions and ionizing radiation may play a significant role in the production of ozone-destroying NO. Secondary electrons with an energy range of 10-100 electron volts are produced in the upper regions of the atmosphere by an intense flux of charged particles and cosmic rays. These electrons can reach only the upper regions of the stratosphere before being destroyed, but they produce more highly penetrating bremsstrahlung X-rays which can reach down to 30-km altitude. Both the energetic electrons and X-rays can bring about the dissociation of stratospheric N_2,

$$N + h\nu \rightarrow N + N \qquad (9.5.22)$$

to produce nitrogen atoms which react with O_2 to yield NO:

$$O_2 + N \rightarrow NO + O \qquad (9.5.23)$$

The production of secondary electrons needed to initiate this reaction sequence depends upon solar radiation, such as energetic solar protons. On the basis of this mechanism, the ozone layer should decrease during periods of maximum solar activity.

Ozone is an undesirable pollutant in the troposphere. It is toxic, and a mild over-dose causes labored breathing, a feeling of chest pressure, cough, and irritated eyes. In addition to its toxicological effects, which are discussed in Chapter 20, ozone damages materials, such as rubber.

9.6. REACTIONS OF ATMOSPHERIC NITROGEN

The 78% by volume of nitrogen contained in the atmosphere constitutes an inexhaustible reservoir of that essential element. The nitrogen cycle and nitrogen fixation by microorganisms were discussed in Chapter 6. A small amount of nitrogen is thought to be fixed in the atmosphere by lightning, and some is also fixed by combustion processes, as in the internal combustion engine.

Before the use of synthetic fertilizers reached its current high levels, chemists were concerned that denitrification processes in the soil would lead to nitrogen depletion on the Earth. Now, with millions of tons of synthetically fixed nitrogen being added to the soil each year, major concern has shifted to possible excess accumulation of nitrogen in soil, fresh water, and the oceans.

Unlike oxygen, which is almost completely dissociated to the monatomic form in higher regions of the thermosphere, molecular nitrogen is not readily dissociated by ultraviolet radiation. However, at altitudes exceeding approximately 100 km, atomic nitrogen is produced by photochemical reactions:

$$N_2 + h\nu \rightarrow N + N \qquad (9.6.1)$$

Other reactions which may produce monatomic nitrogen are:

$$N_2^+ + O \rightarrow NO^+ + N \qquad (9.6.2)$$

$$NO^+ + e^- \rightarrow N + O \qquad (9.6.3)$$

$$O^+ + N_2 \rightarrow NO^+ + N \qquad (9.6.4)$$

Most stratospheric ozone is probably removed by the action of nitric oxide, which reacts as follows:

$$O_3 + NO \rightarrow NO_2 + O_2 \qquad (9.6.5)$$

$$NO_2 + O \rightarrow NO + O_2 \qquad (9.6.6)$$

In the so-called E region of the ionosphere, NO^+ is one of the predominant ions. A plausible sequence of reactions by which NO^+ is formed is the following:

$$N_2 + h\nu \rightarrow N_2^+ + e^- \qquad (9.6.7)$$

$$N_2^+ + O \rightarrow NO^+ + N \qquad (9.6.8)$$

In the lowest region of the ionosphere, the D region, which extends from approximately 50 km in altitude to approximately 85 km, NO^+ is produced directly by ionizing radiation:

$$NO + h\nu \rightarrow NO^+ + e^- \qquad (9.6.9)$$

In the lower part of the D region, the ionic species N_2^+ is formed through the action of galactic cosmic rays:

$$N_2 + h\nu \rightarrow N_2^+ + e^- \qquad (9.6.10)$$

Pollutant oxides of nitrogen, particularly NO_2, are key species involved in air pollution and the formation of photochemical smog. For example, NO_2 is readily dissociated photochemically to NO and reactive atomic oxygen:

$$NO_2 + h\nu \rightarrow NO + O \qquad (9.6.11)$$

This reaction is the most important primary photochemical process involved in smog formation. The roles played by nitrogen oxides in smog formation and other forms of air pollution are discussed in Chapters 11–13.

9.7. ATMOSPHERIC CARBON DIOXIDE

Although only about 0.035% (350 ppm) of air consists of carbon dioxide, it is the atmospheric "nonpollutant" species of most concern. As mentioned in Section 9.3, carbon dioxide, along with water vapor, is primarily responsible for the absorption of infrared energy re-emitted by the Earth such that some of this energy is reradiated back to the Earth's surface. Current evidence suggests that changes in the atmospheric carbon dioxide level will substantially alter the Earth's climate through the greenhouse effect.

Valid measurements of overall atmospheric CO_2 can only be taken in areas remote from industrial activity. Such areas include Antarctica and the top of Mauna

Loa Mountain in Hawaii. Measurements of carbon dioxide levels in these locations over the last 30 years suggest an annual increase in CO_2 of about 1 ppm per year.

The most obvious factor contributing to increased atmospheric carbon dioxide is consumption of carbon-containing fossil fuels. In addition, release of CO_2 from the biodegradation of biomass and uptake by photosynthesis are important factors determining overall CO_2 levels in the atmosphere. The role of photosynthesis is illustrated in Figure 9.8, which shows a seasonal cycle in carbon dioxide levels in the northern hemisphere. Maximum values occur in April and minimum values in late September or October. These oscillations are due to the "photosynthetic pulse," influenced most strongly by forests in middle latitudes. Forests have a much greater influence than other vegetation because forests carry out more photosynthesis. Furthermore, forests store enough fixed, but readily oxidizable carbon in the form of wood and humus to have a marked influence on atmospheric CO_2 content. Thus, during the summer months, forests carry out enough photosynthesis to reduce the atmospheric carbon dioxide content markedly. During the winter, metabolism of biota, such as bacterial decay of humus, releases a significant amount of CO_2. Therefore, the current worldwide trend toward destruction of forests and conversion of forest lands to agricultural uses will contribute substantially to a greater overall increase in atmospheric CO_2 levels.

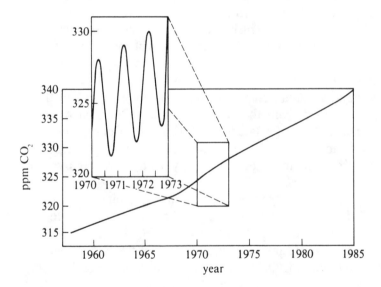

Figure 9.8. Increases in atmospheric CO_2 levels in recent years. The inset illustrates seasonal variations in the northern hemisphere.

With current trends, it is likely that global CO_2 levels will double by the middle of the next century, which may well raise the Earth's mean surface temperature by 1.5–4.5°C. Such a change might have more potential to cause massive irreversible environmental changes than any other disaster short of global nuclear war.

Chemically and photochemically, carbon dioxide is a comparatively insignificant species because of its relatively low concentrations and low photochemical reactivity. However, calculations based on known photochemical reactions, carbon dioxide levels, and ultraviolet radiation intensity indicate that photodissociation of CO_2 by solar ultraviolet radiation should occur in the upper atmosphere:

$$CO_2 + h\nu \rightarrow CO + O \tag{9.7.1}$$

This reaction could be a major source of CO at higher altitudes. The infrared radiation absorbed by carbon dioxide is not energetic enough to cause photochemical reactions to occur.

9.8. ATMOSPHERIC WATER

The water vapor content of the troposphere is normally within a range of 1–3% by volume with a global average of about 1%. However air can contain as little as 0.1% or as much as 5% water. The percentage of water in the atmosphere decreases rapidly with increasing altitude. Water circulates through the atmosphere in the hydrologic cycle as shown in Figure 2.1.

Water vapor absorbs infrared radiation even more strongly than does carbon dioxide, thus greatly influencing the Earth's heat balance. Clouds formed from water vapor reflect light from the sun and have a temperature-lowering effect. On the other hand, water vapor in the atmosphere acts as a kind of "blanket" at night, retaining heat from the Earth's surface by absorption of infrared radiation.

Gaseous water in the upper atmosphere is involved in the formation of hydroxyl and hydroperoxyl radicals as discussed in Section 9.4. Condensed water vapor in the form of very small droplets is of considerable concern in atmospheric chemistry. The harmful effects of some air pollutants—for instance, the corrosion of metals by acid-forming gases—requires the presence of water which may come from the atmosphere. Atmospheric water vapor has an important influence upon pollution-induced fog formation under some circumstances. Water vapor interacting with pollutant particulate matter in the atmosphere may reduce visibility to undesirable levels through the formation of aerosol particles (see Section 9.9).

When ice particles in the atmosphere change to liquid droplets or when these droplets evaporate, heat is absorbed from the surrounding air. Reversal of these processes results in heat release to the air (as latent heat). This may occur many miles from the place where heat was absorbed and is a major mode of energy transport in the atmosphere. It is the predominant type of energy transition involved in thunderstorms, hurricanes, and tornadoes.

On a global basis, rivers drain only about one-third of the precipitation that falls on the Earth's continents. This means that two-thirds of the precipitation is lost as combined evaporation and transpiration (**evapotranspiration**, Figure 2.1). During the summer, evapotranspiration may exceed precipitation because of the large quantities of water stored in the root zone of the soil. In some cases, evapotranspiration furnishes atmospheric water vapor necessary for cloud formation and precipitation. It is probable, therefore, that large-scale deforestation, soil damage (such as by plowing up grasslands in semi-arid areas), and irrigation could have an effect on regional climate and rainfall. As noted in Section 9.2, the cold tropopause serves as a barrier to the movement of water into the stratosphere. The main source of water in the stratosphere is the photochemical oxidation of methane:

$$CH_4 + 2O_2 + h\nu \xrightarrow[\text{(several steps)}]{} CO_2 + H_2O \qquad (9.8.1)$$

The water thus produced serves as a source of stratospheric hydroxyl radical as shown by the following reaction:

$$H_2O + h\nu \rightarrow HO\bullet + H \qquad (9.8.2)$$

9.9. ATMOSPHERIC PARTICLES

Particles are common significant components of the atmosphere, particularly the troposphere. Colloidal-sized particles in the atmosphere are called **aerosols**. Most aerosols from natural sources have a diameter of less than 0.1 μm. These particles originate in nature from sea sprays, smokes, dusts, and the evaporation of organic materials from vegetation. Other typical particles of natural origin in the atmosphere are bacteria, fog, pollen grains, and volcanic ash.

Many important atmospheric phenomena involve aerosol particles, including electrification phenomena, cloud formation, and fog formation. Particles help determine the heat balance of the Earth's atmosphere by reflecting light. Probably the most important function of particles in the atmosphere is their action as nuclei for the formation of ice crystals and water droplets. Current efforts at rain-making are centered around the addition of condensing particles to atmospheres supersaturated with water vapor. Dry ice was used in early attempts; now silver iodide, which forms huge numbers of very small particles, is used.

Particles are involved in many chemical reactions in the atmosphere. Neutralization reactions, which occur most readily in solution, may take place in water droplets suspended in the atmosphere. Small particles of metal oxides and carbon have a catalytic effect on oxidation reactions. Particles may also participate in oxidation reactions induced by light.

LITERATURE CITED

1. Finlayson-Pitts, Barbara J., and James N. Pitts, *Atmospheric Chemistry*, John Wiley and Sons, Inc., New York, 1986, p. 478.

SUPPLEMENTAL REFERENCES

Seinfeld, John H., *Atmospheric Chemistry and Physics of Air Pollution*, John Wiley and Sons, Inc., New York, 1986.

Makhijani, Arjun, Annie Makhijani, and Amanda Bickel, *Saving Our Skins: Technical Potential and Policies for the Elimination of Ozone-Depleting Chlorine Compounds*, Environmental Policy Institute and the Institute for Energy & Environmental Research, Washington, DC, 1988.

Watson, Ann Y, Richard R. Bates, and Donald Kennedy, Eds., *Air Pollution, the Automobile, and Public Health*, National Academy Press, Washington, DC, 1988.

Warneck, Peter, *Chemistry of the Natural Atmosphere*, Academic Press, San Diego, CA, 1988.

Regens, James L., and Robert W. Rycroft, *The Acid Rain Controversy*, University of Pittsburgh Press, Pittsburgh, PA, 1988.

Leifer, Asa, *The Kinetics of Environmental Aquatic Photochemistry: Theory and Practice*, American Chemical Society, Washington, DC, 1988.

Ferraudi, G. J., *Elements of Inorganic Photochemistry*, John Wiley and Sons, New York, NY, 1988.

White, James C., Ed., *Acid Rain: The Relationship Between Sources and Receptors*, Elsevier, New York, NY, 1988.

Park, Chris C., *Acid Rain: Rhetoric and Reality*, Methuen/Chapman & Hall, New York, 1987.

Singer, S. Fred, Ed., *Global Climate Change: Human and Natural Influences*, Paragon House, New York, 1989.

Lioy, Paul J., and Joan M. Daisey, Eds., *Toxic Air Pollution: A Comprehensive Study of Non-Criteria Air Pollutants*, Lewis Publishers, Chelsea, MI, 1987.

Lodge, James P. Jr., Ed., *Methods of Air Sampling and Analysis*, 3rd ed., Lewis Publishers, Inc., Chelsea, MI, 1989.

Licht, Willliam, *Air Pollution Control Engineering*, 2nd Ed., Marcel Dekker, New York, NY, 1988.

Nazaroff, William W., and Anthony V. Nero, Jr., Eds. *Radon and Its Decay Products in Indoor Air*, John Wiley and Sons, New York, 1988.

Schneider, T., Ed., *Atmospheric Ozone Research and its Policy Implications*, Elsevier Science Publishing Co. New York, 1989.

Moomaw, William R., and Irfint M. Mintzer, *Strategies for Limiting Global Climate Change*, World Resources Institute Publications, Washington, DC, 1989.

Brown, Lester R., *State of the World 1990*, Worldwatch Institute, Washington, DC, 1989.

Botkin, Daniel B., Ed., *Changing the Global Environment: Perspectives on Human Involvement*, Academic Press, San Diego, CA, 1989.

Jorgensen, S. E., and I. Johnsen, *Principles of Environmental Science and Technology*, second ed., Elsevier Science Publishers, Amsterdam, The Netherlands, 1989.

Miller, E. Willard, and Ruby M. Miller, *Environmental Hazards: Air Pollution*, ABC-CLIO, Santa Barbara, CA, 1989.

Johnson, Russell W., and Glen E. Gordon, Eds., *The Chemistry of Acid Rain: Sources and Atmospheric Processes*, American Chemical Society, Washington, DC, 1987.

Glantz, Ed., Michael H., *Societal Responses to Regional Climatic Change: Forecasting by Analogy*, Westview Press, Boulder, CO, 1988.

Schneider, Stephen H., *Global Warming: Are We Entering The Greenhouse Century?*, Sierra Club Books, San Francisco, CA, 1989.

Vincent, James H., *Aerosol Samplings: Science and Practice*, John Wiley and Sons, New York, NY, 1989.

Knap, Anthony H., and Mary-Scott Kaiser, Eds.,*The Long-Range Atmospheric Transport of Natural and Contaminant Substances*, Kluwer Academic Publishers, Norwell, MA, 1989.

Bryce-Smith, D., and A. Gilbert, *Photochemistry*, Vol. 20, Royal Society of Chemistry, Letchworth, England, 1989.

Ware, George W., Ed., *Reviews of Environmental Contamination and Toxicology*, Vol. 103, Springer-Verlag, New York, 1988.

Watson, Ann Y., Richard R. Bates, and Donald Kennedy, Eds., *Air Pollution, the Automobile, and Public Health*, National Academy Press, Washington, DC, 1988.

Andersen, N. R., and A. Malahoff, *The Fate of Fossil Fuel CO_2 in the Oceans*, Plenum Publishing Corp., New York, 1977.

Bach, W., J. Pankrath, and W. Kellogg, Eds., *Man's Impact on Climate*, Elsevier, New York, 1979.

Barry, R. J., and R. J. Chorley, *Atmosphere, Weather, and Climate*, 3rd ed., John Wiley and Sons, Inc., New York, 1979.

Budyko, M. I., *The Earth's Climate*, Academic Press, New York, 1982.

Gaskell, T. F., and M. Morris, *World Climate*, Thames and Hudson, Inc., New York, 1980.

Hobbs, J., *Applied Climatology*, Westview Press, Inc., Boulder, Colorado, 1980.

McEwan, M. J., and L. F. Phillips, *Chemistry of the Atmosphere*, John Wiley and Sons, Inc., New York, 1975.

Slater, L. E., and S. K. Levin, Eds., *Climate's Impact on Food Supplies*, Westview Press, Inc., Boulder, Colorado, 1981.

Smith, C. D., and M. Parry, Eds., *Consequences of Climatic Change*, University of Nottingham, Department of Geography, Nottingham, England, 1981.

Walker, J. C. G., *Evolution of the Atmosphere*, Macmillan, Inc., New York, New York, 1977.

Wigley, T. M. L., M. J. Ingram, and G. Farmer, *Climate and History: Studies in Past Climates and their Impact on Man*, Cambridge University Press, New York, 1982.

QUESTIONS AND PROBLEMS

1. What phenomenon is responsible for the temperature maximum at the boundary of the stratosphere and the mesosphere?

2. What function does a third body serve in an atmospheric chemical reaction?

3. Why does the lower boundary of the ionosphere lift at night.

4. Why might it be expected that the reaction of a free radical with NO_2 is a chain-terminating reaction (consider the total number of electrons in NO_2.)

5. The average atmospheric pressure at sea level is 1.012 dynes/m^2. The value of g (acceleration of gravity) at sea level is 980 cm/sec^2. What is the mass in kg of the column of air having a cross-sectional area of 1.00 cm^2 at the Earth's surface and extending to the limits of the atmosphere? (Recall that the dyne is a unit of force and force = mass x acceleration of gravity.)

6. Suppose that 22.4 liters of air at STP is used to burn 1.50 g of carbon to form CO_2, and that the gaseous product is adjusted to STP. What is the volume and the average molecular mass of the resulting mixture?

7. If the pressure is 0.01 atm at an altitude of 38 km and 0.001 at 57 km, what is it at 19 km (ignoring temperature variations)?

8. Measured in μm, what are the lower wavelength limits of solar radiation reaching the Earth; the wavelength at which maximum solar radiation reaches the earth; and the wavelength at which maximum energy is radiated back into space?

9. Of the species O, HO*•, NO_2*, H_3C•, and N$^+$, which could most readily revert to a nonreactive, "normal" species in total isolation?

10. Of the gases neon, sulfur dioxide, helium, oxygen, and nitrogen, which shows the most variation in its atmospheric concentration?

11. A 12.0-liter sample of air at 25°C and 1.00 atm pressure was collected and dried. After drying, the volume of the sample was exactly 11.50 L. What was the percentage *by weight* of water in the original air sample?

12. The sunlight incident upon a 1 square meter area perpendicular to the line of transmission of the solar flux just above the Earth's atmosphere provides energy at a rate most closely equivalent to: (a) that required to power a pocket calculator, (b) that required to provide a moderate level of lighting for a 40-person capacity classroom illuminated with fluorescent lights, (c) that required to propel a 2500 pound automobile at 55 mph, (d) that required to power a 100-watt incandescent light bulb, (e) that required to heat a 40-person classroom to 70°F when the outside temperature is -10°F.

13. At an altitude of 50 km, the average atmospheric temperature is essentially 0°C. What is the average number of air molecules per cubic centimeter of air at this altitude?

14. What two types of condensation nuclei originate with bursting sea-foam bubbles?

15. State two factors that make the stratosphere particularly important in terms of acting as a region where atmospheric trace contaminants are converted to other, chemically less reactive, forms.

16. What two chemical species are most generally responsible for the removal of hydroxyl radical from the unpolluted troposphere.

17. What is the distinction between the symbols * and · in discussing chemically active species in the atmosphere?

18. What is the distinction between chemiluminescence and luminescence caused when light is absorbed by a molecule or atom?

Particles in the Atmosphere

10.1. PARTICLES IN THE ATMOSPHERE

Particles in the atmosphere, which range in size from about one-half millimeter (the size of sand or drizzle) down to molecular dimensions, are made up of an amazing variety of materials and discrete objects that may consist of either solids or liquid droplets. **Particulates** is a term that has come to stand for particles in the atmosphere, although *particulate matter* or simply *particles*, is preferred usage. Particulate matter makes up the most visible and obvious form of air pollution. Atmospheric **aerosols** are solid or liquid particles smaller than 100 μm in diameter. Pollutant particles in the 0.001 to 10 μm range are commonly suspended in the air near sources of pollution, such as the urban atmosphere, industrial plants, highways, and power plants. Some of the terms commonly used to describe atmospheric particles are summarized in Table 10.1.

Table 10.1. Important Terms Describing Atmospheric Particles

Term	Meaning
Aerosol	Colloidal-sized atmospheric particle
Condensation aerosol	Formed by condensation of vapors or reactions of gases
Dispersion aerosol	Formed by grinding of solids, atomization of liquids, or dispersion of dusts
Fog	Term denoting high level of water droplets
Haze	Denotes decreased visibility due to the presence of particles
Mists	Liquid particles
Smoke	Particles formed by incomplete combustion of fuel

Very small, solid particles include carbon black, silver iodide, combustion nuclei, and sea-salt nuclei (see Figure 10.10). Larger particles include cement dust, wind-blown soil dust, foundry dust, and pulverized coal. Liquid particulate matter, **mist**, includes raindrops, fog, and sulfuric acid mist. Some particles are of biological origin, such as viruses, bacteria, bacterial spores, fungal spores, and pollen. Particulate matter may be organic or inorganic; both types are very important atmospheric contaminants.

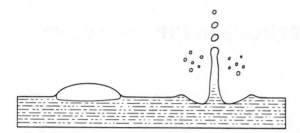

Figure 10.1. Bursting bubbles in seawater form small liquid aerosol particles. Evaporation of water from aerosol particles results in the formation of small solid particles of sea-salt nuclei.

As discussed later in this chapter, particulate matter originates from a wide variety of sources and processes, ranging from simple grinding of bulk matter to com-plicated chemical or biochemical syntheses. The effects of particulate matter are also widely varied. Possible effects on climate are discussed in Chapter 11. Either by itself, or in combination with gaseous pollutants, particulate matter may be detrimental to human health. Atmospheric particles may damage materials, reduce visibility, and cause undesirable esthetic effects.

For the most part, aerosols consist of carbonaceous material, metal oxides and glasses, dissolved ionic species (electrolytes), and ionic solids.[1] The predominant constituents are carbonaceous material, water, sulfate, nitrate, ammonium nitrogen, and silicon. The compossition of aerosol particles varies significantly with size. The very small particles tend to be acidic and often originate from gases, such as from the conversion of SO_2 to H_2SO_4. Larger particles tend to consist of materials generated mechanically, such as by the grinding of limestone, and have a greater tendency to be basic.

10.2. PHYSICAL BEHAVIOR OF PARTICLES IN THE ATMOSPHERE

As shown in Figure 10.2, atmospheric particles undego a number of processes in the atmosphere. Small colloidal particles are subject to *diffusion processes*. Smaller particles *coagulate* together to form larger particles. *Sedimentation* and *scavenging* by raindrops and other forms of precipitation are the major mechanisms for particle removal from the atmosphere. Particles also react with atmospheric gases.

Particle size usually expresses the diameter of a particle, though sometimes it is used to denote the radius. The rate at which a particle settles is a function of particle diameter and density. The settling rate is important in determining the effect of the particle in the atmosphere. For spherical particles greater than approximately 1 μm in diameter, Stokes' law applies,

$$v = \frac{gd^2(\rho_1 - \rho_2)}{18n} \tag{10.2.1}$$

where v is the settling velocity in cm/sec, g is the acceleration of gravity in cm/sec^2, ρ_1 is the density of the particle in g/cm^3, ρ_2 is the density of air in g/cm^3, and n is the viscosity of air in poise. Stokes' law can also be used to express the effective diameter of an irregular nonspherical particle. These are called **Stokes diameters** (aerodynamic diameters) and are normally the ones given when particle diameters are expressed. Furthermore, since the density of a particle is often not known, an arbitrary density of 1 g/cm^3 is conventionally assigned to ρ_1; when this is done, the diameter calculated from Equation 10.2.1 is called the **reduced sedimentation diameter**.

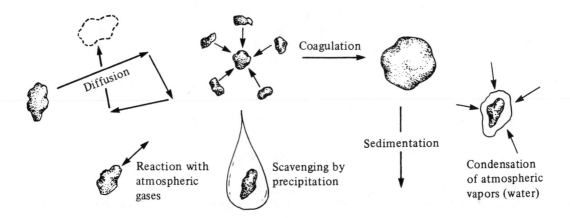

Figure 10.2. Processes that particles undergo in the atmosphere.

Size and Settling of Atmospheric Particles

Most kinds of aerosol particles have unknown diameters and densities and occur over a range of sizes. For such particles the term **mass median diameter (MMD)** may be used to describe aerodynamically equivalent spheres having an assigned density of 1 g/cm^3 at a 50% mass collection efficiency, as determined in samplers calibrated with spherical aerosol particles having a known, uniform size. (Polystyrene latex commonly is used as a material for the preparation of such standard aerosols.) The determination of MMD is accomplished by plotting the log of particle size as a function of the percentage of particles smaller than the given size on a probability scale. Two such plots are shown in Figure 10.3. It is seen from the plot that particles of aerosol X have a mass median diameter of 2.0 µm (ordinate corresponding to 50% on the abscissa). In the case of aerosol Y, linear extrapolation to sizes below the lower measurable size limit of about 0.7 µm gives an estimated value of 0.5 µm for the MMD.

The settling characteristics of particles smaller than about 1 µm in diameter deviate from Stokes' Law because the settling particles "slip between" air molecules. Extremely small particles are subject to **Brownian motion** resulting from random movement due to collisions with air molecules and do not obey Stokes' Law. Deviations are also observed for particles above 10 µm in diameter because they settle rapidly and generate turbulence as they fall.

10.3. PHYSICAL PROCESSES FOR PARTICLE FORMATION

Dispersion aerosols, such as dusts, formed from the disintegration of larger particles are usually above 1 µm in size. Typical processes for forming dispersion aerosols include evolution of dust from coal grinding, formation of spray in cooling towers, and blowing of dirt from dry soil.

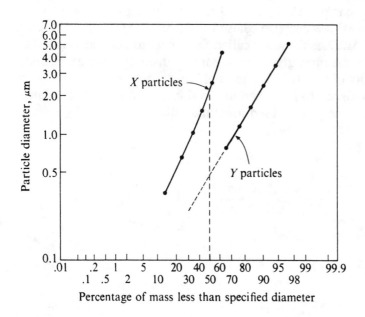

Figure 10.3. Particle size distribution for particles of X (MMD = 2.0 μm) and Y (MMD = 0.5 μm).

Many dispersion aerosols originate from natural sources, such as sea spray, windblown dust, and volcanic dust. However, a vast variety of human activities break up material and disperse it to the amosphere. "All terrain" vehicles churn across desert lands, coating fragile desert plants with layers of dispersed dust. Quarries and rock crushers spew out plumes of ground rock. Cultivation of land has made it much more susceptible to dust-producing wind erosion.

However, since much more energy is required to break material down into small particles than is required for or released by the synthesis of particles through chemical synthesis or the adhesion of smaller particles, most dispersion aerosols are relatively large. Larger particles tend to have fewer harmful effects than smaller ones. As examples, larger particles are less *respirable* in that they do not penetrate so far into the lungs as smaller ones and larger particles are relatively more easy to remove from air pollution effluent sources.

10.4. CHEMICAL PROCESSES FOR PARTICLE FORMATION

Inorganic Particles

Metal oxides constitute a major class of inorganic particles in the atmosphere. These are formed whenever fuels containing metals are burned. For example, particulate iron oxide is formed during the combustion of pyrite-containing coal:

$$3FeS_2 + 8O_2 \rightarrow Fe_3O_4 + 6SO_2 \tag{10.4.1}$$

Organic vanadium in residual fuel oil is converted to particulate vanadium oxide. Part of the calcium carbonate in the ash fraction of coal is converted to calcium oxide and is emitted to the atmosphere through the stack:

$$CaCO_3 + heat \rightarrow CaO + CO_2 \qquad (10.4.2)$$

A common process for the formation of aerosol mists involves the oxidation of atmospheric sulfur dioxide to sulfuric acid, a hygroscopic substance that accumulates atmospheric water to form small liquid droplets:

$$2SO_2 + O_2 + 2H_2O \rightarrow 2H_2SO_4 \qquad (10.4.3)$$

In the presence of basic air pollutants, such as ammonia or calcium oxide, the sulfuric acid reacts to form salts:

$$H_2SO_4(droplet) + 2NH_3(g) \rightarrow (NH_4)_2SO_4 \ (droplet) \qquad (10.4.4)$$

$$H_2SO_4(droplet) + CaO(s) \rightarrow CaSO_4(droplet) \qquad (10.4.5)$$

Under low-humidity conditions water is lost from these droplets and a solid aerosol is fomred.

The preceding examples show several ways in which solid or liquid inorganic aerosols are formed by chemical reactions. Such reactions constitute an important general process for thc formation of aerosols, particularly the smaller particles.

Organic Particles

A significant portion of organic particulate matter is produced by internal combustion engines in complicated processes that involve pyrosynthesis and nitrogenous compounds. These products may include nitrogen-containing compounds and oxidized hydrocarbon polymers. Lubricating oil and its additives may also contribute to organic particulate matter.

PAH Synthesis

The organic particles of greatest concern are PAH hydrocarbons, which consist of condensed ring aromatic molecules. The most often cited example of a PAH compound is benzo(a)pyrene, a compound that the body can metabolize to a carcinogenic form:

Benzo(a)pyrene

PAHs may be synthesized from saturated hydrocarbons under oxygen-deficient conditions. Hydrocarbons with very low molecular masses, including even methane, may act as precursors for the polycyclic aromatic compounds. Low-molecular-mass hydrocarbons form PAHs by **pyrosynthesis**. This happens at temperatures exceeding approximately 500°C at which carbon-hydrogen and carbon-carbon bonds are broken to form free radicals. These radicals undergo dehydrohalogenation and combine chemically to form aromatic ring structures, which are resistant to thermal degradation. The basic process for the formation of such rings from pyrosynthesis starting with ethane is,

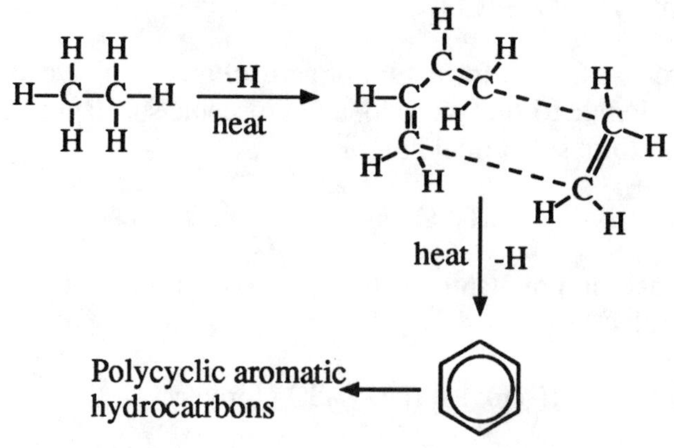

$$\text{H}_3\text{C–CH}_3 \xrightarrow[\text{heat}]{-\text{H}} \cdots \xrightarrow[\text{heat}]{-\text{H}}$$

Polycyclic aromatic
hydrocatrbons ←

stable PAH structures. The tendency of hydrocarbons to form PAHs by pyrosynthesis varies in the order aromatics > cycloolefins > olefins > paraffins. The existing ring structure of cyclic compounds is conducive to PAH formation. Unsaturated compounds are especially susceptible to the addition reactions involved in PAH formation.

Polycyclic aromatic compounds may be formed from higher alkanes present in fuels and plant materials by the process of **pyrolysis**, the "cracking" of organic compounds to form smaller and less stable molecules and radicals.

10.5. THE COMPOSITION OF INORGANIC PARTICLES

Figure 10.4 illustrates the basic factors responsible for the composition of inorganic particulate matter. In general, the proportions of elements in atmospheric particulate matter reflect relative abundances of elements in the parent material. The source of particulate matter is reflected in its elemental compositon, taking into consideration chemical reactions that may change the composition. For example, particulate matter largely from ocean spray origin in a coastal area receiving sulfur dioxide pollution may show anomalously high sulfate and corresponding low chloride content. The sulfate comes from atmospheric oxidation of sulfur dioxide to form nonvolatile ionic sulfate, whereas some chloride originally from the NaCl in the seawater may be lost from the solid aerosol as volatile HCl:

$$2SO_2 + O_2 + 2H_2O \rightarrow 2H_2SO_4 \tag{10.5.1}$$

$$H_2SO_4 + 2NaCl(\text{particulate}) \rightarrow 2Na_2SO_4(\text{particulate}) + 2HCl \tag{10.5.2}$$

The chemical composition of atmospheric particulate matter is quite diverse. Among the constituents of inorganic particulate matter found in polluted atmospheres are salts, oxides, nitrogen compounds, sulfur compounds, various metals, and radionuclides. In coastal areas sodium and chlorine get into atmospheric particles as sodium chloride from sea spray. The major trace elements that typically occur at levels above 1 $\mu g/m^3$ in particulate matter are aluminum, calcium, carbon, iron, potassium, sodium, and silicon; note that most of these tend to originate from terrestrial sources. Lesser quantities of copper, lead, titanium, and zinc and even lower levels of antimony, beryllium, bismuth, cadmium, cobalt, chromium, cesium, lithium,

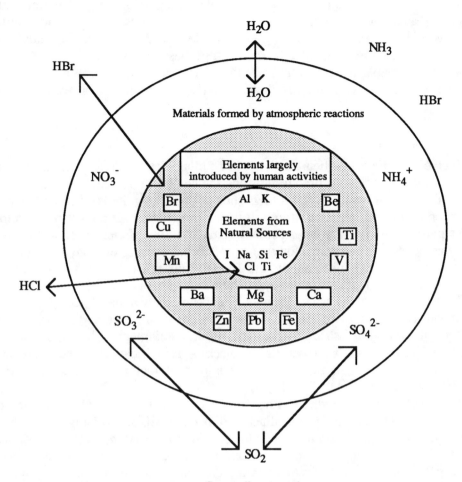

Figure 10.4. Some of the components of inorganic particulate matter and their origins.

manganese, nickel, rubidium, selenium, strontium, and vanadium are commonly observed. The likely sources of some of these elements are given below:

- **Al, Fe, Ca, Si**: Soil erosion, rock dust, coal combustion

- **C**: Incomplete combustion of carbonaceous fuels

- **Na, Cl**: Marine aerosols, chloride from incineration of organohalide polymer wastes

- **Sb, Se**: Very volatile elements, possibly from the combustion of oil, coal, or refuse

- **V**: Combustion of residual petroleum (present at very high levels in residues from Venezuelan crude oil)

- **Zn**: Tends to occur in small particles, probably from combustion

- **Pb**: Combustion of leaded fuels and wastes containing lead

Particulate carbon as soot, carbon black, coke, and graphite originates from auto and truck exhausts, heating furnaces, incinerators, power plants, and steel and foundry operations and composes one of the more visible and troublesome particulate air pollutants. Because of its good adsorbent properties, carbon can be a carrier of gaseous and other particulate pollutants. Particulate carbon surfaces may catalyze some heterogeneous atmospheric reactions, including the important conversion of SO_2 to sulfate.

Fly Ash

Much of the mineral particulate matter in a polluted atmosphere is in the form of oxides and other compounds produced during the combustion of high-ash fossil fuel. Much of the mineral matter in fossil fuels such as coal or lignite is converted during combustion to a fused, glassy bottom ash which presents no air pollution problems. Smaller particles of **fly ash** enter furnace flues and are efficiently collected in a properly equipped stack system. However, some fly ash escapes through the stack and enters the atmosphere. Unfortunately, the fly ash thus released tends to consist of smaller particles that do the most damage to human health, plants, and visibility.

The composition of fly ash varies widely, depending upon the source of fuel. The predominant constituents are oxides of aluminum, calcium, iron, and silicon. Other elements that occur in fly ash are magnesium, sulfur, titanium, phosphorus, potassium, and sodium. Elemental carbon (soot, carbon black) is a significant fly ash constituent.

The size of fly ash particles is a very important factor in determining their removal from stack gas and their ability to enter the body through the respiratory tract. Fly ash from coal-fired utility boilers has shown a bimodal (two peak) distribution of size, with a peak at about 0.1 μm as illustrated in Figure 10.5.[2] Although only about 1-2 percent of the total fly ash mass is in the smaller size fraction, it includes the vast majority of the total number of particles and particle surface area. Submicrometer particles probably result from a volatilization-condensation process during combustion as reflected in a higher concentration of more volatile elements, such as As, Sb,

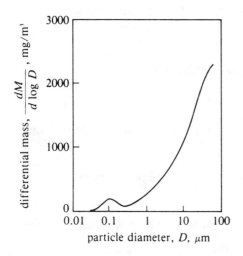

Figure 10.5. General appearance of particle-size distribution in coal-fired power plant ash. The data are given on differential mass coordinates, where M is the mass, so that the area under the curve in a given size range is the mass of the particles in that size range.

Hg, and Zn. Furthermore, the very small particles are the most difficult to remove by electrostatic precipitators and bag houses (see Section 10.10).

Asbestos

Asbestos is the name given to a group of fibrous silicate minerals, typically those of the serpentine group, for which the approximate formula is $Mg_3P(Si_2O_5)(OH)_4$. The tensile strength, flexibility, and nonflammability of asbestos have led to many uses including structural materials, brake linings, insulation, and pipe manufacture.[3] In 1979, 560,000 metric tons of asbestos were used in the U.S. By 1988 annual consumption had dropped to 85,000 metric tons, most of it used for brake linings and pads, roofing products, cement/asbestos pipe, gaskets, heat-resistant packing, and specialty papers. In 1989, the U.S. Environmental Protection Agency announced regulations that would phase out most uses of asbestos by 1996. However, global production of asbestos increased in 1988, largely because of increased demand from India.

Asbestos is of concern as an air pollutant because when inhaled it may cause asbestosis (a pneumonia condition), mesothelioma (tumor of the mesothelial tissue lining the chest cavity adjacent to the lungs), and bronchogenic carcinoma (cancer originating with the air passages in the lungs).[4] Therefore, uses of asbestos have been severely curtailed and widespread programs have been undertaken to remove the material from buildings.

10.6. TOXIC METALS

Some of the metals found predominantly as particulate matter in polluted atmospheres are known to be hazardous to human health. All of these except beryllium are so-called "heavy metals." Lead is the toxic metal of greatest concern in the urban atmosphere because it comes closest to being present at a toxic level; mercury ranks second. Others include beryllium, cadmium, chromium, vanadium, nickel, and arsenic (a metalloid).

Atmospheric Mercury

Atmospheric mercury is of concern because of its toxicity, volatility, and mobility. Some atmospheric mercury is associated with particulate matter. Much of the mercury entering the atmosphere does so as volatile elemental mercury from coal combustion and volcanoes. Volatile organomercury compounds such as dimethylmercury, $(CH_3)_2Hg$, and monomethylmercury salts, such as $(CH_3)HgBr$, are also encountered in the atmosphere.

Atmospheric Lead

With the reduction of leaded fuels, atmospheric lead is of less concern than it used to be. However, during the decades that leaded gasoline containing tetraethyllead was the predominant automotive fuel, particulate lead halides were emitted in large quantities. This occurs through the action of dichlorethane and dibromoethane added as halogenated scavengers to prevent the accumulation of lead oxides inside engines. The lead halides formed,

$$Pb(C_2H_5)_4 + O_2 + \text{halogenated scavengers} \rightarrow CO_2 + H_2O +$$
$$PbCl_2 + PbClBr + PbBr_2 \quad (10.6.1)$$

are volatile enough to exit through the exhaust system but condense in the air to form particles. During the period of peak usage of leaded gasoline in the early 1970's about 200,000 tons of lead were entering the atmosphere each year by this route in the U.S.

Atmospheric Beryllium

Only about 350 metric tons of beryllium are used each year in the U.S. for the formulation of specialty alloys used in electrical equipment, electronic instrumentation, space gear, and nuclear reactor components, so that distribution of beryllium is by no means comparable to that of other toxic metals such as lead or mercury. However, because of its "high tech" applications, consumption of beryllium may increase in the future.

During the 1940s and 1950s, the toxicity of beryllium and beryllium compounds became widely recognized; it has the lowest allowable limit in the atmosphere of all the elements. One of the main results of the recognition of beryllium toxicity hazards was the elimination of this element from phosphors (coatings which produce visible light from ultraviolet light) in fluorescent lamps.

10.7. RADIOACTIVE PARTICLES

A significant natural source of radionuclides in the atmosphere is **radon**, a noble gas product of radium decay.[5] Radon may enter the atmosphere as either of two isotopes, ^{222}Rn (half-life 3.8 days) and ^{220}Rn (half-life 54.5 seconds). Both are alpha emitters in decay chains that terminate with stable isotopes of lead. The initial decay products, ^{218}Po and ^{216}Po are nongaseous and adhere readily to atmospheric particulate matter. Therefore, some of the radioactivity detected in these particles is of natural origin. Furthermore, cosmic rays act on nuclei in the atmosphere to produce other radionuclides, including ^{7}Be, ^{10}Be, ^{14}C, ^{39}Cl, ^{3}H, ^{22}Na, ^{32}P, and ^{33}P.

One of the more serious problems in connection with radon is that of radioactivity originating from uranium mine tailings that have been used in some areas as backfill, soil conditioner, and a base for building foundations. Radon produced by the decay of radium exudes from foundations and walls constructed on tailings. Higher than normal levels of radioactivity have been found in some structures in the city of Grand Junction, Colorado, where uranium mill tailings have been used extensively in construction. Some medical authorities have suggested that the rate of birth defects and infant cancer in areas where uranium mill tailings have been used in residential construction are significantly higher than normal. The combustion of fossil fuels introduces radioactivity into the atmosphere in the form of radionuclides contained in fly ash. Large coal-fired power plants lacking ash-control equipment may introduce up to several hundred milliCuries of radionuclides into the atmosphere each year, far more than either an equivalent nuclear or oil-fired power plant.

The radioactive noble gas ^{85}Kr (half-life 10.3 years) is emitted into the atmosphere by the operation of nuclear reactors and the processing of spent reactor fuels. In general, other radionuclides produced by reactor operation are either chemically reactive and can be removed from the reactor effluent, or have such short half-lives that a short time delay prior to emission prevents their leaving the reactor. Widepread use of fission power will inevitably result in an increased level of ^{85}Kr in the atmosphere. Fortunately, biota cannot concentrate this chemically unreactive element.

The above-ground detonation of nuclear weapons can add large amounts of radioactive particulate matter to the atmosphere. Among the radioisotopes that can be detected in rainfall falling after atmospheric nuclear weapon detonation are ^{91}Y, ^{141}Ce, ^{144}Ce, ^{147}Nd, ^{147}Pm, ^{149}Pm, ^{151}Sm, ^{153}Sm, ^{155}Eu, ^{156}Eu, ^{89}Sr, ^{90}Sr, ^{115m}Cd, ^{129m}Te, ^{131}I, ^{132}Te, and ^{140}Ba. (Note that "m" denotes a metastable state that decays by gamma-ray emission to an isotope of the same element.) The rate of travel of radioactive particles through the atmosphere is a function of particle size. Appreciable fractionation of nuclear debris is observed because of differences in the rates at which various components of nuclear variety move through the atmosphere.

10.8. THE COMPOSITION OF ORGANIC PARTICLES

Organic atmospheric particles occur in a wide variety of compounds. For analysis, such particles can be collected onto a filter; extracted with organic solvents; fractionated into neutral, acid, and basic groups; and analyzed for specific constituents by chromatography and mass spectrometry. The neutral group contains predominantly hydrocarbons, including aliphatic, aromatic, and oxygenated fractions. The aliphatic fraction of the neutral group contains a high percentage of long-chain hydrocarbons, predominantly those with 16-28 carbon atoms. These relatively unreactive compounds are not particularly toxic and do not participate strongly in atmospheric chemical reactions. The aromatic fraction, however, contains carcinogenic polycyclic aromatic hydrocarbons, which are discussed below. Aldehydes, ketones, epoxides, peroxides, esters, quinones, and lactones are found among the oxygenated neutral components, some of which may be mutagenic or carcinogenic. The acidic group contains long-chain fatty acids and nonvolatile phenols. Among the acids recovered from air-pollutant particulate matter are lauric, myristic, palmitic, stearic, behenic, oleic, and linoleic acids. The basic group consists largely of alkaline N-heterocyclic hydrocarbons, such as acridine:

Acridine

Polycyclic Aromatic Hydrocarbons

Polycyclic aromatic hydrocarbons (PAH) in atmospheric particles have received a great deal of attention because of the known carcinogenic effects of some of these compounds, which are discussed in greater detail in Chapter 20. Prominent among these compounds are benzo(a)pyrene, benz(a)anthracene, chrysene, benzo-(e)pyrene, benz(e)acephenanthrylene, benzo(j)fluoranthene, and indenol. Some representative structures of PAH compounds are given below:

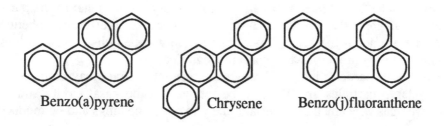

Benzo(a)pyrene   Chrysene   Benzo(j)fluoranthene

Elevated levels of PAH compounds of up to about 20 µg/m^3 are found in the atmosphere. Elevated levels of PAHs are most likely to be encountered in polluted urban atmospheres and in the vicinity of natural fires, such as forest and prairie fires. Coal furnace stack gas may contain over 1000 µg/m^3 of PAH compounds and cigarette smoke almost 100 µg/m^3.

Atmospheric polycyclic aromatic hydrocarbons are found almost exclusively in the solid phase, largely sorbed to soot particles. Soot itself is a highly condensed product of PAHs. Soot contains 1–3% hydrogen and 5–10% oxygen, the latter due to partial surface oxidation. Benzo(a)pyrene adsorbed on soot disappears very rapidly in the presence of light yielding oxygenated products; the large surface area of the particle contributes to the high rate of reaction. Oxidation products of benzo(a) pyrene include epoxides, quinones, phenols, aldehydes, and carboxylic acids[6] as illustrated by the composite structures shown below:

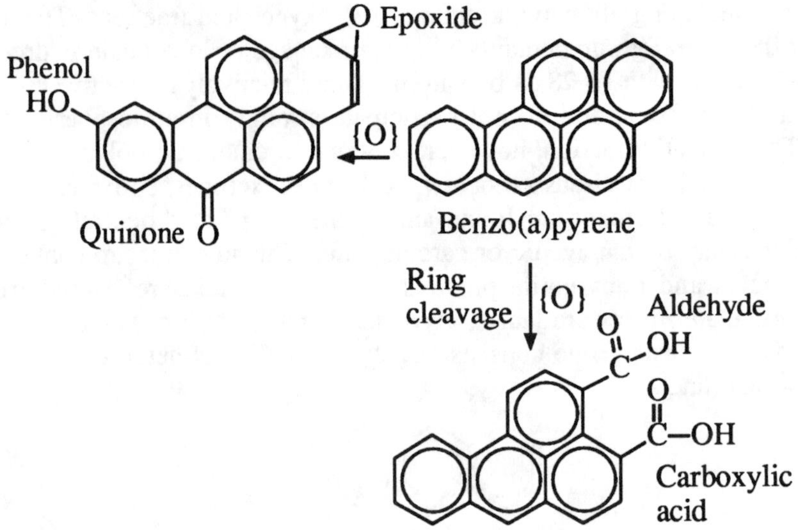

10.9. EFFECTS OF PARTICLES

Atmospheric particles have numerous effects. The most obvious of these is reduction and distortion of visibility. They provide active surfaces upon which heterogeneous atmospheric chemical reactions can occur and nucleation bodies for the condensation of atmospheric water vapor, thereby exerting a significant influence upon weather and air pollution phenomena.

The most visible effects of aerosol particles upon air quality result from their optical effects. Particles smaller than about 0.1 µm in diameter scatter light much like molecules, that is, Rayleigh scattering. Generally, such particles have an insignificant effect upon visibility in the atmosphere. The light-scattering and intercepting properties of particles larger than 1 µm are approximately proportional to the particle's cross-sectional area. Particles of 0.1 µm – 1 µm cause interference phenomena because they are about the same dimensions as the wavelengths of visible light, so their light-scattering properties are especially significant.

Atmospheric particles inhaled through the respiratory tract may damage health. Relatively large particles are likely to be retained in the nasal cavity and in the pharynx, whereas very small particles are likely to reach the lungs and be retained by them. The respiratory system possesses mechanisms for the expulson of inhaled particles. In the ciliated region of the respiratory system, particles are carried as far as the entrance to the gastrointestinal tract by a flow of mucus. Macrophages in the nonciliated pulmonary regions carry particles to the ciliated region.

The repiratory system may be damaged directly by particulate matter that enters the blood system or lymph system through the lungs. In addition, the particulate material or soluble components of it may be transported to organs some distance from the lungs and have a detrimental effect on these organs. Particles cleared from the respiratory tract are to a large extent swallowed into the gastrointestinal tract.

A strong correlation has been found between increases in the daily mortality rate and acute episodes of air pollution. In such cases, high levels of particulate matter are accompanied by elevated concentrations of SO_2 and other pollutants, so that any conclusions must be drawn with caution.

10.10. CONTROL OF PARTICULATE EMISSIONS

The removal of particulate matter from gas streams is the most widely practiced means of air pollution control. A number of devices have been developed for this purposes, which differ widely in effectiveness, complexity, and cost. The selection of a particle removal system for a gaseous waste stream depends upon the particle loading, nature of particles (size distribution), and type of gas scrubbing system used.

Particle Removal by Sedimentation and Inertia

The simplest means of particulate matter removal is **sedimentation**, a phenomenon that occurs continuously in nature. Gravitational settling chambers may be employed for the removal of particles from gas streams by simply settling under the influence of gravity. These chambers take up large amounts of space and have low collection efficiencies, particularly for small particles.

Gravitational settling of particles is enhanced by increased particle size, which occurs spontaneously by coagulation. Thus, over time, the size of particles increases and the number of particles decreases in a mass of air that contains particles. Brownian motion of particles less than about 0.1 μm in size is primarily responsible for their contact, enabling coagulation to occur. Particles greater than about 0.3 μm in radius do not diffuse appreciably and serve primarily as receptors of smaller particles.

Inertial mechanisms are effective for particle removal. These depend upon the fact that the radius of the path of a particle in a rapidly moving, curving air stream is larger than the path of the stream as a whole. Therefore, when a gas stream is spun by vanes, a fan, or a tangential gas inlet, the particulate matter may be collected on a separator wall because the particles are forced outward by centrifugal force. Devices utlizing this mode of operation are called **dry centrifugal collectors**.

Particle filtration

Fabric filters, as their name implies, consist of fabrics that allow the passage of gas but retain particulate matter. These are used to collect dust in bags contained in structures called *baghouses*. Periodically the fabric composing the filter is shaken to remove the particles and to reduce back-pressure to acceptable levels. Typically the bag is in a tubular configuration as shown in Figure 10.6. Numerous other configurations are possible. Collected particulate matter is removed from bags by mechanical agitation, blowing air on the fabric, or rapid expansion and contraction of the bags.

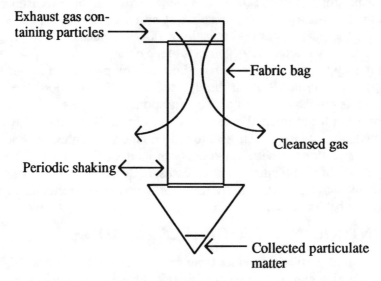

Figure 10.6. Baghouse collection of particulate emissions.

Although simple, baghouses are generally effective in removing particles from exhaust gas. Particles as small as 0.01 μm in diameter are removed, and removal efficiency is relatively high for particles down to 0.5 μm in diameter.

Scrubbers

A venturi scrubber passes gas through a converging section, throat, and diverging section as shown in Figure 10.7. Injection of the scrubbing liquid at right angles to incoming gas breaks the liquid into very small droplets, which are ideal for scavenging particles from the gas stream. In the reduced-pressure (expanding) region of the venturi, some condensation can occur, adding to the scrubbing efficiency. In addition to removing particles, venturis may serve as quenchers to cool exhaust gas and as scrubbers for pollutant gases.

Ionizing wet scrubbers place an electrical charge on particles upstream from a wet scrubber. Larger particles and some gaseous contaminants are removed by scrubbing action. Smaller particles tend to induce opposite charges in water droplets in the scrubber and in its packing material and are removed by attraction of the opposite charges.

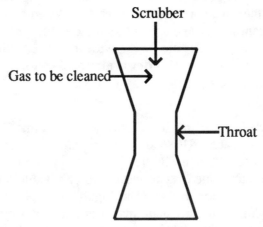

Figure 10.7. Venturi scrubber.

Electrostatic Removal

Aerosol particles may acquire electrical charges. In an electric field, such particles are subjected to a force, F (dynes) given by

$$F = eq \qquad (10.10.1)$$

where E is the voltage gradient (statvolt/cm) and q is the electrostatic charge charge on the particle (in esu). This phenomenon has been widely used in highly efficient **electrostatic precipitators,** as shown in Figure 10.8. The particles acquire a charge when the gas stream is passed through a high-voltage, direct-current corona. Because of the charge, the particles are attracted to a grounded surface, from which they may be later removed. Ozone may be produced by the corona discharge. Similar devices used as household dust collectors may produce toxic oxone if not operated properly.

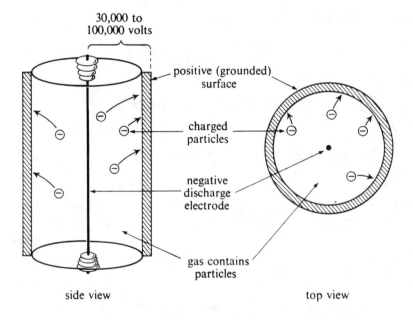

Figure 10.8. Schematic diagram of an electrostatic precipitator

LITERATURE CITED

1. Stelson, A. W., and J. H. Seinfeld, "Chemical Mass Accounting of Urban Aerosol," *Environmental Science and Technology*, **15**, (1981) 671-679.

2. McElroy, M. W., R. C. Carr, D. S. Ensor, and G. R. Markowski, "Size Distribution of Fine Particles from Coal Combustion," *Science*, **215**, (1982) 13-19.

3. Steinway, Daniel M., "Scope and Numbers of Regulations for Asbestos-Containing Materials, Abatement Continue to Grow," *Hazmat World*, April, 1990, pp. 32-58.

4. Fisher, Gerald L., and Michael A. Gallo, Eds., *Asbestos Toxicity*, Marcel Dekker, New York, 1988.

5. Nazaroff, William W., and Anthony V. Nero, Jr., *Radon and its Decay Products in Indoor Air*, John Wiley and Sons, New York, 1988.

6. Finlayson-Pitts, Barbara J., and James N. Pitts, Jr., *Atmospheric Chemistry: Fundamentals and Experimental Techniques*, John Wiley and Sons, New York, 1986.

SUPPLEMENTARY REFERENCES

Seinfeld, John H., *Atmospheric Chemistry and Physics of Air Pollution*, John Wiley and Sons, Inc., Somerset, N. J., 1986.

Botkin, Daniel B., Ed., *Changing the Global Environment: Perspectives on Human Involvement*, Academic Press, San Diego, CA, 1989.

Airborne Particles, University Park Press, Baltimore, MD, 1979.

Control Techniques for Particulate Air Pollutants, National Air Pollution Control Administration Publicaation No. AP-51, 1969.

Air Quality Criteria for Particulate Matter, National Air Pollution Control Administration Publicaation No. AP-49, 1969.

Bjørseth, A., Ed., 1983, *Handbook of Polycyclic Aromatic Hydrocarbons*, Van Nostrand Reinhold Co., New York, 1980.

Bricard, J., "Aerosol Production in the Atmosphere," Chap. 11 in *Environmental Chemistry*, J. O'M. Bockris, Ed., Plenum Publishing Corp., New York, pp. 313-30.

Cheremisinoff, P. N., Ed., *Air/Particulate — Instrumentation and Analysis*, Butterworth Publishers, Woburn, Mass., 1981.

Hinds, W. C., *Aerosol Technology*, John Wiley and Sons, Inc., 1982.

Kneip, T. J., and P. J. Lioy, Eds., *Aerosols: Anthropogenic and Natural Sources and Transport*, New York Academy of Sciences, New York, 1980.

Wolff, G. T., and R. L. Klimisch, Eds., *Particulate Carbon: Atmospheric Life Cycle*, Plenum Publishing Corp., New York, 1982.

QUESTIONS AND PROBLEMS

1. The maximum electrical charge that an atmospheric particle may attain in dry air is about 8 esu/cm^2. How many electrons per square centimeter of surface area is this?

2. For small charged particles, those that are 0.1 μm or less in size, an average charge of 4.77 x 10^{-10} esu, equal to the *elementary quantum of electricity* (the charge on 1 electron or proton), is normally assumed for the whole particle. What is the surface charge in esu/cm^2 for a charged spherical particle with a radius of 0.1 μm?

3. What is the settlling velocity of a particle having a Stoke's diameter of 10 μm and a density of 1 g/cm^3 in air at 1.00 atm pressure and 0°C temperature? (The viscosity of air at 0°C is 170.8 micropoise. The density of air under these conditions is 1.29 g/L.)

4. A freight train that included a tank car containing anhydrous NH_3 and one containing concentrated HCl was wrecked, causing both of the tank cars to leak. In the region between the cars a white aerosol formed. What was it, and how was it produced?

5. Examination of aerosol fume particles produced by a welding process showed that 2% of the particles were greater than 7 μm in diameter and 2% were less than 0.5 μm. What is the mass median diameter of the particles?

6. What two vapor forms of mercury might be found in the atmosphere?

7. Analysis of particulate matter collected in the atmosphere near a seashore shows considerablly more Na than Cl on a molar basis. What does this indicate?

8. What type of process results in the formation of very small aerosol particles?

9. Which size range encompasses most of the particulate matter mass in the atmosphere?

10. Why are aerosols in the 0.1–1 μm size range especially effective in scattering light?

11. Per unit mass, why are smaller particles more effective catalyssts for atmospheric chemical reactions?

12. In terms of origin, what are the three major categories of elements found in atmospheric particles?

13. What are the five major classes of material making up the composition of atmospheric aerosol particles?

14. The size distribution of particles emitted from coal-fired power plants is bimodal. What are some of the properties of the smaller fraction in terms of potential environmental implications?

Gaseous Inorganic Air Pollutants

11.1 INORGANIC POLLUTANT GASES

A number of gaseous inorganic pollutants enter the atmosphere as the result of human activities. Those added in the greatest quantities are CO, SO_2, NO, and NO_2. (These quantities are relatively small, compared to the amount of CO_2 in the atmosphere. The possible environmental effects of increased atmospheric CO_2 levels are discussed in Chapter 14.) Other inorganic pollutant gases include NH_3, N_2O, N_2O_5, H_2S, Cl_2, HCl, and HF. Substantial quantities of some of these gases are added to the atmosphere each year by human activities. Globally, atmospheric emissions of carbon monoxide, sulfur oxides, and nitrogen are of the order of one to several hundred million tons per year.

11.2. PRODUCTION AND CONTROL OF CARBON MONOXIDE

Carbon monoxide, CO, causes problems in cases of locally high concentrations. The toxicity of carbon monoxide is discussed in Chapter 20. The overall atmospheric concentration of carbon monoxide is about 0.1 ppm corresponding to a burden in the Earth's atmosphere of approximately 500 million metric tons of CO with an average residence time ranging from 36 to 110 days. Much of this CO is present as an intermediate in the oxidation of methane by hydroxyl radical. From Table 9.1 it may be seen that the methane content of the atmosphere is about 1.6 ppm, more than 10 times the concentration of CO. Therefore, any oxidation process for methane that produces carbon monoxide as an intermediate is certain to contribute substantially to the overall carbon monoxide burden, probably around two-thirds of the total CO.

Degradation of chlorophyll during the autumn months releases CO, amounting to perhaps as much as 20% of the total annual release. Anthropogenic sources account for about 6% of CO emissions. The remainder of atmospheric CO comes from largely unknown sources. These include some plants and marine organisms known as siphonophores, an order of Hydrozoa. Carbon monoxide is also produced by decay of plant matter other than chlorophyll.

Because of carbon monoxide emissions from internal combustion engines, highest levels of this toxic gas tend to occur in congested urban areas at times when the maximum number of people are exposed, such as during rush hours. At such times, carbon monoxide levels in the atmosphere may become as high as 50-100 ppm.

Atmospheric levels of carbon monoxide in urban areas show a positive correlation with the density of vehicular traffic, and a negative correlation with wind speed. Whereas urban atmospheres may show average carbon monoxide levels of the order of several ppm, data taken in remote areas show much lower levels.

Control of Carbon Monoxide Emissions

Since the internal combustion engine is the primary source of localized pollutant carbon monoxide emissions, control measures have been concentrated on the automobile. Carbon monoxide emissions may be lowered by employing a leaner air-fuel mixture, that is, one in which the weight ratio of air to fuel is relatively high. At air-fuel (weight:weight) ratios exceeding approximately 16:1, an internal combustion engine emits virtually no carbon monoxide.

Modern automobiles use catalytic exhaust reactors to cut down on carbon monoxide emissions. Excess air is pumped into the exhaust gas, and the mixture is passed through a catalytic converter in the exhaust system, resulting in oxidation. of CO to CO_2.

11.3. FATE OF ATMOSPHERIC CO

It is known that the residence time of carbon monoxide in the atmosphere, τ_{CO}, is not long, perhaps of the order of 4 months. It is generally agreed that carbon monoxide is removed from the atmosphere by reaction with hydroxyl radical, $HO^\bullet$:

$$CO + HO^\bullet \rightarrow CO_2 + H \tag{11.3.1}$$

The reaction produces hydroperoxyl radical as a product:[1]

$$O_2 + H + M \rightarrow HOO^\bullet + M \tag{11.3.2}$$

$HO^\bullet$ is regenerated from $HOO^\bullet$ by the following reactions:

$$HOO^\bullet + NO \rightarrow HO^\bullet + NO_2 \tag{11.3.3}$$

$$HOO^\bullet + HOO^\bullet \rightarrow H_2O_2 \tag{11.3.4}$$

The latter reaction is followed by photochemical dissociation of H_2O_2 to regenerate $HO^\bullet$:

$$H_2O_2 + h\nu \rightarrow 2HO^\bullet \tag{11.3.5}$$

Methane is also involved through the atmospheric CO-$HO^\bullet$-CH_4 cycle.

Soil microorganisms act to remove CO from the atmosphere. Therefore, soil is a sink for carbon monoxide.

11.4. SULFUR DIOXIDE SOURCES AND THE SULFUR CYCLE

Figure 11.1 shows the main aspects of the global sulfur cycle. This cycle involves primarily H_2S, SO_2, SO_3, and sulfates. There are many uncertainties regarding the sources, reactions, and fates of these atmospheric sulfur species. On a global basis,

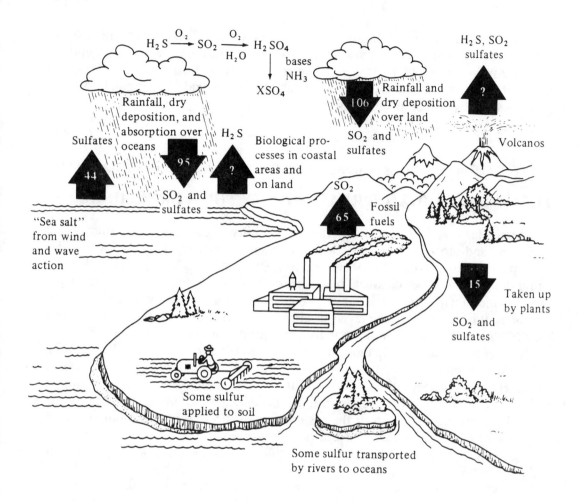

Figure 11.1. The atmospheric sulfur cycle.

sulfur compounds enter the atmosphere to a very large extent through human activities. Approximately 100 million metric tons of sulfur per year enters the global atmosphere through anthropogenic activities, primarily as SO_2 from the combustion of coal and residual fuel oil. The greatest uncertainties in the cycle have to do with nonanthropogenic sulfur, which enters the atmosphere largely as H_2S from volcanoes and from the biological decay of organic matter and reduction of sulfate. The quantity added from biological processes may be as low as 1 million metric tons per year. Any H_2S that does get into the atmosphere is converted rapidly to SO_2 by the following overall process:

$$H_2S + \tfrac{3}{2}O_2 \rightarrow SO_2 + H_2O \qquad (11.4.1)$$

The intial reaction is hydrogen ion abstraction by hydroxyl radical,

$$H_2S + HO^{\bullet} \rightarrow HS^{\bullet} + H_2O \qquad (11.4.2)$$

followed by the following two reactions to give SO_2:

$$HS\cdot + O_2 \rightarrow HO\cdot + SO \qquad\qquad (11.4.3)$$

$$SO + O_2 \rightarrow SO_2 + O \qquad\qquad (11.4.4)$$

The primary source of anthropogenic sulfur dioxide is coal, from which sulfur must be removed at great expense to keep sulfur dioxide emissions at acceptable levels. Approximately half of the sulfur in coal is in some form of pyrite, FeS_2, and the other half is organic sulfur. The production of sulfur dioxide by the combustion of pyrite is given by the following reaction:

$$4FeS_2 + 11O_2 \rightarrow 2Fe_2O_3 + 8SO_2 \qquad\qquad (11.4.5)$$

Essentially all of the sulfur is converted to SO_2; only 1 or 2% leaves the stack as SO_3.

11.5. SULFUR DIOXIDE REACTIONS IN THE ATMOSPHERE

Many factors, including temperature, humidity, light intensity, atmospheric transport, and surface characteristics of particulate matter, may influence the atmospheric chemical reactions of sulfur dioxide. Like many other gaseous pollutants, sulfur dioxide undergoes chemical reactions resulting in the formation of particulate matter, which then settles or is scavenged from the atmosphere by rainfall or other processes. It is known that high levels of air pollution normally are accompanied by a marked increase in aerosol particles and a consequent reduction in visibility. Reaction products of sulfur dioxide are thought to be responsible for some aerosol formation. Whatever the processes involved, much of the sulfur dioxide in the atmosphere ultimately is oxidized to sulfuric acid and sulfate salts, particularly ammonium sulfate and ammonium hydrogen sulfate. In fact, it is likely that these sulfates account for the turbid haze that covers much of the eastern part of the U.S. under all atmospheric conditions except those characterized by massive intrusions of Arctic air masses during the winter months. The potential of sulfates to induce climatic change is high and must be taken into account when considering control of sulfur dioxide.

Some of the possible ways in which sulfur dioxide may react in the atmosphere are (1) photochemical reactions; (2) photochemical and chemical reactions in the presence of nitrogen oxides and/or hydrocarbons, particularly alkenes; (3) chemical processes in water droplets, particularly those containing metal salts and ammonia; and (4) reactions on solid particles in the atmosphere. It should be kept in mind that the atmosphere is a highly dynamic system with great variations in temperature, composition, humidity, and intensity of sunlight; therefore, different processes may predominate under various atmospheric conditions.

Photochemical reactions are probably involved in some of the processes resulting in the atmospheric oxidation of SO_2. Light with wavelengths above 218 nm is not sufficiently energetic to bring about the photodissociation of SO_2, so direct photochemical reactions in the troposphere are of no significance. The oxidation of sulfur dioxide at the parts-per-million level in an otherwise unpolluted atmosphere is a slow process. Therefore, other pollutant species must be involved in the process in atmospheres polluted with SO_2.

The presence of hydrocarbons and nitrogen oxides greatly increases the oxidation rate of atmospheric SO_2. As discussed in Chapter 13, hydrocarbons, nitrogen oxides, and ultraviolet light are the ingredients necessary for the formation of photochemical smog. This disagreeable condition is characterized by high levels of various oxidizing species (photochemical oxidants) capable of oxidizing SO_2. In the smog-prone Los

Angeles area, the oxidation of SO_2 ranges up to 5-10% per hour. Among the oxidizing species present which could bring about this fast reaction are $HO^\bullet$, $HOO^\bullet$, O, O_3, NO_3, N_2O_5, $ROO^\bullet$, and $RO^\bullet$. As discussed in Chapters 12 and 13, the latter two species are reactive, organic free radicals containing oxygen. Although ozone, O_3, is an important product of photochemical smog, it is believed that the oxidation of SO_2 by ozone in the gas phase is too slow to be significant, but it is probably significant in water droplets.

The most important gas-phase reaction leading to the oxidation of SO_2 is the addition of $HO^\bullet$ radical,[2]

$$HO^\bullet + SO_2 \rightarrow HOSO_2^\bullet \tag{11.5.1}$$

forming a reactive free radical which is eventually converted to a form of sulfate.

In all but relatively dry atmospheres, it is probable that sulfur dioxide is oxidized by reactions occurring inside water aerosol droplets. The overall process of sulfur dioxide oxidation in the aqueous phase is rather complicated. It involves the transport of gaseous SO_2 and oxidant to the aqueous phase, diffusion of species in the aqueous droplet, hydrolysis and ionization of SO_2, and oxidation of SO_2 by the following overall process, where $\{O\}$ represents an oxidizing agent such as H_2O_2, $HO^\bullet$, or O_3 and S(IV) is $SO_2(aq)$, $HSO_3^-(aq)$, and $SO_3^{2-}(aq)$.

$$\{O\}(aq) + S(IV)(aq) \rightarrow 2H^+ + SO_4^{2-} \tag{11.5.2}$$

In the absence of catalytic species, the reaction with dissolved molecular O_2,

$$\tfrac{1}{2}O_2(aq) + SO_2(aq) + H_2O \rightarrow H_2SO_4(aq) \tag{11.5.3}$$

is too slow to be significant.[3] Hydrogen peroxide is an important oxidizing agent in the atmosphere.[4] It reacts with dissolved sulfur dioxide through the overall reaction,

$$SO_2(aq) + H_2O_2(aq) \rightarrow H_2SO_4(aq) \tag{11.5.4}$$

to produce sulfuric acid. The major reaction is thought to be with HSO_3^- ion with peroxymonosulfurous acid, SO_2OOH^-, as an intermediate.

Ozone, O_3, oxidizes sulfur dioxide in water. The fastest reaction is with sulfite ion;

$$SO_3^{2-}(aq) + O_3(aq) + H_2O \rightarrow H_2SO_4(aq) + O_2 \tag{11.5.5}$$

reactions are slower with $HSO_3^-(aq)$ and $SO_2(aq)$. The rate of oxidation of aqueous SO_2 species by ozone increases with increasing pH. The oxidation of sulfur dioxide in water droplets is faster in the presence of ammonia, which reacts with sulfur dioxide to produce bisulfite ion and sulfite ion in solution:

$$NH_3 + SO_2 + H_2O \rightarrow NH_4^+ + HSO_3^- \tag{11.5.6}$$

Some solutes dissolved in water catalyze the oxidation of aqueous SO_2. Both iron(III) and Mn(II) have this effect. The reactions catalyzed by these two ions are faster with increasing pH. Dissolved nitrogen species, NO_2 and HNO_2, oxidize aqueous sulfur dioxide in the laboratory. However, the reaction is probably too slow to be significant in the atmosphere.

Heterogeneous reactions on solid particles may also play a role in the removal of sulfur dioxide from the atmosphere. In atmospheric photochemical reactions, such particles may function as nucleation centers. Thus, they act as catalysts and grow in size by accumulating reaction products. The final result would be production of an aerosol with a composition unlike that of the original particle. Soot particles, which consist of elemental carbon contaminated with polynuclear aromatic hydrocarbons (see Chapter 12) produced in the incomplete combustion of carbonaceous fuels, can catalyze the oxidation of sulfur dioxide to sulfate as indicated by the presence of sulfate on the soot particles. Soot particles are very common in polluted atmospheres, so it is very likely that they are strongly involved in catalyzing the oxidation of sulfur dioxide.

Oxides of metals such as aluminum, calcium, chromium, iron, lead, or vanadium may also be catalysts for the heterogenous oxidation of sulfur dioxide. These oxides may also adsorb sulfur dioxide. However, the total surface area of oxide particulate matter matter in the atmosphere is very low so that the fraction of sulfur dioxide oxidized on metal oxide surfaces is relatively small.

Effects of Atmospheric Sulfur Dioxide

Though not terribly toxic to most people, low levels of sulfur dioxide in air do have some health effects. Its primary effect is upon the respiratory tract, producing irritation and increasing airway resistance, especially to people with respiratory weaknesses and sensitized asthmatics. Therefore, exposure to the gas may increase the effort required to breathe. Mucus secretion is also stimulated by exposure to air contaminated by sulfur dioxide. Although SO_2 causes death in humans at 500 ppm, it has not been found to harm laboratory animals at 5 ppm.

Sulfur dioxide has been at least partially implicated in several acute incidents of air pollution. One of these incidents occurred in December, 1930, in the Meuse River Valley of Belgium when a thermal inversion trapped waste products from a number of industrial sources in the narrow valley. Sulfur dioxide levels reached 38 ppm. Approximately 60 people died in the episode, and some cattle were killed. In October, 1948, a similar incident caused illness in over 40% of the population of Donora, Pennsylvania, and 20 people died. Sulfur dioxide concentrations of 2 ppm were recorded. During a five-day period marked by a temperature inversion and fog in London in December, 1952, approximately 3500-4000 deaths in excess of normal occurred. SO_2 levels reached 1.3 ppm. Autopsies revealed irritation of the respiratory tract, and high levels of sulfur dioxide were suspected of contributing to excess mortality.

Atmospheric sulfur dioxide is harmful to plants. Acute exposure to high levels of the gas kills leaf tissue (leaf necrosis). The edges of the leaves and the areas between the leaf veins are particularly damaged. Chronic exposure of plants to sulfur dioxide causes chlorosis, a bleaching or yellowing of the normally green portions of the leaf. Plant injury increases with increasing relative humidity. Plants incur most injury from sulfur dioxide when their stomata (small openings in plant surface tissue that allow interchange of gases with the atmosphere) are open. For most plants, the stomata are open during the daylight hours, and most damage from sulfur dioxide occurs then. Long-term, low-level exposure to sulfur dioxide can reduce the yields of grain crops, such as wheat or barley. Sulfur dioxide in the atmosphere is converted to sulfuric acid, so that in areas with high levels of sulfur dioxide pollution, plants may be damaged by sulfuric acid aerosols. Such damage appears as small spots where sulfuric acid droplets have impinged in leaves.

Sulfur Dioxide Removal

A number of processes are being used to remove sulfur and sulfur oxides from fuel before combustion and from stack gas after combustion. Most of these efforts concentrate on coal, since it is the major source of sulfur oxides pollution. Physical separation techniques may be used to remove discrete particles of pyritic sulfur from coal. Chemical methods may also be employed for removal of sulfur from coal. Fluidized bed combustion of coal promises to eliminate SO_2 emissions at the point of combustion. The process consists of burning granular coal in a bed of finely divided limestone or dolomite maintained in a fluid-like condition by air injection. Heat calcines the limestone,

$$CaCO_3 \rightarrow CaO + CO_2 \tag{11.5.7}$$

and the lime produced absorbs SO_2:

$$CaO + SO_2 + \tfrac{1}{2}O_2 \rightarrow CaSO_3 \tag{11.5.8}$$

Many processes have been proposed or studied for the removal of sulfur dioxide from stack gas. Table 11.1 summarizes major stack gas scrubbing systems. These include throwaway and recovery systems as well as wet and dry systems. A dry throwaway system used with only limited success involves injection of dry limestone or dolomite into the boiler, followed by recovery of dry lime, sulfites, and sulfates. The overall reaction, shown here for dolomite, is

$$CaCO_3 \cdot MgCO_3 + SO_2 + \tfrac{1}{2}O_2 \rightarrow CaSO_4 + MgO + 2CO_2 \tag{11.5.9}$$

The solid sulfate and oxide products are removed by electrostatic precipitators or cyclone separators. The process has an efficiency of 50% or less for the removal of sulfur oxides.

As may be noted from the chemical reactions shown in Table 11.1, all sulfur dioxide removal processes, except for catalytic oxidation, depend upon absorption of SO_2 by an acid-base reaction. The first two processes listed are throwaway processes yielding large quantities of wastes; the others provide for some sort of sulfur product recovery.

Lime or limestone slurry scrubbing for SO_2 removal involves acid-base reactions with SO_2. When sulfur dioxide dissolves in water, equilibrium is established between SO_2 gas and dissolved SO_2:

$$SO_2(g) \leftrightarrow SO_2(aq) \tag{11.5.10}$$

This equilibrium is described by Henry's Law (Section **),

$$[SO_2(aq)] = K \times P_{SO_2} \tag{11.5.11}$$

where $[SO_2(aq)]$ is the concentration of dissolved molecular sulfur dioxide; K is the Henry's Law constant for SO_2; and P_{SO_2} is the partial pressure of sulfur dioxide gas. In the presence of base, Reaction 11.5.10 is shifted strongly to the right by the following reactions:

$$H_2O + SO_2(aq) \leftrightarrow H^+ + HSO_3^- \tag{11.5.12}$$

$$HSO_3^- \leftrightarrow H^+ + SO_3^{2-} \tag{11.5.13}$$

Table 11.1. Major Stack Gas Scrubbing Systems

Process	Chemical reactions	Major advantages or disadvantages
Lime slurry scrubbing[2]	$Ca(OH)_2 + SO_2 \rightarrow CaSO_3 + H_2O$	Up to 200 kg of lime are needed per metric ton of coal, producing huge quantities of waste product.
Limestone slurry scrubbing[2]	$CaCO_3 + SO_2 \rightarrow CaSO_3 + CO_2(g)$	Lower pH than lime slurry, and not so efficient.
Magnesium oxide scrubbing	$Mg(OH)_2(\text{slurry}) + SO_2 \rightarrow MgSO_3 + 2 H_2O$	The sorbent can be regenerated, and this need not be done on site.
Sodium-base scrubbing	$Na_2SO_3 + H_2O + SO_2 \rightarrow 2 NaHSO_3$ $2 NaHSO_3 + \text{heat} \rightarrow Na_2SO_3 + H_2O + SO_2$ (regeneration)	There are no major technological limitations. Annual costs are relatively high.
Double alkali[2]	$2 NaOH + SO_2 \rightarrow Na_2SO_3 + 2 H_2O$ $Ca(OH)_2 + Na_2SO_3 \rightarrow CaSO_3(s) + 2 NaOH$ (regeneration of NaOH)	Allows for regeneration of expensive sodium alkali solution with inexpensive lime.

[1] For details regarding these and more advanced processes, see the following two books. Satriana, M. 1981. *New developments in flue gas desulfurization technology.* Park Ridge, N.J.: Noyes Data Corp. Hudson, J.L., and Rochelle, G.T., eds. 1982. *Flue gas desulfurization.* Washington, D.C.: American Chemical Society.

[2] These processes have also been adapted to produce a gypsum product by oxidation of $CaSO_3$ in the spent scrubber medium:

$$CaSO_3 + \tfrac{1}{2} O_2 + 2 H_2O \rightarrow CaSO_4 \cdot 2 H_2O(s)$$

Gypsum has some commercial value, such as in the manufacture of plasterboard, and makes a relatively settleable waste product.

In the presence of calcium carbonate slurry (as in limestone slurry scrubbing), hydrogen ion is taken up by the reaction

$$CaCO_3 + H^+ \leftrightarrow Ca^{2+} + HCO_3^- \tag{11.5.14}$$

The reaction of calcium carbonate with carbon dioxide from stack gas,

$$CaCO_3 + CO_2 + H_2O \leftrightarrow Ca^{2+} + 2HCO_3^- \tag{11.5.15}$$

results in some sorption of CO_2. The reaction of sulfite and calcium ion to form highly insoluble calcium sulfite hemihydrate

$$Ca^{2+} + SO_3^{2-} + \tfrac{1}{2}H_2O \leftrightarrow CaSO_3 \cdot \tfrac{1}{2}H_2O(s) \tag{11.5.16}$$

also shifts Reactions 11.5.12 and 11.5.13 to the right. Gypsum is formed in the scrubbing process by the oxidation of sulfite,

$$SO_3^{2-} + \tfrac{1}{2}O_2 \rightarrow SO_4^{2-} \tag{11.5.17}$$

followed by reaction of sulfate ion with calcium ion:

$$Ca^{2+} + SO_4^{2-} + 2H_2O \xleftarrow{\rightarrow} CaSO_4 \cdot 2H_2O(s) \tag{11.5.18}$$

Formation of gypsum in the scrubber is undesirable because it creates scale in the scrubber equipment. However, gypsum is sometimes produced deliberately in the spent scrubber liquid downstream from the scrubber.

When lime, $Ca(OH)_2$, is used in place of limestone (lime slurry scrubbing), a source of hydroxide ions is provided for direct reaction with H:

$$H^+ + OH^- \rightarrow H_2O \tag{11.5.19}$$

The reactions involving sulfur species in a lime slurry scrubber are essentially the same as those just discussed for limestone slurry scrubbing. The pH of a lime slurry is higher than that of a limestone slurry, so that the former has more of a tendency to react with CO_2, resulting in the absorption of that gas:

$$CO_2 + OH^- \rightarrow HCO_3^- \tag{11.5.20}$$

Current practice with lime and limestone scrubber systems calls for injection of the slurry into the scrubber loop beyond the boilers. A number of power plants are now operating with this kind of system. Experience to date has shown that these scrubbers remove well over 90% of both SO_2 and fly ash when operating properly. (Fly ash is fuel combustion ash normally carried up the stack with flue gas, see Chapter 10.) In addition to corrosion and scaling problems, disposal of lime sludge poses formidable obstacles. The quantity of this sludge may be appreciated by considering that approximately 1 ton of limestone is required for each 5 tons of coal. The sludge is normally disposed of in large ponds, which can present some disposal problems. Water seeping through the sludge beds becomes laden with calcium sulfate and other salts. It is difficult to stabilize this sludge as a structurally stable, nonleachable solid.

Recovery systems in which sulfur dioxide or elemental sulfur are removed from the spent sorbing material, which is recycled, are much more desirable from an environmental viewpoint than are throwaway systems. Many kinds of recovery processes have been investigated, including those that involve scrubbing with magnesium oxide slurry, sodium sulfite solution, ammonia solution, or sodium citrate solution.

Sulfur dioxide trapped in a stack-gas-scrubbing process can be converted to hydrogen sulfide by reaction with synthesis gas (H_2, CO, CH_4),

$$SO_2 + (H_2, CO, CH_4) \xleftarrow{\rightarrow} H_2S + CO_2 \tag{11.5.21}$$

The Claus reaction is then employed to produce elemental sulfur:

$$2H_2S + SO_2 \xleftarrow{\rightarrow} 2H_2O + 3S \tag{11.5.22}$$

11.6. NITROGEN OXIDES IN THE ATMOSPHERE

The three oxides of nitrogen normally encountered in the atmosphere are nitrous oxide (N_2O), nitric oxide (NO), and nitrogen dioxide (NO_2). Nitrous oxide, a commonly used anesthetic known as "laughing gas," is produced by microbiological processes and is a component of the unpolluted atmosphere at a level of approximately 0.3 ppm (see Table 9.1). This gas is relatively unreactive and probably does not significantly influence important chemical reactions in the lower atmosphere. Its concentration decreases rapidly with altitude in the stratosphere due to the photochemical reaction

$$N_2O + h\nu \rightarrow N_2 + O \tag{11.6.1}$$

and some reaction with singlet atomic oxygen:

$$N_2O + O \rightarrow N_2 + O_2 \tag{11.6.2}$$

$$N_2O + O \rightarrow NO + NO \tag{11.6.3}$$

These reactions are significant in terms of depletion of the ozone layer. Increased global fixation of nitrogen, accompanied by increased microbial production of N_2O, could contribute to ozone layer depletion.

Colorless, odorless nitric oxide (NO) and pungent red-brown nitrogen dioxide (NO_2) are very important in polluted air. Collectively designated NO_x, these gases enter the atmosphere from natural sources, such as lightning and biological processes, and from pollutant sources. The latter are much more significant because of regionally high NO_2 concentrations, which can cause severe air quality deterioration. Practically all anthropogenic NO_2 enters the atmosphere as a result of the combustion of fossil fuels in both stationary and mobile sources. Globally, somewhat less than 100 million metric tons of nitrogen oxides are emitted to the atmosphere from these sources each year, compared to several times that much from widely dispersed natural sources. United States production of nitrogen oxides is of the order of 20 million metric tons per year. The contribution of automobiles to nitric oxide production in the U.S. has become somewhat lower in the last decade as newer automobiles with nitrogen oxide pollution controls have become more common.

Most NO_2 entering the atmosphere from pollution sources does so as NO generated from internal combustion engines. At very high temperatures, the following reaction occurs:

$$N_2 + O_2 \rightarrow 2NO \tag{11.6.4}$$

The speed with which this reaction takes place increases steeply with temperature. The equilibrium concentration of NO in a mixture of 3% O_2 and 75% N_2, typical of that which occurs in the combustion chamber of an internal combustion engine, is shown as a function of temperature in Figure 11.2. At room temperature (27°C) the equilibrium concentration of NO is only 1.1×10^{-10} ppm, whereas at high temperatures it is much higher. Therefore, high temperatures favor both a high equilibrium concentration and a rapid rate of formation of NO. Rapid cooling of the exhaust gas from combustion "freezes" NO at a relatively high concentration because equilibrium is not maintained. Thus, by its very nature, the combustion process both in the internal combustion engine and in furnaces produces high levels of NO in the combustion products.

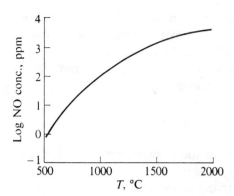

Figure 11.2. Log of equilibrium NO concentration as a function of temperature in a mixture containing 75% N_2 and 3% O_2.

The mechanism for formation of nitrogen oxides from N_2 and O_2 during combustion is a complicated process. Both oxygen and nitrogen atoms are formed at the very high combustion temperatures by the reactions

$$O_2 + M \rightarrow O + O + M \tag{11.6.5}$$

$$N_2 + M \rightarrow N^{\bullet} + N^{\bullet} + M \tag{11.6.6}$$

where M is a highly energetic third body that imparts enough energy to the molecular N_2 and O_2 to break their chemical bonds. The energies required for these reactions are quite high because breakage of the oxygen bond requires 118 kcal/mole and breakage of the nitrogen bond requires 225 kcal/mole. Once formed, O and N atoms participate in the following chain reaction for the formation of nitric oxide from nitrogen and oxygen:

$$N_2 + O \rightarrow NO + N \tag{11.6.7}$$

$$\underline{N + O_2 \rightarrow NO + O} \tag{11.6.8}$$
$$N_2 + O_2 \rightarrow 2NO \tag{11.6.9}$$

There are, of course, many other species present in the combustion mixture besides those shown. The oxygen atoms are especially reactive toward hydrocarbon fragments by reactions such as the following.

$$RH + O \rightarrow R^{\bullet} + HO^{\bullet} \tag{11.6.10}$$

where RH represents a hydrocarbon fragment with an extractable hydrogen atom. These fragments compete with N_2 for oxygen atoms. It is partly for this reason that the formation of NO is appreciably higher at air/fuel ratios exceeding the stoichiometric ratio (lean mixture), as shown in Figure 13.3.

The hydroxyl radical itself can participate in the formation of NO. The reaction is

$$N + HO^\bullet \rightarrow NO + H^\bullet \tag{11.6.11}$$

Nitric oxide, NO, is a product of the combustion of coal and petroleum containing chemically bound nitrogen. Production of NO by this route occurs at much lower temperatures than those required for "thermal" NO, discussed previously.

Atmospheric chemical reactions convert NO_x to nitric acid, inorganic nitrate salts, organic nitrates, and peroxyacetyl nitrate (see Chapter 13).

Atmospheric Reactions of NO_x

The principal reactive nitrogen oxide species in the troposphere are NO, NO_2, and HNO_2. These species cycle among each other, as shown in Figure 11.3. Although NO is the primary form in which NO_x is released to the atmosphere, the conversion of NO to NO_x is relatively rapid in the troposphere.

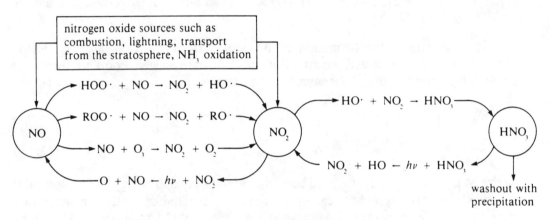

Figure 11.3. Principal reactions among NO, NO_2, and HNO_3 in the atmosphere. ROO• represents an organic peroxyl radical, such as the methylperoxyl radical, $CH_3OO\bullet$.

Nitrogen dioxide is a very reactive and significant species in the atmosphere. It absorbs light throughout the ultraviolet and visible spectrum penetrating the troposphere. At wavelengths below 398 nm, photodissociation occurs,

$$NO_2 + hv \rightarrow NO + O \tag{11.6.12}$$

to produce ground state oxygen atoms. Above 430 nm, only excited molecules are formed,

$$NO_2 + hv \rightarrow NO_2^* \tag{11.6.13}$$

whereas at wavelengths between 398 nm and 430 nm, either process may occur. Photodissociation at these wavelengths requires input of rotational energy from rotation of the NO_2 molecule. The reactivity of NO_2 to photodissociation is shown clearly by the fact that in direct sunlight the half-life of NO_2 is much shorter than that of any other atmospheric component (only 85 seconds at 40° latitude).

The photodissociation of nitrogen dioxide can give rise to the following significant inorganic reactions, in addition to a host of atmospheric reactions involving organic species:

$$O + O_2 + M(\text{third body}) \rightarrow O_3 + M \qquad (11.6.14)$$

$$NO + O_3 \rightarrow NO_2 + O_2 \qquad (11.6.15)$$

$$NO_2 + O_3 \rightarrow NO_3 + O_2 \qquad (11.6.16)$$

$$O + NO_2 \rightarrow NO + O_2 \qquad (11.6.17)$$

$$O + NO_2 + M \rightarrow NO_3 + M \qquad (11.6.18)$$

$$NO_2 + NO_3 \rightarrow N_2O_5 \qquad (11.6.19)$$

$$NO + NO_3 \rightarrow 2NO_2 \qquad (11.6.20)$$

$$O + NO + M \rightarrow NO_2 + M \qquad (11.6.21)$$

Nitrogen dioxide ultimately is removed from the atmosphere as nitric acid, nitrates, or (in atmospheres where photochemical smog is formed) as organic nitrogen. Dinitrogen pentoxide formed in Reaction 11.6.19 is the anhydride of nitric acid, which it forms by reacting with water:

$$N_2O_5 + H_2O \rightarrow 2HNO_3 \qquad (11.6.22)$$

In the stratosphere, nitrogen dioxide reacts with hydroxyl radicals to produce nitric acid:

$$HO\cdot + NO_2 \rightarrow HNO_3 \qquad (11.6.23)$$

In this region, the nitric acid can also be destroyed by hydroxyl radicals,

$$HO\cdot + HNO_3 \rightarrow H_2O + NO_3 \qquad (11.6.24)$$

or by a photochemical reaction,

$$HNO_3 + h\nu \rightarrow HO\cdot + NO_2 \qquad (11.6.25)$$

so that HNO_3 serves as a temporary sink for NO_2 in the stratosphere. Nitric acid produced from NO_2 is removed as precipitation, or reacts with bases (ammonia, particulate lime) to produce particulate nitrates.

Harmful Effects of Nitrogen Oxides

Nitric oxide, NO, is biochemically less active and less toxic than NO_2. Like carbon monoxide and nitrite, NO attaches to hemoglobin and reduces oxygen transport efficiency. However, in a polluted atmosphere, the concentration of nitric oxide normally is much lower than that of carbon monoxide so that the effect on hemoglobin is much less.

Acute exposure to NO_2 can be quite harmful to human health. For exposures ranging from several minutes to one hour, a level of 50-100 ppm of NO_2 causes inflammation of lung tissue for a period of 6-8 weeks, after which time the subject normally recovers. Exposure of the subject to 150-200 ppm of NO_2 causes *bronchiolitis fibrosa obliterans*, a condition fatal within 3-5 weeks after exposure. Death generally results within 2-10 days after exposure to 500 ppm or more of NO_2. "Silo-filler's disease," caused by NO_2 generated by the fermentation of ensilage containing nitrate, is a particularly striking example of nitrogen dioxide poisoning. Deaths have resulted from the inhalation of NO_2-containing gases from burning celluloid and nitrocellulose film and from spillage of NO_2 oxidant (used with liquid hydrazine fuel) from missile rocket motors.

Although extensive damage to plants is observed in areas receiving heavy exposure to NO_2, most of this damage probably comes from secondary products of nitrogen oxides, such as PAN formed in smog (see Chapter 14). Exposure of plants to several parts per million of NO_2 in the laboratory causes leaf spotting and breakdown of plant tissue. Exposure to 10 ppm of NO causes a reversible decrease in the rate of photosynthesis. The effect on plants of long-term exposure to a few tenths of a part per million of NO_2 is less certain.

Nitrogen oxides are known to cause fading of dyes and used in some textiles. This has been observed in gas clothes dryers and is due to NO_x formed in the dryer flame. Much of the damage to materials caused by NO_x comes from secondary nitrates and nitric acid. For example, stress-corrosion cracking of springs used in telephone relays occurs far below the yield strength of the nickel-brass spring metal because of the action of particulate nitrates and aerosol nitric acid formed from NO_x.

Concern has been expressed about the possibility that NO_x emitted to the atmosphere by supersonic transport planes could catalyze the partial destruction of the stratospheric ozone layer, which absorbs damaging short-wavelength (240-300 nm) ultraviolet radiation (see Chapter 11). Detailed consideration of this effect is quite complicated, and only the main features are considered here.

In the upper stratosphere and in the mesosphere, molecular oxygen is photodissociated by ultraviolet light of less than 242-nm wavelength:

$$O_2 + h\nu \rightarrow O + O \tag{11.6.26}$$

In the presence of energy-absorbing third bodies, the atomic oxygen reacts with molecular oxygen to produce ozone:

$$O_2 + O + M \rightarrow O_3 + M \tag{11.6.27}$$

Ozone can be destroyed by reaction with atomic oxygen,

$$O_3 + O \rightarrow O_2 + O_2 \tag{11.6.28}$$

and its formation can be prevented by recombination of oxygen atoms:

$$O + O + M \rightarrow O_2 + M \tag{11.6.29}$$

Addition of the reaction of nitric oxide with ozone,

$$NO + O_3 \rightarrow NO_2 + O_2 \tag{11.6.30}$$

to the reaction of nitrogen dioxide with atomic oxygen,

$$NO_2 + O \rightarrow NO + O_2 \tag{11.6.31}$$

results in a net reaction for the destruction of ozone:

$$O + O_3 \rightarrow O_2 + O_2 \tag{11.6.32}$$

Along with NO_x, water vapor is also emitted into the atmosphere by aircraft exhausts, which could accelerate ozone depletion by the following two reactions:

$$O + H_2O \rightarrow HO^\bullet + HO^\bullet \tag{11.6.33}$$

$$HO^\bullet + O_3 \rightarrow HOO^\bullet + O_2 \tag{11.6.34}$$

However, there are many natural stratospheric buffering reactions which tend to mitigate the potential ozone destruction from those reactions outlined above. Atomic oxygen capable of regenerating ozone is produced by the photochemical reaction,

$$NO_2 + h\nu \rightarrow NO + O \quad (\lambda < 420\,nm) \tag{11.6.35}$$

A competing reaction removing catalytic NO is

$$NO + HOO^\bullet \rightarrow NO_2 + HO^\bullet \tag{11.6.36}$$

Current belief is that supersonic aircraft emissions will not cause nearly as much damage to the ozone layer as chlorofluorocarbons.

Control of Nitrogen Oxides

The level of NO_x emitted from stationary sources such as power plant furnaces generally falls within the range of 50-1000 ppm. NO production is favored both kinetically and thermodynamically by high temperatures and by high excess oxygen concentrations. These factors must be considered in reducing NO emissions from stationary sources. Reduction of flame temperature to prevent NO formation is accomplished by adding recirculated exhaust gas, cool air, or inert gases. Unfortunately, this decreases the efficiency of energy conversion as calculated by the Carnot Equation (see Chapter 21).

Low-excess-air firing is effective in reducing NO_x emissions during the combustion of fossil fuels. As the term implies, low-excess-air firing uses the minimum amount of excess air required for oxidation of the fuel, so that less oxygen is available for the reaction

$$N_2 + O_2 \rightarrow 2NO_2 \tag{11.6.37}$$

in the high temperature region of the flame. Incomplete fuel burnout, with the emission of hydrocarbons, soot, and CO, is an obvious problem with low-excess-air firing. This may be oevercome by a two-stage combustion process consisting of the following steps:

1. A first stage in which the fuel is fired at a relatively high temperature with a substoichiometric amount of air, for example, 90-95% of the stoichiometric requirement. NO formation is limited by the absence of excess oxygen.

2. A second stage in which fuel burnout is completed at a relatively low temperature in excess air. The low temperature prevents formation of NO.

In some power plants fired with gas, the emission of NO has been reduced by as much as 90% by a two-stage combustion process.

Removal of NO_x from stack gas presents some formidable problems. Possible approaches to NO_x removal are catalytic decomposition of nitrogen oxides, catalytic reduction of nitrogen oxides, and sorption of NO_x by liquids or solids.

A typical catalytic reduction of NO in stack gas involves methane:

$$CH_4 + 4NO \rightarrow 2N_2 + CO_2 + 2H_2O \tag{11.6.38}$$

Production of undesirable byproducts is a major concern in these processes. For example, sulfur dioxide reacts with carbon monoxide used to reduce NO to produce toxic carbonyl sulfide, COS:

$$SO_2 + 3CO \rightarrow 2CO_2 + COS \tag{11.6.39}$$

Most sorption processes have been aimed at the simultaneous removal of both nitrogen oxides and sulfur oxides. Sulfuric acid solutions or alkaline scrubbing solutions containing $Ca(OH)_2$ or $Mg(OH)_2$ may be used. The species N_2O_3 produced by the reaction

$$NO_2 + NO \rightarrow N_2O_3 \tag{11.6.40}$$

is most efficiently absorbed. Therefore, since NO is the primary combustion product, the introduction of NO_2 into the flue gas is required to produce the N_2O_3, which is absorbed efficiently.

11.7. ACID RAIN

As discussed in this chapter, much of the sulfur and nitrogen oxides entering the atmosphere are converted to sulfuric and nitric acids, respectively. When combined with hydrochloric acid arising from hydrogen chloride emissions, these acids cause acidic precipitation (acid rain) that is now a major pollution problem in some areas.

Headwater streams and high-altitude lakes are especially susceptible to the effects of acid rain and may sustain loss of fish and other aquatic life. Other effects include reductions in forest and crop productivity; leaching of nutrient cations and heavy metals from soils, rocks, and the sediments of lakes and streams; dissolution of metals such as lead and copper from water distribution pipes; corrosion of exposed metal; and dissolution of the surfaces of limestone buildings and monuments.

As a result of its widespread distribution and effects, acid rain is an air pollutant that may pose a threat to the global atmosphere. Therefore, it is discussed in greater detail in Chapter 14.

11.8 AMMONIA IN THE ATMOSPHERE

Ammonia is present even in unpolluted air as a result of natural biochemical and chemical processes. Among the various sources of atmospheric ammonia are micro-organisms, decay of animal wastes, sewage treatment, coke manufacture, ammonia manufacture, and leakage from ammonia-based refrigeration systems. High concentrations of ammonia gas in the atmosphere are generally indicative of accidental release of the gas.

Ammonia is removed from the atmosphere by its affinity for water and by its action as a base. It is a key species in the formation and neutralization of nitrate and sulfate aerosols in polluted atmospheres. Ammonia reacts with these acidic aerosols to form ammonium salts:

$$NH_3 + HNO_3 \rightarrow NH_4NO_3 \tag{11.8.1}$$

$$NH_3 + H_2SO_4 \rightarrow NH_4HSO_3 \tag{11.8.2}$$

Ammonium salts are among the more corrosive salts in atmospheric aerosols.

11.9. FLUORINE, CHLORINE, AND THEIR GASEOUS COMPOUNDS

Fluorine, hydrogen fluoride, and other volatile fluorides are produced in the manufacture of aluminum, and hydrogen fluoride is a byproduct in the conversion of fluorapatite (rock phosphate) to phosphoric acid, superphosphate ferilizers, and other phosphorus products. The wet process for the production of phosphoric acid involves the reaction of fluorapatite, $Ca_5F(PO_4)_3$, with sulfuric acid:

$$Ca_5F(PO_4)_3 + 5H_2SO_4 + 2H_2O \rightarrow CaSO_4 \cdot 2H_2O + HF + 3H_3PO_4 \tag{11.9.1}$$

It is necessary to recover most of the by-product fluorine from rock phosphate processing to avoid severe pollution problems. Recovery as fluosilicic acid, H_2SiF_6, is normally practiced.

Hydrogen fluoride gas is a dangerous substance that is so corrosive that it even reacts with glass. It is irritating to body tissues, and the respiratory tract is very sensitive to it. Brief exposure to HF vapors at the part-per-thousand level may be fatal. The acute toxicity of F_2 is even higher than that of HF. Chronic exposure to high levels of fluorides causes fluorosis, the symptoms of which include mottled teeth and pathological bone conditions.

Plants are particularly susceptible to the effects of gaseous fluorides. Fluorides from the atmosphere appear to enter the leaf tissue through the stomata. Fluoride is a cumulative poison in plants, and exposure of sensitive plants to even very low levels of fluorides for prolonged periods results in damage. Characteristic symptoms of fluoride poisoning are chlorosis (fading of green color due to conditions other than the absence of light), edge burn, and tip burn. Conifers (such as pine trees) afflicted with fluoride poisoning may have reddish-brown necrotic needle tips. The sensitivity of some conifers to fluoride poisoning is illustrated by the fact that fluorine produced by aluminum plants in Norway has destroyed forests of *Pinus sylvestris* up to 8 miles distant; trees were damaged at distances as great as 20 miles from the plant.

Silicon tetrafluoride gas, SiF_4, is another gaseous fluoride pollutant produced during some steel and metal smelting operations that employ CaF_2, fluorspar. Fluorspar reacts with silicon dioxide (sand), releasing SiF_4 gas:

$$2CaF_2 + 3SiO_2 \rightarrow 2CaSiO_3 + SiF_4 \tag{11.9.2}$$

Another gaseous fluorine compound, sulfur hexafluoride, SiF_6, occurs in the atmosphere at levels of about 0.3 parts per trillion. It is extremely unreactive and is used as an atmospheric tracer. It does not absorb ultraviolet light in either the troposphere or stratosphere and is probably destroyed above 50 km by reactions beginning with its capture of free electrons.

Chlorofluorocarbons

The fluorine-containing air pollutants with the greatest potential for damage to the atmosphere are the **chlorofluorocarbons (CFC)**, commonly called Freons, which are used as fluids in refrigeration mechanisms, as blowing agents in the fabrication of flexible and rigid foams, and, until several years ago, as propellants in spray cans containing deodorants, hair spray, and many other products. Chlorofluorocarbons are volatile 1- and 2-carbon compounds that contain Cl and F bonded to carbon. These compounds are notably stable and non-toxic. The most widely manufactured of these compounds are CCl_3F (CFC-11, bp 24°C), CCl_2F_2 (CFC-12, bp - 28°C), $C_2Cl_3F_3$ (CFC-113), $C_2Cl_2F_4$ (CFC-114), and C_2ClF_5 (CFC-115).

Halons are related compounds that contain bromine and are used in fire extinguisher systems. The major commercial halons are $CBrClF_2$ (Halon-1211), $CBrF_3$ (Halon-1301), and $C_2Br_2F_4$ (Halon-2402), where the sequence of numbers denotes the number of carbon, fluorine, chlorine, and bromine atoms, respectively, per molecule. Halons are particularly effective fire extinguishing agents because of the way in which they stop combustion. Halons act by chain reactions (see Section 3.11) that destroy hydrogen atoms which sustain combustion. The basic sequence of reactions involved is outlined below:

$$CBrClF_2 + H\cdot \rightarrow CClF_2\cdot + HBr \tag{11.9.3}$$

$$HBr + H\cdot \rightarrow Br\cdot + H_2 \tag{11.9.4}$$

Chain reaction

$$Br\cdot + H\cdot \rightarrow HBr \tag{11.9.5}$$

Halons are used in automatic fire extinguishing systems, particularly those located in flammable solvent storage areas, and in specialty fire extinguishers, such as those on aircraft.

The nonreactivity of CFC compounds, combined with worldwide production of approximately one-half million metric tons per year and deliberate or accidental release to the atmosphere, has resulted in CFCs becoming homogeneous components of the global atmosphere. In 1974 it was convincingly suggested[5] that chlorofluoromethanes could catalyze the destruction of stratospheric ozone, which filters out cancer-causing ultraviolet radiation from the sun. Although quite inert in the lower atmosphere, CFCs undergo photodecomposition by the action of high-energy ultraviolet radiation in the stratosphere, which is energetic enough to break their very strong C-Cl bonds through reactions such as,

$$Cl_2CF_2 + h\nu \rightarrow Cl\cdot + ClCF_2\cdot \tag{11.9.6}$$

thereby releasing Cl atoms. These atoms react with ozone, destroying it and producing ClO:

$$Cl + O_3 \rightarrow ClO + O_2 \tag{11.9.7}$$

In the stratosphere, there is an appreciable concentration of atomic oxygen, by virtue of the reaction

$$O_3 + hv \rightarrow O_2 + O \tag{11.9.8}$$

Nitric oxide, NO, is also present. The ClO species may react with either O or NO, regenerating Cl atoms and resulting in chain reactions that cause the net destruction of ozone:

$$ClO + O \rightarrow Cl + O_2 \tag{11.9.9}$$

$$\underline{Cl + O_2 \rightarrow ClO + O_2} \tag{11.9.10}$$

$$O_2 + O \rightarrow 2O_2 \tag{11.9.11}$$

$$ClO + NO \rightarrow Cl + NO_2 \tag{11.9.12}$$

$$\underline{O_3 + Cl \rightarrow ClO + O_2} \tag{11.9.13}$$

$$O_3 + NO \rightarrow NO_2 + O_2 \tag{11.9.14}$$

Both ClO and Cl involved in the above chain reactions have been detected in the 25-45-km region.[6]

The effects of CFCs on the ozone layer may be the single greatest threat to the global atmosphere and is discussed as such in Chapter 14. U. S. Environmental Protection Agency regulations imposed in accordance with the 1986 Montreal Protocol on Substances that Deplete the Ozone Layer, curtailed production of CFCs and halocarbons in the U. S. starting in 1989. The most likely substitutes for these halocarbons are hydrogen-containing chlorofluorocarbons (HCFCs) and hydrogen-containing fluorocarbons (HFCs). The substitute compounds most likely to be produced commercially first are CH_2FCF_3 (HFC-134a, a substitute for CFC-12 in automobile air conditioners and refrigeration equipment), $CHCl_2CF_3$ (HCFC-123, substitute for CFC-11 in plastic foam-blowing), CH_3CCl_2F (HCFC-141b, substitute for CFC-11 in plastic foam-blowing), $CHClF_2$ (HCFC-22, air conditioners and manufacture of plastic foam food containers). Because of the more readily broken H-C bonds that they contain, these compounds are more easily destroyed by atmospheric chemical reactions (particularly with hydroxyl radical, see Section 3.11) before they reach the stratosphere. Relative to a value of 1.0 for CFC-11, the ozone-depletion potentials of these substitutes are HFC-134a, 0; HCFC-123, 0.016; HCFC-141b, 0.081; and HCFC-22, 0.053. In 1991, Du Pont began marketing Freon substitutes for refrigeration applications under the trade name Suva.[7] Like the CFCs that they have been synthesized to replace, the four leading Freon substitutes have shown no evidence of causing skin or eye irritation, birth defects, or other short term toxic effects.[8]

Industrial concerns have undertaken a program to replace CFCs in manufacturing processes.[9] In August, 1989, The American Telephone and Telegraph Company, which consumes 3 million pounds of CFCs per year to manufacture circuit boards and electronic chips, stated that it will cut consumption in half by 1991 and totally by 1994. Other major companies that have announced similar intentions include Japan's Seiko Epson Company and Canada's Northern Telecom.

The Du Pont Company, which introduced chlorofluorocarbons in the 1930s and is the largest manufacturer of them, has announced that it intends to cease production shortly after the year 2000. The company has plans to manufacture HFC and HCFC substitutes during the early 1990s, including HFC-134a in a plant in Corpus Christi, Texas, HCFC-123 in a plant in Maitland, Ontario, and HCFC-141b in a plant in Montague, Michigan. ICI intends to produce HFC-134a in St. Gabriel, Louisiana, in the U.S. and in Runcorn, England.

Chlorine and Hydrogen Chloride

Chlorine gas, Cl_2, does not occur as an air pollutant on a large scale but can be quite damaging on a local scale. Chlorine was the first poisonous gas deployed in World War I. It is widely used as a manufacturing chemical, in the plastics industry, for example, as well as for water treatment and as a bleach. Therefore, possibilities for its release exist in a number of locations. Chlorine is quite toxic and is a mucous-membrane irritant. It is very reactive and a powerful oxidizing agent. Chlorine dissolves in atmospheric water droplets, yielding hydrochloric acid and hypochlorous acid, an oxidizing agent:

$$H_2O + Cl_2 \rightarrow H^+ + Cl^- + HOCl \tag{11.9.15}$$

Spills of chlorine gas have caused fatalities among exposed persons. For example, the rupture of a derailed chlorine tank car at Youngstown, Florida, on February 25, 1978, resulted in the deaths of 8 people who inhaled the deadly gas, and a total of 89 people were injured.

Hydrogen chloride, HCl, is emitted from a number of sources. Incineration of chlorinated plastics, such as polyvinylchloride, releases HCl as a combustion product.

$$\cdots\cdots \underset{\displaystyle \overset{|}{\underset{H}{}}}{\overset{\displaystyle \overset{Cl}{|}}{C}} - \underset{H}{\overset{H}{C}} - \underset{Cl}{\overset{H}{C}} - \underset{H}{\overset{H}{C}} - \underset{H}{\overset{Cl}{C}} - \underset{H}{\overset{H}{C}} - \underset{Cl}{\overset{H}{C}} - \underset{H}{\overset{H}{C}} - \underset{H}{\overset{Cl}{C}} - \underset{H}{\overset{H}{C}} \cdots\cdots \text{ Polyvinylchloride}$$

Some compounds released to the atmosphere as air pollutants hydrolyze to form HCl. One such incident occurred on April 26, 1974, when a storage tank containing 750,000 galons of liquid silicon tetrachloride, $SiCl_4$, began to leak in South Chicago, Illinois. This compound reacted with water in the atmosphere to form a choking fog of hydrochloric acid droplets:

$$SiCl_4 + 2H_2O \rightarrow SiO_2 + 4HCl \tag{11.9.16}$$

Many people became ill from inhaling the vapor.

In February, 1981, in Stroudsburg, Pennsylvania, a wrecked truck dumped 12 tons of powdered aluminum chloride during a rainstorm. This compound produces HCl gas when wet,

$$AlCl_3 + 3H_2O \rightarrow Al(OH)_3 + 3HCl \tag{11.9.17}$$

and more than 1,200 residents had to be evacuated from their homes because of the fumes generated.

11.10. HYDROGEN SULFIDE, CARBONYL SULFIDE, AND CARBON DISULFIDE

Hydrogen sulfide is produced by microbial decay of sulfur compounds and microbial reduction of sulfate (see Chapter 6), from geothermal steam, from wood pulping, and from a number of miscellaneous natural and anthropogenic sources. Most atmospheric hydrogen sulfide is rapidly converted to SO_2 and to sulfates. The organic homologs of hydrogen sulfide, the mercaptans, enter the atmosphere from decaying organic matter and have particularly objectionable odors.

Hydrogen sulfide pollution from artificial sources is not as much of an overall air pollution problem as sulfur dioxide pollution. However, there have been several acute incidents of hydrogen sulfide emissions resulting in damage to human health and even fatalities. The most notorious such incident occurred in Poza Rica, Mexico, in 1950. Accidental release of hydrogen sulfide from a plant used for the recovery of sulfur from natural gas caused the deaths of 22 people and the hospitalization of over 300. The symptoms of poisoning included irritation of the respiratory tract and damage to the central nervous system. Unlike sulfur dioxide, which appears to affect older people and those with respiratory weaknesses, there was little evidence of correlation between the observed hydrogen sulfide poisoning and the age or physical condition of the victim.

In a tragic incident which occurred in February 1975, hydrogen sulfide gas leaking from an experimental secondary-recovery oil well near Denver City, Texas, killed nine people trying to flee the lethal fumes. A process was being tried in which carbon dioxide, rather than water, was injected under high pressure to recover petroleum. Leakage from the well released deadly hydrogen sulfide present in the oil-bearing formation. Efforts to tap very deep natural gas formations have increased the hazard from hydrogen sulfide. A pocket of H_2S was struck at 15,000 feet while drilling such a well near Athens, Texas, in 1978. Leakage of hydrogen sulfide on May 12, 1978, forced the evacuation of 50 families. As an emergency measure in such cases, the gas may be ignited to form less toxic SO_2.

Hydrogen sulfide at levels well above ambient concentrations destroys immature plant tissue. This type of plant injury is readily distinguished from that due to other phytotoxins. More sensitive species are killed by continuous exposure to around 3000 ppb H_2S, whereas other species exhibit reduced growth, leaf lesions, and defoliation.

Damage to certain kinds of materials is a very expensive effect of hydrogen sulfide pollution. Paints containing lead pigments, $2PbCO_3 \cdot Pb(OH)_2$ (no longer used), are particularly susceptible to darkening by H_2S. Darkening results from exposure over several hours to as little as 50 ppb H_2S. The lead sulfide originally produced by reaction of the lead pigment with hydrogen sulfide eventually may be converted to white lead sulfate by atmospheric oxygen after removal of the source of H_2S, thus partially reversing the damage.

A black layer of copper sulfide forms on copper metal exposed to H_2S. Eventually, this layer is replaced by a green coating of basic copper sulfate such as $CuSO_4 \cdot 3Cu(OH)_2$. The green "patina," as it is called, is very resistant to further corrosion. Such layers of corrosion can seriously impair the function of copper contacts on electrical equipment. Hydrogen sulfide also forms a black sulfide coating on silver.

Carbonyl sulfide, COS, is now recognized as a component of the atmosphere at a tropospheric concentration of approximately 500 parts per trillion by volume, corresponding to a global burden of about 2.4 teragrams. It is, therefore, a significant sulfur species in the atmosphere. It is possible that the HO• radical-initiated oxidation of COS and carbon disulfide (CS_2) would yield 8-12 teragrams as S in atmospheric sulfur dioxide per year. Though this is a small yield compared to pollution sources, the HO•-initiated process could account for much of the SO_2 burden in the remote troposphere.

Both COS and CS_2 are oxidized in the atmosphere by reactions initiated by the hydroxyl radical. The initial reactions are

$$HO• + COS \rightarrow CO_2 + HS• \tag{11.10.1}$$

$$HO• + CS_2 \rightarrow COS + HS• \tag{11.10.2}$$

The sulfur-containing products undergo further reactions to sulfur dioxide and, eventually, to sulfate species.

LITERATURE CITED

1. Finlayson-Pitts, Barbara J., and James N. Pitts, Jr., "Sources, Atmospheric Lifetimes, and Chemical Fates of Species in the Natural Troposphere," Chapter 14 in *Atmospheric Chemistry: Fundamentals and Experimental Techniques*, John Wiley and Sons, New York, 1986, pp. 961-1007.

2. Seinfeld, John H., "Gas-Phase Atmospheric Chemistry," Chapter 4 in *Atmospheric Chemistry and Physics of Air Pollution*, John Wiley and Sons, New York, 1986, pp. 111-194.

3. Seinfeld, John H., "Aqueous-Phase Atmospheric Chemistry," Chapter 5 in *Atmospheric Chemistry and Physics of Air Pollution*, John Wiley and Sons, New York, 1986, pp. 195-249.

4. Sakugawa, Hiroshi, Isaac R. Kaplan, Wangteng Tsai, and Yoram Cohen, "Atmospheric Hydrogen Peroxide," *Environmental Science and Technology*, **24**, 1990, pp. 1452-1461.

5. Molina, M. J., and F. S. Rowland, *Nature*, **249**, p. 810

6. Anderson, J. G., J. J. Margitan, and Donald H. Stedman, "Atomic Chlorine and Chlorine Monoxide in the Stratosphere: Three *In Situ* Observations," *Science*, **198**, 1977, pp. 501-3.

7. "Du Pont Introduces New Refrigerants," *New York Times*, January 22, 1991, p. C4.

8. Zurer, Pamela S., "CFC Substitutes: Candidates Pass Early Toxicity Tests," *Chemical and Engineering News*, October 9, 1989, p. 4.

9. Shabecoff, Philip, "A.T. & T. Barring Chemicals Depleting Ozone Layer," *New York Times*, Aug. 2, 1989, p. 8

SUPPLEMENTARY REFERENCES

Cogan, Douglas G., *Stones in a Glass House: CFCs and Ozone Depletion*, Investor Responsibility Research Center Inc., Washington, DC, 1988.

Makhijani, Arjun, Annie Makhijani, and Amanda Bickel, *Saving Our Skins: Technical Potential and Policies for the Elimination of Ozone-Depleting Chlorine Compounds*, Environmental Policy Institute and the Institute for Energy & Environmental Research, Washington, DC, 1988.

Watson, Ann Y., Richard R. Bates, and Donald Kennedy, Eds., *Air Pollution, the Automobile, and Public Health*, National Academy Press, Washington, DC, 1988.

Warneck, Peter, *Chemistry of the Natural Atmosphere*, Academic Press, San Diego, CA, 1988.

Regens, James L., and Robert W. Rycroft, *The Acid Rain Controversy*, University of Pittsburgh Press, Pittsburgh, PA, 1988.

White, James C., Ed., *Acid Rain: The Relationship Between Sources and Receptors*, Elsevier, New York, NY, 1988.

Park, Chris C., *Acid Rain: Rhetoric and Reality*, Methuen, New York, NY, 1988.

Licht, William, *Air Pollution Control Engineering*, 2nd Ed., Marcel Dekker, New York, NY, 1988.

Benarie, M. M., Ed., *Atmospheric Pollution*, Elsevier Publlishing Co., New York, 1982.

Bragg, G. M., and W. Strauss, Eds., *Air Pollution Control*, Part 4, John Wiley and Sons, Inc., New York, 1981.

Glassman, I., *Combustion*, Academic Press, Inc., New York, 1977.

Lee, S. D., Ed., *Nitrogen Oxides and their Effects on Health*, Ann Arbor Science Publishers, Inc., Ann Arbor, Michigan, 1980.

Nriagu, J. O., *Sulfur in the Environment*, John Wiley and Sons, Inc., New York, 1978.

Theodore, L., and A. J. Buonicore, *Air Pollution Control Equipment: Selection, Design, Operation, and Maintenance*, Prentice-Hall, Inc., Englewood Cliffs, NJ, 1982.

Shriner, D. S., C. R. Richmond, and S. E. Lindberg, Eds., *Atmospheric Sulfur Deposition: Environmental Impact and Health Effects*, Ann Arbor Science Publishers,Inc., Ann Arbor, Michigan, 1980.

Schryer, D. R., Ed., *Heterogeneous Atmospheric Chemistry*, AGU Publications, Inc., Washington, DC, 1982.

QUESTIONS AND PROBLEMS

1. Why is it that "highest levels of carbon monoxide tend to occur in congested urban areas at times when the maximum number of people are exposed?"

2. Which unstable, reactive species is responsible for the removal of CO from the atmosphere?

3. Which of the following fluxes in the atmospheric sulfur cycle is smallest: (a) Sulfur species washed out in rainfall over land, (b) sulfates entering the atmosphere as "sea salt," (c) sulfur species entering the atmosphere from volcanoes, (d) sulfur species entering the atmosphere from fossil fuels, (e) hydrogen sulfide entering the atmosphere from biological processes in coastal areas and on land.

4. Of the following agents, the one that would not favor conversion of sulfur dioxide to sulfate species in the atmosphere is: (a) Ammonia, (b) water, (c) contaminant reducing agents, (d) ions of transition metals such as manganese, (e) sunlight.

5. Of the stack gas scrubber processes discussed in this chapter, which is the least effficient for the removal of SO_2?

6. The air inside a garage was found to contain 10 ppm CO by volume at standard temperature and pressure (STP). What is the concentration of CO in mg/L and in ppm by mass?

7. Assume that an incorrectly adjusted lawn mower is operated in a garage such that the combustion reaction in the engine is

$$C_8H_{18} + {}^{17}/_2O_2 \rightarrow 8CO + 9H_2O$$

If the dimensions of the garage are 5 x 3 x 3 meters, how many grams of gasoline must be burned to raise the level of CO in the air to 1000 ppm by volume at STP?

8. A 12.0-L sample of waste air from a smelter process was collected at 25°C and 1.00 atm pressure, and the sulfur dioxide was removed. After SO_2 removal, the volume of the air sample was 11.50 L. What was the percentage by weight of SO_2 in the original sample?

9. What is the oxidant in the Claus reaction?

10. Carbon monoxide is present at a level of 10 ppm by volume in an air sample taken at 15°C and 1.00 atm pressure. At what temperature (at 1.00 atm pressure) would the sample also contain 10 mg/m^3 of CO?

11. How many metric tons of 5%–S coal would be needed to yield the H_2SO_4 required to produce a 3.00–cm rainfall of pH 2.00 over a 100 km^2 area?

12. In what major respect is NO_2 a more significant species than SO_2 in terms of participation in atmospheric chemical reactions?

13. How many metric tons of coal containing an average of 2% S are required to produce the SO emitted by fossil fuel combustion shown in Figure 11.1? (Note that the values given in the figure are in terms of elemental sulfur, S.) How many metric tons of SO are emitted?

14. Assume that the wet limestone process requires 1 metric ton of $CaCO_3$ to remove 90% of the sulfur from 4 metric tons of coal containing 2% S. Assume that the sulfur product is $CaSO_4$. Calculate the percentage of the limestone converted to calcium sulfate.

15. Referring to the two preceding problems, calculate the number of metric tons of $CaCO_3$ required each year to remove 90% of the sulfur from 1 billion metric tons of coal (approximate annual U.S. consumption), assuming an average of 2% sulfur in the coal.

16. If a power plant burning 10,000 metric tons of coal per day with 10% excess air emits stack gas containing 100 ppm by volume of NO, what is the daily output of NO?

17. How many cubic kilometers of air at 25°C and 1 atm pressure would be contaminated to a level of 0.5 ppm NO_x from the power plant discussed in the preceding question?

Organic Air Pollutants

12.1. ORGANIC COMPOUNDS IN THE ATMOSPHERE

Organic pollutants may have a strong effect upon atmospheric quality. The effects of organic pollutants in the atmosphere may be divided into two major catergories. The first consists of **direct effects**, such as cancer caused by exposure to vinyl chloride. The second is the formation of **secondary pollutants**, especially photochemical smog (discussed in detail in Chapter 13). In the case of pollutant hydrocarbons in the atmosphere, the latter is the more important effect. In some localized situations, particularly the workplace, direct effects of organic air pollutants may be equally important.

This chapter discusses the nature and distribution of organic compounds in the atmosphere. Chapter 13 deals with photochemical smog and addresses the mechanisms by which organic compounds undergo photochemical reactions in the atmosphere.

12.2. ORGANIC COMPOUNDS FROM NATURAL SOURCES

Natural sources are the most important contributors of organics in the atmosphere. For instance, hydrocarbons generated and released by human activities consitute only about 1/7 of the total hydrocarbons in the atmosphere. This ratio is primarily the result of the huge quantities of methane produced by anaerobic bacteria in the decomposition of organic matter in water, sediments, and soil:

$$2\{CH_2O\} \text{ (bacterial action)} \rightarrow CO_2(g) + CH_4(g) \qquad (12.2.1)$$

Flatulent emissions from domesticated animals, arising from bacterial decomposition of food in their digestive tracts, add about 85 million metric tons of methane to the atmosphere each year. Methane is a natural constituent of the atmosphere and is present at a level of about 1.4 parts per million (ppm) in the troposphere.

Methane in the troposphere contributes to the photochemical production of carbon monoxide and ozone. The photochemical oxidation of methane is a major source of water vapor in the stratosphere.

Vegetation is the most important natural source of atmospheric hydrocarbons. A compilation of organic compounds in the atmosphere[1] lists a total of 367 different compounds that are released to the atmosphere from vegetation sources. Other natural sources include microorganisms, forest fires, animal wastes, and volcanoes.

One of the simplest organic compounds given off by plants is ethylene, C_2H_4. This compound is produced by a variety of plants and released to the atmosphere. Because of its double bond, ethylene is highly reactive with hydroxyl radical, $HO^•$, and with oxidizing species in the atmosphere. Ethylene from vegetation sources should be considered as an active participant in atmospheric chemical processes.

Most of the hydrocarbons emitted by plants are **terpenes**, which constitute a large class of organic compounds found in essential oils. Essential oils are obtained when parts of some types of plants are subjected to steam distillation. Most of the plants that produce terpenes belong to the family *Coniferae*, the family *Myrtaceace*, and the genus *Citrus*. One of the most common terpenes emitted by trees is α-pinene, a principal component of turpentine. The terpene limonene, found in citrus fruit and pine needles, is encountered in the atmosphere around these sources. Isoprene (2-methyl-1,3-butadiene), a hemiterpene, has been identified in the emissions from cottonwood, eucalyptus, oak, sweetgum and white spruce trees. Oather terpenes known to be given off by trees include β-pinene, myrcene, ocimene, and α-terpinene.

As exemplified by the structures of α-pinene, isoprene, and limonene,

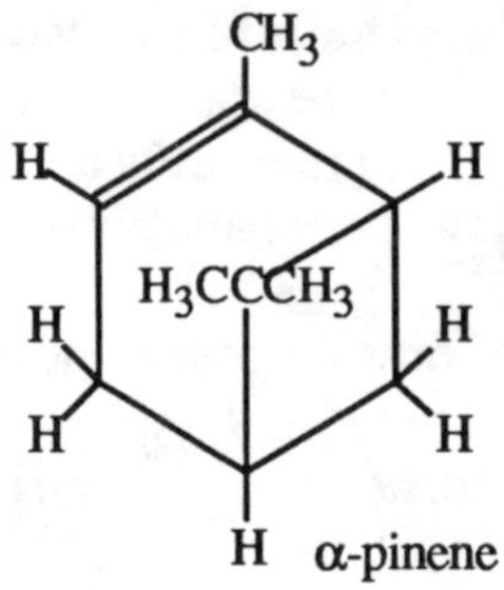

isoprene

α-pinene

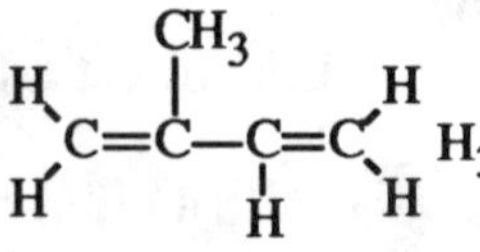

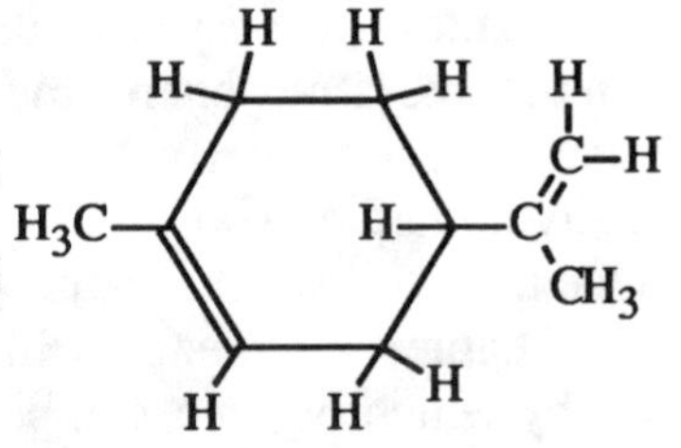

limonene

terpenes contain alkenyl (olefinic) bonds, usually two or more per molecule. Because of these and other structural features, terpenes are among the most reactive compounds in the atmosphere. The reaction of terpenes with hydroxyl radical, $HO^•$, is very rapid, and terpenes also react with other oxidizing agents in the atmosphere, particularly ozone, O_3. Turpentine, a common mixture of terpenes, has been widely used in paint because it reacts with atmospheric oxygen to form a peroxide, then a hard resin. It is likely that compounds such as α-pinene and isoprene undergo similar reactions in the atmosphere to form particulate matter. The resulting Aitken nuclei aerosols (see Chapter 10) probably cause the blue haze in the atmosphere above some heavy growths of vegetation.

Smog-chamber experiments have been performed in an effort to determine the fate of atmospheric terpenes. For example, when a mixture of α -pinene with NO and NO_2 in air is irradiated with ultraviolet light, pinonic acid is formed:

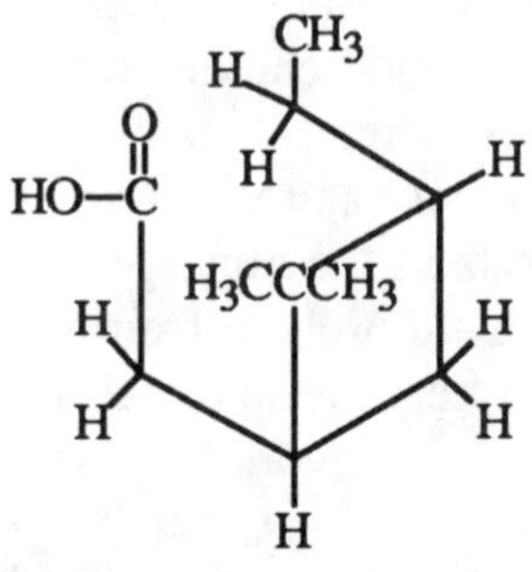

Found in forest aerosol particles, this compound is almost certainly produced by photochemical processes acting upon α-pinene.

In an investigation of the influence of α-pinene on the smog-forming tendencies of a synthetic mixture of hydrocarbons,[2] it was found that substitution of the terpene for up to 20% of the synthetic mixture of low-molecular-weight alkanes and alkenes did not appreciably increase the overall reactivity of the mixture as evidenced by the behavior of NO, NO_2, and O_3. Therefore, despite the higher reactivity of α-pinene as measured by its reaction with hydroxyl radical, its presence in moderate quantities along with synthetic hydrocarbons does not dramatically increase the smog-forming tendencies of a contaminated atmosphere.

Perhaps the greatest variety of compounds emitted by plants consist of **esters**. However, they are released in such small quantities that they have little influence upon atmospheric chemistry. Esters are primarily responsible for the fragrances associated with much vegetation. Some typical esters that are released by plants to the atmosphere are shown below:

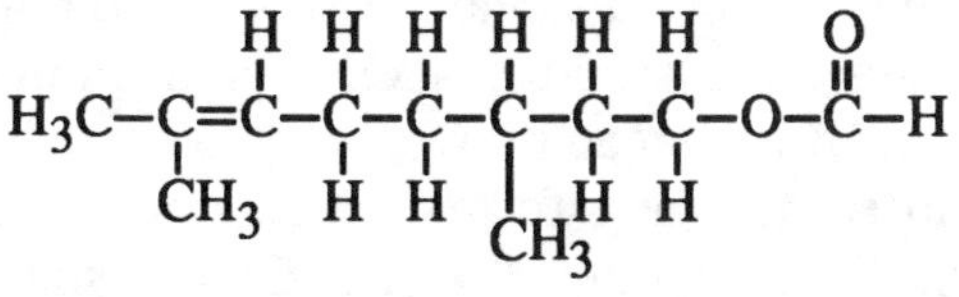

Citronellyl formate

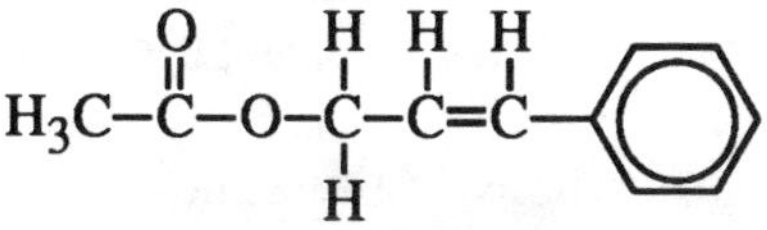

Cinnamyl acetate 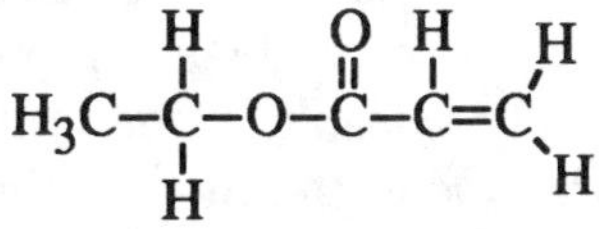   Ethyl acrylate

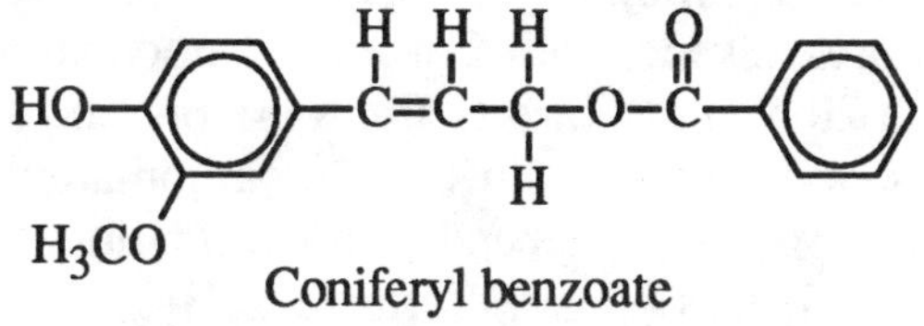

Coniferyl benzoate

12.3. POLLUTANT HYDROCARBONS

Ethylene and terpenes, which were discussed in the preceding section, are **hydrocarbons**, organic compounds containing only hydrogen and carbon. The major classes of hydrocarbons are **alkanes** (formerly called paraffins), such as 2,2,3-trimethylbutane;

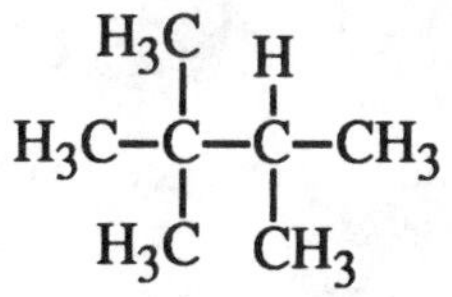

alkenes (olefins, compounds with double bonds between adjacent carbon atoms), such as ethylene; **alkynes** (compounds with triple bonds), such as acetylene;

$$H-C\equiv C-H$$

and **aromatic compounds**, such as naphthalene:

Because of their widespread use in fuels, hydrocarbons predominate among organic atmospheric pollutants. Petroleum products, primarily gasoline, are the source of most of the anthropogenic pollutant hydrocarbons found in the atmosphere. Hydrocarbons may enter the atmosphere either directly or as by-products of the partial combustion of other hydrocarbons. The latter are particularly important because they tend to be unsaturated and relatively reactive (see Chapter 13 for a discussion of hydrocarbon reactivity in photochemical smog formation). Most hydrocarbon pollutant sources produce about 15% reactive hydrocarbons, whereas those from gasoline are about 45% reactive. It has been shown that the hydrocarbons in uncontrolled automobiles exhausts are only about 1/3 alkanes, with the remainder divided approximately equally between more reactive alkenes and aromatic hydrocarbons, thus accounting for the relatively high reactivity of automotive exhaust hydrocarbons.

Investigators who study smog formation in smog chambers have developed synthetic mixtures of hydrocarbons that mimic the smog-forming behavior of hydrocarbons in a polluted atmosphere. The composition of one such mixture and the structures of the constituent hydrocarbons are given in Table 12.1. The compounds shown in this table provide a simplified idea of the composition of pollutant hydrocarbons likely to lead to smog formation.

Alkanes are among the more stable hydrocarbons in the atmosphere. Straight-chain alkanes with 1 to more than 30 carbon atoms and branched-chain alkanes with 6 or fewer carbon atoms are commonly present in polluted atmospheres. Because of their high vapor pressures, alkanes with 6 or fewer carbon atoms are normally present as gases, alkanes with 20 or more carbon atoms are present as aerosols or sorbed to atmospheric particles, and alkanes with 6 to 20 carbon atoms per molecule may be present either as vapor or particles, depending upon conditions.

In the atmosphere, alkanes (general formula C_xH_{2x+2}) are attacked primarily by hydroxyl radical, HO$\cdot$, resulting in the loss of a hydrogen atom and formation of an **alkyl radical**,

$$C_xH_{2x+1}\cdot$$

Subsequent reaction with O_2 causes formation of **alkylperoxyl radical**,

$$C_xH_{2x+1}O_2\cdot$$

These radicals may act as oxidants, losing oxygen to produce **alkoxyl radicals**:

$$C_xH_{2x+1}O\cdot$$

As a result of these and subsequent reactions, lower-molecular-mass alkanes are eventually oxidized to species that can be precipitated from the atmosphere with particulate matter to eventually undergo biodegradation in soil.

Table 12.2. Hydrocarbons Composing a Typical Synthetic Mixture in Smog Chamber Experiments.

Hydrocarbon name	Hydrocarbon structure	Mole percent in hydrocarbon mixture
Alkanes:		
2-Methylbutane	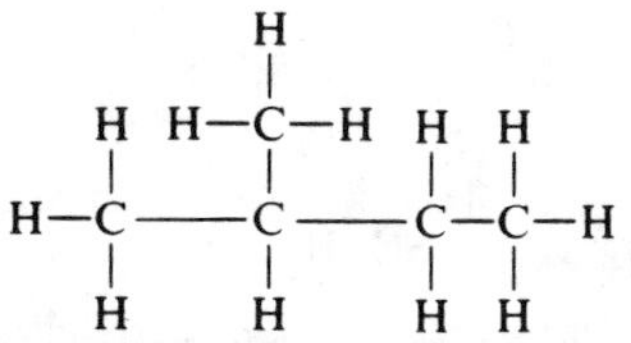	0.127
n-Pentane	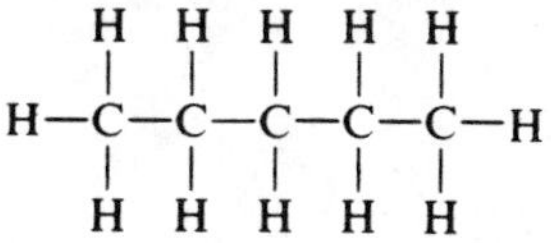	0.241
2-Methylpentane	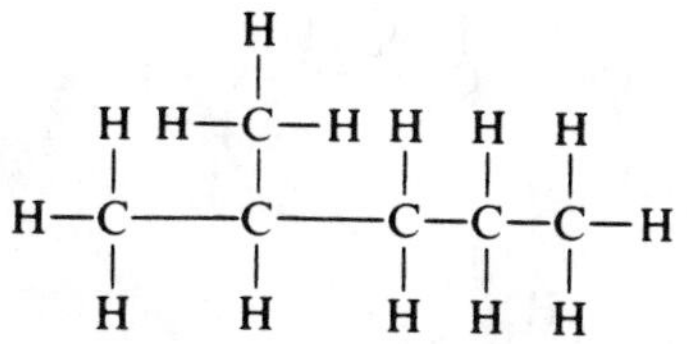	0.071
2,4-Dimethylpentane	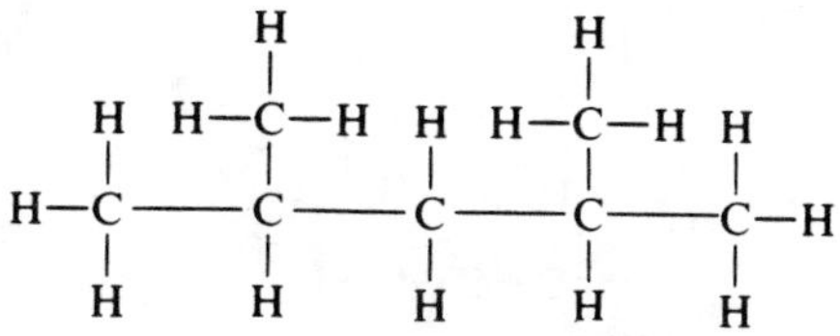	0.053
2,2,4-Trimethylpentane	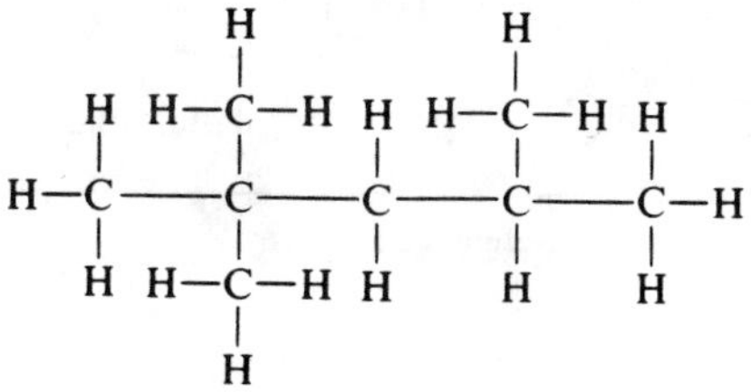	0.064
	Total alkanes:	0.556
Alkenes:		
1-Butene	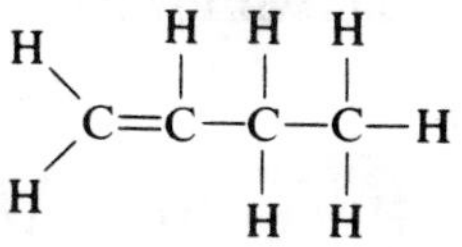	0.027

Hydrocarbon name	Hydrocarbon structure	Mole percent in hydrocarbon mixture[1]
cis-2-Butene	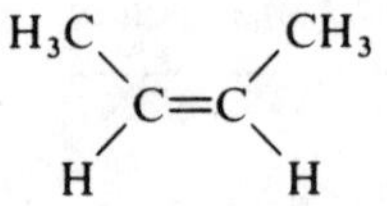	0.034
2-Methyl-1-butene	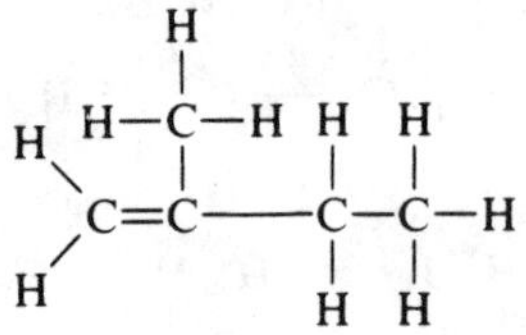	0.030
2-Methyl-2-butene	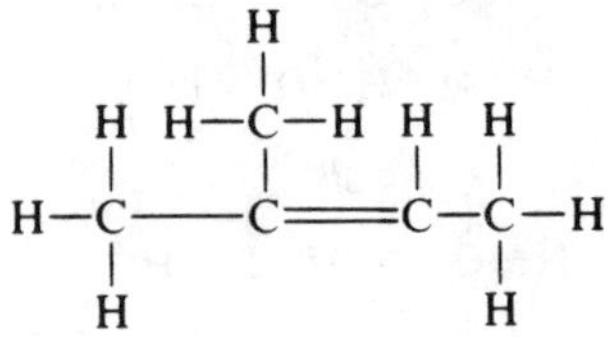	0.027
Ethylene	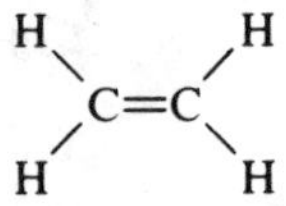	0.250
Propylene	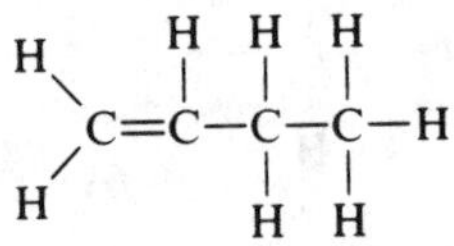	0.075
	Total alkenes:	0.443

[1]Synthetic mixture of hydrocarbons cited in Reference 5.

Alkenes enter the atmosphere from a variety of processes, including emissions from internal combustion engines and turbines, foundry operations, and petroleum refining. Several alkenes, including the ones shown below, are among the top 50 chemicals produced each year, with annual worldwide production of several billion kg:

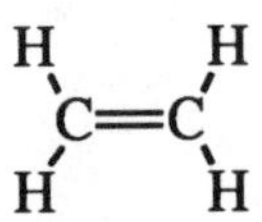

Ethylene 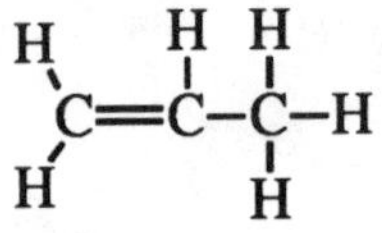   Propylene 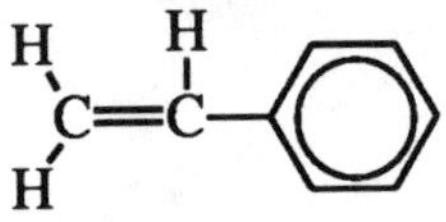   Styrene

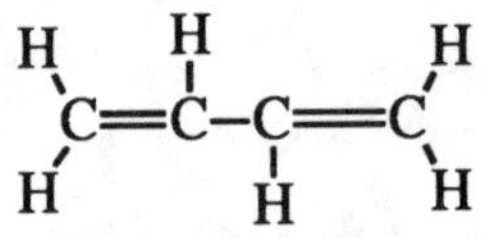

Butadiene

These compounds are used primarily as monomers, which are polymerized to create polymers for plastics (polyethylene, polypropylene, polystyrene), synthetic rubber (styrenebutadiene, polybutadiene), latex paints (styrenebutadiene), and other applications. All of these compounds, as well as others manufactured in lesser quantities, are released to the atmosphere. In addition to the direct release of alkenes, these hydrocarbons are commonly produced by the partial combustion and "cracking" at high temperatures of alkanes, particularly in the internal combustion engine.

Alkynes occur much less commonly in the atmosphere than do alkenes. Detectable levels are sometimes found of acetylene, used as a fuel for welding torches, and 1-butyne, used in synthetic rubber manufacture:

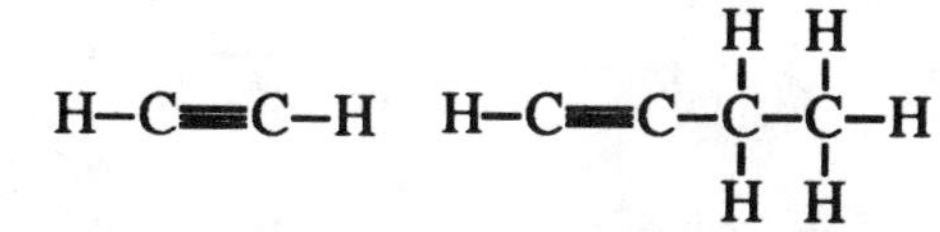

$$H-C\equiv C-H \qquad H-C\equiv C-\underset{\underset{H}{|}}{\overset{\overset{H}{|}}{C}}-\underset{\underset{H}{|}}{\overset{\overset{H}{|}}{C}}-H$$

Acetylene   1-Butyne

Unlike alkanes, alkenes are highly reactive in the atmosphere, especially in the presence of NO_x and sunlight. Hydroxyl radical reacts readily with alkenes, either by abstracting a hydrogen atom or by adding to the double bond. If hydroxyl radical adds to the double bond in propylene, for example, the product is:

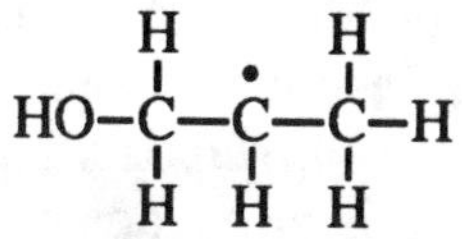

$$HO-\underset{\underset{H}{|}}{\overset{\overset{H}{|}}{C}}-\underset{\underset{H}{|}}{\overset{\overset{\bullet}{}}{C}}-\underset{\underset{H}{|}}{\overset{\overset{H}{|}}{C}}-H$$

Addition of molecular O_2 to this radical results in the formation of a peroxyl radical:

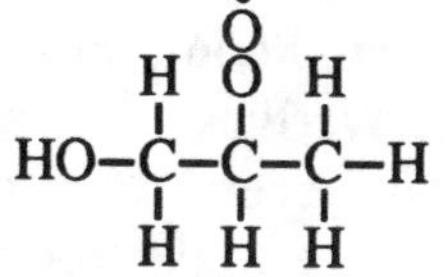

$$HO-\underset{\underset{H}{|}}{\overset{\overset{H}{|}}{C}}-\underset{\underset{H}{|}}{\overset{\overset{\overset{\bullet}{O}}{\underset{|}{O}}}{C}}-\underset{\underset{H}{|}}{\overset{\overset{H}{|}}{C}}-H$$

These radicals then participate in reaction chains, such as those discussed for the formation of photochemical smog in Chapter 13.

Ozone, O_3, adds across double bonds and is rather reactive with alkenes. The exact chemistry of ozone-alkene reactions in the atmosphere is not well known, but aldehydes are among the products.

12.4. AROMATIC HYDROCARBONS

Aromatic hydrocarbons may be divided into the two major classes of those that have only one benzene ring and those with multiple rings. As discussed in Chapter 10, the latter are *polycyclic aromatic hydrocarbons, PAH*. Aromatic hydrocarbons with two rings, such as naphthalene, are intermediate in their behavior. Some typical aromatic hydrocarbons are:

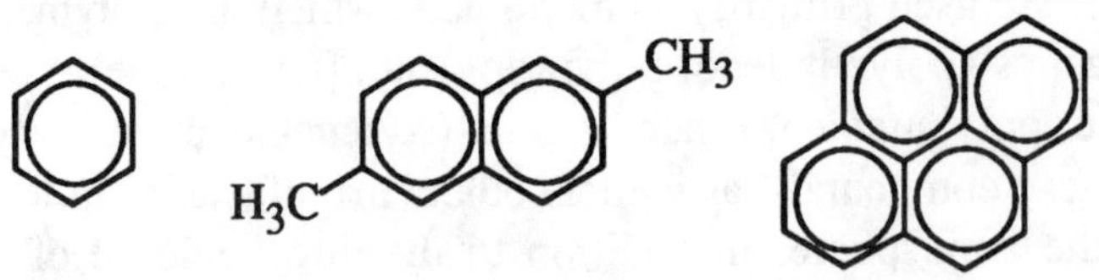

Benzene   2,6-Dimethylnaphthalene   Pyrene

The following aromatic hydrocarbons are among the top 50 chemicals manufactured each year:

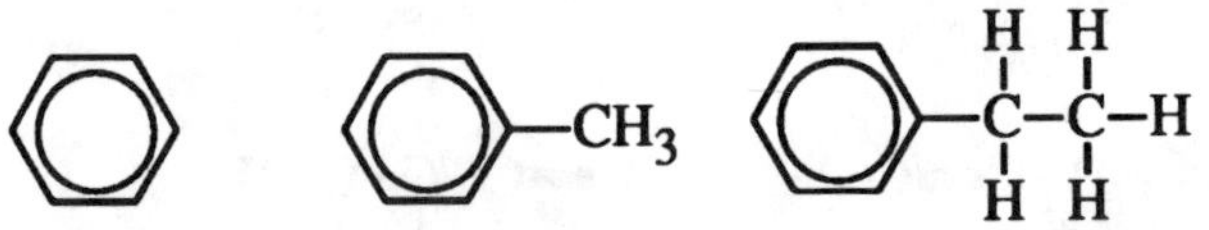

Benzene   Toluene   Ethylbenzene

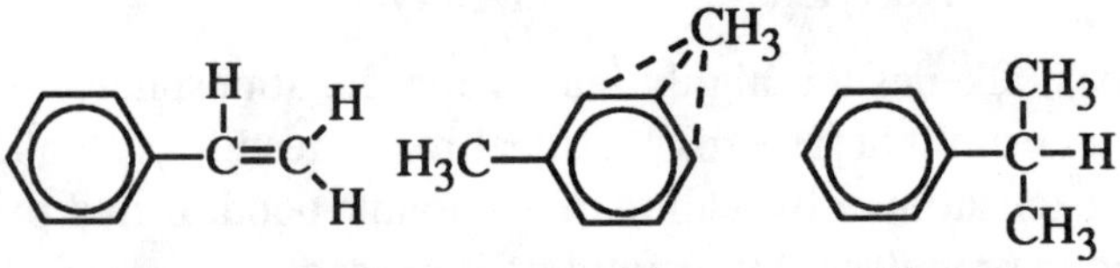

Styrene   Xylene (3 isomers)   Cumene

Single-ring aromatic compounds are important constituents of lead-free gasoline, which has largely replaced leaded gasoline. Aromatic solvents are widely used in industry. Aromatic hydrocarbons are raw materials for the manufacture of monomers and plasticizers in polymers. Styrene is a monomer in plastics and synthetic rubber. Cumene is oxidized to produce phenol and valuable acetone as a byproduct. With all of these applications, plus production of these compounds as combustion byproducts, it is no surprise that aromatic compounds are common atmospheric pollutants.

Approximately 55 hydrocarbons containing a single benzene ring and approximately 30 hydrocarbon derivatives of naphthalene have been found as atmospheric pollutants.[1] In addition, several compounds containing two or more *unconjugated* rings (not sharing the same π electron cloud between rings) have been detected as atmospheric pollutants. One such compound is biphenyl,

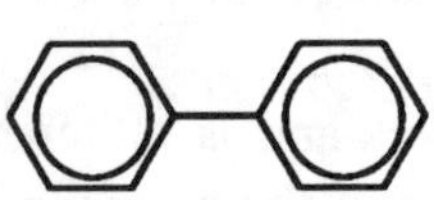

Biphenyl

detected in diesel smoke. It should be pointed out that many of these aromatic hydrocarbons have been detected primarily as ingredients of tobacco smoke and are, therefore, of much greater significance in an indoor environment than in an outdoor one.

Reactions of Atmospheric Aromatic Hydrocarbons

As with most atmospheric hydrocarbons, the most likely reaction of benzene and its derivatives is with hydroxyl radical. Addition of HO· to the benzene ring results in the formation of an unstable radical species,

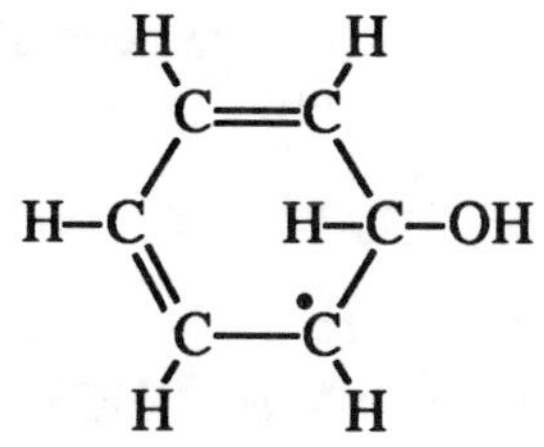

$$
\begin{array}{c}
\text{H} \quad\quad \text{H} \\
\text{C}=\text{C} \\
\text{H}-\text{C} \quad\quad \text{H}-\text{C}-\text{OH} \\
\text{C}-\overset{\bullet}{\text{C}} \\
\text{H} \quad\quad \text{H}
\end{array}
$$

where the dot denotes an unpaired electron in the radical. The electron is not confined to one atom; therefore, it is **delocalized** and may be represented in the aromatic radical structure by a half-circle with a dot in the middle. Using this notation for the radical above, its reaction with O_2 is,

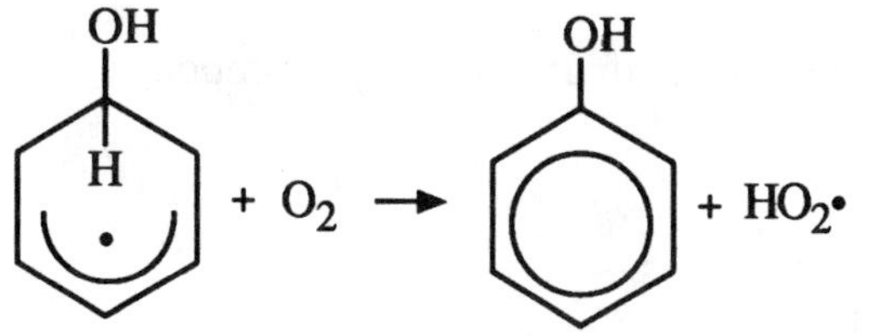

$$(12.4.1)$$

to form stable phenol and reactive hydroperoxyl radical, $HOO^{\bullet}$. Alkyl-substituted aromatics may undergo reactions involving the alkyl group. For example, abstraction of alkyl H by $HO^{\bullet}$ from a compound such as *p*-xylene can result in the formation of a radical,

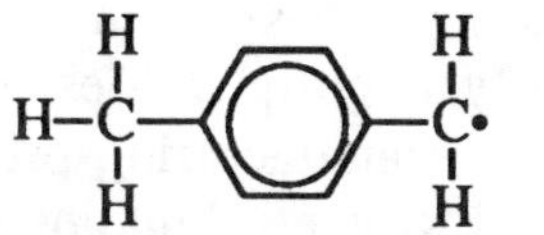

which can react further with O_2 to form a peroxyl radical, then enter chain reactions involved in the formation of photochemical smog (Chapter 13).

As discussed in Section 10.8 polycyclic aromatic hydrocarbons are present as aerosols in the atmosphere because of their extremely low vapor pressures. These compounds are the most stable form of hydrocarbons having low hydrogen-to-carbon ratios and are formed by the combustion of hydrocarbons under oxygen-deficient conditions. The partial combustion of coal, which has a hydrogen-to-carbon ratio less than 1, is a major source of PAH compounds.

12.5. ALDEHYDES AND KETONES

Carbonyl compounds, consisting of aldehydes and ketones, are often the first species formed, other than unstable reaction intermediates, in the photochemical oxidation of atmospheric hydrocarbons. The general formulas of aldehydes and ketones are represented by the following, where R and R' represent the hydrocarbon *moieties* (portions), such as the –CH₃ group.

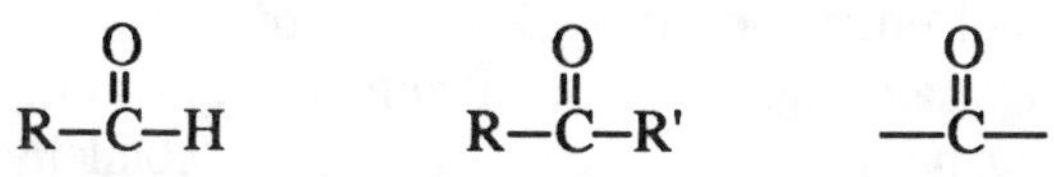

$$
\begin{array}{ccc}
\overset{\text{O}}{\overset{\|}{\text{R}-\text{C}-\text{H}}} & \quad \overset{\text{O}}{\overset{\|}{\text{R}-\text{C}-\text{R}'}} & \quad \overset{\text{O}}{\overset{\|}{-\text{C}-}} \\
\text{Aldehyde} & \text{Ketone} & \text{Carbonyl moiety}
\end{array}
$$

Carbonyl compounds are byproducts of the generation of hydroxyl radicals from organic peroxyl radicals (see Section 12.3) by reactions such as the following:

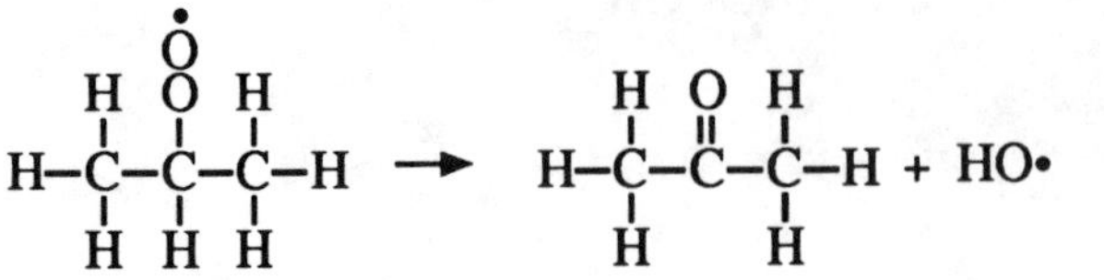

$$\begin{array}{ccccc}
& \overset{\displaystyle\cdot}{\underset{}{O}} & & & \\
H & O & H & & \\
| & | & | & & \\
H\text{--}C\text{--}C\text{--}C\text{--}H & \longrightarrow & H\text{--}C\text{--}C\text{--}C\text{--}H + HO\cdot & & (12.5.1)\\
| & | & | & & \\
H & H & H & &
\end{array}$$

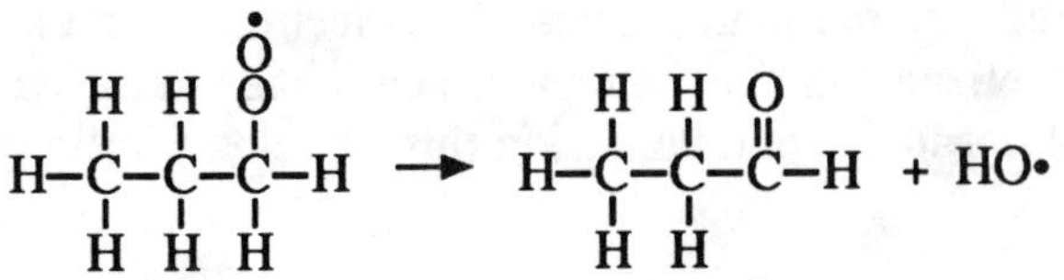

$$\begin{array}{ccccc}
& & \overset{\displaystyle\cdot}{\underset{}{O}} & & \\
H & H & O & & \\
| & | & | & & \\
H\text{--}C\text{--}C\text{--}C\text{--}H & \longrightarrow & H\text{--}C\text{--}C\text{--}C\text{--}H + HO\cdot & & (12.5.2)\\
| & | & | & & \\
H & H & H & &
\end{array}$$

The simplest and most widely produced of the carbonyl compounds is **formaldehyde**,

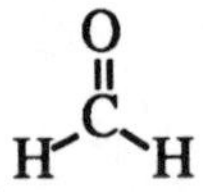

$$\begin{array}{c}
O \\
\| \\
C \\
H \diagup \diagdown H
\end{array} \quad \text{Formaldehyde}$$

with annual global production exceeding 1 billion kg. Used in the manufacture of plastics, resins, lacquers, dyes, and explosives, formaldehyde is uniquely important because of its widespread distribution and toxicity. Humans may be exposed to formaldehyde in the manufacture and use of phenol, urea, and melamine resin plastics and from formaldehyde-containing adhesives in pressed wood products, such as particle board, used in especially large quantities in mobile home construction.[3] However, significantly improved manufacturing processes have greatly reduced formaldehyde emissions from these synthetic building materials. Formaldehyde occurs in the atmosphere primarily in the gas phase.

The structures of some important aldehydes and ketones are shown below:

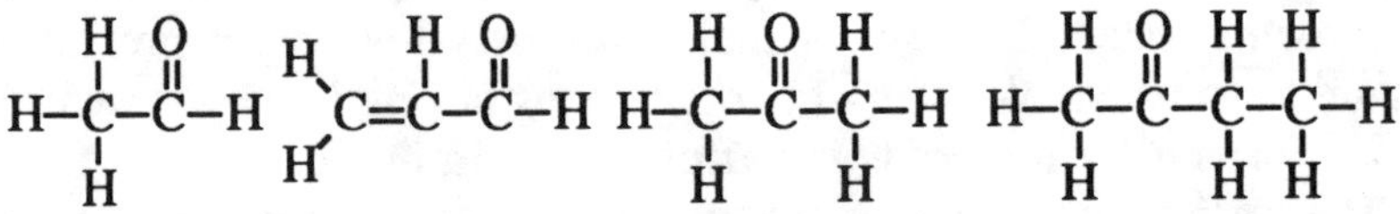

$$\begin{array}{cccc}
\text{Acetaldehyde} & \text{Acrolein} & \text{Acetone} & \text{Methylethyl ketone}
\end{array}$$

Acetaldehyde is a widely produced organic chemical used in the manufacture of acetic acid, plastics, and raw materials. Of the order of a billion kg of acetone are produced each year as a solvent and for applications in the rubber, leather, and plastics industries. Methylethyl ketone is employed as a low-boiling solvent for coatings and adhesives and for the synthesis of other chemicals.

In addition to their production from hydrocarbons by photochemical oxidation, carbonyl compounds enter the atmosphere from a large number of sources and processes. These include direct emissions from internal combustion engine exhausts, incinerator emissions, spray painting, polymer manufacture, printing, petrochemicals manufacture, and lacquer manufacture. Formaldehyde and acetaldehyde are produced by microorganisms and acetaldehyde is emitted by some kinds of vegetation.[4]

Aldehydes are second only to NO_2 as atmospheric sources of free radicals produced by the absorption of light. This is because the carbonyl group is a **chromophore**, a molecular group that readily absorbs light. It absorbs well in the near-ultraviolet region of the spectrum. The activated compound produced when a photon is absorbed by an aldehyde dissociates into a formyl radical,

$$\overset{\bullet}{H}CO$$

and an alkyl radical. The photodissociation of acetaldehyde illustrates this two-step process:

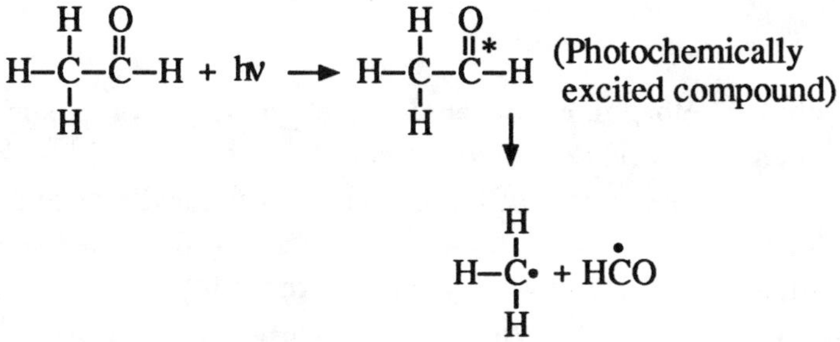

$$
\underset{\underset{H}{|}}{\overset{\overset{H}{|}}{H-C}}-\overset{O}{\overset{||}{C}}-H + h\nu \longrightarrow \underset{\underset{H}{|}}{\overset{\overset{H}{|}}{H-C}}-\overset{O}{\overset{||*}{C}}-H \quad \text{(Photochemically excited compound)}
$$

$$
\underset{\underset{H}{|}}{\overset{\overset{H}{|}}{H-C}}\bullet + H\overset{\bullet}{C}O \qquad (12.5.3)
$$

Photolytically excited formaldehyde, CH_2O, may dissociate in two ways. The first of these produces an H atom and $H\overset{\bullet}{C}O$ radical; the second produces chemically stable H_2 and CO.

Because of the presence of both double bonds and carbonyl groups, olefinic aldehydes are especially reactive in the atmosphere. The most common of these found in the atmosphere is acrolein,

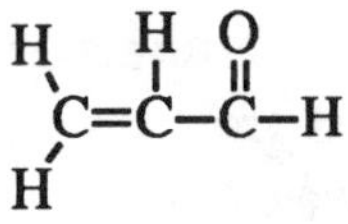

$$
\underset{H}{\overset{H}{\diagdown}}C=\underset{\underset{H}{|}}{\overset{\overset{H}{|}}{C}}-\overset{O}{\overset{||}{C}}-H
$$

a powerful lachrymator (tear producer) which is used as an industrial chemical and produced as a combustion byproduct.

Ketones commonly undergo photochemical dissociation in the atmosphere at one of the bonds joining the carbonyl group to the hydrocarbon moieties:

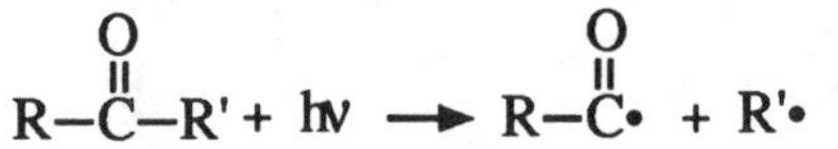

$$
R-\overset{O}{\overset{||}{C}}-R' + h\nu \longrightarrow R-\overset{O}{\overset{||}{C}}\bullet + R'\bullet \qquad (12.5.4)
$$

The radicals produced then undergo subsequent reactions with O_2 and other chemical species in the atmosphere.

12.6. MISCELLANEOUS OXYGEN-CONTAINING COMPOUNDS

Oxygen-containing aldehydes, ketones, and esters in the atmosphere were covered in preceding sections. This section discusses the oxygen-containing organic compounds consisting of **aliphatic alcohols, phenols, ethers,** and **carboxylic acids**. These compounds have the general formulas given below, where R and R' represent hydrocarbon moieties and ϕ stands specifically for an aromatic moiety, such as the phenyl group (benzene less an H atom):

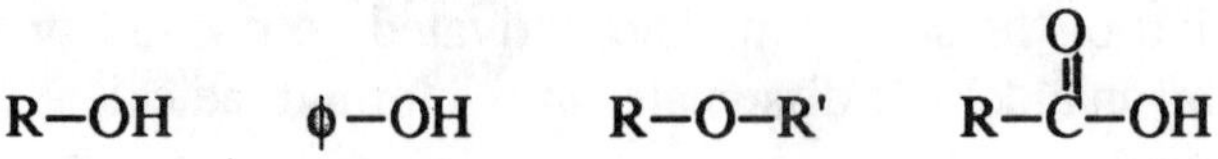

$$R{-}OH \qquad \phi{-}OH \qquad R{-}O{-}R' \qquad R{-}\overset{\displaystyle O}{\overset{\|}{C}}{-}OH$$

Aliphatic   Phenols   Ethers   Carboxylic
alcohols   acids

These classes of compounds include many important organic chemicals.

Alcohols

Of the alcohols, methanol, ethanol, isopropanol, and ethylene glycol rank among the top 50 chemicals with annual worldwide production of the order of a billion kg or more. The most common of the many uses of these chemicals is for the manufacture of other chemicals. Methanol is widely used in the manufacture of formaldehyde (see Section 12.5), as a solvent, and mixed with water as an antifreeze formulation. Ethanol is used as a solvent and as the starting material for the manufacture of acetaldehyde, acetic acid, ethyl ether, ethyl chloride, ethyl bromide, and several important esters. Both methanol and ethanol can be used as motor vehicle fuels, usually in mixtures with gasoline. Ethylene glycol is a common antifreeze compound.

A number of aliphatic alcohols have been reported in the atmosphere. Because of their volatility, the lower alcohols, especially methanol and ethanol, predominate as atmospheric pollutants. Among the other alcohols released to the atmosphere are 1-propanol, 2-propanol, propylene glycol, 1-butanol, and even octadecanol, $CH_3(CH_2)_{16}CH_2OH$, a volatile material evolved by plants. Alcohols can undergo photochemical reactions, beginning with abstraction of hydrogen by hydroxyl radical. Mechanisms for scavenging alcohols from the atmosphere are relatively efficient because the lower alcohols are quite water soluble and the higher ones have low vapor pressures.

Some alkenyl alcohols have been found in the atmosphere, largely as by-products of combustion. Typical of these is 2-buten-1-ol,

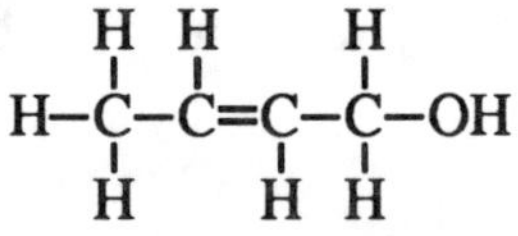

$$\begin{array}{cccccccc} & H & H & & & H & & \\ & | & | & & & | & & \\ H{-} & C & {-} & C & {=}C{-} & C & {-}OH \\ & | & & & & | & & \\ & H & & H & & H & & \end{array}$$

which has been detected in automobile exhausts.

Phenols

Phenols are aromatic alcohols that have an –OH group bonded to an aromatic ring. They are more noted as water pollutants than as air pollutants. Some typical phenols that have been reported as atmospheric contaminants are shown at the top of the following page:

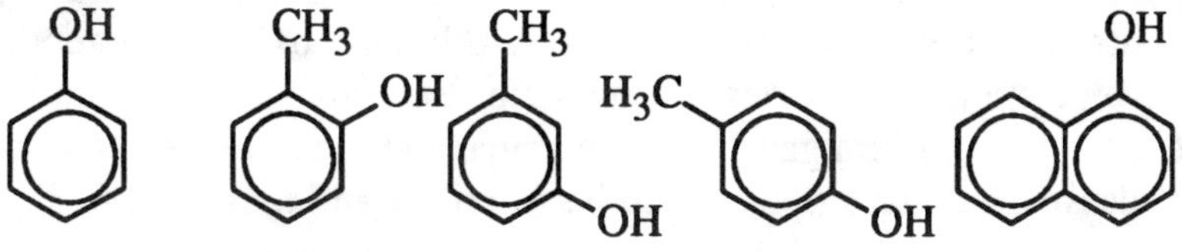

phenol   *o*- cresol   *m*- cresol   *p*- cresol   1-naphthol

Phenol is among the top 50 chemicals produced. It is most commonly used in the manufacture of resins and polymers, such as Bakelite, a phenol-formaldehyde copolymer. Phenols are produced by the pyrolysis of coal and are major by-products of coking. Thus, in local situations involving coal coking and similar operations phenols can be troublesome air pollutants.

Ethers

Ethers are relatively uncommon atmospheric pollutants; however, the flammability hazard of diethyl ether vapor in an enclosed work space is well known. In addition to aliphatic ethers, such as dimethyl ether and diethyl ether, several alkenyl ethers, including vinylethyl ether are produced by internal combustion engines. A cyclic ether and important industrial solvent, tetrahydrofuran, occurs as an air contaminant. Methyltertiarybutyl ether, MTBE, has become the octane booster of choice to replace tetraethyllead in gasoline. Because of its widespread distribution, MTBE has the potential to be an air pollutant, although its hazard is limited by its low vapor pressure. The structural formulas of the ethers mentioned above are given below:

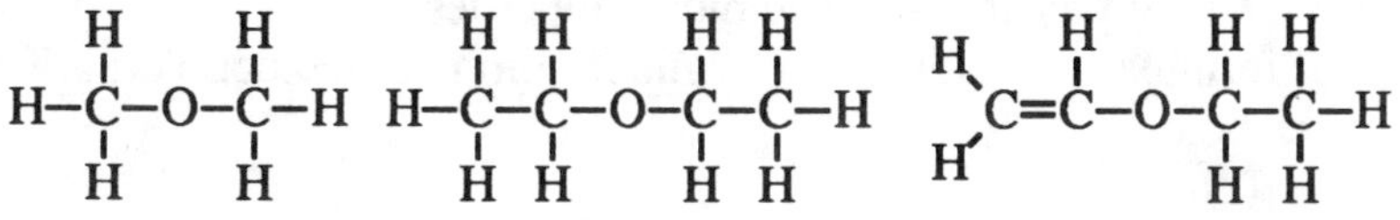

Dimethyl ether   Diethyl ether   Vinylethyl ether

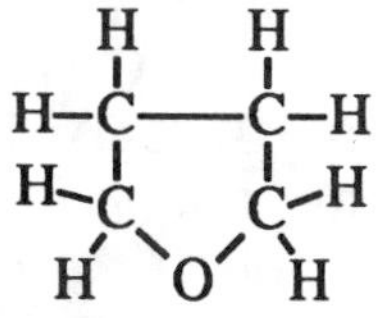

Tetrahydrofuran 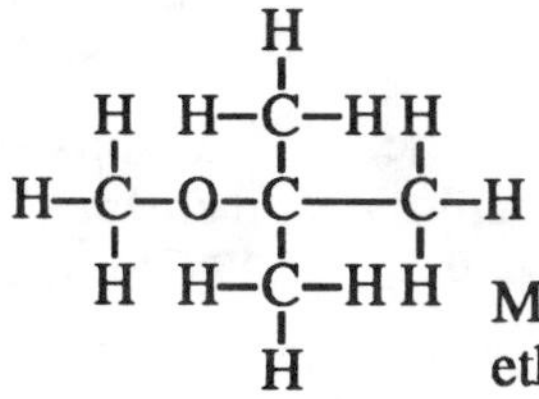   Methyltertiarybutyl ether (MTBE)

Oxides

Ethylene oxide and propylene oxide,

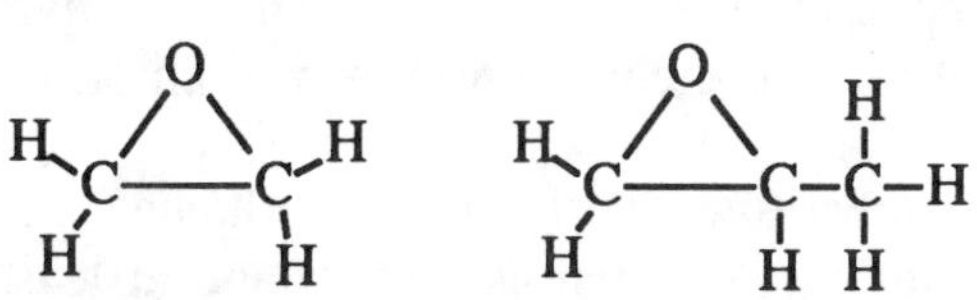

Ethylene oxide   Propylene oxide

rank among the 50 most widely produced industrial chemicals and have a limited potential to enter the atmosphere as pollutants. Ethylene oxide is a moderately to highly toxic sweet-smelling, colorless, flammable, explosive gas used as a chemical intermediate, sterilant, and fumigant. It is a mutagen and a carcinogen to experimental animals. It is classified as hazardous for both its toxicity and ignitability.

Carboxylic Acids

Carboxylic acids have at least one of the functional groups,

$$\begin{matrix} O \\ \| \\ -C-OH \end{matrix}$$

attached to an alkane, alkene, or aromatic hydrocarbon moiety. A carboxylic acid, pinonic acid, produced by the photochemical oxidation of naturally-produced α-pinene, was discussed in Section 12.2. Most of the many carboxylic acids found in the atmosphere are probably the result of the photochemical oxidation of other organic compounds through gas-phase reactions or by reactions of other organic compounds dissolved in aqueous aerosols. These acids are often the end products of photochemical oxidation because their low vapor pressures and water solubilities make them susceptible to scavenging from the atmosphere.

12.7. ORGANOHALIDE COMPOUNDS

Organohalides consist of halogen-substituted hydrocarbon molecules, each of which contains at least one atom of F, Cl, Br, or I. They may be saturated (**alkyl halides**), unsaturated (**alkenyl halides**), or aromatic (**aryl halides**). Organohalides exhibit a wide range of physical and chemical properties.

Structural formulas of several alkyl halides commonly encountered in the atmosphere are given below:

$$\begin{matrix} H \\ | \\ H-C-Cl \\ | \\ H \end{matrix} \qquad \begin{matrix} H \\ | \\ Cl-C-Cl \\ | \\ H \end{matrix} \qquad \begin{matrix} F \\ | \\ Cl-C-Cl \\ | \\ F \end{matrix} \qquad \begin{matrix} Cl\ \ H \\ |\ \ \ | \\ Cl-C-C-H \\ |\ \ \ | \\ Cl\ \ H \end{matrix}$$

Chloromethane   Dichloromethane   Dichlorodifluoro-   1,1,1-Trichloroeth-
(bp-24°C)   (methylene chloride,   methane ("Freon-   ane (methyl chloro-
             fp -97°C, bp 40°C)   12," bp -29°C)   form, bp 74°C)

Volatile **chloromethane** (methyl chloride) is consumed in the manufacture of silicones. **Dichloromethane** is a volatile liquid with excellent solvent properties for nonpolar organic solutes. It has been used as a solvent for the decaffeination of coffee, in paint strippers, as a blowing agent in urethane polymer manufacture, and to depress vapor pressure in aerosol formulations. **Dichlorodifluoromethane** is one of the chlorofluorocarbon compounds used as a refrigerant and involved in stratospheric ozone depletion. One of the more common industrial chlorinated solvents is **1,1,1-trichloroethane**.

Viewed as halogen-substituted derivatives of alkenes, the **alkenyl** or **olefinic organohalides** contain at least one halogen atom and at least one carbon–carbon double bond. The most significant of these are the lighter chlorinated compounds, such as those shown at the top of page 309.

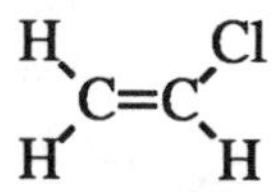

$$H_2C=CHCl \quad Cl_2C=CHCl \quad H_2C=CH-CH_2-Cl$$

Monochloroethylene   Trichloroethylene   3-Chloropropene
(vinyl chloride)   (TCE)   (allyl chloride)

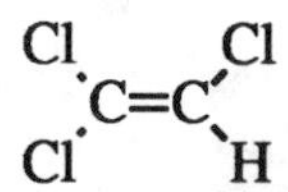

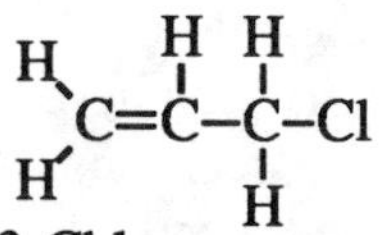

Vinyl chloride is consumed in large quantities as a raw material to manufacture pipe, hose, wrapping, and other products fabricated from polyvinylchloride plastic. This highly flammable, volatile, sweet-smelling gas is known to cause angiosarcoma, a rare form of liver cancer. **Trichloroethylene** is a clear, colorless, nonflammable, volatile liquid. It is an excellent degreasing and drycleaning solvent and has been used as a household solvent and for food extraction (for example, in decaffeination of coffee). **Allyl chloride** is an intermediate in the manufacture of allyl alcohol and other allyl compounds, including pharmaceuticals, insecticides, and thermosetting varnish and plastic resins.

Some commonly used aryl halide derivatives of benzene and toluene are shown below:

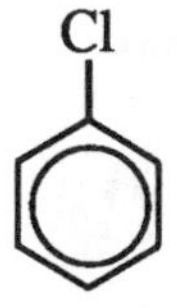

Monochlor-   Monobrom-   Hexachlor-   1-Chloro-2-
obenzene   obenzene   obenzene   methylbenzene

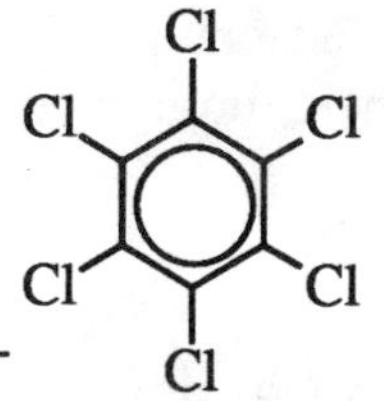

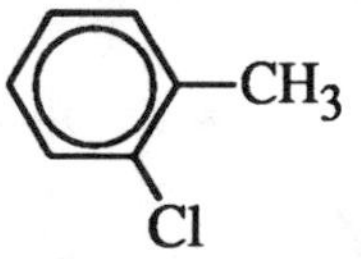

Aryl halide compounds have many uses, which have resulted in substantial human exposure and environmental contamination. Polychlorinated biphenyls, PCBs, a group of compounds formed by the chlorination of biphenyl,

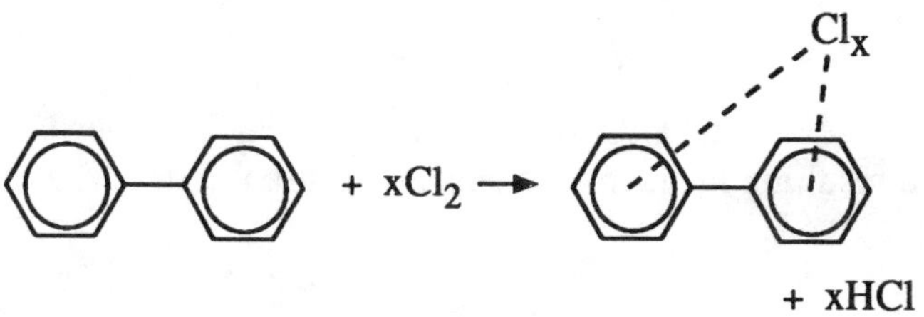

$$\text{biphenyl} + xCl_2 \rightarrow \text{PCB (Cl}_x) + xHCl \tag{12.7.1}$$

have extremely high physical and chemical stabilities and other qualities that have led to their being used in many applications, including heat transfer fluids, hydraulic fluids, and dielectrics.

As expected from their low vapor pressures, the lighter organohalide compounds are the most likely to be found in the atmosphere. On a global basis, the three most abundant organochlorine compounds in the atmosphere are methyl chloride, methyl chloroform, and carbon tetrachloride, which have tropospheric concentrations ranging from a tenth to several tenths of a part per billion. Methyl chloroform is relatively persistent in the atmosphere, with residence times of several years. Therefore, it may pose a threat to the stratospheric ozone layer in the same way as chlorofluorocarbons. Also found are methylene chloride; methyl bromide, CH_3Br; bromform, $CHBr_3$; assorted chlorofluorcarbons; and halogen-substituted ethylene compounds, such as trichloroethylene, vinyl chloride, perchloroethylene, (CCl_2=CCl_2) and solvent ethylene dibromide ($CHBr$=$CHBR$).

Chlorofluorocarbons (CFCs), such as dichlorodifluoromethane, are volatile 1- and 2-carbon compounds that contain Cl and F bonded to carbon. These compounds are notably stable and non-toxic. They have been widely used in recent decades in the fabrication of flexible and rigid foams and as fluids for refrigeration and air conditioning. The most widely manufactured of these compounds are CCl_3F (CFC-11), CCl_2F_2 (CFC-12), $C_2Cl_3F_3$ (CFC-113), $C_2Cl_2F_4$ (CFC-114), and C_2ClF_5 (CFC-115). **Halons** are related compounds that contain bromine and are used in fire extinguisher systems. The major commercial halons are $CBrClF_2$ (Halon-1211), $CBrF_3$ (Halon-1301), and $C_2Br_2F_4$ (Halon-2402), where the sequence of numbers denotes the number of carbon, fluorine, chlorine, and bromine atoms, respectively, per molecule.

All of the chlorofluorocarbons and halons discussed above have been implicated in the halogen-atom-catalyzed destruction of atmospheric ozone, which filters out cancer-causing ultraviolet radiation from the sun. Although quite inert in the lower atmosphere, CFCs undergo photodecomposition by the action of high-energy ultraviolet radition in the stratosphere through reactions such as

$$CCl_2F_2 + h\nu \rightarrow Cl\cdot + \cdot CClF_2 \tag{12.7.1}$$

which release Cl• atoms, denoted below simply as Cl. These atoms react with ozone, destroying it and producing ClO:

$$Cl + O_3 \rightarrow ClO + O_2 \tag{12.7.2}$$

In this region of the atmosphere, there is an appreciable concentration of atomic oxygen, by virtue of the reaction

$$O_3 + h\nu \rightarrow O_2 + O \tag{12.7.3}$$

Nitric oxide, NO, is also present. The ClO species may react with either O or NO, regenerating Cl atoms and resulting in chain reactions that cause the net destruction of ozone:

$$ClO + O \rightarrow Cl + O_2 \tag{12.7.4}$$

$$\underline{Cl + O_3 \rightarrow ClO + O_2} \tag{12.7.5}$$
$$O + O_3 \rightarrow 2O_2 \tag{12.7.6}$$

$$ClO + NO \rightarrow Cl + NO_2 \tag{12.7.7}$$

$$\underline{Cl + O_3 \rightarrow ClO + O_2} \tag{12.7.8}$$
$$O_3 + NO \rightarrow NO_2 + O_2 \tag{12.7.9}$$

The potential of the overall process discussed above to deplete atmospheric ozone and to pose a major threat to the global environment is discussed further in Chapter 14. As a result of U. S. Environmental Protection Agency regulations imposed in accordance with the 1986 Montreal Protocol on Substances that Deplete the Ozone Layer, production of CFCs and halocarbons in the U. S. was curtailed starting in 1989.[5] The most likely substitutes for these halocarbons are hydrogen-containing chlorofluorocarbons (HCFCs) and hydrogen-containing fluorocarbons (HFCs). The substitute compounds most likely to be produced commercially first are CH_2FCF_3 (HFC-134a, a substitute for CFC-12 in automobile air conditioners and refrigeration equipment), $CHCl_2CF_3$ (HCFC-123, substitute for CFC-11 in plastic foam-blowing), CH_3CCl_2F (HCFC-141b, substitute for CFC-11 in plastic foam-blowing), $CHClF_2$ (HCFC-22, air conditioners and manufacture of plastic foam food containers). Because of the more readily broken H-C bonds that they contain, these compounds are more easily destroyed by atmospheric chemical reactions (particularly with hydroxyl radical) before they reach the stratosphere.

12.8. ORGANOSULFUR COMPOUNDS

Substitution of alkyl or aryl hydrocarbon groups such as phenyl and methyl for H on hydrogen sulfide, H_2S, leads to a number of different organosulfur thiols (mercaptans, R–SH) and sulfides, also called thioethers (R–S–R). Structural formulas of examples of these compounds are shown below:

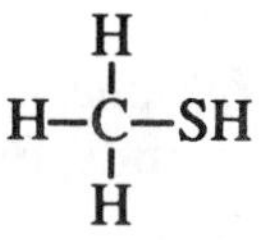

Methanethiol 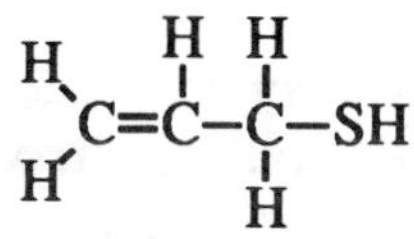   2-Propene-1-thiol 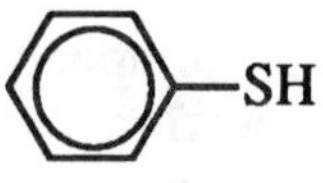   Benzenethiol

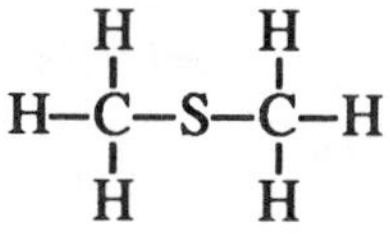

Dimethylsulfide 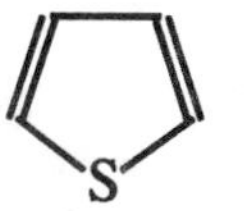   Thiophene 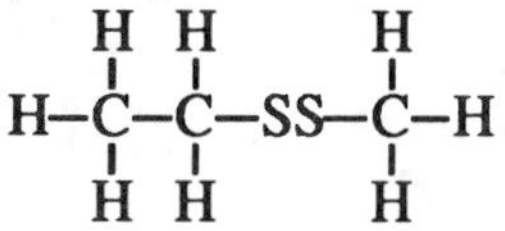   Ethylmethyldisulfide

Methanethiol and other lighter alkyl thiols are fairly common air pollutants that have "ultragarlic" odors; both 1- and 2-butanethiol are associated with skunk odor. Gaseous methanethiol and volatile liquid ethanethiol are used as odorant leak-detecting additives for natural gas, propane, and butane and are also employed as intermediates in pesticide synthesis. Allyl mercaptan (2-propene-1-thiol) is a toxic, irritating volatile liquid with a strong garlic odor. Benzenethiol (phenyl mercaptan), is the simplest of the aryl thiols. It is a toxic liquid with a severely "repulsive" odor.

Alkyl sulfides or thioethers contain the C-S-C functional group. The lightest of these compounds is dimethyl sulfide, a volatile liquid (bp 38°C) that is moderately toxic by ingestion. Cyclic sulfides contain the C-S-C group in a ring structure. The most common of these compounds is thiophene, a heat-stable liquid (bp 84°C) with a solvent action much like that of benzene, that is used in the manufacture of pharmaceuticals, dyes, and resins.

Although not highly significant as atmospheric contaminants on a large scale, organic sulfur compounds can cause local air pollution problems because of their bad odors. Major sources of organosulfur compounds in the atmosphere include microbial degradation, wood pulping, volatile matter evolved from plants, animal wastes, packing house and rendering plant wastes, starch manufacture, sewage treatment, and petroleum refining.

Although the impact of organosulfur compounds on atmospheric chemistry is minimal in areas such as aerosol formation or production of acid precipitation components, these compounds are the worst of all in producing odor. Therefore, it is important to prevent their release to the atmosphere.

As with all hydrogen-containing organic species in the atmosphere, reaction of organosulfur compounds with hydroxyl radical is a first step in their atmospheric photochemical reactions. The sulfur from both mercaptans and sulfides ends up as SO_2. In both cases there is thought to be a readily oxidized SO intermediate, and HS· radical may also be an intermediate in the oxidation of mercaptans. Another possibility is the addition of O atoms to S, resulting in the formation of free radicals as shown below for methyl mercaptan:

$$CH_3SH + O \rightarrow H_3C\cdot + HSO\cdot \tag{12.8.1}$$

The HSO· radical is readily oxidized by atmospheric O_2 to SO_2.

12.9. ORGANONITROGEN COMPOUNDS

Organic nitrogen compounds that may be found as atmospheric contaminants may be classified as **amines, amides, nitriles, nitro compounds,** or **heterocyclic nitrogen compounds**. Structures of common examples of each of these five classes of compounds reported as atmospheric contaminants are:

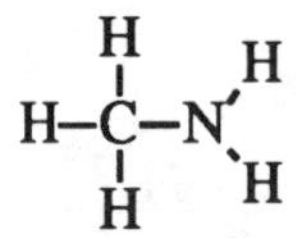

Methylamine 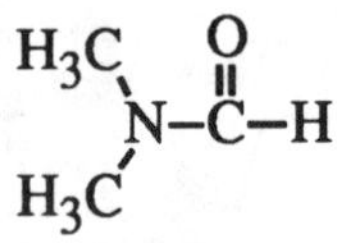   Dimethyl formamide 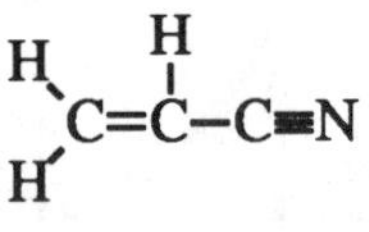   Acrylonitrile

Nitrobenzene    Pyridine    Aniline

Amines consist of compounds in which one or more of the hydrogen atoms in NH_3 has been replaced by a hydrocarbon moiety. Lower-molecular-mass amines are volatile. These are prominent among the compounds giving rotten fish their characteristic odor—an obvious reason why air contamination by amines is undesirable. The simplest and most important aromatic amine is aniline, used in the manufacture of dyes, amides, photographic chemicals, and drugs. A number of amines are widely used industrial chemicals and solvents, so that industrial sources have the potential to contaminate the atmosphere with these chemicals. Decaying organic matter, especially protein wastes, produce amines, so that rendering plants, packing houses, and sewage treatment plants are important sources of these substances.

Aromatic amines are of particular concern as atmospheric pollutants, particularly in the workplace, because some are known to cause urethral tract cancer (particularly of the bladder) in exposed individuals.[6] Aromatic amines are widely used as chemical intermediates, antioxidants, and curing agents in the manufacture of polymers (rubber and plastics), drugs, pesticides, dyes, pigments, and inks. In addition to aniline, some aromatic amines of potential concern are the following:

Benzidine   3,3'-Dichlorobenzidine

1-Naphthylamine   2-Naphtylamine   Phenyl-2-naph-thylamine

In the atmosphere, amines can be attacked by hydroxyl radical and undergo further reactions. Amines are bases (electron-pair donors). Therefore, their acid-base chemistry in the atmosphere may be important, particularly in the presence of acids in acidic precipitation.

The amide most likely to be encountered as an atmospheric pollutant is dimethylformamide. It is widely used commercially as a solvent for the synthetic polymer, polyacrylonitrile (Orlon, Dacron). Most amides have relatively low vapor pressures, which limits their entry into the atmosphere.

Nitriles, which are characterized by the $—C\!\equiv\!N$ group, have been reported as air contaminants, particularly from industrial sources. Both acrylonitrile and acetonitrile, CH_3CN, have been reported in the atmosphere as a result of synthetic rubber manufacture. As expected from their volatilities and levels of industrial production, most of the nitriles reported as atmospheric contaminants are low-molecular-mass aliphatic or olefinic nitriles, or aromatic nitriles with only one benzene ring. Acrylonitrile, used to make polyacrylonitrile polymer, is the only nitrogen-containing organic chemical among the top 50 chemicals, with annual worldwide production exceeding 1 billion kg.

Among the nitro compounds, RNO_2, reported as air contaminants are nitromethane, nitroethane, and nitrobenzene. These compounds are produced from industrial sources. Highly oxygenated compounds containing the NO_2 group, particularly peroxyacetyl nitrate (PAN, discussed in Chapter 13), are end products of the photochemical oxidation of hydrocarbons in urban atmospheres.

A large number of **heterocyclic nitrogen compounds** have been reported in tobacco smoke, and it is inferred that many of these compounds can enter the atmosphere from burning vegetation. Coke ovens are another major source of these compounds. In addition to the derivatives of pyridine, some of the heterocyclic nitrogen compounds are derivatives of pyrrole:

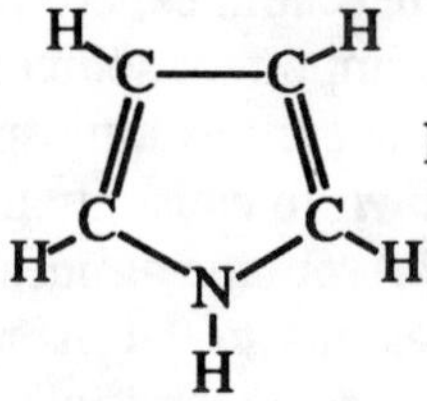

Pyrrole

Heterocyclic nitrogen compounds occur almost entirely in association with aerosols in the atmosphere.

Nitrosamines, general formula,

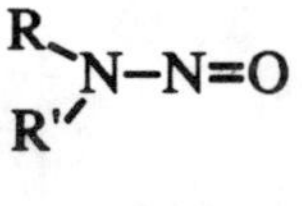

$$\begin{array}{c} R \\ \diagdown \\ N-N=O \\ \diagup \\ R' \end{array}$$

deserve special mention as atmospheric contaminants because some are known carcinogens. Both N-nitrosodimethylamine and N-nitrosodiethylamine have been detected in the atmosphere.

LITERATURE CITED

1. Graedel, T. E., *Chemical Compounds in the Atmosphere*. Academic Press, New York, 1978

2. Kamens, R. M., et al., "The impact of α-pinene on urban smog formation: an Outdoor Smog Chamber Study," *Atmospheric Environment*, **15** (1981), 969-981.

3. Gammage, R. G., and C. C. Travis, "Formaldehyde Exposure and Risk in Mobile Homes," Chapter 17 in *The Risk Assessment of Environmental and Human Health Hazards: A Textbook of Case Studies*, John Wiley and Sons, New York, 1989, pp. 601-611.

4. Nicholas, H. J., "Miscellaneous Volatile Plant Products," *Photochemistry*, **2**, L. P. Miller, Ed., John Wiley and Sons, New York, 1973, pp. 381-399.

5. Zurer, Pamela S., "Producers, Users Grapple with Realities of CFC Phaseout," *Chemical and Engineering News*, July 24, 1989, pp. 7-13.

6. Fishbein, Lawrence, "Aromatic Amines," in *The Handbook of Environmental Chemistry*, Vol. 3, Part C, Otto Hutzinger, Ed., Springer-Verlag, Berlin, 198

SUPPLEMENTARY REFERENCES

Seinfeld, John H., *Atmospheric Chemistry and Physics of Air Pollution*, John Wiley and Sons, Inc., Somerset, NJ, 1986.

Warneck, Peter, *Chemistry of the Natural Atmosphere*, Academic Press, San Diego, CA, 1988.

Howard, Philip H., *Handbook of Environmental Fate and Exposure Data for Organic Chemicals* (Three Volumes), Lewis Publishers, Chelsea, MI, 1989.

Graedel, T. E., "Atmospheric Photochemistry," in *The Handbook of Environmental Chemistry*, Otto Hutzinger, Ed., Springer-Verlag, Berlin, 1980, pp. 107-143.

Monteith, J. L., Ed., *Vegetation and the Atmosphere*, Academic Press, Inc., New York, 1977.

Vapor-Phase Organic Pollutants, National Academy of Sciences, Washington, DC, 1976.

QUESTIONS AND PROBLEMS

1. Match each organic pollutant in the left column with its expected effect in the right column, below:

   (a) CH_3SH

   (b) $CH_3CH_2CH_2CH_3$

   (c)
   $$\begin{array}{c} H \\ \diagdown \\ \diagup \\ H \end{array} C = C - C - C - H \text{ (with H substituents)}$$

   1. Most likely to have a secondary effect in the atmosphere
   2. Most likely to have a direct effect
   3. Should have the least effect of these three

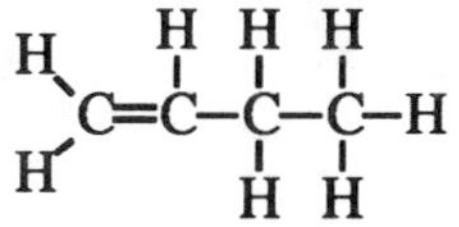

2. Why are hydrocarbon emissions from uncontrolled automobiles exhaust particularly reactive?

3. Assume an accidental release of a mixture of gaseous alkanes and alkenes into an urban atmosphere early in the morning. If the atmosphere at the release site is monitored for these compounds, what can be said about their total and relative concentrations at the end of the day? Explain.

4. Match each radical in the left column with its type in the right column, below:

   (a) $H_3C^\bullet$

   (b) $CH_3CH_2O^\bullet$

   (c) $H\overset{\bullet}{C}O$

   (d) $CH_xCH_{2x+1}O_2{}^\bullet$

   1. Formyl radical
   2. Alkylperoxyl radical
   3. Alkyl radical
   4. Alkoxyl radical

5. When reacting with hydroxyl radical, alkenes have a reaction mechanism not available to alkanes, which makes the alkenes much more reactive. What is this mechanism?

6. What is the most stable type of hydrocarbon that has a very low hydrogen-to-carbon ratio?

7. In the sequence of reactions leading to the oxidation of hydrocarbons in the atmosphere, what is the first stable class of compounds generally produced?

8. Give a sequence of reactions leading to the formation of acetaldehyde from ethane starting with the reaction of hydroxyl radical.

9. What important photochemical property do carbonyl compounds share with NO_2?

Photochemical Smog

13.1. INTRODUCTION

This chapter discusses the **oxidizing smog** or **photochemical smog** that permeates atmospheres in Los Angeles, Mexico City, Zurich, and many other urban areas. Although **smog** is the term used in this book to denote a photochemically oxidizing atmosphere, the word originally was used to describe the unpleasant combination of smoke and fog laced with sulfur dioxide, which was formerly prevalent in London when high-sulfur coal was the primary fuel used in that city. This mixture is characterized by the presence of sulfur dioxide, a reducing compound; therefore it is a **reducing smog** or **sulfurous smog**. In fact, readily oxidized sulfur dioxide has a short lifetime in an atmosphere where oxidizing photochemical smog is present.

Smog has a long history. Exploring what is now southern California in 1542, Juan Rodriguez Cabrillo named San Pedro Bay "The Bay of Smokes" because of the heavy haze that covered the area. Complaints of eye irritation from anthropogenically polluted air in Los Angeles were recorded as far back as 1868. Characterized by reduced visibility, eye irritation, cracking of rubber, and deterioration of materials, smog became a serious nuisance in the Los Angeles area during the 1940s. It is now recognized as a major air pollution problem in many areas of the world.

Air-quality officials define a smoggy day as one having moderate to severe eye irritation or visibility below 3 miles when the relative humidity is below 60%. The formation of oxidants in the air, particularly ozone, is indicative of smog formation. Serious levels of photochemical smog may be assumed to be present when the oxidant level exceeds 0.15 ppm for more than one hour. The three ingredients required to generate photochemical smog are ultraviolet light, hydrocarbons, and nitrogen oxides.

13.2. SMOG-FORMING AUTOMOTIVE EMISSIONS

Internal combustion engines used in automobiles and trucks produce reactive hydrocarbons and nitrogen oxides, two of the three key ingredients required for smog to form. Therefore, automotive air emissions are discussed next.

The production of nitrogen oxides was discussed in Section 11.6. At the high temperature and pressure conditions in an internal combustion engine, products of incompletely burned gasoline undergo chemical reactions which produce several hundred different hydrocarbons. Many of these are highly reactive in forming photo chemical smog. As shown in Figure 13.1, the automobile has several potential sources of hydrocarbon emissions other than the exhaust. The first of these to be controlled was the crankcase mist of hydrocarbons composed of lubricating oil and "blowby." The latter consists of exhaust gas and unoxidized carbureted mixture that enters the crankcase from the combustion chambers from around the pistons. This mist is destroyed by recirculating it through the engine intake manifold by way of the positive crankcase ventilation (PVC) valve.

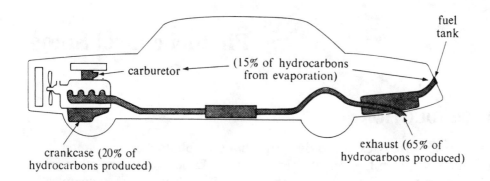

Figure 13.1. Potential sources of pollutant hydrocarbons from an automobile without pollution control devices.

A second major source of automotive hydrocarbon emissions is the fuel system, from which hydrocarbons are emitted through fuel tank and carburetor vents. When the engine is shut off and the engine heat warms up the fuel system, gasoline may be evaporated and emitted to the atmosphere. In addition, heating during the daytime and cooling at night causes the fuel tank to breathe and emit gasoline fumes. Such emissions are reduced by fuel formulated to reduce volatility. Automobiles are equipped with canisters of carbon which collect evaporated fuel from the fuel tank and fuel system, to be purged and burned when the engine is operating.

Control of Exhaust Hydrocarbons

In order to understand the production and control of automotive hydrocarbon exhaust products, it is helpful to understand the basic principles of the internal combustion engine. As shown in Figure 13.2, the four steps involved in one complete cycle of the four-cycle engine used in most vehicles are the following:

1. **Intake:** An air-gasoline mixture produced in the carburetor (or air in the case of a fuel-injection engine) is drawn into the cylinder through the open intake valve.

2. **Compression:** The combustible mixture is compressed at a ratio of about 7:1. Higher compression ratios favor thermal efficiency and complete combustion of hydrocarbons. However, higher temperatures, premature combustion ("pinging"), and high production of nitrogen oxides also result from higher combustion ratios.

3. **Ignition and power stroke:** As the fuel-air mixture is ignited by the spark plug near top-dead-center, a temperature of about 2,500°C is reached very rapidly at pressures up to 40 atm. As the gas volume increases with downward movement of the piston, the temperature decreases in a few milliseconds. This rapid cooling "freezes" nitric oxide in the form of NO without allowing it time to dissociate to

N_2 and O_2, which are thermodynamically favored at the normal temperatures and pressures of the atmosphere.

1. **Exhaust:** Exhaust gases consisting largely of N_2 and CO_2, with traces of CO, NO, hydrocarbons, and O_2, are pushed out through the open exhaust valve, thus completing the cycle.

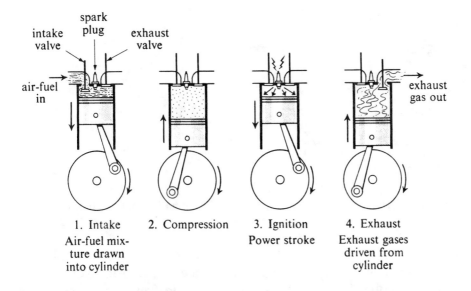

Figure 13.2. Steps in one complete cycle of a four-cycle internal combustion engine.

The primary cause of unburned hydrocarbons in the engine cylinder is wall quench, wherein the relatively cool wall in the combustion chamber of the internal combustion engine causes the flame to be extinguished within several thousandths of a centimeter from the wall. Part of the remaining hydrocarbon may be retained as residual gas in the cylinder, and part may be oxidized in the exhaust system. The remainder is emitted to the atmosphere as pollutant hydrocarbons. Engine misfire due to improper adjustment and deceleration greatly increases the emission of hydrocarbons. Turbine engines are not subject to the wall quench phenomenon because their surfaces are always hot.

Several engine design characteristics favor lower exhaust hydrocarbon emissions. Wall quench, which is mentioned above, is diminished by design that decreases the combustion chamber surface/volume ratio through reduction of compression ratio, more nearly spherical combustion chamber shape, increased displacement per engine cylinder, and increased ratio of stroke relative to bore.

Spark retard also reduces exhaust hydrocarbon emissions. For optimum engine power and economy, the spark should be set to fire appreciably before the piston reaches the top of the compression stroke and begins the power stroke. Retarding the spark to a point closer to top-dead-center reduces the hydrocarbon emissions markedly. One reason for this reduction is that the effective surface/volume ratio of the combustion chamber is reduced, thus cutting down on wall quench. Second, when the spark is retarded, the combustion products are purged from the cylinders sooner after combustion. Therefore, the exhaust gas is hotter, and reactions consuming hydrocarbons are promoted in the exhaust system.

As shown in Figure 13.3, the air/fuel ratio in the internal combustion engine has a marked effect upon the emission of hydrocarbons.

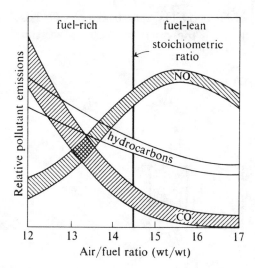

Figure 13.3. Effects of air/fuel ratio on pollutant emissions from an internal combustion piston engine.

As the air/fuel ratio becomes richer in fuel than the stoichiometric ratio, the emission of hydrocarbons increases sharply. There is a moderate decrease in hydrocarbon emissions when the mixture becomes appreciably leaner in fuel than the stoichiometric ratio requires. The lowest level of hydrocarbon emissions occurs at an air/fuel ratio somewhat leaner in fuel than the stoichiometric ratio. This behavior is the result of a combination of factors, including minimum quench layer thickness at an air/fuel ratio somewhat richer in fuel than the stoichiometric ratio, decreasing hydrocarbon concentration in the quench layer with a leaner mixture, increasing oxygen concentration in the exhaust with a leaner mixture, and a peak exhaust temperature at a ratio slightly leaner in fuel than the stoichiometric ratio. Catalytic converters are now used to destroy pollutants in exhaust gases. A reduction catalyst is employed to reduce NO in the exhaust gas and an oxidation catalyst to oxidize hydrocarbons and CO. A dual catalyst system involves running the engine slightly rich in fuel and passing the exhaust gas first over a reduction catalyst to reduce nitrogen oxides. Air is pumped into the exhaust downstream from this device and the exhaust stream passes through an oxidation catalyst where hydrocarbons and carbon monoxide are oxidized.

Noble metals (Pd, Pt, Ru) and nonstoichiometric oxides, such as Fe_2O_3 or $CoO \cdot Cr_2O_3$, may be employed as oxidation catalysts. The latter have the property of being able to gain oxygen in their lattice structure, followed by loss of oxygen to the substance being oxidized.

The objective of a reduction catalyst is to reduce NO to harmless N_2. The reduction may be catalyzed by noble metals (Pd, Pt, Ru, Rh), base metals (Co, Ni, Cu), or oxides (CuO, $CuCrO_4$). With carbon monoxide as the reducing agent, the two major reactions are

$$2NO + CO \rightarrow N_2O + CO_2 \tag{13.2.1}$$

$$N_2O + CO \rightarrow N_2 + CO_2 \tag{13.2.2}$$

which add up to the following net reaction:

$$2NO + 2CO \rightarrow N_2 + 2CO_2 \qquad (13.2.3)$$

Several undesirable side-reactions, such as the formation of ammonia,

$$2NO + 5H_2 \rightarrow 2H_2O + 2NH_3 \qquad (13.2.4)$$

can occur with a reduction catalyst.

Since lead can poison auto exhaust catalysts, automobiles equipped with catalytic exhaust-control devices require lead-free gasoline, which has become the standard motor fuel.

The 1990 U. S. Clean Air Act calls for reformulating gasoline by adding more oxygenated compounds to reduce emissions of hydrocarbons and carbon monoxide. A study released by the Auto/Oil Air Quality Improvement Research Program[1] has indicated that the reductions were not as substantial as had been hoped and that for 1983-85 automobiles emissions of hydrocarbons actually increased with oxygenated fuels.

13.3. SMOG-FORMING REACTIONS OF ORGANIC COMPOUNDS IN THE ATMOSPHERE

Hydrocarbons are eliminated from the atmosphere by a number of chemical and photochemical reactions.These reactions are responsible for the formation of many noxious secondary pollutant products and intermediates from relatively innocuous hydrocarbon precursors. These pollutant products and intermediates make up photo-chemical smog.

Hydrocarbons and most other organic compounds in the atmosphere are thermodynamically unstable toward oxidation and tend to be oxidized through a series of steps. The oxidation process terminates with formation of CO_2, solid organic particulate matter which settles from the atmosphere, or water-soluble products (for example, acids, aldehydes) which are removed by rain. Inorganic species such as ozone or nitric acid are byproducts of these reactions.

Photochemical Reactions of Methane

Some of the major reactions involved in the oxidation of atmospheric hydrocarbons may be understood by considering the oxidation of methane, the most common and widely dispersed atmospheric hydrocarbon. Like other hydrocarbons, methane reacts with oxygen atoms (generally produced by the photochemical dissoc-iation of NO_2, Reaction 13.5.11) to generate the all-important hydroxyl radical and an alkyl (methyl) radical

$$CH_4 + O \rightarrow H_3C^\bullet + HO^\bullet \qquad (13.3.1)$$

The methyl radical produced reacts rapidly with molecular oxygen to form very reactive peroxyl radicals,

$$H_3C^\bullet + O_2 + M \rightarrow H_3COO^\bullet + M \qquad (13.3.2)$$

which with methane as the hydrocarbon is the methoxyl radical, $H_3COO\cdot$. Such radicals participate in a variety of subsequent chain reactions, including those leading to smog formation. The hydroxyl radical reacts rapidly with hydrocarbons to form reactive hydrocarbon radicals,

$$CH_4 + HO\cdot \rightarrow H_3C\cdot + H_2O \qquad (13.3.3)$$

in this case, the methyl radical, $H_3C\cdot$. The following are more reactions involved in the overall oxidation of methane:

$$H_3COO\cdot + NO \rightarrow H_3CO\cdot + NO_2 \qquad (13.3.4)$$

(This is a very important kind of reaction in smog formation because the oxidation of NO by peroxyl radicals is the predominant means of regenerating NO_2 in the atmosphere, after it has been photochemically dissociated to NO.)

$$H_3CO\cdot + O_3 \rightarrow \text{various products} \qquad (13.3.5)$$

$$H_3CO\cdot + O_2 \rightarrow CH_2O + HOO\cdot \qquad (13.3.6)$$

$$H_3COO\cdot + NO_2 + M \rightarrow CH_3OONO_2 + M \qquad (13.3.7)$$

(The species CH_3OONO_2 is peroxyacetyl nitrate, PAN, a very strong oxidant.)

$$H_2CO + h\nu \rightarrow \text{photodissociation products} \qquad (13.3.8)$$

As will be seen throughout this chapter, hydroxyl radical, $HO\cdot$, and hydroperoxyl radical, $HOO\cdot$, are ubiquitous intermediates in photochemical chain-reaction processes. These two species are known collectively as odd hydrogen radicals.

Reactions such as (13.3.1) and (13.3.3) are **abstraction reactions** involving the removal of an atom, usually hydrogen, by reaction with an active species. **Addition reactions** of organic compounds are also common. Typically, hydroxyl radical reacts with an alkene such as propylene to form another reactive free radical:

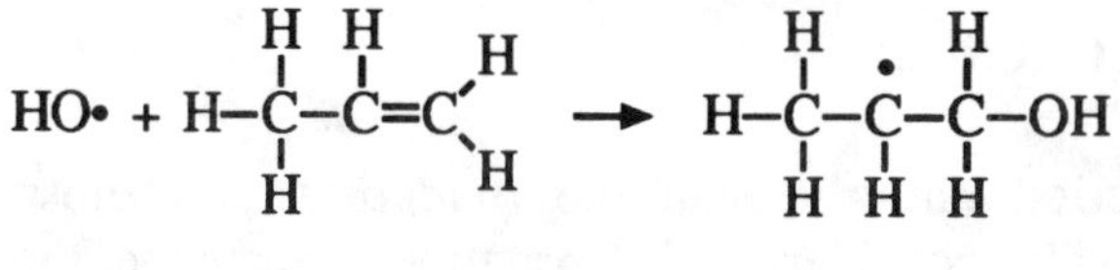

$$(13.3.9) \qquad (13.3.9)$$

Ozone adds to unsaturated compounds to form reactive ozonides:

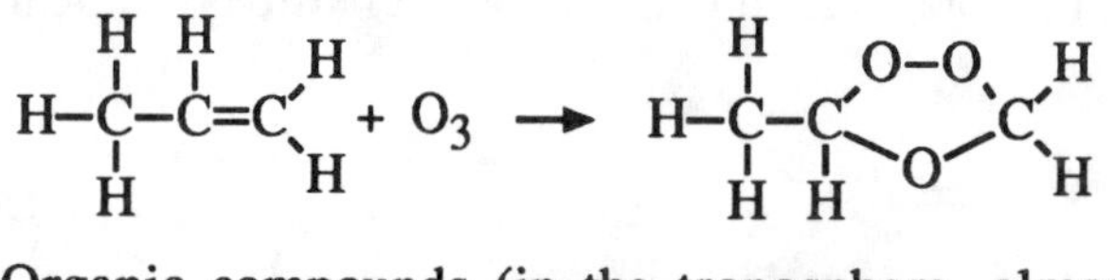

$$(13.3.10) \qquad (13.3.10)$$

Organic compounds (in the troposphere, almost exclusively carbonyls) can undergo primary photochemical reactions resulting in the direct formation of free radicals. By far the most important of these is the photochemical dissociation of aldehydes:

$$H_3C-\overset{\overset{\displaystyle O}{\|}}{C}-H + h\nu \longrightarrow H_3C\cdot + H\overset{\displaystyle \cdot}{C}O \qquad (13.3.11)$$

Organic free radicals undergo a number of chemical reactions. Hydroxyl radicals may be generated from organic peroxyl reactions such as,

$$H_3C-\overset{\overset{\displaystyle \cdot}{\overset{\displaystyle O}{\underset{\displaystyle |}{O}}}}{\underset{\displaystyle H}{C}}-CH_3 \longrightarrow H_3C-\overset{\overset{\displaystyle O}{\|}}{C}-CH_3 + HO\cdot \qquad (13.3.12)$$

leaving an aldehyde or ketone. The hydroxyl radical may react with other organic compounds, maintaining the chain reaction. Gas-phase reaction chains commonly have many steps. Furthermore, chain-branching reactions take place in which a free radical reacts with an excited molecule, causing it to produce two new radicals. Chain termination may occur in several ways, including reaction of two free radicals,

$$2HO\cdot \rightarrow H_2 + O_2 \qquad (13.3.13)$$

adduct formation with nitric oxide or nitrogen dioxide (which, because of their odd numbers of electrons, are themselves stable free radicals),

$$HO\cdot + NO_2 + M \rightarrow HNO_3 + M \qquad (13.3.14)$$

or reaction of the radical with a solid particle surface.

Hydrocarbons may undergo heterogeneous reactions on particles in the atmosphere. Dusts composed of metal oxides or charcoal have a catalytic effect upon the oxidation of organic compounds. Metal oxides may enter into photochemical reactions. For example, zinc oxide photosensitized by exposure to light promotes oxidation of organic compounds.

The kinds of reactions just discussed are involved in the formation of photochemical smog in the atmosphere. Next, consideration is given to the smog-forming process.

13.4. OVERVIEW OF SMOG FORMATION

This section addresses the conditions that are characteristic of a smoggy atmosphere and the overall processes involved in smog formation. In atmospheres that receive hydrocarbon and NO pollution accompanied by intense sunlight and stagnant air masses, oxidants tend to form. In air-pollution parlance, **gross photochemical oxidant** is a substance in the atmosphere capable of oxidizing iodide ion to elemental iodine. Sometimes other reducing agents are used to measure oxidants. The primary oxidant in the atmosphere is ozone. Other atmospheric oxidants include H_2O_2, organic peroxides (ROOR'), organic hydroperoxides (ROOH), and peroxyacyl nitrates, such as PAN mentioned in the preceding section.

Nitrogen dioxide, NO_2, is not regarded as a gross photochemical oxidant. However, it is about 15% as efficient as O_3 in oxidizing iodide to iodine(0), and a correction is made in measurements for the positive interference of NO_2. Sulfur dioxide is oxidized by O_3 and produces a negative interference, for which a measurement correction must also be made.

The formation of peroxyacetyl nitrate, PAN, was shown in Equation 13.3.7. PAN and its homologs, such as peroxybenzoyl nitrate,

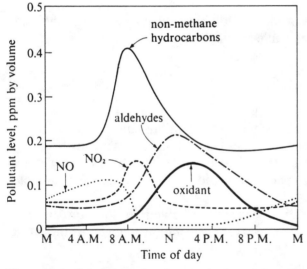

a powerful eye irritant and lachrymator, are produced photochemically in atmospheres containing alkenes and NO_x. PAN, especially, is a notorious organic oxidant. In addition to PAN and PBN, some other specific organic oxidants that may be important in polluted atmospheres are peroxypropionyl nitrate (PPN); peracetic acid, $CH_3(CO)OOH$; acetylperoxide, $CH_3(CO)OO(CO)CH_3$; ethyl hydroperoxide, $CH_3CH_2CH_2CH_2OOH$; and tert-butylhydroperoxide, $(CH_3)_3COOH$.

As shown in Figure 13.4, smoggy atmospheres show characteristic variations with time of day in levels of NO, NO_2, hydrocarbons,

Figure 13.4. Generalized plot of atmospheric concentrations of species involved in smog formation as a function of time of day.

aldehydes, and oxidants. Examination of the figure shows that, shortly after sunrise, the level of NO in the atmosphere decreases markedly, a decrease that is accompanied by a peak in the concentration of NO_2. During midday (significantly, after the concentration of NO has fallen to a very low level), the levels of aldehydes and oxidants become relatively high. The concentration of total hydrocarbons in the atmosphere peaks sharply in the morning, then decreases during the remaining daylight hours.

An overview of the processes responsible for the behavior just discussed is summarized in Figure 13.5. The chemical basis for the processes illustrated in this figure are explained in the following section:

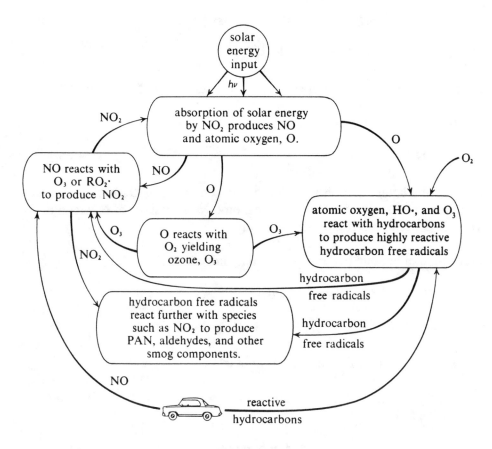

Figure 13.5. Generalized scheme for the formation of photochemical smog.

13.5. MECHANISMS OF SMOG FORMATION

Here are discussed some of the primary aspects of photochemical smog formation. For more details the reader is referred to books on atmospheric chemistry.[2,3] Since the exact chemistry of photochemical smog formation is very complex, many of the reactions are given as plausible illustrative examples rather than proven mechanisms.

The kind of behavior summarized in Figure 13.4 contains several apparent anomalies which puzzled scientists for many years. The first of these was the rapid increase in NO_2 concentration and decrease in NO concentration under conditions where it was known that photodissociation of NO_2 to O and NO was occurring. Furthermore, it could be shown that the disappearance of olefins and other hydrocarbons was much more rapid than could be explained by their relatively slow reactions with O_3 and O. These anomalies are now explained by chain reactions involving the interconversion of NO and NO_2, the oxidation of hydrocarbons, and the generation of reactive intermediates, particularly hydroxyl radical ($HO^\bullet$).

Figure 13.5 shows the overall reaction scheme for smog formation, which is based upon the photochemically initiated reactions that occur in an atmosphere containing nitrogen oxides, reactive hydrocarbons, and oxygen. The time variations in levels of hydrocarbons, ozone, NO, and NO_2 are explained by the following overall reactions:

1. Primary photochemical reaction producing oxygen atoms:

$$NO_2 + h\nu \, (\lambda < 420 \, nm) \rightarrow NO + O \qquad (13.5.1)$$

2. Reactions involving oxygen species (M is an energy-absorbing third body):

$$O_2 + O + M \rightarrow O_3 + M \qquad (13.5.2)$$

$$O_3 + NO \rightarrow NO_2 + O_2 \qquad (13.5.3)$$

Because the latter reaction is rapid, the concentration of O_3 remains low until that of NO falls to a low value. Automotive emissions of NO tend to keep O_3 concentrations low along freeways.

3. Production of organic free radicals from hydrocarbons, RH:

$$O + RH \rightarrow R\cdot + \text{other products} \qquad (13.5.4)$$

$$O_3 + RH \rightarrow R\cdot + \text{and/or other products} \qquad (13.5.5)$$

($R\cdot$ is a free radical which may or may not contain oxygen.)

4. Chain propagation, branching, and termination by a variety of reactions such as the following:

$$NO + ROO\cdot \rightarrow NO_2 + \text{and/or other products} \qquad (13.5.6)$$

$$NO_2 + R\cdot \rightarrow \text{products (for example, PAN)} \qquad (13.5.7)$$

The latter kind of reaction is the most common chain-terminating process in smog because NO_2 is a stable free radical present at high concentrations. Chains may terminate also by reaction of free radicals with NO or by reaction of two $R\cdot$ radicals, although the latter is uncommon because of the relatively low concentrations of radicals compared to molecular species. Chain termination by radical sorption on a particle surface is also possible and may contribute to aerosol particle growth.

A large number of specific reactions are involved in the overall scheme for the formation of photochemical smog. The formation of atomic oxygen by a primary photochemical reaction (Reaction 13.5.1) leads to several reactions involving oxygen and nitrogen oxide species:

$$O + O_2 + M \rightarrow O_3 + M \qquad (13.5.8)$$

$$O + NO + M \rightarrow NO_2 + M \qquad (13.5.9)$$

$$O + NO_2 \rightarrow NO + O_2 \qquad (13.5.10)$$

$$O_3 + NO \rightarrow NO_2 + O_2 \qquad (13.5.11)$$

$$O + NO_2 + M \rightarrow NO_3 + M \qquad (13.5.12)$$

$$O_3 + NO_2 \rightarrow NO_3 + O_2 \qquad (13.5.13)$$

There are a number of significant atmospheric reactions involving nitrogen oxides, water, nitrous acid, and nitric acid:

$$NO_3 + NO_2 \rightarrow N_2O_5 \tag{13.5.14}$$

$$N_2O_5 \rightarrow NO_3 + NO_2 \tag{13.5.15}$$

$$NO_3 + NO \rightarrow 2NO_2 \tag{13.5.16}$$

$$N_2O_5 + H_2O \rightarrow 2HNO_3 + O_2 \tag{13.5.17}$$

(This reaction is slow in the gas phase but may be fast on surfaces.)

Very reactive $HO^\bullet$ radicals can be formed by the reaction of excited atomic oxygen with water,

$$O^* + H_2O \rightarrow 2HO^\bullet \tag{13.5.18}$$

by photodissociation of hydrogen peroxide,

$$H_2O_2 + hv\,(\lambda < 350\,nm) \rightarrow 2HO^\bullet \tag{13.5.19}$$

or by the photolysis of nitrous acid,

$$HNO_2 + hv \rightarrow 2HO^\bullet + NO \tag{13.5.20}$$

Among the inorganic species with which the hydroxyl radical reacts are oxides of nitrogen,

$$HO^\bullet + NO_2 \rightarrow HNO_3 \tag{13.5.21}$$

$$HO^\bullet + NO + M \rightarrow HNO_2 + M \tag{13.5.22}$$

and carbon monoxide,

$$CO + HO^\bullet + O_2 \rightarrow CO_2 + HOO^\bullet \tag{13.5.23}$$

The last reaction is significant in that it is responsible for the disappearance of much atmospheric CO (see Section 11.3) and because it produces the hydroperoxyl radical $HOO^\bullet$. One of the major inorganic reactions of the hydroperoxyl radical is the oxidation of NO:

$$HOO^\bullet + NO \rightarrow HO^\bullet + NO_2 \tag{13.5.24}$$

For purely inorganic systems, kinetic calculations and experimental measurements cannot explain the rapid transformation of NO to NO_2 that occurs in an atmosphere undergoing photochemical smog formation and predict that the concentration of NO_2 should remain very low. However, in the presence of reactive hydrocarbons, NO_2 accumulates very rapidly by a reaction process beginning with its photodissociation! It may be concluded, therefore, that the organic compounds form species which react with NO directly, rather than with NO_2.

A number of chain reactions have been shown to result in the general type of species behavior shown in Figure 13.4. When aliphatic hydrocarbons, RH, react with O, O_3, or or HO• radical,

$$RH + O + O_2 \rightarrow ROO• + HO• \tag{13.5.25}$$

$$RH + HO• + O_2 \rightarrow ROO• + H_2O \tag{13.5.26}$$

reactive oxygenated organic radicals, ROO• are produced. Alkenes (R=R') are much more reactive, undergoing overall reactions such as

$$R=R' + O + 2O_2 \rightarrow ROO• + \overset{\overset{\displaystyle O}{\|}}{R''\text{--}C}\text{--}OO• \tag{13.5.27}$$

$$R=CH_2 + O + 2O_2 \rightarrow ROO• + HOO• + CO \tag{13.5.28}$$

$$R=R' + HO• + O_2 \rightarrow ROO• + \overset{\overset{\displaystyle O}{\|}}{H\text{--}C}\text{--}R'' \tag{13.5.29}$$
$$\text{(an aldehyde)}$$

Aromatic hydrocarbons, φH, may also react with O and HO•. Addition reactions of aromatics with HO• are favored. The product is phenol, as shown by the following reaction sequence:

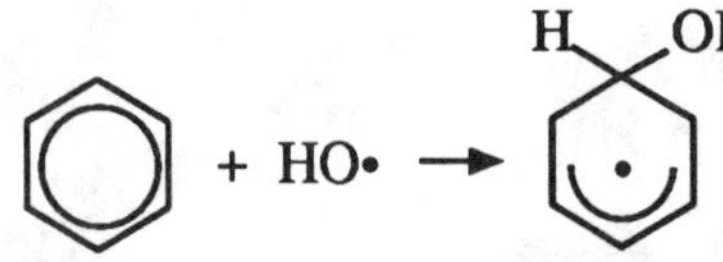

$$\tag{13.5.30}$$

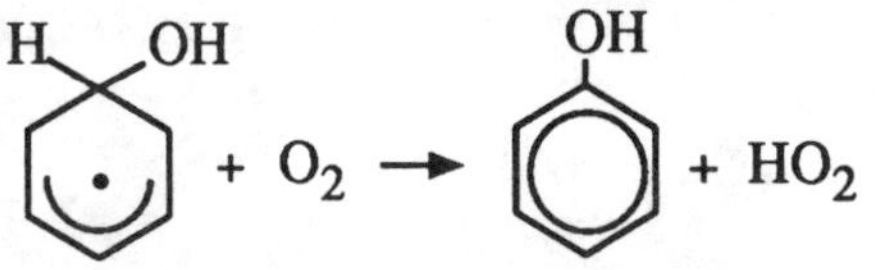

$$\tag{13.5.31}$$

In the case of alkyl benzenes, such as toluene, the hydroxyl radical attack may occur on the alkyl group, leading to reaction sequences such as those of alkanes.

Aldehydes react with HO•,

$$\overset{\overset{\displaystyle O}{\|}}{R\text{--}C}\text{--}H + HO• + O_2 \rightarrow \overset{\overset{\displaystyle O}{\|}}{R\text{--}C}\text{--}OO• + H_2O \tag{13.5.32}$$

$$\overset{H}{\underset{H}{}}C=O + HO• + \tfrac{3}{2}O_2 \rightarrow CO_2 + HOO• + H_2O \tag{13.5.33}$$

and undergo photochemical reactions:

$$\overset{\overset{\displaystyle O}{\parallel}}{R-C-H} + h\nu + 2O_2 \longrightarrow ROO\bullet + CO + HOO\bullet \tag{13.5.34}$$

$$\overset{\displaystyle H}{\underset{\displaystyle H}{\diagdown}}C=O + h\nu + 2O_2 \longrightarrow CO + 2HOO\bullet \tag{13.5.35} \tag{13.5.35}$$

Hydroxyl radical (HO•), which reacts with some hydrocarbons at rates that are almost diffusion-controlled, is the predominant reactant in early stages of smog formation. Significant contributions are made by hydroperoxyl radical (HOO•) and O_3 after smog formation is well underway.

One of the most important reaction sequences in the smog-formation process begins with the abstraction by HO• of a hydrogen atom from a hydrocarbon and leads to the oxidation of NO to NO_2 as follows:

$$RH + HO\bullet \rightarrow R\bullet + H_2O \tag{13.5.36}$$

The alkyl radical, R•, reacts with O_2 to produce a peroxyl radical, ROO•:

$$R\bullet + O_2 \rightarrow ROO\bullet \tag{13.5.37}$$

This strongly oxidizing species very effectively oxidizes NO to NO_2,

$$ROO\bullet + NO \rightarrow RO\bullet + NO_2 \tag{13.5.38}$$

thus explaining the once-puzzling rapid conversion of NO to NO_2 in an atmosphere in which the latter is undergoing photodissociation. The alkoxyl radical product, RO•, is not so stable as ROO•. In cases where the oxygen atom is attached to a carbon atom that is also bonded to H, a carbonyl compound is likely to be formed by the following type of reaction:

$$H_3CO\bullet + O_2 \longrightarrow \overset{\overset{\displaystyle O}{\parallel}}{H-C-H} + HOO\bullet \tag{13.5.39}$$

The rapid production of photosensitive carbonyl compounds from alkoxyl radicals is an important stimulant for further atmospheric photochemical reactions. In the absence of extractable hydrogen, cleavage of a radical containing the carbonyl group occurs:

$$\overset{\overset{\displaystyle O}{\parallel}}{H_3C-C-O\bullet} \longrightarrow H_3C\bullet + CO_2 \tag{13.5.40}$$

Another reaction that can lead to the oxidation of NO is of the following type:

$$\overset{\overset{\displaystyle O}{\parallel}}{R-C-OO\bullet} + NO + O_2 \longrightarrow ROO\bullet + NO_2 + CO_2 \tag{13.5.41}$$

Peroxyacyl nitrates (PAN) are highly significant air pollutants formed by an addition reaction with NO_2:

$$R-\overset{\overset{\displaystyle O}{\|}}{C}-OO\bullet + NO_2 \longrightarrow R-\overset{\overset{\displaystyle O}{\|}}{C}-OO-NO_2 \qquad (13.5.42)$$

When R is the methyl group, the product is peroxyacetyl nitrate, mentioned in Section 14.4. Alkyl nitrates and alkyl nitrites may be formed by the reaction of alkoxyl radicals ($RO\bullet$) with nitrogen dioxide and nitric oxide, respectively:

$$RO\bullet + NO_2 \rightarrow RONO_2 \qquad (13.5.43)$$

$$RO\bullet + NO \rightarrow RONO \qquad (13.5.44)$$

Addition reactions with NO_2 such as these are important in terminating the reaction chains involved in smog formation. Since NO_2 is involved both in the chain-initiation step (Reaction 13.5.1) and the chain termination step, only moderate reductions in NO_x emissions alone may not curtail smog formation and in some circumstances may even increase it.

As shown in Reaction 13.5.39, the reaction of oxygen with alkoxyl radicals produces hydroperoxyl radical. Peroxyl radicals can react with one another to produce reactive hydrogen peroxide, alkoxyl radicals, and hydroxyl radicals:

$$HOO\bullet + HOO\bullet \rightarrow H_2O_2 + O_2 \qquad (13.5.45)$$

$$HOO\bullet + ROO\bullet \rightarrow RO\bullet + HO\bullet + O_2 \qquad (13.5.46)$$

$$ROO\bullet + ROO\bullet \rightarrow 2RO\bullet + O_2 \qquad (13.5.47)$$

Photolyzable Compounds in the Atmosphere

It may be useful at this time to review the types of compounds capable of undergoing photolysis in the troposphere and thus initiating chain reaction. Under most tropospheric conditions, the most important of these is NO_2:

$$NO_2 + h\nu \ (\lambda < 420 \text{ nm}) \rightarrow NO + O \qquad (13.5.1)$$

In relatively polluted atmospheres, the next most important photodissociation reaction is that of carbonyl compounds, particularly formaldehyde:

$$CH_2O + h\nu \ (\lambda < 335 \text{ nm}) \rightarrow H\bullet + H\overset{\bullet}{C}O \qquad (13.5.48)$$

Hydrogen peroxide photodissociates to produce two hydroxyl radicals:

$$HOOH + h\nu \ (\lambda < 350 \text{ nm}) \rightarrow 2HO\bullet \qquad (13.5.49)$$

Finally, organic peroxides may be formed and subsequently dissociate by the following reactions, starting with a peroxyl radical:

$$H_3COO^\bullet + HOO^\bullet \rightarrow H_3COOH + O_2 \tag{13.5.50}$$

$$H_3COOH + h\nu\,(\lambda < 350\,nm) \rightarrow H_3CO^\bullet + HO^\bullet \tag{13.5.51}$$

It should be noted that each of the last three photochemical reactions gives rise to two free radical species per photon absorbed. Ozone undergoes photochemical dissociation to produce excited oxygen atoms at wavelengths less than 315 nm. These atoms may react with H_2O to produce hydroxyl radicals.

Singlet Oxygen

Recall from Section 9.5 that ground state triplet molecular oxygen, $^3O_2$, can be excited to singlet molecular oxygen, $^1O_2$. Although it was once believed that singlet molecular oxygen was reactive enough with alkenes to play a significant role in photochemical smog formation, current evidence suggests that such is not the case.

13.6. REACTIVITY OF HYDROCARBONS

The reactivity of hydrocarbons in the smog formation process is an important consideration in understanding the process and in developing control strategies. It is useful to know which are the most reactive hydrocarbons so that their release can be minimized. Less reactive hydrocarbons, of which propane is a good example, may cause smog formation far downwind from the point of release.

Hydrocarbon reactivity is best based upon the interaction of hydrocarbons with hydroxyl radical. Methane, which is perhaps the least reactive gas-phase hydrocarbon and has an atmospheric half-life exceeding 10 days, is assigned a reactivity of 1.0. (Despite its low reactivity, methane is so abundant in the atmosphere that it accounts for a significant fraction of total hydroxyl radical reactions.) In contrast, β-pinene produced by conifer trees and other vegetation, is almost 9000 times as reactive as methane and d-limonene, produced by orange rind, is almost 19000 times as reactive. Relative to their rates of reaction with hydroxyl radical, hydrocarbon reactivities may be classified from I through V as shown in Table 13.1.

13.7. INORGANIC PRODUCTS FROM SMOG

Two major classes of inorganic products from smog are sulfates and nitrates. Inorganic sulfates and nitrates, along with sulfur and nitrogen oxides can contribute to acidic precipitation, corrosion, reduced visibility, and adverse health effects.

Although the oxidation of SO_2 to sulfate species is relatively slow in a clean atmosphere, it is much faster under smoggy conditons. During severe photochemical smog conditions, oxidation rates of 5-10% per hour may occur, as compared to only a fraction of a percent per hour under normal atmospheric conditions. Thus, sulfur dioxide exposed to smog can produce very high local concentrations of sulfate, which can aggravate already bad atmospheric conditions.

Table 13.1. Relative Reactivities of Hydrocarbons and CO with HO• Radical[1]

Reactivity class	Reactivity range[2]	Approximate half-life in the atmosphere	Compounds in increasing order of reactivity
I	<10	>10 days	methane
II	10–100	24 h–10 d	CO, acetylene, ethane
III	100–1000	2.4–24 h	benzene, propane, *n*-butane, isopentane, methyl ethyl ketone, 2-methylpentane, toluene, *n*-propylbenzene, isopropylbenzene, ethene, *n*-hexane, 3-methylpentane, ethylbenzene
IV	1,000–10,000	15 min–2.4 h	*p*-xylene, *p*-ethyltoluene, *o*-ethyltoluene, *o*-xylene, methyl isobutyl ketone, *m*-ethyltoluene, *m*-xylene, 1,2,3-trimethylbenzene, propene, 1,2,4-trimethylbenzene, 1,3,5-trimethylbenzene, *cis*-2-butene, β-pinene, 1,3-butadiene
V	>10,000	<15 min	2-methyl-2-butene, 2,4-dimethyl-2-butene, *d*-limonene

[1] Based on data from K.R. Darnall, A.C. Lloyd, A.M. Winer, and J.N. Pitts, Jr., "Reactivity Scale for Atmospheric Hydrocarbons Based on Reaction with Hydroxyl Radical," *Environmental Science and Technology* 10: 692–6 (1976).

[2] Based on an assigned reactivity of 1.0 for methane reacting with hydroxyl radical.

Several oxidant species in smog can oxidize SO_2. Among the oxidants are compounds, including O_3, NO_3, and N_2O_5, as well as reactive radical species, particularly HO•, HOO•, O, RO•, and ROO•. The two major primary reactions are oxygen transfer,

$$SO_2 + O \text{ (from O, RO•, RO•,)} \rightarrow SO_3 \rightarrow H_2O_4, \text{ sulfates} \qquad (13.7.1)$$

or addition. As an example of the latter, HO• adds to SO_2 to form a reactive species which can further react with oxygen, nitrogen oxides, or other species to yield sulfates, other sulfur compounds, or compounds of nitrogen:

$$HO• + SO_2 \rightarrow HOSOO• \qquad (13.7.2)$$

The presence of HO• (typically at a level of 3 x 10^6 radicals/cm^3, but appreciably higher in smoggy atmosphere), makes this a likely route. Addition of SO_2 to RO• or ROO• can yield organic sulfur compounds.

It should be noted that the reaction of H_2S with HO• is quite rapid. As a result, the normal atmospheric half-life of H_2S of about one-half day becomes much shorter in the presence of photochemical smog.

Inorganic nitrates or nitric acid are formed by several reactions in smog. Among the important reactions forming nitric acid are the reaction of N_2O_5 with water

(Reaction 13.5.17) and the addition of hydroxyl radical to NO_2 (Reaction 13.5.21). The oxidation of NO or NO_2 to nitrate species may occur after absorption of gas by an aerosol droplet. Nitric acid formed by these reactions reacts with ammonia in the atmosphere to form ammonium nitrate:

$$NH_3 + HNO_3 \rightarrow NH_4NO_3 \tag{13.7.3}$$

Other nitrate salts may also be formed.

Nitric acid and nitrates are among the more damaging end-products of smog. In addition to possible adverse effects on plants and animals, they cause severe corrosion problems. Electrical relay contacts and small springs associated with electrical switches are especially susceptible to damage from nitrate-induced corrosion.

13.8. EFFECTS OF SMOG

The harmful effects of smog occur mainly in the areas of (1) human health and comfort, (2) damage to materials, (3) effects on the atmosphere, and (4) toxicity plants. The exact degree to which exposure to smog affects human health is not known, although substantial adverse effects are suspected. Pungent-smelling, smog-produced ozone is known to be toxic. Ozone at 0.15 ppm causes coughing, wheezing, bronchial constriction, and irritation to the respiratory mucous system in healthy, exercising individuals. Peroxyacyl nitrates and aldehydes found in smog are eye irritants. Materials are adversely affected by some smog components. Rubber has a high affinity for ozone and is cracked and aged by it. Indeed, the cracking of rubber used to be employed as a test for the presence of ozone.

Ozone attacks natural rubber and similar materials by oxidizing and breaking double bonds in the polymer according to the following

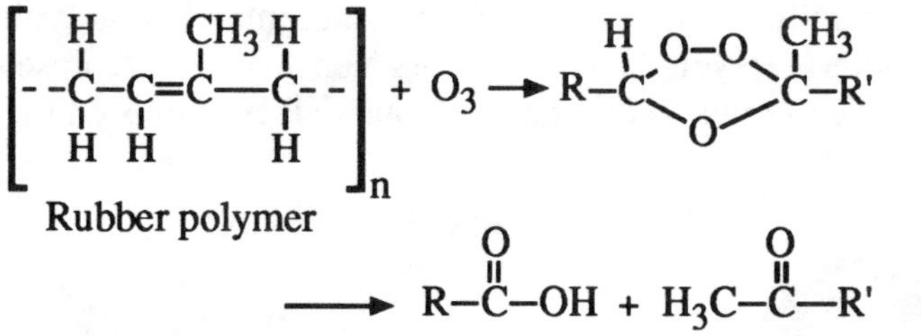

$$(13.8.1)$$

This oxidative scission type of reaction causes bonds in the polymer structure to break and results in deterioration of the polymer.

Aerosol particles that reduce visibility are formed by the polymerization of the smaller moleules produced in smog-forming reactions. Since these reactions largely involve the oxidation of hydrocarbons, it is not suprising that oxygen-containing organics make up the bulk of the particulate matter produced from smog. Ether-soluble aerosols collected from the Los Angeles atmosphere have shown an empirical formula of approximately CH_2O. Among the specific kinds of compounds identified in organic smog aerosols are alcohols, aldehydes, ketones, organic acids, esters, and organic nitrates.

Smog aerosols likely form by condensation on existing nuclei rather than by self-nucleation of smog reaction product molecules. In support of this view are electron micrographs of these aerosols showing that smog aerosol particles in the micrometer-size region consist of liquid droplets with an inorganic electron-opaque core (Figure 13.6). Thus, particulate matter from a source other than smog may have some influence on the formation and properties of smog aerosols.

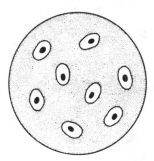

Figure 13.6. Representation of an electron micrograph of smog aerosol particles collected by a jet inertial impactor, showing electron-opaque nuclei in the centers of the impacted droplets.

In view of worldwide shortages of food, the known harmful effects of smog on plants is of particular concern. These effects are largely due to oxidants in the smoggy atmosphere. The three major oxidants invovled are ozone, PAN, and nitrogen oxides. Of these, PAN has the highest toxicity to plants, attacking younger leaves and causing "bronzing" and "glazing" of their surfaces. Exposure for several hours to an atmosphere containing PAN at a level of only 0.02-0.05 ppm will damage vegetation. The sulfhydryl group of proteins in organisms is susceptible to damage by PAN, which reacts with such groups as both an oxidizing agent and an acetylating agent. Fortunately, PAN is usually present at only low levels. Nitrogen oxides occur at relatively high concentrations during smoggy conditions, but their toxicity to plants is relatively low. The low toxicity of nitrogen oxides and the usually low levels of PAN leave ozone as the greatest smog-produced threat to plant life.

Typical of the phytotoxicity of O_3, ozone damage to a lemon leaf is typified by chlorotic stippling (characteristic yellow spots on a green leaf), as represented in Figure 13.7. Reduction in plant growth may occur without visible lesions on the plant.

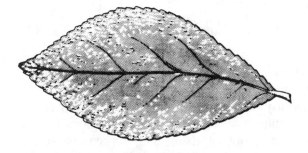

Figure 13.7. Representation of ozone damage to a lemon leaf. In color, the spots appear as yellow chlorotic stippling on the green upper surface caused by ozone exposure.

Brief exposure to approximately 0.06 ppm of ozone may temporarily cut photosynthesis rates in some plants in half. Crop damage from ozone and other photochemical air pollutants in California alone is estimated to cost millions of dollars each year. The geographic distribution of damage to plants in California is illustrated in Figure 13.8.

San
Francisco

Los Angeles

Figure 13.8. Geographic distribution of plant damage from smog in California.

LITERATURE CITED

1. Wald, Matthew L., "For Some Cars, 'Clean Gas' is found to Raise Emissions," *New York Times*, December 12, 1990, p. C1.

2. Finlayson-Pitts, Barbara J., and James N. Pitts, *Atmospheric Chemistry*, John Wiley and Sons, Inc., New York, 1986, p. 478.

3. Seinfeld, John H., *Atmospheric Chemistry and Physics of Air Pollution*, John Wiley and Sons, Inc., New York, 1986.

SUPPLEMENTARY REFERENCES

Fawell, J. K., and S. Hunt, *Environmental Toxicology: Organic Pollutants*, John Wiley and Sons, New York, NY, 1988.

Watson, Ann Y., Richard R. Bates, Donald Kennedy, Eds., *Air Pollution, the Automobile, and Public Health*, National Academy Press, Washington, DC, 1988.

Warneck, Peter, *Chemistry of the Natural Atmosphere*, Academic Press, San Diego, CA, 1988.

Lodge, James P. Jr., Ed., *Methods of Air Sampling and Analysis*, 3rd ed., Lewis Publishers, Inc., Chelsea, MI, 1989.

Schneider, T., Ed., *Atmospheric Ozone Research and its Policy Implications*, Elsevier Science Publishing Co., New York, NY, 1989.

IARC, *Diesel and Gasoline Engine Exhausts and Some Nitroarenes*, World Health Organization. WHO Publications Center USA, Albany, NY, 1989.

Howard, Philip H., *Handbook of Environmental Fate and Exposure Data for Organic Chemicals*, Vol. 1: *Large Production and Priority Pollutants*, Lewis Publishers, Chelsea, MI, 1989.

Bryce-Smith, D., and A. Gilbert, *Photochemistry*, Vol. 20, Royal Society of Chemistry, Letchworth, England, 1989.

Chameides, W. L., and D. D. Davis, "Chemistry in the Troposphere," *Chemical and Engineering News*, October 4, 1982, pp. 38-52.

Hidy, G. M., Eds., *The Character and Origins of Smog Particles*, John Wiley and Sons, Inc., New York, 1982.

Logan, J. A., M. J. Prather, S. C. Wofsy, and M. B. McElroy, "Tropospheric Chemistry: A Global Perspective," *Journal of Geophysical Research*, **85**, 7210 (1981)

Ozone and Other Photochemical Oxidants, National Academy of Sciences, Washington, DC, 1977.

Photochemical Oxidant Air Pollution, OECD Publications Center, Washington, DC, 1975.

Photochemical Oxidants, WHO Publications Center USA, Albany, NY, 1979.

Vapor-Phase Organic Pollutants, National Academy of Sciences, Washington, DC, 1976.

QUESTIONS AND PROBLEMS

1. Of the following species, the one which is the least likely product of the absorption of a photon of light by a molecule of NO_2 is: (a) O, (b) a free radical species, (c) NO, (d) NO_2^*, (e) N atoms.

2. Which of the following statements is true: (a) $RO^{\bullet}$ reacts with NO to form alkyl nitrates, (b) $RO^{\bullet}$ is a free radical, (c) $RO^{\bullet}$ is not a very reactive species, (d) $RO^{\bullet}$ is readily formed by the action of stable hydrocarbons and ground state NO_2, (e) $RO^{\bullet}$ is not thought to be an intermediate in the smog-forming process.

3. Of the following species, the one most likely to be bound in reducing smogs is: ozone, relatively high levels of atomic oxygen, SO_2, PAN, PBN.

4. Why are automotive exhaust pollutant hydrocarbons even more damaging to the environment than their quantities would indicate?

5. At what point in the smog-producing chain reaction is PAN formed?

6. What particularly irritating product is formed in the laboratory by the irradiation of a mixture of benzaldelhyde and NO_2 with ultraviolet light?

7. Which of the following species reaches its peak value last on a smog-forming day: NO, oxidants, hydrocarbons, NO_2?

8. What is the main species responsible for the oxidation of NO to NO_2 in a smoggy atmosphere?

9. Give two reasons that a turbine engine should have lower hydrocarbon emissions than an internal combustion engine.

10. What pollution problem does a lean mixture aggravate when employed to control hydrocarbon emissions from an internal combustion engine?

11. Why are two catalytic reactors necessary to control all major automotive exhaust pollutants?

12. What is the distinction between *reactivity* and *instability* as applied to some of the chemically active species in a smog-forming atmosphere?

13. Why might carbon monoxide be chosen as a standard against which to compare automotive hydrocarbon emissions in atmospheres where smog is formed? What are some pitfalls created by this choice?

14. What is an undesirable byproduct of an oxidation catalyst? Of a reduction catalyst?

15. Some atmospheric chemical reactions are abstraction reactions and others are addition reactions. Which of these applies to thereaction of hydroxyl racical with propane? With propene (propylene)?

16. How might oxidants be detected in the atmosphere?

17. Although NO_x is necessary for smog formation, there are circumstances where it is plausible that moderate reductions of NO_x levels might actually increase the rate of smog formation. Suggest a reason why this might be so.

18. Why is ozone especially damaging to rubber?

19. Show how hydroxyl radical, HO•, might react differently with ethylene, $H_2C=CH_2$, and methane, CH_4.

20. Name the stable product that results from an initial addition reaction of hydroxyl radical, HO•, with benzene.

The Endangered Global Atmosphere

14.1. ANTHROPOGENIC CHANGE IN THE ATMOSPHERE

Ever since life first appeared on Earth, the atmosphere has been influenced by the metabolic processes of living organisms. When the first primitive life molecules were formed approximately 3.5 billion years ago, the atmosphere was very different from its present state. At that time, it was chemically reducing, consisting primarily of methane, ammonia, water vapor, and hydrogen. These gases and water in the sea were bombarded by intense, bond-breaking ultraviolet radiation, which, along with lightning and radiation from radionuclides, provided the energy to bring about chemical reactions that resulted in the production of relatively complicated molecules, including even amino acids and sugars. From this rich chemical mixture, life molecules evolved. Initially, these very primitive life forms derived their energy from fermentation of organic matter formed by chemical and photochemical processes, but eventually they gained the capability to produce organic matter, "$\{CH_2O\}$," by photosynthesis,

$$CO_2 + H_2O + h\nu \rightarrow \{CH_2O\} + O_2(g) \tag{14.1.1}$$

and the stage was set for the massive biochemical transformation that resulted in the production of almost all the atmosphere's oxygen.

The oxygen initially produced by photosynthesis was probably quite toxic to primitive life forms. However, much of this oxygen was converted to iron oxides by reaction with soluble iron(II):

$$2Fe^{2+} + O_2 + 4H_2O \rightarrow 2Fe_2O_3 + 8H^+ \tag{14.1.2}$$

The enormous deposits of iron oxides thus formed provide major evidence for the liberation of free oxygen in the primitive atmosphere.

Eventually enzyme systems developed that enabled organisms to mediate the reaction of waste-product oxygen with oxidizable organic matter in the sea. Later, this mode of waste-product disposal was utilized by organisms to produce energy for respiration, which is now the mechanism by which nonphotosynthetic organisms obtain energy.

In time, oxygen accumulated in the atmosphere, providing an abundant source of O_2 for respiration. It had an additional benefit in that it enabled the formation of an ozone shield against solar ultraviolet radiation (see Section 11.8). With this shield in place, the Earth became a much more hospitable environment for life, and life forms were enabled to move from the protective surroundings of the sea to the more exposed environment of the land.

Other instances of climatic change and regulation induced by organisms can be cited. An example is the maintenance of atmospheric carbon dioxide at low levels through the action of photosynthetic organisms (note from Reaction 14.1.1 that photosynthesis removes CO_2 from the atmosphere). But, at an ever accelerating pace during the last 200 years, another organism, humankind, has engaged in a number of activities that are altering the atmosphere profoundly.[1] These are summarized below:

- Industrial activities, which emit a variety of atmospheric pollutants including SO_2, particulate matter, photochemically reactive hydrocarbons, chlorofluorocarbons, and inorganic substances (such as toxic heavy metals)
- Burning of large quantities of fossil fuel, which can introduce CO_2, CO, SO_2, NO_x, hydrocarbons (including CH_4), and particulate soot, polycyclic aromatic hydrocarbons, and fly ash into the atmosphere
- Transportation practices, which emit CO_2, CO, NO_x, photochemically reactive (smog forming) hydrocarbons (including CH_4,), and polycyclic aromatic hydrocarbons
- Alteration of land surfaces, including deforestation
- Burning of biomass and vegetation, including tropical and subtropical forests and savanna grasses, which produces atmospheric CO_2, CO, NO_x, and particulate soot and polycyclic aromatic hydrocarbons
- Agricultural practices, which produce methane (from the digestive tracts of domestic animals and from the cultivation of rice in waterlogged anaerobic soils) and N_2O from bacterial denitrification of nitrate-fertilized soils

These kinds of human activities have significantly altered the atmosphere, particularly in regard to its composition of minor constituents and trace gases. Major effects have been the following:

- Increased acidity in the atmosphere
- Production of pollutant oxidants in localized areas of the lower troposphere (see Photochemical Smog, Chapter 13)
- Elevated levels of infrared-absorbing gases (greenhouse gases)
- Threats to the ultraviolet-filtering ozone layer in the stratosphere
- Increased corrosion of materials induced by atmospheric pollutants

In 1957 photochemical smog was only beginning to be recognized as a serious problem, acid rain and the greenhouse effect were scientific curiosities, and the ozone-destroying potential of chlorofluorocarbons had not even been imagined. In that year, Revelle and Suess[2] prophetically referred to human perturbations of the Earth and its climate as a massive "geophysical experiment." The effects that this experiment may have on the global atmosphere are discussed in this chapter. For additional information the reader is referred to a series of excellent articles from *Science* dealing with urban air pollution (smog),[3] acid precipitation,[4] threats to the stratospheric ozone layer,[5] and global warming.[6]

14.2. GREENHOUSE GASES AND GLOBAL WARMING

This section deals with infrared-absorbing trace gases (other than water vapor) in the atmosphere that contribute to global warming — the "greenhouse effect" — by allowing incoming solar radiant energy to penetrate to the Earth's surface while reabsorbing infrared radiation emanating from it. Levels of these "greenhouse gases" have increased at a rapid rate during recent decades and are continuing to do so. Concern over this phenomenon has intensified since about 1980. This is because ever since accurate temperature records have been kept, the 1980s have been the warmest 10-year period recorded. On an annual basis, 1988 was the warmest year ever recorded, 1987 was second and 1981 third.[7] In addition to being a scientific issue, greenhouse warming of the atmosphere is also becoming a major policy and political issue.

There are many uncertainties surrounding the issue of greenhouse warming. However, several things about the phenomenon are certain. It is known that CO_2 and other greenhouse gases, such as CH_4, absorb infrared radiation by which Earth loses heat. The levels of these gases have increased markedly since about 1850 as nations have become industrialized and as forest lands and grasslands have been converted to agriculture. Chlorofluorocarbons, which also are greenhouse gases, were not even introduced into the atmosphere until the 1930s. Although trends in levels of these gases are well known, their effects on global temperature and climate are much less certain. The phenomenon has been the subject of much computer modelling. Most models predict global warming of 1.5-5°C, about as much again as has occurred since the last ice age. Such warming would have profound effects on rainfall, plant growth, and sea levels, which might rise as much as 0.5-1.5 meters.

Carbon dioxide is the gas most commonly thought of as a greenhouse gas; it is responsible for about half of the atmospheric heat retained by trace gases. It is produced primarily by burning of fossil fuels and deforestation accompanied by burning and biodegradation of biomass. On a molecule-for-molecule basis, methane, CH_4, is 20–30 times more effective in trapping heat than is CO_2. Other trace gases that contribute are chlorofluorocarbons and N_2O.

Analyses of gases trapped in polar ice samples indicate that pre-industrial levels of CO_2 and CH_4 in the atmosphere were approximately 260 parts per million and 0.70 ppm, respectively. Over the last 300 years these levels have increased to current values of around 350 ppm, and 1.7 ppm, respectively; most of the increase by far has taken place at an accelerating pace over the last 100 years. (A note of interest is the observation based upon analyses of gases trapped in ice cores that the atmospheric level of CO_2 at the peak of the last ice age about 18,000 years past was 25 percent below preindustrial levels.) About half of the increase in carbon dioxide in the last 300 years can be attributed to deforestation, which still accounts for approximately 20 percent of the annual increase in this gas. Carbon dioxide is increasing by about 1 ppm per year. Methane is going up at a rate of almost 0.02 ppm/year.[8,9] The comparatively very rapid increase in methane levels is attributed to a number of factors resulting from human activities. Among these are direct leakage of natural gas, byproduct emissions from coal mining and petroleum recovery, and release from the burning of savannas and tropical forests. Biogenic sources resulting from human activities produce large amounts of atmospheric methane. These include methane from bacteria degrading organic matter, such as municipal refuse in landfills; methane evolved from anaerobic biodegradation of organic matter in rice paddies; and methane emitted as the result of bacterial action in the digestive tracts of ruminant animals.

Both positive and negative feedback mechanisms may be involved in determining the rates at which carbon dioxide and methane build up in the atmosphere. Laboratory studies indicate that increased CO_2 levels in the atmosphere cause accelerated uptake of this gas by plants undergoing photosynthesis, which tends to slow buildup of atmospheric CO_2. Given adequate rainfall, plants living in a warmer climate that would result from the greenhouse effect would grow faster and take up more CO_2. This could be an especially significant effect of forests, which have a high CO_2-fixing ability. However, the projected rate of increase in carbon dioxide levels is so rapid that forests would lag behind in their ability to fix additional CO_2. Similarly, higher atmospheric CO_2 concentrations will result in accelerated sorption of the gas by oceans. The amount of dissolved CO_2 in the oceans is about 60 times the amount of CO_2 gas in the atmosphere. However, the times for transfer of carbon dioxide from the atmosphere to the ocean are of the order of years. Because of low mixing rates, the times for transfer of oxygen from the upper approximately 100-meter layer of the oceans to ocean depths is much longer, of the order of decades. Therefore, like the uptake of CO_2 by forests, increased absorption by oceans will lag behind the emissions of CO_2. Severe drought conditions resulting from climatic warming could cut down substantially on CO_2 uptake by plants. Warmer conditions would accelerate release of both CO_2 and CH_4 by microbial degradation of organic matter. (It is important to realize that about twice as much carbon is held in soil in dead organic matter — necrocarbon — potentially degradable to CO_2 and CH_4 as is present in the atmosphere.) Global warming might speed up the rates at which biodegradation adds these gases to the atmosphere.

It is certain that atmospheric CO_2 levels will continue to increase significantly. The degree to which this occurs depends upon future levels of CO_2 production and the fraction of that production that remains in the atmosphere. Given plausible projections of CO_2 production and a reasonable estimate that half of that amount will remain in the atmosphere, projections can be made that indicate that sometime during the middle part of the next century the concentration of this gas will reach 600 ppm in the atmosphere. This is well over twice the levels estimated for pre-industrial times. Much less certain are the effects that this change will have on climate. It is virtually impossible for the elaborate computer models used to estimate these effects to accurately take account of all variables, such as the degree and nature of cloud cover. Clouds both reflect incoming light radiation and absorb outgoing infrared radiation, with the former effect tending to predominate. The magnitude of these effects depend upon the degree of cloud cover, brightness, altitude, and thickness. In the case of clouds, too, feedback phenomena occur; for example, warming induces formation of more clouds, which reflect more incoming energy. Most computer models predict global warming of at least 3.0°C and as much as 5.5°C occurring over a period of just a few decades. These estimates are sobering because they correspond to the approximate temperature increase since the last ice age 18,000 years past, which took place at a much slower pace of only about 1 or 2°C per 1,000 years.

Drought is one of the most serious problems that could arise from major climatic change resulting from greenhouse warming. Typically, a 3-degree warming would be accompanied by a 10 percent decrease in precipitation. Water shortages would be aggravated, not just from decreased rainfall, but from increased evaporation, as well. Increased evaporation results in decreased runoff, thereby reducing water available for agricultural, municipal, and industrial use. Water shortages, in turn, lead to increased demand for irrigation and to the production of lower quality, higher salinity runoff water and wastewater. In the U. S., such a problem would be especially intense in the

Colorado River basin, which supplies much of the water used in the rapidly growing U. S. Southwest. The magnitude of this problem has been emphasized by a long-term drought in California, which caused the state's governor to propose drastic mandatory water conservation measures in February of 1991.

A variety of other problems, some of them unforeseen as of now, could result from global warming. An example is the effect of warming on plant and animal pests — insects, weeds, diseases, and rodents. Many of these would certainly thrive much better under warmer conditions.

Interestingly, another air pollutant, acid-rain-forming sulfur dioxide (see Section 14.3), may have a counteracting effect on greenhouse gases. This is because sulfur dioxide is oxidized in the atmosphere to sulfuric acid, forming a light-reflecting haze. Furthermore, the sulfuric acid and resulting sulfates act as condensation nuclei (Sections 9.9 and 10.9) that increases the extent, density, and brightness of light-reflecting cloud cover.

14.3. ACID RAIN

Precipitation made acidic by the presence of acids stronger than $CO_2(aq)$ is commonly called **acid rain**; the term applies to all kinds of acidic aqueous precipitation, including fog, dew, snow, and sleet. In a more general sense, **acid deposition** refers to the deposition on the Earth's surface of aqueous acids, acid gases (such as SO_2), and acidic salts (such as NH_4HSO_4).[4] According to this definition, deposition in solution form is *acid precipitation*, and deposition of dry gases and compounds is *dry deposition*. Sulfur dioxide, SO_2, contributes more to the acidity of precipitation than does CO_2 present at higher levels in the atmosphere for two reasons. The first of these is that sulfur dioxide is significantly more soluble in water, as indicated by its Henry's Law constant (Section 5.3) of 1.2 mol x L^{-1} x atm^{-1} compared to 3.38 x 10^{-2} mol x L^{-1} x atm^{-1} for CO_2. Secondly, the value of K_{a1} for $SO_2(aq)$,

$$SO_2(aq) + H_2O \xleftarrow{\hspace{0.3cm}} \rightarrow H^+ + HSO_3^- \tag{14.3.1}$$

$$K_{a1} = \frac{[H^+][HSO_3^-]}{[SO_2]} = 1.7 \times 10^{-2} \tag{14.3.2}$$

is more than four orders of magnitude higher than the value of 4.45 x 10^{-7} for CO_2.

Although acid rain can originate from the direct emission of strong acids, such as HCl gas or sulfuric acid mist, most of it is a secondary air pollutant produced by the atmospheric oxidation of acid-forming gases such as the following:

$$SO_2 + \tfrac{1}{2}O_2 + H_2O \xrightarrow[\text{ing of several steps}]{\text{Overall reaction consist-}} \{2H^+ + SO_4^{2-}\}(aq) \tag{14.3.3}$$

$$2NO_2 + \tfrac{1}{2}O_2 + H_2O \xrightarrow[\text{ing of several steps}]{\text{Overall reaction consist-}} 2\{H^+ + NO_3^-\}(aq) \tag{14.3.4}$$

Chemical reactions such as these play a dominant role in determining the nature, transport, and fate of acid precipitation. As the result of such reactions the chemical properties (acidity, ability to react with other substances) and physical properties (volatility, solubility) of acidic atmospheric pollutants are altered drastically. For example, even the small fraction of NO that does dissolve in water does not react significantly.

However, its ultimate oxidation product, HNO_3, though volatile, is highly water-soluble, strongly acidic, and very reactive with other materials. Therefore, it tends to be removed readily from the atmosphere and to do a great deal of harm to plants, corrodable materials, and other things that it contacts.

Although emissions from industrial operations and fossil fuel combustion are the major sources of acid-forming gases, acid rain has also been encountered in areas far from such sources. This is due in part to the fact that acid-forming gases are oxidized to acidic constituents and deposited over several days, during which time the air mass containing the gas may have moved as much as several thousand km.. It is likely that the burning of biomass, such as is employed in "slash-and-burn" agriculture evolves the gases that lead to acid formation in more remote areas. In arid regions, dry acid gases or acids sorbed to particles may be deposited, with effects similar to those of acid rain deposition.

Acid rain spreads out over areas of several hundred to several thousand kilometers. This classifies it as a *regional* air pollution problem compared to a *local* air pollution problem for smog and a *global* one for ozone-destroying chlorofluoro-carbons and greenhouse gases. Other examples of regional air pollution problems are those caused by soot, smoke, and fly ash from combustion sources and fires (forest fires). Nuclear fallout from weapons testing or from reactor fires (of which, fortunately, their has been only one major one to date — the one at Chernobyl in the Soviet Union) may also be regarded as a regional phenomenon.

Acid precipitation shows a strong geographic dependence, as illustrated in Figure 14.1, representing the pH of precipitation in the continental U.S. The preponderance of acidic rainfall in the northeastern U.S. is obvious. Analyses of the movements of air masses have shown a correlation between acid precipitation and prior movement of an air mass over major sources of anthropogenic sulfur and nitrogen oxides emissions. This is particularly obvious in southern Scandinavia, which receives a heavy burden of air pollution from densely populated, heavily industrialized areas in Europe.

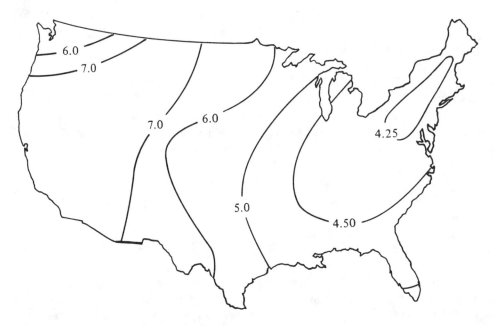

Figure 14.1. Hypothetical precipitation-pH pattern in the continental United States. Actual values found may vary with the time of year and climatic conditions.

Acid rain is not a new phenomenon; it has been observed for well over a century, with many of the older observations from Great Britain. The first manifestations of this phenomenon were elevated levels of SO_4^{2-} in precipitation collected in industrialized areas. More modern evidence was obtained from analyses of precipitation in Sweden in the 1950s and of U.S. precipitation a decade or so later. A vast research effort on acid rain was conducted in North America by the National Acid Precipitation Assessment Program, which resulted from the U.S. Acid Precipitation Act of 1980.

The longest-term experimental study of acid precipitation in the U.S. has been conducted at the U.S. Forest Service Hubbard Brook Experimental Forest in New Hampshire's White Mountains. It is downwind from major U.S. urban and industrial centers and is, therefore, a prime candidate to receive acid precipitation. This is reflected by mean annual pH values ranging from 4.0 to 4.2 during the 1964-74 period. During this period, the annual hydrogen ion input ($[H^+]$ x volume) increased by 36%.

Table 14.1 shows typical major cations and anions in pH-4.25 precipitation. Although actual values encountered vary greatly with time and location of collection, this table does show some major features of ionic solutes in precipitation. From the predominance of sulfate anion, it is apparent that sulfuric acid is the major contributor to acid precipitation. Nitric acid makes up a smaller but growing contribution to the acid present. Hydrochloric acid ranks third.

An important factor in the study of acid rain and sulfur pollution involves the comparison of primary sulfate species (those emitted directly by point sources) and secondary sulfate species (those formed from gaseous sulfur compounds, primarily by the atmospheric oxidation of SO_2). A low primary-sulfate content indicates transport of the pollutant from some distance, whereas a high primary-sulfate content indicates local emissions. This information can be useful in determining the effectiveness of SO_2 control in reducing atmospheric sulfate, including sulfuric acid. Primary and secondary sulfates can be measured using the oxygen-18 content of the sulfates. This content is higher in sulfate emitted directly from a power plant than it is in sulfate formed by the oxidation of SO_2. This technique can yield valuable information on the origins and control of acid rain.

Table 14.1. Typical Values of Ion Concentrations in Acidic Precipitation

Cations		Anions	
Ion	Concentration equivalents/L x 10^6	Ion	Concentration equivalents/L x 10^6
H^+	56	SO_4^{2-}	51
NH_4^+	10	NO_3^-	20
Ca^{2+}	7	Cl^-	12
Na^+	5	Total	83
Mg^{2+}	3		
K^+	2		
Total	83		

Ample evidence exists of the damaging effects of acid rain. The major such effects are the following:

- Direct phytotoxicity to plants from excessive acid concentrations. (Evidence of direct or indirect phytoxicity of acid rain is provided by the declining health of Eastern U.S. and Scandinavian forests and especially by damage to Germany's Black Forest.)
- Phytotoxicity from acid-forming gases, particularly SO_2 and NO_2, that accompany acid rain
- Indirect phytotoxicity, such as from Al^{3+} liberated from soil
- Destruction of sensitive forests
- Respiratory effects on humans and other animals
- Acidification of lake water with toxic effects to lake flora and fauna, especially fish fingerlings
- Corrosion to exposed structures, electrical relays, equipment, and ornamental materials. Because of the effect of hydrogen ion,

$$2H^+ + CaCO_3(s) \rightarrow Ca^{2+} + CO_2(g) + H_2O$$

  limestone, $CaCO_3$, is especially susceptible to damage from acid rain
- Associated effects, such as reduction of visibility by sulfate aerosols and the influence of sulfate aerosols on physical and optical properties of clouds. (As mentioned in Section 14.2, intensification of cloud cover and changes in the optical properties of cloud droplets — specifically, increased reflectance of light — resulting from acid sulfate in the atmosphere may even have a mitigating effect on greenhouse warming of the atmosphere.) A significant association exists between acidic sulfate in the atmosphere and haziness.

Soil sensitivity to acid precipitation can be estimated from cation exchange capacity (CEC, see Chapter 5). Soil is generally insensitive if free carbonates are present or if it is flooded frequently. Soils with a cation exchange capacity above 15.4 milliequivalents/100 g are also insensitive. Soils with cation exchange capacities between 6.2 meq/100 g and 15.4 meq/100 g are slightly sensitive. Soils with cation exchange capacities below 6.2 meq/100 g normally are sensitive if free carbonates are absent and the soil is not frequently flooded.

Forms of precipitation other than rainfall, such as snow, are susceptible to excess acidity. Acidic fog can be especially damaging because it is very penetrating. In early December, 1982, Los Angeles experienced a severe, two-day episode of acid fog. This fog consisted of a heavy concentration of acidic mist particles at ground level, which reduced visibility and were very irritating to breathe. The pH of the water in these particles was 1.7, much lower than ever before recorded for acid precipitation.

14.4. OZONE LAYER DESTRUCTION

Recall from Section 9.5 that stratospheric ozone, O_3, serves as a shield to absorb harmful ultraviolet radiation in the stratosphere, protecting living beings on the Earth from the effects of excessive amounts of such radiation. The two reactions by which stratospheric ozone are produced are,

$$O_2 + h\nu \rightarrow O + O \qquad (\lambda < 242.4\,\text{nm}) \tag{14.4.1}$$

$$O + O_2 + M \quad \rightarrow O_3 + M \text{ (energy-absorbing } N_2 \text{ or } O_2) \tag{14.4.2}$$

and it is destroyed by photodissociation,

$$O_3 + h\nu \rightarrow O_2 + O \qquad (\lambda < 325\,\text{nm}) \tag{14.4.3}$$

and a series of reactions from which the net result is the following:

$$O + O_3 \rightarrow 2O_2 \tag{14.4.4}$$

The concentration of ozone in the stratosphere is a steady-state concentration resulting from the balance of ozone production and destruction by the above processes. The quantities of ozone involved are interesting. A total of about 350,000 metric tons of ozone are formed and destroyed daily. Ozone never makes up more than a small fraction of the gases in the ozone layer. In fact, if all the atmosphere's ozone were in a layer at 273 K and 1 atm, it would be only 3 mm thick!

Ozone absorbs ultraviolet radiation very strongly in the region 220-330 nm. Therefore, it is effective in filtering out dangerous UV-B radiation , 290 nm $< \lambda <$ 320 nm. (UV-A radiation, 320 nm-400 nm, is relatively less harmful and UV-C radiation, < 290 nm does not penetrate to the troposphere.) If UV-B were not absorbed by ozone, severe damage would result to exposed forms of life on the Earth. Absorption of electromagnetic radiation by ozone converts the radiation's energy to heat and is responsible for the temperature maximum encountered at the boundary between the stratosphere and the mesosphere at an altitude of approximately 50 km. The reason that the temperature maximum occurs at a higher altitude than that of the maximum ozone concentration arises from the fact that ozone is such an effective absorber of ultraviolet light, so that most of this radiation is absorbed in the upper stratosphere, where it generates heat, and only a small fraction reaches the lower altitudes, which remain relatively cool.

Increased intensities of ground-level ultraviolet radiation caused by stratospheric ozone destruction would have some significant adverse consequences. One major effect would be on plants, including crops used for food. The destruction of microscopic plants that are the basis of the ocean's food chain (phytoplankton) could severely reduce the productivity of the world's seas. Human exposure would result in an increased incidence of cataracts. The effect of most concern to humans is the elevated occurrence of skin cancer in individuals exposed to ultraviolet radiation. This is because UV-B radiation is absorbed by cellular DNA (see Chapter 20) resulting in photochemical reactions that alter the function of DNA so that the genetic code is improperly translated during cell division. This can result in uncontrolled cell division leading to skin cancer. People with light complexions lack protective melanin, which absorbs UV-B radiation, and are especially susceptible to its effects. The most common type of skin cancer resulting from ultraviolet exposure is squamous cell carcinoma, which forms lesions that are readily removed and has little tendency to spread (metastasize). Readily metastasized malignant melanoma caused by absorption of UV-B radiation is often fatal. Fortunately, this form of skin cancer is relatively uncommon.

The major culprit in ozone depletion consists of chlorofluorocarbon (CFC) compounds, commonly known as "Freons." These volatile compounds have been used and released to a very large extent in recent decades. The major use associated with CFCs is as refrigerant fluids. Other applications have included solvents, aerosol propellants, and blowing agents in the fabrication of foam plastics. The same extreme chemically stability that makes CFCs nontoxic enables them to persist for years in the atmosphere and to enter the stratosphere. In the stratosphere, as discussed in Section 11.8, the photochemical dissociation of CFCs by intense ultraviolet radiation,

$$CF_2Cl_2 + h\nu \ \rightarrow \ Cl^\bullet + CCl_2F^\bullet \tag{14.4.5}$$

yields chlorine atoms each of which can go through chain reactions, particularly the following:

$$Cl^\bullet + O_3 \ \rightarrow \ ClO^\bullet + O_2 \tag{14.4.6}$$

$$ClO^\bullet + O \rightarrow \ Cl^\bullet + O_2 \tag{14.4.7}$$

The net effect of these reactions is catalysis of the destruction of several thousand molecules of O_3 for each Cl atom produced. Because of their widespread use and persistency, the two CFCs of most concern in ozone destruction are CFC-11 and CFC-12, $CFCl_3$ and CF_2Cl_2, respectively. Even in the intense ultraviolet radiation of the stratosphere the most persistent chlorofluorocarbons have lifetimes of the order of 100 years.

The most prominent instance of ozone layer destruction is the so-called "Antarctic ozone hole" that has been documented in recent years.[10] This phenomenon is manifested by the appearance during the Antarctic's late winter and early spring of severely depleted stratospheric ozone (up to 50%) over the polar region. The reasons why this occurs are related to the normal effect of NO_2 in limiting Cl-atom-catalyzed destruction of ozone by combining with ClO,

$$ClO + NO_2 \ \rightarrow \ ClONO_2 \tag{14.4.8}$$

In the polar regions, particularly Antarctica, NO_x gases are removed along with water by freezing in polar stratospheric clouds at temperatures below -70°C as compounds such as $HNO_3 \cdot 3H_2O$. Furthermore, chlorine species can be liberated from $ClONO_2$ and other chlorine compounds (HCl) by reactions in the cloud ice, followed by photodissociation to yield ozone-destroying atomic chlorine:

$$ClONO_2 + H_2O \rightarrow HOCl + HNO_3 \tag{14.4.9}$$

$$ClONO_2 + HCl \rightarrow Cl_2 + HNO_3 \tag{14.4.10}$$

$$HOCl + h\nu \rightarrow HO^\bullet + Cl \tag{14.4.11}$$

$$Cl_2 + h\nu \rightarrow Cl + Cl \tag{14.4.12}$$

14.5. PHOTOCHEMICAL SMOG

Photochemical smog is a major air pollution phenomenon discussed in Chapter 13. It occurs in urban areas where the combination of pollution-forming emissions and appropriate atmospheric conditions are right for its formation. In order for high levels of smog to form, relatively stagnant air must be subjected to sunlight under low humidity conditions in the presence of pollutant nitrogen oxides and hydrocarbons. Although the automobile is the major source of these pollutants, hydrocarbons may come from biogenic sources, of which α-pinene and isoprene from trees are the most abundant. Stated succinctly, "The urban atmosphere is a giant chemical reactor in which pollutant gases such as hydrocarbons and oxides of nitrogen and sulfur react under the influence of sunlight to create a variety of products."[3] Although not as great a threat to the global atmosphere as some of the other air pollutants discussed in this chapter, smog does pose significant hazards to living things and materials in local urban areas in which millions of people are exposed.

Ironically, ozone, which serves an essential protective function in the stratosphere, is the major culprit in tropospheric smog. In fact, surface ozone levels are used as a measure of smog. Ozone's phytotoxicity raises particular concern in respect to trees and crops. Ozone is the smog constituent responsible for most of the respiratory system distress and eye irritation characteristic of human exposure to smog. Breathing is impaired at ozone levels approaching only about 0.1 ppm[11] (the U.S. National Ambient Air Quality Standard for ozone is 0.12 ppm allowed to be exceeded only once per year). Ozone is the "criteria" air pollutant that has been most resistant to control measures. Because of its strongly oxidizing nature ozone attacks unsaturated bonds in fatty acid constituents of cell membranes. Other oxidants, such as PAN (Section 13.3), also contribute to the toxicity of smog, as do aldehydes produced as reactive intermediates in smog formation.

Smog is a secondary air pollutant that forms some time after, and some distance from the injection into the atmosphere of the primary pollutant nitrogen oxides and reactive hydrocarbons required for its formation. The U.S. Environmental Protection Agency's Empirical Kinetic Modeling Approach uses the concept of an **air parcel** to model smog formation. This model utilizes the concept of a "parcel" of relatively unpolluted air moving across an urban area in which it becomes contaminated with smog-forming gases. When the upper boundary of this parcel is restricted to about 1,000 meters by a temperature inversion and subjected to sunlight, the primary pollutants react to form smog in a system that involves photochemical reaction processes, transport, mixing, and dilution. As the hydrocarbons are consumed by photochemical oxidation processes in the air and as nitrogen oxides are removed as nitrates and nitric acid (especially at nightime), ozone levels reach a peak concentration at a time and place some distance removed from the source of pollutants.

The most visible manifestation of smog is the **urban aerosol**, which greatly reduces visibility in smoggy urban aerosols. Many of the particles composing this aerosol are made from gases by chemical processes (see Section 10.5) and are therefore quite small, usually less than 2 μm. Particles of such a size are especially harmful because they scatter light most efficiently and are the most respirable. Aerosol particles formed from smog often contain toxic constituents, such as respiratory tract irritants and mutagens. The urban aerosol also contains particle constituents that originate from processes other than smog formation. Oxidation of pollutant sulfur dioxide by the strongly oxidizing conditions of photochemical smog,

$$SO_2 + \text{ }^1/_2O_2 + H_2O \rightarrow H_2SO_4 \quad \text{(overall process)} \qquad (14.5.1)$$

produces sulfuric acid and sulfate particles. Nitric acid and nitrates are produced at night when sunlight is absent, which readily decomposes intermediate NO_3 radical:

$$O_3 + NO_2 \rightarrow O_2 + NO_3 \qquad (14.5.2)$$

$$NO_3 + NO_2 + M \text{ (energy-absorbing third body)} \rightarrow N_2O_5 + M \qquad (14.5.3)$$

$$N_2O_5 + H_2O \rightarrow 2HNO_3 \qquad (14.5.4)$$

$$HNO_3 + NH_3 \rightarrow NH_4NO_3 \qquad (14.5.5)$$

As indicated by the last reaction above, ammonium salts are common constituents of urban aerosol particles; they tend to be particularly corrosive. Metals, which may contribute to the toxicity of urban aerosol particles and which may catalyze reactions on their surfaces, occur in the particles. Water is always present, even in low humidity atmospheres, and is usually present in urban aerosol particles. Carbon and polycyclic aromatic hydrocarbons from partial combustion and diesel engine emissions are usually abundant constituents; particulate elemental carbon is usually the particulate constituent most responsible for absorbing light in the urban aerosol. If the air parcel originates over the ocean, it contains sea salt particles consisting largely of NaCl, from which some of the chloride may be lost as volatile HCl by the action of less volatile strong acids produced by smog. This phenomenon is responsible for Na_2SO_4 and $NaNO_3$ found in the urban aerosol.

Polycyclic aromatic hydrocarbons, PAH (see Section 10.8), are among the urban aerosol particle constituents of most concern, particularly because metabolites of some of these compounds (see the 7,8-diol-9,10-epoxide of benzo(a)pyrene in Figure 20.14) are carcinogenic. PAHs include unsubstituted compounds, as well as those with alkyl, oxygen, or nitrogen substituents or O or N hetero atoms:

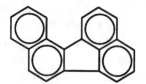

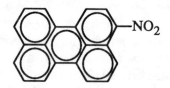

Benzo(j)fluoranthene   3-Nitroperylene (mutagenic)

These kinds of compounds are emitted by internal combustion engine exhausts and occur in both the gas and particulate phases. Numerous mechanisms exist for their destruction and chemical alteration, particularly reaction with oxidant species — HO•, O_3, NO_2, N_2O_5, and HNO_3. Direct photolysis is also possible. PAH compounds in the vapor phase are destroyed relatively rapidly by these means, whereas PAHs sorbed to particles are much more resistant to reaction.

Another kind of urban aerosol particulate matter of considerable concern is **acid fog**, which may have pH values below 2 due to the presence of H_2SO_4 or HNO_3. Seinfeld[3] has pointed out that acid fog formation covers a wide range of atmospheric chemical and physical phenomena. The gas-phase oxidation of SO_2 and NO_x

produces strong acids, which form very small aerosol particles. These, in turn, act as condensation nuclei for water vapor. Acid-base phenomena occur in the droplets, and they act as scavengers to remove ionic species from air. Because fog aerosol particles form in areas of intense acid gas pollution near the surface, the concentrations of acids and ionic species in fog aerosol droplets tend to be much higher than in cloud aerosol droplets at higher altitudes.

In addition to health effects and damage to materials, one of the greater problems caused by smog is destruction of crops and reduction of crop yields. The annual cost of these effects in California, alone, is about $15 billion.[12]

Even lightly populated non-industrial areas are subject to the effects of smog brought about by human activities. Particularly, the practice of burning savanna grasses for agricultural purposes causes smog. This burning produces NO_x and reactive hydrocarbons that are required for smog formation. Furthermore, these grasses grow in tropical regions which have the intense sunlight required for smog formation. The net result is rapid development of smoggy conditions as manifested by ozone levels several times normal background values.

14.6. NUCLEAR WINTER

Nuclear winter is a term used to describe a catastrophic atmospheric effect that might occur after a massive exchange of nuclear firepower between major powers. The heat from the nuclear blasts and from resulting fires would result in powerful updrafts carrying combustion products to stratospheric regions.[13] The reflection and scattering of sunlight by particles carried into the stratosphere would result in several years of much lower temperatures and freezing temperatures even during summertime,[14] such as occurred in 1816, "the year without a summer," following the astoundingly massive Tambora, Indonesia volcanic explosion of 1915. Brutally cold years around 210 B.C. that followed a similar volcanic incident in Iceland were recorded in ancient China. In addition to the direct suffering caused, massive starvation would result from crop failures accompanying years of nuclear winter. The incidents cited above clearly illustrate the climatic effects of huge quantities of particulate matter ejected high into the atmosphere.

Evidence exists to suggest that military explosives can result in the introduction of large quantities of particulate matter into the atmosphere. For example, carpet bombings of cities, such as the tragic bombing of Dresden, Germany, near the end of World War II, have caused huge firestorms that created their own wind causing a particle-laden updraft into the atmosphere. Of course, the effect of a full scale nuclear exchange would be many-fold higher.

An idea of the potential climatic effect resulting from a full-scale nuclear exchange may be obtained by considering the magnitude of the blasts that might be involved. Only two nuclear bombs have been used in warfare, both dropped on cities in Japan in 1945. The Hiroshima fission bomb had the explosive force of 12,000 kilotons of TNT explosive. Its blast, fireball, and instantaneous emissions of neutrons and gamma radiation, followed by fires and exposure to radioactive fission products, killed about 100,000 people and destroyed the city on which it was dropped. By comparison with this 12 kiloton bomb, modern fusion bombs are typically rated at 500 kilotons, and 10 <u>mega</u>ton weapons are common. A full-scale nuclear exchange might involve a total of the order of 5,000 megatons of nuclear explosives. As a result, unimaginable quantities of soot from the partial combustion of wood, plastics, paving asphalt, petroleum, forests, and other combustibles would be carried to the stratosphere. At such high altitudes tropospheric removal mechanisms (see Figure 10.2) are not effective because there is not enough water in the stratosphere to produce rainfall to wash particles from the air and convection processes are very limited. Much of the particulate matter would be in the μm size range in which light is reflected, scattered, and absorbed most effectively and settling is very slow. Therefore, vast areas of the Earth would be overlain by a stable cloud of particles and the fraction of sunlight reaching the Earth's surface would be drastically reduced, resulting in a dramatic cooling effect. There would be other effects as well. The extreme heat and pressure in the fireball would result in fixation of nitrogen,

$$O_2 + N_2 \rightarrow 2NO \tag{14.6.1}$$

as ozone-destroying nitrogen oxides.

The timing and location of nuclear blasts are very important in determining their climatic effects. Atmospheric testing of nuclear weapons, including a 58-ton monster detonated by the Soviet Union, have had little atmospheric effect. Such tests were carried out at widely spaced intervals on deserts, small tropical islands, and other locations with minimal combustible matter. In contrast, military use of nuclear weapons would involve a high concentration of firepower, both in time and in space, on industrial and military targets consisting largely of combustibles. Furthermore, destruction of hardened military sites requires blasts that disrupt large quantities of soil, rock, and concrete, which are pulverized, vaporized, and blown into the atmosphere.

On a hopeful note, the East-West conflict that has dominated world politics and threatened nuclear war since the mid-1900s has now abated and the probability of nuclear warfare seems to have diminished. However, the outbreak of war in the Middle East in 1991, nuclear proliferation, disintegration of great powers with vast nuclear arsenals, racial hatred, and a "trigger-happy" state of mind among even educated people who should know better should still give us cause for concern in respect to the prospect of "nuclear winter."

14.7. WHAT IS TO BE DONE?

Of all environmental hazards, there is little doubt that major disruptions in the atmosphere and climate have the greatest potential for catastrophic and irreversible environmental damage. If levels of greenhouse gases and reactive trace gases continue to increase at present rates, major environmental effects are virtually certain. On a hopeful note, the bulk of these emissions arise from industrialized nations, which — in principle — can apply the resources needed to reduce them substantially. The best example to date has been the 1987 "Montreal Protocol on Substances that Deplete the Ozone Layer," an international treaty through which a large number of nations agreed to cut chlorofluorocarbon emissions by 50% by the year 2000. Developments since then indicate that this goal may even be exceeded. This agreement may pave the way for more encompassing agreements covering carbon dioxide and other trace gases.

More ominous, however, is the combination of population pressure and desire for better living standards on a global basis. Consider, for example, the demand that these two factors place on energy resources, and the environmental disruption that may result. In many highly populated developing nations, high-sulfur coal is the most readily available, cheapest source of energy. It is understandably difficult to persuade populations faced with real hunger to forego short-term economic gain for the sake of long-term environmental quality. Destruction of rain forests by "slash-and-burn" agricultural methods does make economic sense to those engaged in subsistence farming to obtain badly needed hard currency, which can be earned by converting forest to pasture land and exporting fast-food-hamburger beef to wealthier nations.

What is to be done? First of all, it is important to keep in mind that the atmosphere has a strong ability to cleanse itself of pollutant species. Water-soluble gases, including greenhouse-gas CO_2, acid-gas SO_2, and fine particulate matter are removed with precipitation. For most gaseous contaminants, oxidation precedes or accompanies removal processes. To a degree, oxidation is carried out by O_3. To a larger extent, the most active atmospheric oxidant is hydroxyl radical, $HO^\bullet$. As illustrated in Figure 9.6, this atmospheric scavenger species reacts with all important trace gas species, except for CO_2 and chlorofluorocarbons. It is now generally recognized that $HO^\bullet$ is an almost universal atmospheric cleansing agent. Given this crucial role of $HO^\bullet$ radical, any pollutants that substantially reduce its concentration in the atmosphere are potentially troublesome. One concern over carbon monoxide emissions to the atmosphere is the reactivity of $HO^\bullet$ with CO,

$$CO + HO^\bullet \rightarrow CO_2 + H \tag{14.7.1}$$

which could result in removal of $HO^\bullet$ from the atmosphere.

Of all the major threats to the global climate, it is virtually certain that humankind will have to try to cope with greenhouse warming and the climatic effects thereof. The measures to be taken in dealing with this problem fall into the three following categories:

- **Minimization** by reducing emissions of greenhouse gases, switching to alternate energy sources, increasing energy conservation, and reversing deforestation. It is especially sensible to use measures that have major benefits in addition to reduction of greenhouse warming. Such measures include, as examples, reforestation, restoration of grasslands, increased energy conservation, and a massive shift to solar energy sources. On the other hand, shifting to nuclear-based energy sources just to prevent possible greenhouse warming may not be such a good idea.

- **Counteracting measures**, such as injecting light-reflecting particles into the upper atmosphere.

- **Adaptation**, particularly through increased efficiency and flexibility of the distribution and use of water, which might be in very short supply in many parts of the world as a consequence of greenhouse warming. Important examples are implementation of more efficient irrigation practices and changes in agriculture to grow crops that require less irrigation. Emphasis on adaptation is favored by those who contend that not enough is known about the types and severity of global warming to justify massive expenditures on minimization and counteractive measures. In any case, adaptation will certainly have to be employed as a means of coping with global warming.

A common measure taken against the effects of another atmospheric hazard, ultraviolet radiation, provides an example of adaptation. This measure is the use of sunscreens placed on the skin as lotions to filter out UV-B radiation. The active ingredient of sunscreen must absorb ultraviolet light effectively. But this is not enough because it is the absorption of ultraviolet light by skin that makes it so dangerous in the first place. Therefore, active compounds in sunscreen formulations must also dissipate the absorbed energy in a harmless way. The way in which this is done is illustrated below for *o*-hydroxybenzophenone contained in sunscreens, which reacts as follows:

$$\text{(o-hydroxybenzophenone)} + h\nu \xrightarrow{\ (1)\ } \text{(enol form)} \xrightarrow{\ (2)\ } \text{(keto form)} \qquad (14.7.2)$$

The first step in the above reaction sequence can be regarded as *intramolecular transfer* of energy and *internal isomerization* (see Section 9.4) by which absorbed ultraviolet energy is accomodated within the molecule, which reacts to produce the more energized enol form. In Step 2 the molecule reverts back to the more stable keto form, losing energy thermally in the process. The net result is that energy is absorbed, the excited absorbing species reacts only with itself, then the energy is dissipated harmlessly as thermal energy, which does not cause additional reactions to occur.

The "tie-in strategy" has been proposed as a sensible approach to dealing with the kinds of global environmental problems discussed in this chapter. This approach was first enunciated in 1980.[15] It advocates taking measures consisting of "high-leverage actions" which are designed to prevent problems from occurring and which have substantial merit even if the major problems that they are designed to avoid do not materialize. An example is implementation of environmentally sound substitutes for fossil fuels to lower atmospheric CO_2 output and prevent greenhouse warming. Even if it turns out that the greenhouse effect is exaggerated, such substitutes would save the Earth from other kinds of environmental damage, such as disruption of land by strip mining coal or preventing oil spills from petroleum transport. Definite economic and political benefits would also accrue from lessened dependence on uncertain, volatile petroleum supplies. Increased energy efficiency would diminish both greenhouse gas and acid rain production, while lowering costs of production and reducing the need for expensive and environmentally disruptive new power plants. The implementation of these kinds of tie-in strategies requires some degree of incentive beyond normal market forces, and, therefore, is opposed by some on ideological grounds. A good example is opposition to mandatory fuel mileage standards for automobiles. However, to quote Schneider,[6] "a market that does not include the costs of environmental disruptions can hardly be called a free market."

LITERATURE CITED

1. Graedel, Thomas E., and Paul J. Crutzen, "The Changing Atmosphere," in *Managing Planet Earth*, special issue of *Scientific American*, September, 1989, pp. 58-68.

2. Revelle, R., and H. Suess, *Tellus*, **9**, 18 (1957).

3. Seinfeld, John H., "Urban Air Pollution: State of the Science," *Science*, **243**, February 10, 1989, pp. 745-752.

4. Schwartz, Stephen E., "Acid Deposition: Unraveling a Regional Phenomenon," *Science*, **243**, February 10, 1989, pp. 753-763.

5. McElroy, Michael B., and Ross J. Salawitch, "Changing Composition of the Global Stratosphere," *Science*, **243**, February 10, 1989, pp. 763-770.

6. Schneider, Stephen H., "The Greenhouse Effect: Science and Policy," **243**, February 10, 1989, pp. 751-781.

7. Schneider, Stephen, H., "The Changing Climate," in *Managing Planet Earth*, special issue of *Scientific American*, September, 1989, pp. 70-79.

8. Pearce, F., "Methane, the Hidden Greenhouse Gas," *New Scientist*, May 6, 1989, pp. 37-41.

9. Khalil, M. A. K., and R. A. Rasmussen, "Atmospheric Methane: Recent Global Trends," *Environmental Science and Technology*, **24**, 549-553 (1991).

10. Stolarski, Richard S., and Paul J. Crutzen, "The Antarctic Ozone Hole," *Scientific American*, January, 1988.

11. Lippman, M., "Health Effects of Ozone," *J. Air Pollut. Control Assoc.*, **39**, 672-695 (1989).

12. Anderson, I., *New Scientist*, April 1, 1991, p. 21.

13. Crutzen, P. J., and J. W. Birks, "The Atmosphere after a Nuclear War: Twilight at Noon," *Ambio*, **11**, 114-125 (1982).

14. Sagan, Carl, and Richard Turco, *A Path Where No Man Thought: Nuclear Winter and the End of the Arms Race*, Random House, New York, 1990.

15. E. Boulding, in *Carbon Dioxide Effects, Research and Assessment Program: Workshop on Environmental and Societal Consequence of a Possible CO_2-Induced Climatic Change*, Report 009, CONF-7904143, U. S. Department of Energy, U. S. Government Printing Office, Washington, DC, October 1980, pp. 79-10.

SUPPLEMENTARY REFERENCES

Seinfeld, John H., *Atmospheric Chemistry and Physics of Air Pollution*, John Wiley and Sons, Inc., Somerset, NJ, 1986.

Makhijani, Arjun, Annie Makhijani, and Amanda Bickel, *Saving Our Skins: Technical Potential and Policies for the Elimination of Ozone-Depleting Chlorine Compounds*, Environmental Policy Institute and the Institute for Energy and Environmental Research, Washington, DC, 1988.

Warneck, Peter, *Chemistry of the Natural Atmosphere*, Academic Press, San Diego, CA, 1988.

Regens, James L., and Robert W. Rycroft, *The Acid Rain Controversy*, University of Pittsburgh Press, Pittsburgh, PA, 1988.

Ferraudi, G. J., *Elements of Inorganic Photochemistry*, John Wiley and Sons, New York, NY, 1988.

James C. White, Ed., *Acid Rain: The Relationship Between Sources and Receptors*, Elsevier, New York, NY, 1988.

Park, Chris C., *Acid Rain: Rhetoric and Reality*, Methuen/Chapman and Hall, New York, NY, 1987.

Singer, Fred S., Ed., *Global Climate Change: Human and Natural Influences*, Paragon House, New York, NY, 1989.

Schneider, T., Ed., *Atmospheric Ozone Research and its Policy Implications*, Elsevier Science Publishing Co. New York, NY, 1989.

Moomaw, William R., and Irfint M. Mintzer, *Strategies for Limiting Global Climate Change*, World Resources Institute Publications, Washington, DC, 1989.

Kemp, David D., *Global Environmental Issues: A Climatological Approach*, Routledge, Chapman and Hall, Inc., New York, NY, 1990.

Brower, Michael, *Cool Energy: The Renewable Solution to Global Warming*, Union of Concerned Scientists, Cambridge, MA, 1989.

Brown, Lester R., *State of the World 1990*, Worldwatch Institute, Washington, DC, 1989.

Park, Chris C., *Chernobyl: The Long Shadow*, Routledge, London and New York, 1989.

Schneider, Stephen H., *Global Warming: Are We Entering The Greenhouse Century?*, Sierra Club Books, San Francisco, CA, 1989.

Oppenheimer, Michael, and Robert H. Boyle, *Dead Heat: The Race Against the Greenhouse Effect*, Basic Books, New York, NY, 1990.

Lyman, Francesca, *The Greenhouse Trap*, Beacon Press, Boston, MA, 1990.

Knap, Anthony H., and Mary-Scott Kaiser, Eds.,*The Long-Range Atmospheric Transport of Natural and Contaminant Substances*, Kluwer Academic Publishers, Norwell, MA, 1989.

QUESTIONS AND PROBLEMS

1. How do modern transportation problems contribute to the kinds of atmospheric problems discussed in this chapter?

2. What is the rationale for classifying most acid rain as a secondary pollutant?

3. Distinguish among UV-A, UV-B, and UV-C radiation. Why does UV-B pose the greatest danger in the troposphere?

4. How does the extreme cold of stratospheric clouds in Antarctic regions contribute to the Antarctic ozone hole?

5. How does the oxidizing nature of ozone from smog contribute to the damage that it does to cell membranes?

6. What may be said about the time and place of the occurrence of maximum ozone levels from smog in respect to the origin of the primary pollutants that result in smog formation?

7. What is the basis for "nuclear winter"?

8. What is meant by a "tie-in strategy"?

15

The Geosphere and Geochemistry

15.1. INTRODUCTION

The **geosphere**, or solid Earth, is that part of the Earth upon which humans live and from which they extract most of their food, minerals, and fuels. Once thought to have an almost unlimited buffering capacity against the perturbations of humankind, the geosphere is now known to be rather fragile and subject to harm by human activities. For example, some billions of tons of Earth material are mined or otherwise disturbed each year in the extraction of minerals and coal. Two atmospheric pollutant phenomena — excess carbon dioxide and acid rain (see Chapter 14) — have the potential to cause major changes in the geosphere. Too much carbon dioxide in the atmosphere may cause global heating ("greenhouse effect"), which could significantly alter rainfall patterns and turn currently productive areas of the Earth into desert regions. The low pH characteristic of acid rain can bring about drastic changes in the solubilities and oxidation-reduction rates of minerals. Erosion caused by intensive cultivation of land is washing away vast quantities of topsoil from fertile farmlands each year. In some areas of industrialized countries, the geosphere has been the dumping ground for toxic chemicals (see the discussion of hazardous wastes in Chapters 17-19). Ultimately, the geosphere must provide disposal sites for the nuclear wastes of the approximately 400 nuclear reactors now operating worldwide, as well as those yet to be completed. It may be readily seen that the preservation of the geosphere in a form suitable for human habitation is one of the greatest challenges facing humankind.

Human activities on the Earth's surface may have an effect upon climate. The most direct such effect is through the change of surface albedo, defined as the percentage of incident solar radiation reflected by a land or water surface. For example, if the sun radiates 100 units of energy per minute to the outer limits of the atmosphere, and the Earth's surface receives 60 units per minute of the total, then reflects 30 units upward, the albedo is 50 percent. Some typical albedo values for different areas on the Earth's surface are[1]: evergreen forests, 7-15%; dry, plowed fields, 10-15%; deserts, 25-35%; fresh snow, 85-90%; asphalt, 8%. In some heavily developed areas, anthropogenic heat release is comparable to the solar input. The anthropogenic (human-produced) energy release over the 60 square kilometers of Manhattan Island averages about 4 times the solar energy falling on the area; over the 3,500 km^2 of Los Angeles the anthropogenic energy release is about 13% of the solar flux.

One of the greater impacts of humans upon the geosphere is the creation of desert areas through abuse of land with marginal amounts of rainfall. This process, called **desertification** is manifested by declining groundwater tables, salinization of topsoil and water, reduction of surface waters, unnaturally high soil erosion, and desolation of native vegetation. The problem is severe in some parts of the world, particularly Africa's Sahel (southern rim of the Sahara), where the Sahara advanced southward at a particularly rapid rate during the period 1968-73, contributing to widespread starvation in Africa during the 1980s. Large, arid areas of the western U.S. are experiencing at least some of the symptoms of desertification as the result of human activities and a severe drought that has lasted five years as of 1991. As the populations of the Western states increase, one of the greatest challenges facing the residents is to prevent additional conversion of land to desert.

The most important part of the geosphere for life on earth is soil. It is the medium upon which plants grow and virtually all terrestrial organisms depend upon it for their existence. The productivity of soil is strongly affected by environmental conditions and pollutants. Because of the importance of soil, all of Chapter 16 is devoted to its environmental chemistry.

With increasing population and industrialization, one of the more important aspects of our use of the geosphere has to do with the protection of water sources. Mining, agricultural, chemical, and radioactive wastes all have the potential for contaminating both surface water and groundwater. Sewage sludge spread on land may contaminate water by release of nitrate and heavy metals. Landfills may likewise be sources of contamination. Leachates from unlined pits and lagoons containing hazardous liquids or sludges may pollute drinking water.

It should be noted, however, that many soils have the ability to assimilate and neutralize pollutants. Various chemical and biochemical phenomena in soils operate to reduce the harmful nature of pollutants. These phenomena include oxidation-reduction processes, hydrolysis, acid-base reactions, precipitation, sorption, and biochemical degradation. Some hazardous organic chemicals may be degraded to harmless products on soil, and heavy metals may be sorbed by it. In general, however, extreme care should be exercised in disposing of chemicals, sludges, and other potentially hazardous materials on soil, particularly where the possibility of water contamination exists.

15.2. THE NATURE OF SOLIDS IN THE GEOSPHERE

The earth is divided into layers, including the solid iron-rich inner core, molten outer core, mantle, and crust. Environmental chemistry is most concerned with the **lithosphere**, which consists of the outer mantle and the **crust**. The latter is the earth's outer skin that is accessible to humans. It is extremely thin compared to the diameter of the earth, ranging from 5 to 40 km thick.

Most of the solid earth crust consists of rocks. Rocks are composed of minerals, where a **mineral** is a naturally-occurring inorganic solid with a definite internal structure and chemical composition[2]. A **rock** is a mass of pure mineral or an aggregate of two or more minerals.

Structure and Properties of Minerals

Many minerals have very well defined crystalline structures which result from the ways in which the atoms and ions of different kinds and sizes are chemically bonded together in the mineral. Physical properties of minerals can be used to classify them.

The characteristic external appearance of a pure crystalline mineral is its **crystal form**. Because of space constrictions on the ways that minerals grow, the pure crystal form of a mineral is often not expressed. **Color** is an obvious characteristic of minerals, but can vary widely due to the presence of impurities. The appearance of a mineral surface in reflected light describes its **luster**. Minerals may have a metallic luster or appear partially metallic (or submetallic), vitreous (like glass), dull or earthy, resinous, or pearly. The color of a mineral in its powdered form as observed when the mineral is rubbed across an unglazed porcelain plate is known as **streak**. **Hardness** is expressed on Mohs scale, which ranges from 1 to 10 and is based upon 10 minerals that vary from talc, hardness 1, to diamond, hardness 10. **Cleavage** denotes the manner in which minerals break along planes and the angles in which these planes intersect. For example, mica cleaves to form thin sheets. Most minerals **fracture** irregularly, although some fracture along smooth curved surfaces or into fibers or splinters. **Specific gravity**, density relative to that of water, is another important physical characteristic of minerals.

Kinds of Minerals

Although over two thousand minerals are known, only about 25 **rock-forming minerals** make up most of the earth's crust. The nature of these minerals may be better understood with a knowledge of the elemental composition of the crust. Oxygen and silicon make up 46.6% and 27.7% by mass of the earth's crust, respectively. Therefore, most minerals are **silicates** such as quartz, SiO_2, or orthoclase, $KAlSi_3O_8$. In descending order of abundance the other elements in the earth's crust are aluminum (8.1%), iron (5.0%), calcium (3.6%), sodium (2.8%), potassium (2.6%), magnesium (2.1%), and other (1.5%). Table 15.1 summarizes the major kinds of minerals in the earth's crust.

Table 15.1. Major Mineral Groups in the Earth's Crust

Mineral group	Examples	Formula
Silicates	Quartz	SiO_2
	Olivine	$(Mg,Fe)_2SiO_4$
	Potassium feldspar	$KAlSi_3O_8$
Oxides	Corundum	Al_2O_3
	Magnetite	Fe_3O_4
Carbonates	Calcite	$CaCO_3$
	Dolomite	$CaCO_3 \cdot MgCO_3$
Sulfides	Pyrite	FeS_2
	Galena	PbS
Sulfates	Gypsum	$CaSO_4 \cdot 2H_2O$
Halides	Halite	$NaCl$
	Fluorite	CaF_2
Native elements	Copper	Cu
	Sulfur	S

Secondary minerals are formed by alteration of parent mineral matter. **Clays** are silicate minerals, usually containing aluminum, that constitute one of the most significant classes of secondary minerals. Olivine, augite, hornblende, and feldspars all form clays. Clays are discussed in detail in Section 15.4.

Evaporites

Evaporites are soluble salts that precipitate from solution under special arid conditions, commonly as the result of the evaporation of seawater. The most common evaporite is **halite**, NaCl. Other simple evaporite minerals are sylvite (KCl), thenardite (Na_2SO_4), and anhydrite ($CaSO_4$). Many evaporites are hydrates, including bischofite ($MgCl_2 \cdot 6H_2O$), gypsum ($CaSO_4 \cdot 2H_2O$), keiserite ($MgSO_4 \cdot H_2O$), and epsomite ($MgSO_4 \cdot 7H_2O$). Double salts, such as carnallite ($KMgCl_3 \cdot 6H_2O$), kainite ($KMgClSO_4 \cdot ^{11}/_4H_2O$), glaserite ($K_3Na(SO_4)_2$), polyhalite ($K_2MgCa_2(SO_4)_4 \cdot 2H_2O$), and loeweite ($Na_{12}Mg_7(SO_4)_{13} \cdot 15H_2O$), are very common in evaporites.

The precipitation of evaporites from marine and brine sources depends upon a number of factors. Prominent among these are the concentrations of the evaporite ions in the water and the solubility products of the evaporite salts. The presence of a common ion decreases solubility; for example, $CaSO_4$ precipitates more readily from a brine that contains Na_2SO_4 than it does from a solution that contains no other source of sulfate. The presence of other salts that do not have a common ion increases solubility because it decreases activity coefficients. Differences in temperature result in significant differences in solubility.

The nitrate deposits that occur in the hot and extraordinarily dry regions of northern Chile are chemically unique because of the stability of highly oxidized nitrate salts. The dominant salt, which has been mined for its nitrate content for use in explosives and fertilizers, is Chile saltpeter, $NaNO_3$. Traces of highly oxidized $CaCrO_4$ and $Ca(ClO_4)_2$ are also encountered in these deposits, and some regions contain enough $Ca(IO_3)_2$ to serve as a commercial source of iodine.

Volcanic Sublimates

A number of mineral substances are gaseous at the magmatic temperatures of volcanoes and are mobilized with volcanic gases. These kinds of substances condense near the mouths of volcanic fumaroles and are called **sublimates**. Elemental sulfur is a common sublimate. Some oxides, particularly of iron and silicon, are deposited as sublimates. Most other sublimates consist of chloride and sulfate salts. The cations most commonly involved are monovalent cations of ammonium ion, sodium, and potassium; magnesium; calcium; aluminum; and iron. Fluoride and chloride sublimates are sources of gaseous HF and HCl formed by their reactions at high temperatures with water, such as the following:

$$2H_2O + SiF_4 \rightarrow 4HF + SiO_2 \tag{15.2.1}$$

Igneous and Sedimentary Rock

The solidification of molten rock, called magma, produces **igneous rock**. Common igneous rocks are granite, basalt, quartz (SiO_2), pyroxene ($(Mg,Fe)SiO_3$), feldspar ($(Ca,Na,K)AlSi_3O_8$), olivine ($(Mg,Fe)_2SiO_4$), and magnetite (Fe_3O_4). Igneous rocks are formed under water-deficient, chemically reducing conditions of high temperature and high pressure. Exposed igenous rocks, therefore, are not in chemical equilibrium with their surroundings and disintegrate by a process called **weathering**. Weathering tends to be slow because igneous rocks are often hard, non-porous, and of low reactivity. Erosion from wind, water, or glaciers picks up materials from weathering rocks and converts it to **sedimentary rock** and **soil**, which in contrast to the parent igneous rocks are porous, soft, and chemically reactive. Heat and pressure convert sedimentary rock to **metamorphic rock**.

Sedimentary rocks may be **detrital rocks** consisting of solid particles eroded from igneous rocks as a consequence of weathering; quartz is the most likely to survive weathering and transport from its original location chemically intact. A second kind of sedimentary rocks consists of **chemical sedimentary rocks** produced by the precipitation or coagulation of dissolved or colloidal weathering products. **Organic sedimentary rocks** contain residues of plant and animal remains. Carbonate minerals of calcium and magnesium — **limestone** or **dolomite** — are especially abundant in sedimentary rocks. Important examples of sedimentary rocks are the following:

- Sandstone produced from sand-sized particles of minerals such as quartz
- Conglomerates made up of relatively larger particles of variable size
- Shale formed from very fine particles of silt or clay
- Limestone, $CaCO_3$, produced by the chemical or biochemical precipitation of calcium carbonate:

$$Ca^{2+} + CO_3^{2-} \rightarrow CaCO_3(s)$$

$$Ca^{2+} + 2HCO_3^- + h\nu(\text{algal photosynthesis}) \rightarrow \{CH_2O\}(\text{biomass}) + CaCO_3(s) + O_2(g)$$

- Chert consisting of microcrystalline, SiO_2

Stages of Weathering

Weathering can be classified into **early, intermediate**, and **advanced stages**.[3] The stage of weathering to which a mineral is exposed depends upon time; chemical conditions, including exposure to air, carbon dioxide, and water; and physical conditions, such as temperature and mixing with water and air.

Reactive and soluble minerals such as carbonates, gypsum, olivine, feldspars, and iron(II)-rich substances can survive only early weathering. This stage is characterized by dry conditions, low leaching, absence of organic matter, reducing conditions, and limited time of exposure. Quartz, vermiculite, and smectites can survive the intermediate stage of weathering manifested by retention of silica, sodium, potassium, magnesium, calcium, and iron(II) not present in iron(II) oxides. These substances are mobilized in advanced-stage weathering, other characteristics of which are intense leaching by fresh water, low pH, oxidizing conditions (iron(II) → iron(III)), presence of hydoxy polymers of aluminum, and dispersion of silica.

15.3. SEDIMENTS

Vast areas of land, as well as lake and stream sediments, are formed from sedimentary rocks. The properties of these masses of material depend strongly upon their origins and transport. Water is the main vehicle of sediment transport, although wind can also be significant. Enormous amounts of sediment are carried by major rivers, for example, about 750 million tons of per year by the Mississippi River.

The action of flowing water in streams cuts away stream banks and carries sedimentary materials for great distances. Sedimentary materials may be carried by flowing water in streams as the following:

- **Dissolved load** from sediment-forming minerals in solution.
- **Suspended load** from solid sedimentary materials carried along in suspension
- **Bed load** dragged along the bottom of the stream channel

The transport of calcium carbonate as dissolved calcium bicarbonate provides a straightforward example of dissolved load. Water with a high dissolved carbon dioxide content (usually present as the result of bacterial action) in contact with calcium carbonate formations contains Ca^{2+} and HCO_3^- ions. Flowing water containing calcium as such *temporary hardness* may become more basic by loss of CO_2 to the atmosphere, consumption of CO_2 by algal growth, or contact with dissolved base, resulting in the deposition of insoluble $CaCO_3$:

$$Ca^{2+} + 2HCO_3^- \rightarrow CaCO_3(s) + CO_2(g) + H_2O \qquad (15.3.1)$$

Most flowing water that contains dissolved load originates underground, where it dissolves minerals from the rock strata that it flows through.

Most sediments are transported by streams as suspended load, obvious in the observation of "mud" in the flowing water of rivers draining agricultural areas or finely divided rock in Alpine streams fed by melting glaciers. Under normal conditions, finely divided silt, clay, or sand make up most of the suspended load, although larger particles are transported in rapidly flowing water. The degree and rate of movement of suspended sedimentary material in streams are functions of the velocity of water flow and the settling velocity of the particles in suspension.

Bed load is moved along the bottom of a stream by the action of water "pushing" particles along. Particles carried as bed load do not move continuously. The grinding action of such particles is an important factor in stream erosion.

Typically, about 2/3 of the sediment carried by a stream is transported in suspension, about 1/4 in solution, and the remaining relatively small fraction as bed load. The ability of a stream to carry water increases with both the overall rate of flow of the water (mass per unit time) and the velocity of the water. Both of these are higher under flood conditions, so floods are particularly important in the transport of sediments.

Streams mobilize sedimentary materials through **erosion**, **transport** materials along with stream flow, and release them in a solid form during **deposition**. Deposits of stream-borne sediments are called **alluvium**. As conditions such as lowered stream velocity begin to favor deposition, larger, more settleable particles are released first. This results in **sorting** such that particles of a similar size and type tend to occur together in alluvium deposits. Much sediment is deposited in flood plains where streams overflow their banks.

15.4. CLAYS

Clays are extremely common and important in mineralogy. Furthermore, in general (see Section 15.5 and Chapter 16), clays predominate in the inorganic components of most soils and are very important in holding water and in plant nutrient cation exchange. All clays contain silicate and most contain aluminum and water. Physically, clays consist of very fine grains having sheet-like structures. For purposes of discussion here, **clay** is defined as a group of microcrystalline secondary minerals consisting of hydrous aluminum silicates that have sheet-like structures. The most abundant clay minerals are illites, montmorillonites, chlorites, and kaolinites. These clay minerals are distinguished from each other by general chemical formula, structure, and chemical and physical properties. The three major groups of clay minerals are **montmorillonite** $(Al_2(OH)_2Si_4O_{10})$, **illite** $(K_{0-2}Al_4(Si_{8-6}Al_{0-2})O_{20}(OH)_4$, and **kaolinite** $(Al_2Si_2O_5(OH)_4)$.[4] Many clays contain large amounts of sodium, potassium, magnesium, calcium, and iron, as well as trace quantities of other metals. Clays bind cations such as Ca^{2+}, Mg^{2+}, K^+, Na^+, and NH_4^+, which protects the cations from leaching by water but keeps them available in soil as plant nutrients. Since many clays are readily suspended in water as colloidal particles, they may be leached from soil or carried to lower soil layers.

Olivine, augite, hornblende, and feldspars are all parent minerals that form clays. An example is the formation of kaolinite $(Al_2Si_2O_5(OH)_4)$ from potassium feldspar rock $(KAlSi_3O_8)$:

$$KAlSi_3O_8(s) + 2H^+ + 9H_2O \rightarrow Al_2Si_2O_5(OH)_4(s) + 2K^+(aq)$$
$$+ 4H_4SiO_4(aq) \quad (15.4.1)$$

The layered structures of clays consist of sheets of silicon oxide alternating with sheets of aluminum oxide. The silicon oxide sheets are made up of tetrahedra in which each silicon atom is surrounded by four oxygen atoms. Of the four oxygen atoms in each tetrahedron, three are shared with other silicon atoms that are components of other tetrahedra. This sheet is called the **tetrahedral sheet**. The aluminum oxide is contained in an **octahedral sheet**, so named because each aluminum atom is surrounded by six oxygen atoms in an octahedral configuration. The structure is such that some of the oxygen atoms are shared between aluminum atoms and some are shared with the tetrahedral sheet.

Structurally, clays may be classified as either **two-layer clays** in which oxygen atoms are shared between a tetrahedral sheet and an adjacent octahedral sheet, and **three-layer clays**, in which an octahedral sheet shares oxygen atoms with tetrahedral sheets on either side. These layers composed of either two or three sheets are called **unit layers**. A unit layer of a two-layer clay typically is around 0.7 nanometers (nm) thick, whereas that of a three-layer clay exceeds 0.9 nm in thickness. The structure of the two-layer clay kaolinite is represented in Figure 15.1. Some clays, particularly the montmorillonites, may absorb large quantities of water between unit layers, a process accompanied by swelling of the clay.

As described in Section 5.4, clay minerals may attain a net negative charge by **ion replacement**, in which Si(IV) and Al(III) ions are replaced by metal ions of similar size but lesser charge. Compensation must be made for this negative charge by association of cations with the clay layer surfaces. Since these cations need not fit specific sites in the crystalline lattice of the clay, they may be relatively large ions, such as K^+, Na^+, or NH_4^+. These cations are called **exchangeable cations** and are exchangeable for other cations in water. The amount of exchangeable cations, expressed as milliequivalents (of monovalent cations) per 100 g of dry clay, is called the **cation-exchange capacity**, CEC, of the clay and is a very important characteristic of colloids and sediments that have cation-exchange capabilities.

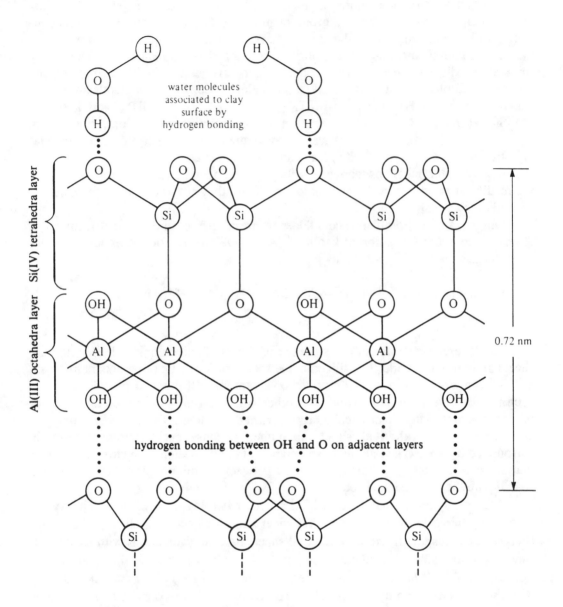

Figure 15.1. Representation of the structure of kaolinite, a two-layer clay.

15.5. SOIL

Insofar as environmental chemistry and life on earth are concerned, the most important part of the earth's crust is soil. The nature of soil is discussed briefly here, and its environmental chemistry is the subject of Chapter 16. **Soil** is a variable mixture of minerals, organic matter, and water, capable of supporting plant life on the earth's surface. It is the final product of the weathering action of physical, chemical, and biological processes on rocks, which largely produces clay minerals. The organic portion of soil consists of plant biomass in various stages of decay. High populations of bacteria, fungi, and animals such as earthworms may be found in soil. Soil contains air spaces and generally has a loose texture (Figure 15.2).

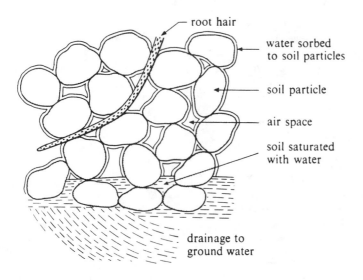

root hair

water sorbed
to soil particles

soil particle

air space

soil saturated
with water

drainage to
ground water

Figure 15.2. Fine structure of soil, showing solid, water, and air phases.

The solid fraction of typical productive soil is approximately 5% organic matter and 95% inorganic matter. Some soils, such as peat soils, may contain as much as 95% organic material. Other soils contain as little as 1% organic matter.

Typical soils exhibit distinctive layers with increasing depth (Figure 15.3). These layers are called **horizons**. Horizons form as the result of complex interactions among processes that occur during weathering. Rainwater percolating through soil carries dissolved and colloidal solids to lower horizons where they are deposited. Biological processes, such as bacterial decay of residual plant biomass, produces slightly acidic CO_2, organic acids, and complexing compounds that are carried by rainwater to lower horizons where they interact with clays and other minerals, altering the properties of the minerals. The top layer of soil, typically several inches in thickness, is known as the A horizon, or **topsoil**. This is the layer of maximum biological activity in the soil and contains most of the soil organic matter. Metal ions and clay particles in the A horizon are subject to considerable leaching. The next layer is the B horizon, or **subsoil**. It receives material such as organic matter, salts, and clay particles leached from the topsoil. The C horizon is composed of weathered parent rocks from which the soil originated.

Soils exhibit a large variety of characteristics that are used for their classification for various purposes, including crop production, road construction, and waste disposal. Soil profiles are discussed above. The parent rocks from which soils are formed obviously play a strong role in determining the composition of soils. Other soil characteristics include strength, workability, soil particle size, permeability, and degree of maturity. One of the more important classes of productive soils is the **podzol** type of soil formed under relatively high rainfall conditions in temperate zones of the world. These generally rich soils tend to be acidic (pH 3.5–4.5) such that alkali and alkaline earth metals and, to a lesser extent aluminum and iron, are leached from their A horizons, leaving kaolinite as the predominant clay mineral. At somewhat higher pH in the B horizons, hydrated iron oxides and clays are redeposited.

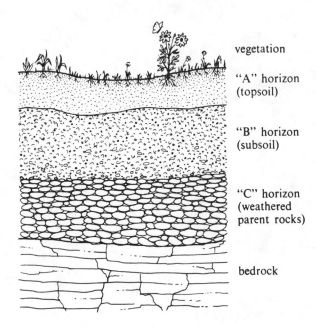

vegetation

"A" horizon
(topsoil)

"B" horizon
(subsoil)

"C" horizon
(weathered
parent rocks)

bedrock

Figure 15.3. Soil profile showing soil horizons.

From the engineering standpoint, especially, the mechanical properties of soil are emphasized.[5] These properties are largely determined by particle size. According to the United Classification System (UCS), the four major categories of soil particle sizes are the following: Gravels (2–60 mm) > sands (0.06–2 mm) > silts (0.06–0.006 mm) > clays (less than 0.002 mm). In the UCS classification scheme clays represent a size fraction rather than a specific class of mineral matter.

Water and Air in Soil

Large quantities of water are required for the production of most plant materials. For example, several hundred kg of water are required to produce one kg of dry hay. Water is part of the three-phase, solid-liquid-gas system making up soil. It is the basic transport medium for carrying essential plant nutrients from solid soil particles into plant roots and to the farthest reaches of the plant's leaf structure (Figure 15.4). The water enters the atmosphere from the plant's leaves, a process called **transpiration**.

Normally, because of the small size of soil particles and the presence of small capillaries and pores in the soil, the water phase is not totally independent of soil solid matter. The availability of water to plants is governed by gradients arising from capillary and gravitational forces. The availability of nutrient solutes in water depends upon concentration gradients and electrical potential gradients. Water present in larger spaces in soil is relatively more available to plants and readily drains away. Water held in smaller pores, or between the unit layers of clay particles (see Chapter 5) is held much more strongly. Soils high in organic matter may hold appreciably more water than other soils, but it is relatively less available to plants because of physical and chemical sorption of the water by the organic matter.

There is a very strong interaction between clays and water in soil. Water is absorbed on the surfaces of clay particles. Because of the high surface/volume ratio of

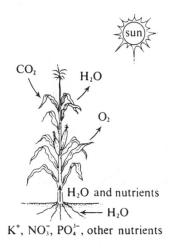

CO₂ H₂O O₂ sun

H₂O and nutrients

H₂O

K⁺, NO₃⁻, PO₄³⁻, other nutrients

Figure 15.4. Plants transport water from the soil to the atmosphere by transpiration. Nutrients are also carried from the soil to the plant extremeties by this process. Plants remove CO_2 from the atmosphere and add O_2 by photosynthesis. The reverse occurs during plant respiration.

colloidal clay particles, a great deal of water may be bound in this manner. Water is also held between the unit layers of the expanding clays, such as the montmorillonite clays.

As soil becomes waterlogged (water-saturated) it undergoes drastic changes in physical, chemical, and biological properties. Oxygen in such soil is rapidly used up by the respiration of microorganisms that degrade soil organic matter. In such soils, the bonds holding soil colloidal particles together are broken, which causes disruption of soil structure. Thus, the excess water in such soils is detrimental to plant growth, and the soil does not contain the air required by most plant roots. Most useful crops, with the notable exception of rice, cannot grow on waterlogged soils.

One of the most marked chemical effects of waterlogging is a reduction of pE by the action of organic reducing agents acting through bacterial catalysts. Thus, the redox condition of the soil becomes much more reducing, and the soil pE may drop from that of water in equilibrium with air (+13.6 at pH 7) to 1 or less. One of the more significant results of this change is the mobilization of iron and manganese as soluble iron(II) and manganese(II) through reduction of their insoluble higher oxides:

$$MnO_2 + 4H^+ + 2e^- \rightarrow Mn^{2+} + 2H_2O \tag{15.5.1}$$

$$Fe_2O_3 + 6H^+ + 2e^- \rightarrow 2Fe^{2+} + 3H_2O \tag{15.5.2}$$

Although soluble manganese generally is found in soil as Mn^{2+} ion, soluble iron(II) frequently occurs as negatively charged iron-organic chelates. Strong chelation of iron(II) by soil fulvic acids (Chapter 3) apparently enables reduction of iron(III) oxides at more positive pE values than would otherwise by possible. This causes an upward shift in the Fe(II)-Fe(OH)₃ boundary shown in Figure 4.4.

Some soluble metal ions such as Fe^{2+} and Mn^{2+} are toxic to plants at high levels. Their oxidation to insoluble oxides may cause formation of deposits of Fe_2O_3 and MnO_2, which clog tile drains in fields.

Roughly 35% of the volume of typical soil is composed of air-filled pores. Whereas the normal dry atmosphere at sea level contains 21% O_2 and 0.03% CO_2 by volume, these percentages may be quite different in soil air because of the decay of organic matter:

$$\{CH_2O\} + O_2 \rightarrow CO_2 + H_2O \tag{15.5.3}$$

This process consumes oxygen and produces CO_2. As a result, the oxygen content of air in soil may be as low as 15%, and the carbon dioxide content may be several percent. Thus, the decay of organic matter is soil increases the equilibrium level of dissolved CO_2 in groundwater. This lowers the pH and contributes to weathering of carbonate minerals, particularly calcium carbonate (see Reaction 3.4.1). As discussed in Chapter 16, CO_2 also shifts the equilibrium of the process by which roots absorb metal ions from soil.

The Inorganic Component of Soil

The weathering of parent rocks and minerals to form the inorganic soil components results ultimately in the formation of inorganic colloids. These colloids are repositories of water and plant nutrients, which may be made available to plants as needed. Inorganic soil colloids often absorb toxic substances in soil, thus playing a role in detoxification of substances that otherwise would harm plants. The abundance and nature of inorganic colloidal material in soil are obviously important factors in determining soil productivity.

The uptake of plant nutrients by roots often involves complex interactions with the water and inorganic phases. For example, a nutrient held by inorganic colloidal material has to traverse the mineral/water interface and then the water/root interface. This process is often strongly influenced by the ionic structure of soil inorganic matter.

As noted in Section 15.2, the most common elements in the earth's crust are oxygen, silicon, aluminum, iron, calcium, sodium, potassium, and magnesium. Therefore, minerals composed of these elements — particularly silicon and oxygen — constitute most of the mineral fraction of the soil. Common soil mineral constituents are finely divided quartz (SiO_2), orthoclase ($KAlSi_3O_8$), albite ($NaAlSi_3O_8$), epidote ($4CaO \cdot 3(AlFe)_2O_3 \cdot 6SiO_2 \cdot H_2O$), geothite ($FeO(OH)$), magnetite ($Fe_3O_4$), calcium and magnesium carbonates ($CaCO_3$, $CaCO_3 \cdot MgCO_3$), and oxides of manganese and titanium

Organic Matter in Soil

Though typically comprising less than 5% of a productive soil, organic matter largely determines soil productivity. It serves as a source of food for microorganisms; undergoes chemical reactions such as ion exchange; and influences the physical properties of soil. Some organic compounds even contribute to the weathering of mineral matter, the process by which soil is formed. For example, $C_2O_4^{2-}$, oxalate ion, produced as a soil fungi metabolite, occurs in soil as the calcium salts whewellite and weddelite. Oxalate in soil water dissolves minerals, thus speeding the weathering process and increasing the availability of nutrient ion species. This weathering process involves oxalate complexation of iron or aluminum in minerals, represented by the reaction

$$3H^+ + M(OH)_3(s) + 2CaC_2O_4(s) \rightarrow M(C_2O_4)_2^-(aq) +$$
$$2Ca^{2+}(aq) + 3H_2O \qquad (15.5.4)$$

in which M is Al or Fe. Some soil fungi produce citric acid, and other chelating organic acids, which react with silicate minerals and release potassium and other nutrient metal ions held by these minerals.

The strong chelating agent 2-ketogluconic acid is produced by some soil bacteria. By solubilizing metal ions, it may contribute to the weathering of minerals. It may also be involved in the release of phosphate from insoluble phosphate compounds.

Biologically active components of the organic soil fraction include polysaccharides, amino sugars, nucleotides, and organic sulfur and phosphorus compounds. Humus, a water-insoluble material that biodegrades very slowly, makes up the bulk of soil organic matter. The organic compounds in soil have been described in detail[6] and are summarized in Table 15.2.

Table 15.2. Major Classes of Organic Compounds in Soil

Compound type	Composition	Significance
Humus	Degradation-resistant residue from plant decay, largely C, H, and O	Most abundant organic component, improves soil physical properties, exchanges nutrients, reservoir of fixed N
Fats, resins, and waxes	Lipids extractable by organic solvents	Generally, only several percent of soil organic matter, may adversely affect soil physical properties by repelling water, perhaps phytotoxic
Saccharides	Cellulose, starches, hemicellulose, gums	Major food source for soil microorganisms, help stabilize soil aggregates
N-containing organics	Nitrogen bound to humus, amino acids, amino sugars, other compounds	Provide nitrogen for soil fertility
Phosphorous compounds	Phosphate esters, inositol phosphates (phytic acid). phospholipids	Sources of plant phosphate

The accumulation of organic matter in soil is strongly influenced by temperature and by the availability of oxygen. Since the rate of biodegradation decreases with decreasing temperature, organic matter does not degrade rapidly in colder climates and tends to build up in soil. In water and in waterlogged soils, decaying vegetation does not have easy access to oxygen, and organic matter accumulates. The organic content may reach 90% in areas where plants grow and decay in soil saturated with water.

Of the organic components listed in Table 15.2, **soil humus** is by far the most significant. Humus, composed of a base-soluble fraction called humic and fulvic acids (described in Section 3.12) and an insoluble fraction called humin, is the residue left when bacteria and fungi biodegrade plant material. The bulk of plant biomass consists of relatively degradable cellulose and degradation-resistant lignin, which is a polymeric substance with a higher carbon content than cellulose. Among lignin's prominent chemical components are aromatic rings connected by alkyl chains, methoxyl groups, and hydroxyl groups. Lignin is the precursor of most soil humus.

An increase in nitrogen/carbon ratio is a significant feature of the transformation of plant biomass to humus. This ratio starts at approximately 1/100 in fresh plant biomass. During the humification process, microorganisms convert organic carbon to CO_2 to obtain energy. Simultaneously, the bacterial action incorporates bound nitrogen with the compounds produced by the decay processes. The result is a nitrogen/carbon ratio of about 1/10 upon completion of humification. As a general rule, therefore, humus is relatively rich in organically bound nitrogen.

Humic materials in soil strongly sorb many solutes in soil water and have a particular affinity for heavy polyvalent cations. Soil humic substances may contain levels of uranium more than 10^4 times that of the water with which they are in equilibrium. Thus, water becomes depleted of its cations (or purified) in passing through humic-rich soils. Humic substances in soils also have a strong affinity for organic compounds with low water-solubility such as DDT or Atrazine, a herbicide widely used to kill weeds in corn fields.

Atrazine

In some cases, there is a strong interaction between the organic and inorganic portions of soil. This is especially true of the strong complexes formed between clays and humic (fulvic) acid compounds. In many soils, 50-100% of soil carbon is complexed with clay. These complexes play a role in determining the physical properties of soil, soil fertility, and stabilization of soil organic matter. One of the mechanisms for the chemical binding between clay colloidal particles and humic organic particles is probably of the flocculation type (see Chapter 5), in which anionic organic molecules with carboxylic acid functional groups serve as bridges in combination with cations to bind clay colloidal particles together as a floc. Support is given to this hypothesis by the known ability of NH_4^+, Al^{3+}, Ca^{2+}, and Fe^{3+} cations to stimulate clay-organic complex formation. The synthesis, chemical reactions, and biodegradation of humic materials are affected by interaction with clays. The lower-molecular-weight fulvic acids may be bound in the spaces in layers in clay particles.

The presence of naturally occurring polynuclear aromatic (PAH) compounds is an interesting feature of soil organic matter. These compounds, some of which are carcinogenic, are discussed as air pollutants in Sections 10.4 and 12.4. PAH compounds found in soil include fluoranthene, pyrene, and chrysene. The origin of PAH compounds in soil is unknown, although the most likely source is combustion. Terpenes also occur in soil organic matter. Extraction of soil with ether and alcohol yields the pigments β-carotene, chlorophyll, and xanthophyll.

15.6. GEOCHEMISTRY

Geochemistry deals with chemical species, reactions, and processes in the lithosphere and their interactions with the atmosphere and hydrosphere. The branch of geochemistry that explores the complex interactions among the rock/water/air/life systems that determine the chemical characteristics of the surface environment is **environmental geochemistry**.[7] Obviously, geochemistry and its environmental subdiscipline are very important in environmental chemistry.

Physical Aspects of Weathering

Defined in Section 15.2, *weathering*, is discussed here as a geochemical phenomenon. Rocks tend to weather more rapidly when there are pronounced differences in physical conditions — alternate freezing and thawing and wet periods alternating with severe drying. Other mechanical aspects are swelling and shrinking of minerals with hydration and dehydration as well as growth of roots through cracks in rocks. Temperature is involved in that the rates of chemical weathering (below) increase with increasing temperature.

Chemical Weathering

As a chemical phenomenon, weathering can be viewed as the result of the tendency of the rock/water/mineral system to attain equilibrium. This occurs through the usual chemical mechanisms of dissolution/precipitation, acid-base reactions, complexation, hydrolysis, and oxidation-reduction.

Weathering occurs extremely slowly in dry air. Water increases the rate of weathering by many orders of magnitude for several reasons. Water, itself, is a chemically active substance in the weathering process. Furthermore, water holds weathering agents in solution such that they are transported to chemically active sites on rock minerals and contact the mineral surfaces at the molecular and ionic level. Prominent among such weathering agents are CO_2, O_2, organic acids (including humic and fulvic acids, see Section 3.12), sulfur acids ($SO_2(aq)$, H_2SO_4), and nitrogen acids (HNO_3, HNO_2). Water provides the source of H^+ ion needed for acid-forming gases to act as acids as shown by the following:

$$CO_2 + H_2O \rightarrow H^+ + HCO_3^- \tag{15.6.1}$$

$$SO_2 + H_2O \rightarrow H^+ + HSO_3^- \tag{15.6.2}$$

Rainwater is essentially free of mineral solutes. It is usually slightly acidic due to the presence of dissolved carbon dioxide or more highly acidic because of acid-rain forming constitutents. Therefore, rainwater is *chemically aggressive* (see Section 8.7) toward some kinds of mineral matter, which it breaks down by a process called **chemical weathering**. Because of this process, river water has a higher concentration of dissolved inorganic solids than does rainwater.

The processes involved in chemical weathering may be divided into the following major categories:[8,9]

- **Hydration/dehydration**, for example:

$$CaSO_4(s) + 2H_2O \rightarrow CaSO_4 \cdot 2H_2O(s)$$

$$Fe(OH)_3 \cdot xH_2O(s) \rightarrow Fe_2O_3(s) + (3+x)H_2O$$

- **Dissolution**, for example:

$$CaSO_4 \cdot 2H_2O(s) \text{ (water)} \rightarrow Ca^{2+}(aq) + SO_4^{2-}(aq) + 2H_2O$$

- **Oxidation**, such as occurs in the dissolution of pyrite:

$$4FeS_2(s) + 15O_2(g) + (8+x)H_2O \rightarrow$$
$$2Fe_2O_3 \cdot xH_2O + 8SO_4^{2-}(aq) + 16H^+(aq)$$

or in the following example in which dissolution of an iron(II) mineral is followed by oxidation of iron(II) to iron(III):

$$Fe_2SiO_4(s) + 4CO_2(aq) + 4H_2O \rightarrow 2Fe^{2+} + 4HCO_3^- + H_4SiO_4$$

$$4Fe^{2+} + 8HCO_3^- + O_2(s) \rightarrow 2Fe_2O_3(s) + 8CO_2 + 4H_2O$$

The second of these two reactions may occur at a site some distance from the first, resulting in a net transport of iron from its original location. Iron, manganese, and sulfur are the major elements that undergo oxidation as part of the weathering process.

- **Dissolution with hydrolysis** as occurs with the hydrolysis of carbonate ion when mineral carbonates dissolve:

$$CaCO_3(s) + H_2O \rightarrow Ca^{2+}(aq) + HCO_3^-(aq) + OH^-(aq)$$

Hydrolysis is the major means by which silicates undergo weathering as shown by the following reaction of forsterite:

$$Mg_2SiO_4(s) + 4CO_2 + 4H_2O \rightarrow 2Mg^{2+} + 4HCO_3^- + H_4SiO_4$$

The weathering of silicates yields soluble silicon as species such as H_4SiO_4, and residual silicon-containing minerals (clay minerals).

- **Acid hydrolysis**, which accounts for the dissolution of significant amounts of $CaCO_3$ and $CaCO_3 \cdot MgCO_3$ in the presence CO_2-rich water:

$$CaCO_3(s) + H_2O + CO_2(aq) \rightarrow Ca^{2+}(aq) + 2HCO_3^-(aq)$$

- **Complexation**, as exemplified by the reaction of oxalate ion, $C_2O_4^{2-}$ on muscovite, $K_2(Si_6Al_2)Al_4O_{20}(OH)_4$:

$$K_2(Si_6Al_2)Al_4O_{20}(OH)_4:(s) + 6C_2O_4^{2-}(aq) + 20H^+ \rightarrow$$
$$6AlC_2O_4^+(aq) + 6Si(OH)_4 + 2K^+$$

Reactions such as these largely determine the kinds and concentrations of solutes in surface water and groundwater. Acid hydrolysis, especially, is the predominant process that releases elements such as Na^+, K^+, and Ca^{2+} from silicate minerals.

15.7. GROUNDWATER IN THE GEOSPHERE

Groundwater (Figure 15.4) is a vital resource in its own right that plays a crucial role in geochemical processes, such as the formation of secondary minerals. The

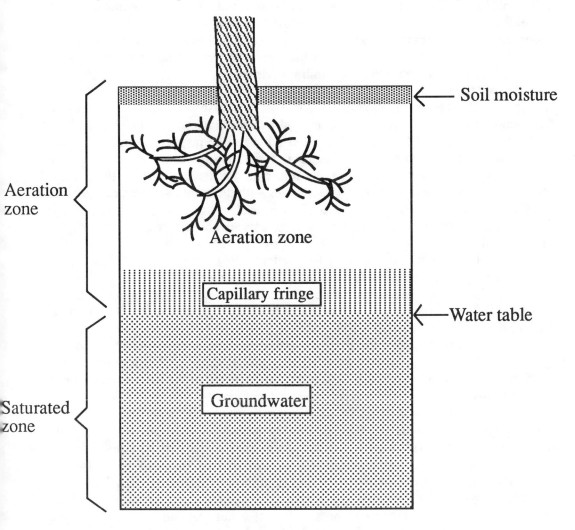

Figure 15.5. Some major features of the distribution of water underground.

nature, quality and mobility of groundwater are all strongly dependent upon the rock formations in which the water is held. Physically, an important characteristic of such formations is their **porosity**, which determines the percentage of rock volume available to contain water. A second important physical characteristic is **permeability**, which describes the ease of flow of the water through the rock. High permeability is usually associated with high porosity. However, clays tend to have low permeability even when a large percentage of the volume is filled with water.

The nature of water underground is illustrated by Figure 15.4. Most groundwater originates as **meteoritic** water from precipitation in the form of rain or snow. If water from this source is not lost by evaporation, transpiration, or to stream runoff, it may infiltrate into the ground. Initial amounts of water from precipitation onto dry soil are held very tightly as a film on the surfaces and in the micropores of soil particles in a **belt of soil moisture**. At intermediate levels, the soil particles are covered with films of water, but air is still present in larger voids in the soil. The region in which such water is held is called the **unsaturated zone** or **zone of aeration** and the water present in it is **vadose water**. At lower depths in the presence of adequate amounts of water, all voids are filled to produce a **zone of saturation**, the upper level of which is the **water table**. Water present in a zone of saturation is called **groundwater**. Because of its surface tension, water is drawn somewhat above the water table by capillary-sized passages in soil in a region called the **capillary fringe**.

The water table is crucial in explaining and predicting the flow of wells and springs and the levels of streams and lakes. It is also an important factor in determining the extent to which pollutant and hazardous chemicals underground are likely to be transported by water. The water table can be mapped by observing the equilibrium level of water in wells, which is essentially the same as the top of the saturated zone. The water table is usually not level, but tends to follow the general contours of the surface topography. It also varies with differences in permeability and water infiltration. The water table is at surface level in the vicinity of swamps and frequently above the surface where lakes and streams are encountered. The water level in such bodies may be maintained by the water table. **Influent** streams or reservoirs are located above the water table; they lose water to the underlying aquifer and cause an upward bulge in the water table beneath the surface water.

Groundwater **flow** is an important consideration in determining the accessibility of the water for use and transport of pollutants from underground waste sites. Various parts of a body of groundwater are in hydraulic contact so that a change in pressure at one point will tend to affect the pressure and level at another point. For example, infiltration from a heavy, localized rainfall may affect water level at a point remote from the infiltration. Groundwater flow occurs as the result of the natural tendency of the water table to assume even levels by the action of gravity.

Groundwater flow is strongly influenced by rock permeability. Porous or extensively fractured rock is relatively highly **pervious**. Because water can be extracted from such a formation, it is called an **aquifer**. By contrast, an **aquiclude** is a rock formation that is too impermeable or unfractured to yield groundwater. Impervious rock in the unsaturated zone may retain water infiltrating from the surface to produce a **perched water table** that is above the main water table and from which water may be extracted. However, the amounts of water that can be extracted from such a formation are limited and the water is vulnerable to contamination.

Water Wells

Most groundwater is tapped for use by water wells drilled into the saturated zone. The use and mis-use of water from this source has a number of environmental implications. In the U.S. about 2/3 of the groundwater pumped is consumed for irrigation; lesser amounts of ground water are used for industrial and municipal applications.

As water is withdrawn, the water table in the vicinity of the well is lowered. This **drawdown** of water creates a **zone of depression**. In extreme cases the groundwater is severely depleted and surface land levels can even subside (which is one reason that Venice, Italy is now very vulnerable flooding). Heavy drawdown can result in infiltration of pollutants from sources such as septic tanks, municipal refuse sites, and hazardous waste dumps. When soluble iron(II) or manganese(II) are present in groundwater, exposure to air at the well/rock formation interface can result in the formation of deposits of insoluble iron(III) and manganese(IV) mineral deposits:

$$4Fe^{2+}(aq) + O_2(aq) + 10H_2O \rightarrow 4Fe(OH)_3(s) + 8H^+ \qquad (15.7.1)$$

$$2Mn^{2+}(aq) + O_2(aq) + (2x + 2)H_2O \rightarrow 2MnO_2 \cdot xH_2O(s) + 4H^+ \qquad (15.7.2)$$

Deposits of iron(III) and manganese(IV) that result from the processes outlined above coat the surfaces from which water flows into the well with a coating that is relatively impermeable to water. The deposits fill the spaces that water must traverse to enter the well. As a result, they can seriously impede the flow of water into the well from the water-bearing aquifer. This creates major water source problems for municipalities using groundwater for water supply. As a result of this problem, chemical or mechanical cleaning, drilling of new wells, or even acquisition of new water sources may be required.

LITERATURE CITED

1. Barney, G. O., *The Global 2000 Report to the President*, 2 vols., U. S. Government Printing Office, Washington, DC, 1980.

2. Tarbuck, Edward J., and Frederick K. Lutgens, *The Earth, an Introduction to Physical Geology*, Charles E. Merrill Publishing Co., Columbus, Ohio, 1984.

3. Sposito, Garrison, *The Chemistry of Soils*, Oxford University Press, New York, 1989.

4. Leet, L. Don, Sheldon Judson, and Marvin E. Kauffman, *Physical Geology*, 6th ed., Prentice-Hall, Inc., Englewood Cliffs, New Jersey, 1982.

5. McLean, Adam, and Colin Gribble, *Geology for Civil Engineers*, George Allen and Unwin, London, 1979

6. Gieseking, J. E., Ed., *Soil Components, Vol. 1, Organic Components*, Springer Verlag, Inc., New York.

7. Plant, Jane A., and Robert Raiswell, "Principles of Environmental Geochemistry," Chapter 1 in *Applied Environmental Geochemistry*, Iain Thornton, Ed., Academic Press, New York, 1983, pp. 1-40.

8. Raiswell, R. W., P. Brimblecombe, D. L. Dent, and P. S. List, *Environmental Chemistry*, Edward Arnold, London, 1980.

9. Krauskopf, Konrad B., "Introduction to Geochemistry," 2nd ed., McGraw-Hill Book Company, New York, 1979

SUPPLEMENTARY REFERENCES

Suffet, I. H., and Patrick MacCarthy, Eds., *Aquatic Humic Substances: Influence on Fate and Treatment of Pollutants*, Advances in Chemistry Series 219. American Chemical Society, Washington, DC, 1989.

Wolf, K., W. J. Van Den Brink, and F. J. Colon, *Contaminated Soil '88*, Vols 1 and 2, Kluwer Academic Publishers, Norwell, MA, 1988.

Brownlow, A. H., *Geochemistry*, Prentice-Hall, Inc., Englewood Cliffs, NJ, 1978.

Craig, P. J., *The Natural Environment and The Biogeochemical Cycles*, Springer-Verlag, Inc., New York, 1980.

Fortescue, A. C., *Environmental Geochemistry*, Springer-Verlag, Inc., New York, 1980.

Krenvolden, K. A., Ed., *Geochemistry of Organic Molecules*, Academic Press, Inc., New York, 1980.

Lerman, A., *Geochemical Processes*, Wiley-Interscience, New York, 1979.

Lindsay, W. L., *Chemical Equilibria in Soils*, Wiley-Interscience, New York, 1979.

Morrill, L. G., B. C. Mahilum, and S. H. Mohiuddin, *Organic Compounds in Soils: Sorption, Degradation, and Persistence*, Ann Arbor Science Publishers, Inc., Ann Arbor, Michigan, 1982.

Nancollas, G. H., Ed., *Biological Mineralization and Demineralization*, Springer-Verlag, Inc., New York, 1982.

Olson, G. W., *Soils and the Environment*, Methuen, Inc., New York, 1981.

Paton, T. R., *The Formation of Soil Material*, Allen and Unwin, Inc., Winchester, Mass., 1979.

Speidel, G. H., and Agnew, A. F., *The Natural Geochemistry of Our Environment*, Westview Press, Boulder, CO, 1982.

Stevenson, F. J., *Humus Chemistry: Genesis, Composition, Reactions*, Wiley-Interscience, New York, 1982.

QUESTIONS AND PROBLEMS

1. Of the following, the one that is **not** a manifestation of desertification is (a) declining groundwater tables, (b) salinization of topsoil and water, (c) production of deposits of MnO_2 and $Fe_2O_3 \cdot H_2O$ from anaerobic processes, (d) reduction of surface waters, (e) unnaturally high soil erosion.

2. Give an example of how each of the following chemical or biochemical phenomena in soils operate to reduce the harmful nature of pollutants: (a) Oxidation-reduction processes, (b) hydrolysis, (c) acid-base reactions, (d) precipitation, (e) sorption, (f) biochemical degradation.

3. Why do silicates and oxides predominate among Earth's minerals?

4. Give the common characteristic of the minerals with the following formulas: $NaCl$, Na_2SO_4, $CaSO_4 \cdot 2H_2O$, $MgCl_2 \cdot 6H_2O$, $MgSO_4 \cdot 7H_2O$, $KMgClSO_4 \cdot ^{11}/_4H_2O$, $K_2MgCa_2(SO_4)_4 \cdot 2H_2O$.

5. Explain how the following are related: weathering, igneous rock, sedimentary rock, soil.

6. Match the following:

1. Metamorphic rock	(a) produced by the precipitation or coagulation of dissolved or colloidal weathering products
2. Chemical sedimentary rocks	
3. Metamorphic rock	(b) Contain residues of plant and animal remains
4. Organic sedimentary rocks	(c) Formed from action of heat and pressure on sedimentary rock
	(d) Formed from solid particles eroded from igneous rocks as a consequence of weathering

7. Where does most flowing water that contains dissolved load originate? Why does it tend to come from this source?

8. What role might be played by water pollutants in the production of dissolved load and in the precipitation of secondary minerals from it?

9. As defined in this chapter, are the ions involved in ion replacement the same as exchangeable cations? If not, why not?

10. Match the following:

1. Subsoil	(a) Weathered parent rocks from which the soil originated
2. Gravels	
3. Topsoil	(b) Largest particle size fraction (2–60 mm) according to the United Classification System
4. C horizon	(c) B horizon of soil
	(d) Layer of maximum biological activity in soil that contains most of the soil organic matter

11. Under what conditions do the reactions,

$$MnO_2 + 4H^+ + 2e^- \rightarrow Mn^{2+} + 2H_2O$$

and

$$Fe_2O_3 + 6H^+ + 2e^- \rightarrow 2Fe^{2+} + 3H_2O$$

occur in soil? Name two detrimental effects that can result from these reactions.

12. What are four important effects of organic matter in soil?

13. How might irrigation water treated with fertilizer potassium and ammonia become depleted of these nutrients in passing through humus-rich soil?

14. Which three elements are most likely to undergo oxidation as part of chemical weathering process? Give example reactions of each.

15. Match the following:

1. Groundwater	(a) Water from from precipitation in the form of rain or snow
2. Vadose water	(b) Water present in a zone of saturation
3. Meteoritic water	(c) Water held in the unsaturated zone or zone of aeration
4. Water in capillary fringe	(d) Water drawn somewhat above the water table by surface tension

Soil Chemistry

16.1. THE NATURE AND IMPORTANCE OF SOIL

The composition and properties of soil were briefly introduced in Chapter 15. In this chapter the environmental chemistry of soil is addressed in greater detail. To humans and most terrestrial organisms, soil is the most important part of the geosphere. Though only a tissue-thin layer compared to the Earth's total diameter, soil is the medium that produces most of the food required by most living things. Good soil and climate are the most valuable assets a nation can have.

In addition to being the site of most food production, soil is the receptor of large quantities of pollutants, such as particulate matter from power plant smokestacks. Fertilizers and some other materials applied to soil often contribute to water and air pollution. Therefore, soil is a key component of environmental chemical cycles.

As discussed in Section 15.5, soils are formed by the weathering of parent rocks as the result of interactive geological, hydrological, and biological processes. Soils are porous and are vertically stratified into horizons.[1] Soils are open systems that undergo continual exchange of matter and energy with the atmosphere, hydrosphere, and biosphere.

The Soil Solution

The **soil solution** is the aqueous portion of soil that contains dissolved matter from soil chemical and biochemical processes and from exchange with the hydrosphere and biosphere.[2] This medium transports chemical species to and from soil particles and provides intimate contact between the solutes and the soil particles. In addition to providing water for plant growth, it is an essential pathway for the exchange of plant nutrients between roots and solid soil.

Obtaining a sample of soil solution is often very difficult because the most significant part of it is bound in capillaries and as surface films. The most straightforward means is collection of drainage water. Displacement with a water-immiscible fluid or mechanical separation by centrifugation or under pressure or vacuum can be employed.[3]

Dissolved mineral matter in soil is largely present as ions. Prominent among the cations are H^+, Ca^{2+}, Mg^{2+}, K^+, Na^+, and usually very low levels of Fe^{2+}, Mn^{2+}, and Al^{3+}. The last three cations may be present in partially hydrolized form, such as $FeOH^+$, or complexed by organic humic substance ligands. Anions that may be present are HCO_3^-, CO_3^{2-}, HSO_4^-, SO_4^{2-}, Cl^-, and F^-. In addition to being bound to H^+ in species such as bicarbonate, anions may be complexed with metal ions, such as in AlF^{2+}. Multivalent cations and anions form ion pairs with each other in soil solutions. Examples of these are $CaSO_4^0$ and $FeSO_4^0$.

16.2. ACID-BASE AND ION EXCHANGE REACTIONS IN SOILS

One of the more important chemical functions of soils is the exchange of cations. As discussed in Chapter 5, the ability of a sediment or soil to exchange cations is expressed as the cation-exchange capacity (CEC), the number of milliequivalents (meq) of monovalent cations that can be exchanged per 100 g of soil (on a dry-weight basis). The CEC should be looked upon as a conditional constant, since it may vary with soil conditions such as pE and pH. Both the mineral and organic portions of soils exchange cations. Clay minerals exchange cations because of the presence of negatively charged sites on the mineral, resulting from the substitution of an atom of lower oxidation number for one of higher number; for example, magnesium for aluminum. Organic materials exchange cations because of the presence of the carboxylate group and other basic functional groups. Humus typically has a very high cation-exchange capacity. The cation-exchange capacity of peat may range from 300-400 meq/100 g. Values of cation-exchange capacity for soils with more typical levels of organic matter are around 10-30 meq/100 g.

Cation exchange in soil is the mechanism by which potassium, calcium, magnesium, and essential trace-level metals are made available to plants. When nutrient metal ions are taken up by plant roots, hydrogen ion is exchanged for the metal ions. This process, plus the leaching of calcium, magnesium, and other metal ions from the soil by water containing carbonic acid, tends to make the soil acidic:

$$\text{Soil}\}Ca^{2+} + 2CO_2 + 2H_2O \rightarrow \text{Soil}\}(H^+)_2 + Ca^{2+}(\text{root}) + 2HCO_3^- \quad (16.2.1)$$

Soil acts as a buffer and resists changes in pH. The buffering capacity depends upon the type of soil.

Production of Mineral Acid in Soil

The oxidation of pyrite in soil causes formation of acid-sulfate soils sometimes called "cat clays":

$$FeS_2 + {}^7/_2O_2 + H_2O \rightarrow Fe^{2+} + 2H^+ + 2SO_4^{2-} \quad (16.2.2)$$

Cat clay soils may have pH values as low as 3.0. These soils, which are commonly found in Delaware, Florida, New Jersey, and North Carolina, are formed when neutral or basic marine sediments containing FeS_2 become acidic upon oxidation of pyrite when exposed to air. For example, soil reclaimed from marshlands and used for citrus groves has developed high acidity detrimental to plant growth. In addition, H_2S released by increased acidity is very toxic to citrus roots.

Soils are tested for potential acid-sulfate formation using a peroxide test. This test consists of oxidizing FeS_2 in the soil with 30% H_2O_2,

$$FeS_2 + {}^{15}/_2H_2O_2 \rightarrow Fe^{3+} + H^+ + 2SO_4^{2-} + 7H_2O_2 \quad (16.2.3)$$

then testing for acidity and sulfate. Appreciable levels of sulfate and a pH below 3.0 indicate potential to form acid-sulfate soils. If the pH is above 3.0, either little FeS_2 is present or sufficient $CaCO_3$ is in the soil to neutralize the sulfuric acid and acidic Fe^{3+}.

Pyrite-containing mine spoils (residue left over from mining) also form soils similar to acid-sulfate soils of marine origin. In addition to high acidity and toxic H_2S, a major chemical species limiting plant growth on such soils is Al(III). Aluminum ion liberated in acidic soils is very toxic to plants.

Adjustment of Soil Acidity

Most common plants grow best in soil with a pH near neutrality. If the soil becomes too acidic for optimum plant growth, it may be restored to productivity by liming, ordinarily through the addition of calcium carbonate:

$$Soil\}(H^+)_2 + CaCO_3 \rightarrow Soil\}Ca^{2+} + CO_2 + H_2O \tag{16.2.4}$$

In areas of low rainfall, soils may become too basic (alkaline) due to the presence of basic salts such as Na_2CO_3. Alkaline soils may be treated with aluminum or iron sulfate, which release acid on hydrolysis:

$$2Fe^{3+} + 3SO_4^{2-} + 6H_2O \rightarrow 2Fe(OH)_3(s)^{3+} + 6H^+ + 3SO_4^{2-} \tag{16.2.5}$$

Sulfur added to soils is oxidized by bacterially mediated reactions to sulfuric acid:

$$S + 3/2 O_2 + H_2O \rightarrow 2H^+ + SO_4^{2-} \tag{16.2.6}$$

and sulfur is used, therefore, to acidify alkaline soils. The huge quantities of sulfur now being removed from fossil fuels to prevent air pollution by sulfur dioxide may make the treatment of alkaline soils by sulfur much more attractive economically.

Ion Exchange Equilibria in Soil

Competition of different cations for cation exchange sites on soil cation exchangers may be described semiquantitatively by exchange constants. For example, soil reclaimed from an area flooded with seawater will have most of its cation exchange sites occupied by Na^+, and restoration of fertility requires binding of nutrient cations such as K^+:

$$Soil\}Na^+ + K^+ \leftrightarrow Soil\}K^+ + Na^+ \tag{16.2.7}$$

The exchange constant is Kc,

$$K_c = \frac{N_K[Na^+]}{N_{Na}[K^+]} \tag{16.2.8}$$

which expresses the relative tendency of soil to retain K^+ and Na^+. In this equation, N_K and N_{Na} are the equivalent ionic fractions of potassium and sodium, respectively, bound to soil, and $[Na^+]$ and $[K^+]$ are the concentrations of these ions in the surrounding soil water. For example, a soil with all cation exchange sites occupied by

Na^+ would have a value of 1.00 for N_{Na}; with one-half of the cation exchange sites occupied by Na^+, N_{Na} is 0.5; etc. The exchange of anions by soil is not nearly so clearly defined as is the exchange of cations. In many cases, the exchange of anions does not involve a simple ion-exchange process. This is true of the strong retention of orthophosphate species by soil. At the other end of the scale, nitrate ion is very weakly retained by the soil.

Anion exchange may be visualized as occurring at the surfaces of oxides in the mineral portion of soil. A mechanism for the acquisition of surface charge by metal oxides is shown in Figure 5.5, using MnO_2 as an example. At low pH, the oxide surface may have a net positive charge, enabling it to hold anions such as chloride by electrostatic attraction:

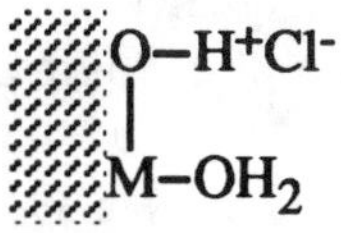

$$O-H^+Cl^-$$
$$|$$
$$M-OH_2$$

At higher pH values, the metal oxide surface has a net negative charge due to the formation of OH^- ion on the surface, caused by loss of H+ from the water molecules bound to the surface:

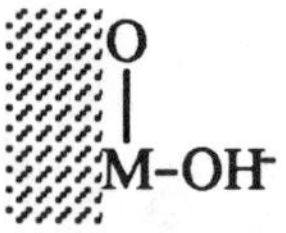

$$O$$
$$|$$
$$M-OH^-$$

In such cases, it is possible for anions such as HPO_4^{2-} to displace hydroxide ion and bond directly to the oxide surface:

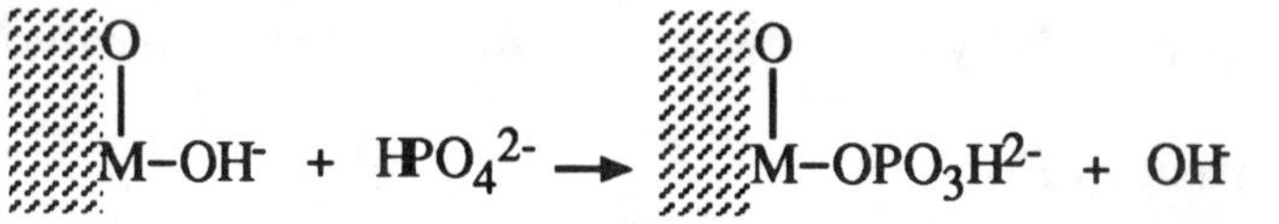

$$\begin{array}{c} O \\ | \\ M-OH^- \end{array} + HPO_4^{2-} \longrightarrow \begin{array}{c} O \\ | \\ M-OPO_3H^{2-} \end{array} + OH^-$$

16.3. MACRONUTRIENTS IN SOIL

One of the most important functions of soil in supporting plant growth is to provide essential plant nutrients — macronutrients and micronutrients. Macronutrients are those elements that occur in substantial levels in plant materials or in fluids in the plant. Micronutrients (Section 16.5) are elements that are essential only at very low levels and generally are required for the functioning of essential enzymes.

The elements generally recognized as essential macronutrients for plants are carbon, hydrogen, oxygen, nitrogen, phosphorus, potassium calcium, magnesium, and sulfur. Carbon, hydrogen, and oxygen are obtained from the atmosphere. The other essential macronutrients must be obtained from soil. Of these, nitrogen, phosphorus, and potassium are the most likely to be lacking are commonly added to soil as fertilizers. Because of their importance, these elements are discussed separately in Section 16.4.

Calcium-deficient soils are relatively uncommon. Liming, a process used to treat acid soils (see Section 16.2), provides a more than adequate calcium supply for plants.

However, calcium uptake by plants and leaching by carbonic acid (Reaction 16.2.1) may produce a calcium deficiency in soil. Acid soils may still contain an appreciable level of calcium which, because of competition by hydrogen ion, is not available to plants. Treatment of acid soil to restore the pH to near-neutrality generally remedies the calcium deficiency. In alkaline soils, the presence of high levels of sodium, magnesium, and potassium sometime produces calcium deficiency because these ions compete with calcium for availability to plants.

Although magnesium makes up 2.1% of the Earth's crust, most of it is rather strongly bound in minerals. Generally, exchangeable magnesium is considered available to plants and is held by ion-exchanging organic matter or clays. The availability of magnesium to plants depends upon the calcium/magnesium ratio. If this ratio is too high, magnesium may not be available to plants and magnesium deficiency results. Similarly, excessive levels of potassium or sodium may cause magnesium deficiency.

Sulfur is assimilated by plants as the sulfate ion, SO_4^{2-}. In addition, in areas where the atmosphere is contaminated with SO_2, sulfur may be absorbed as sulfur dioxide by plant leaves. Atmospheric sulfur dioxide levels have been high enough to kill vegetation in some areas (see Chapter 11). However, some experiments designed to show SO_2 toxicity to plants have resulted in increased plant growth where there was an unexpected sulfur deficiency in the soil used for the experiment.

Soils deficient in sulfur do not support plant growth well, largely because sulfur is a component of some essential amino acids and of thiamin and biotin. Sulfate ion is generally present in the soil as immobilized insoluble sulfate minerals or as soluble salts, which are readily leached from the soil and lost as soil water runoff. Unlike the case of nutrient cations such as K^+, little sulfate is adsorbed to the soil (that is, bound by ion exchange binding) where it is resistant to leaching while still available for assimilation by plant roots.

Soil sulfur deficiencies have been found in a number of regions of the world. Whereas most fertilizers used to contain sulfur, its use in commercial fertilizers is declining. If this trend continues, it is possible that sulfur will become a limiting nutrient in more cases.

As noted in Section 16.2, the reaction of FeS_2 with acid in acid-sulfate soils may release H_2S, which is very toxic to plants and which also kills many beneficial microorganisms. Toxic hydrogen sulfide can also be produced by reduction of sulfate ion through microorganism-mediated reactions with organic matter. Production of hydrogen sulfide in flooded soils may be inhibited by treatment with oxidizing compounds, one of the most effective of which is KNO_3.

16.4. NITROGEN, PHOSPHORUS, AND POTASSIUM IN SOIL

Nitrogen, phosphorus, and potassium are plant nutrients that are obtained from soil. They are so important for crop productivity that they are commonly added to soil as fertilizers. The environmental chemistry of these elements is discussed here and their production as fertilizers in Section 16.6.

Nitrogen

Figure 16.1 summarizes the primary sinks and pathways of nitrogen in soil. In most soils, over 90% of the nitrogen content is organic. This organic nitrogen is primarily the product of the biodegradation of dead plants and animals. It is eventually hydrolyzed to NH_4^+, which can be oxidized to NO_3^- by the action of bacteria in the soil.

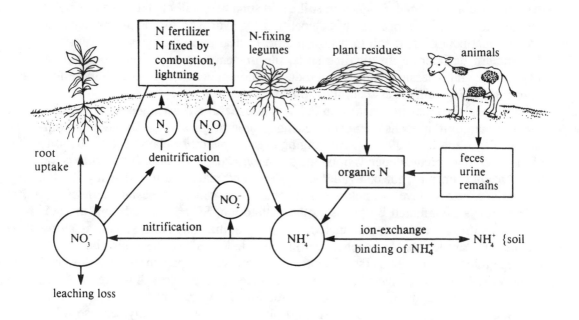

Figure 16.1. Nitrogen sinks and pathways in soil.

Nitrogen bound to soil humus (see Section 15.5) is especially important in maintaining soil fertility. Unlike potassium or phosphate, nitrogen is not a significant product of mineral weathering (see Section 15.6). Nitrogen-fixing organisms ordinarily cannot supply sufficient nitrogen to meet peak demand. Inorganic nitrogen from fertilizers and rainwater is often largely lost by leaching. Soil humus, however, serves as a reservoir of nitrogen required by plants. It has the additional advantage that its rate of decay, hence its rate of nitrogen release to plants, roughly parallels plant growth — rapid during the warm growing season, slow during the winter months.

Nitrogen is an essential component of proteins and other constituents of living matter. Plants and cereals grown on nitrogen-rich soils not only provide higher yields, but are often substantially richer in protein and therefore, more nutritious. Nitrogen is most generally available to plants as nitrate ion, NO_3^-. Some plants such as rice may utilize ammonium nitrogen; however, other plants are poisoned this form of nitrogen. When nitrogen is applied to soils in the ammonium form, nitrifying bacteria perform an essential function in converting it to available nitrate ion.

Plants may absorb excessive amounts of nitrate nitrogen from soil. This phenomenon occurs particularly in heavily fertilized soils under drought conditions. Forage crops containing excessive amounts of nitrate can poison ruminant animals such as cattle or sheep. Plants having excessive levels of nitrate can endanger people when used for ensilage, an animal food consisting of finely chopped plant material such as partially matured whole corn plants, fermented in a structure called a silo. Under the reducing conditions of fermentation, nitrate in ensilage may be reduced to toxic NO_2 gas, which can accumulate to high levels in enclosed silos. There have been many cases reported of persons being killed by accumulated NO_2 in silos.

Nitrogen fixation is the process by which atmospheric N_2 is converted to nitrogen compounds available to plants. Human activities are resulting in the fixation of a great deal more nitrogen than would otherwise be the case. Artificial sources now account for 30-40% of all nitrogen fixed. These include chemical fertilizer manufacture; nitrogen fixed during fuel combustion; combustion of nitrogen-containing fuels; and the increased cultivation of nitrogen-fixing legumes (see the following paragraph). A major concern with this increased fixation of nitrogen is the possible effect upon the atmospheric ozone layer by N_2O released during denitrification of fixed nitrogen.

Prior to the widespread introduction of nitrogen fertilizers, soil nitrogen was provided primarily by legumes. These are plants such as soybeans, alfalfa, and clover, which contain on their root structures bacteria capable of fixing atmospheric nitrogen. Leguminous plants have a symbiotic (mutually advantageous) relationship with the bacteria that provide their nitrogen. Legumes may add significant quantities of nitrogen to soil, up to 10 pounds per acre per year, which is comparable to amounts commonly added as synthetic fertilizers. Soil fertility with respect to nitrogen may be maintained by rotating plantings of nitrogen-consuming plants with plantings of legumes, a fact recognized by agriculturists as far back as the Roman era.

The nitrogen-fixing bacteria in legumes exist in special structures on the roots called root nodules (see Fig. 16.2). The rod-shaped bacteria that fix nitrogen are members of a special genus called Rhizobium. These bacteria may exist independently, but cannot fix nitrogen except in symbiotic combination with plants. Although all species of Rhizobium appear to be very similar, they exhibit a great deal of specificity in their choice of host plants. Curiously, legume root nodules also contain a form of hemoglobin, which must somehow be involved in the nitrogen-fixation process.

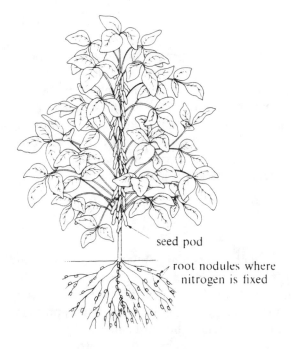

seed pod

root nodules where
nitrogen is fixed

Figure 16.2. A soybean plant, showing root nodules where nitrogen is fixed.

Nitrate pollution of some surface waters and groundwater has become a major problem in some agricultural areas (see Chapter 7). Although fertilizers have been implicated in such pollution, there is evidence that feedlots are a major source of nitrate pollution. The growth of livestock populations and the concentration of livestock in feedlots have aggravated the problem. Such concentrations of cattle, coupled with the fact that a steer produces approximately 18 times as much waste material as a human, have resulted in high levels of water pollution in rural areas with small human populations. Streams and reservoirs in such areas frequently are just as polluted as those in densely populated and highly industrialized areas.

Nitrate in farm wells is a common and especially damaging manifestation of nitrogen pollution from feedlots because of the susceptibility of ruminant animals to nitrate poisoning. The stomach contents of ruminant animals such as cattle and sheep constitute a reducing medium (low pE) and contain bacteria capable of reducing nitrate ion to toxic nitrite ion:

$$NO_3^- + 2H^+ + 2e^- \rightarrow NO_2^- + 2H_2O \qquad (16.4.1)$$

The origin of most nitrate produced from feedlot wastes is amino nitrogen present in nitrogen-containing waste products. Approximately one-half of the nitrogen excreted by cattle is contained in the urine. Part of this nitrogen is proteinaceous and the other part is in the form of urea, NH_2CONH_2. As a first step in the degradation process, the amino nitrogen is probably hydrolyzed to ammonia, or ammonium ion:

$$RNH_2 + H_2O \rightarrow R\text{-}OH + NH_3 (NH_4^+) \qquad (16.4.2)$$

This product is then oxidized through microorganism-catalyzed reactions to nitrate ion:

$$NH_3 + 2O_2 \rightarrow H^+ + NO_3^- + H_2O \qquad (16.4.3)$$

Under some conditions, an appreciable amount of the nitrogen originating from the degradation of feedlot wastes is present as ammonium ion. Ammonium ion is rather strongly bound to soil (recall that soil is a generally good cation exchanger), and a small fraction is fixed as nonexchangeable ammonium ion in the crystal lattice of clay minerals. Because nitrate ion is not strongly bound to soil, it is readily carried through soil formations by water. Many factors, including soil type, moisture, and level of organic matter, affect the production of ammonia and nitrate ion originating from feedlot wastes, and a marked variation is found in the levels and distributions of these materials in feedlot areas.

Phosphorus

Although the percentage of phosphorus in plant material is relatively low, it is an essential component of plants. Phosphorus, like nitrogen, must be present in a simple inorganic form before it can be taken up by plants. In the case of phosphorus, the utilizable species is some form of orthophosphate ion. In the pH range that is present in most soils, $H_2PO_4^-$ and HPO_4^{2-} are the predominant orthophosphate species.

Orthophosphate is most available to plants at pH values near neutrality. It is believed that in relatively acidic soils, orthophosphate ions are precipitated or sorbed by species of Al(III) and Fe(III). In alkaline soils, orthophosphate may react with calcium carbonate to form relatively insoluble hydroxyapatite:

$$3HPO_4^{2-} + 5CaCO_3(s) + 2H_2O \rightarrow Ca_5(PO_4)_3(OH)(s)$$

$$+ 5HCO_3^- + OH^- \qquad (16.4.4)$$

In general, because of these reactions, little phosphorus applied as fertilizer leaches from the soil. This is important from the standpoint of both water pollution and utilization of phosphate fertilizers. More research remains to be done to establish fully the nature of the chemical interactions determining the availability of orthophosphates in soils.

Potassium

Relatively high levels of potassium are utilized by growing plants. Potassium activates some enzymes and plays a role in the water balance in plants. It is also essential for some carbohydrate transformations. Crop yields are generally greatly reduced in potassium-deficient soils. The higher the productivity of the crop, the more potassium is removed from soil. When nitrogen fertilizers are added to soils to increase productivity, removal of potassium is enhanced. Therefore, potassium may become a limiting nutrient in soils heavily fertilized with other nutrients.

Potassium is one of the most abundant elements in the Earth's crust, of which it makes up 2.6%; however, much of this potassium is not easily available to plants. For example, some silicate minerals such as leucite, $K_2O \cdot Al_2O_3 \cdot 4SiO_2$, contain strongly bound potassium. Exchangeable potassium held by clay minerals is relatively more available to plants.

16.5. MICRONUTRIENTS IN SOIL

Boron, chlorine, copper, iron, manganese, molybdenum (for N-fixation), sodium, vanadium, and zinc are considered essential plant **micronutrients**. These elements are needed by plants only at very low levels and frequently are toxic at higher levels. There is some chance that other elements will be added to this list as techniques for growing plants in environments free of specific elements improve. Most of these elements function as components of essential enzymes. Manganese, iron, chlorine, zinc, and vanadium may be involved in photosynthesis.

Iron and manganese occur in a number of soil minerals. Sodium and chlorine (as chloride) occur naturally in soil and are transported as atmospheric particulate matter from marine sprays (see Chapter 10). Some of the other micronutrients and trace elements are found in primary (unweathered) minerals that occur in soil. Boron is substituted isomorphically for Si in some micas and is present in tourmaline, a mineral with the formula $NaMg_3Al_6B_3Si_6O_{27}(OH,F)_4$. Copper is present isomorphically substituted for other elements in feldspars, amphiboles, olivines, pyroxenes, and micas; it also occurs as trace levels of copper sulfides in silicate minerals. Molybdenum occurs as molybdenite (MoS_2). Vanadium is isomorphically substituted for Fe or Al in oxides, pyroxenes, amphiboles, and micas. Zinc is present as the result of isomorphic substitution for Mg, Fe, and Mn in oxides, amphiboles, olivines, and pyroxenes and as traces of zinc sulfide in silicates. Other trace level elements that occur as specific minerals, sulfide inclusions, or by isomorphic substitution for other elements in minerals are chromium, cobalt, arsenic, selenium, nickel, lead, and cadmium.

The trace elements listed above may be coprecipitated with secondary minerals (see Section 15.2) that are involved in soil formation. Such secondary minerals include oxides of aluminium, iron, and manganese (precipitation of hydrated oxides of iron and manganese very efficiently removes many trace metal ions from solution); calcium and magnesium carbonates; smectites; vermiculites; and illites.

Some plants accumulate extremely high levels of specific trace metals. Those accumulating more than 1.00 mg/g of dry weight are called **hyperaccumulators**. Hyperaccumulation of nickel and copper has been described;[4] for example, *Aeolanthus biformifolius DeWild* growing in copper-rich regions of Shaba Province, Zaire, contains up to 1.3% copper (dry weight) and is known as a "copper flower".

16.6. FERTILIZERS

Crop fertilizers contain nitrogen, phosphorus, and potassium as major components. Magnesium, sulfate, and micronutrients may also be added. Fertilizers are designated by numbers, such as 6-12-8, showing the respective percentages of nitrogen expressed as N (in this case 6%), phosphorus as P_2O_5 (12%), and potassium as K_2O (8%). Farm manure corresponds to an approximately 0.5-0.24-0.5 fertilizer. The organic fertilizers such as manure must undergo biodegradation to release the simple inorganic species (NO_3^-, $H_xPO_4^{x-3}$, K^+) assimilable by plants.

Most modern nitrogen fertilizers are made by the Haber process, in which N_2 and H_2 are combined over a catalyst at temperatures of approximately 500°C and pressures up to 1000 atm:

$$N_2 + 3H_2 \rightarrow 2NH_3 \tag{16.6.1}$$

The anhydrous ammonia product has a very high nitrogen content of 82%. It may be added directly to the soil, for which it has a strong affinity because of its water solubility and formation of ammonium ion:

$$NH_3(g) \text{ (water)} \rightarrow NH_3(aq) \tag{16.6.2}$$

$$NH_3(aq) + H_2O \rightarrow NH_4^+ + OH^- \tag{16.6.3}$$

Special equipment is required, however, because of the toxicity of ammonia gas. Aqua ammonia, a 30% solution of NH_3 in water, may be used with much greater safety. It is sometimes added directly to irrigation water. It should be pointed out that ammonia vapor is toxic and NH_3 is reactive with some substances. Improperly discarded or stored ammonia can be a hazardous waste. In 1989, for example, a tank containing 10,000 gallons of liquid ammonia at a closed Nassau County, Long Island, manufacturing plant was drained under emergency conditions by the U.S. Coast Guard to prevent release of ammonia to the surrounding area.[5] The tank was partially under a building which was in danger of collapse, an event that might have ruptured the full tank and caused release of ammonia gas.

Ammonium nitrate, NH_4NO_3, is a common solid nitrogen fertilizer. It is made by oxidizing ammonia over a platinum catalyst, converting the nitric oxide product to nitric acid, and reacting the nitric acid with ammonia. The molten ammonium nitrate product is forced through nozzles at the top of a *prilling tower* and solidifies to form small pellets while falling through the tower. The particles are coated with a water repellent. Ammonium nitrate contains 33.5% nitrogen. Although convenient to apply to soil, it requires considerable care during manufacture and storage because it is explosive. Ammonium nitrate also poses some hazards. It is mixed with fuel oil to form an explosive that serves as a substitute for dynamite in quarry blasting and construction. About 45,000 pounds of this mixture that was detonated as the result of a fire was involved in two massive explosions that killed 6 firefighters at a construction site in Kansas City, Missouri, on November 29, 1988.

Urea,

$$
\begin{array}{c}
\text{O} \\
\parallel \\
\text{H}_2\text{N}-\text{C}-\text{NH}_2
\end{array}
$$

is easier to manufacture and handle than ammonium nitrate. It is now the favored solid nitrogen-containing fertilizer. The overall reaction for urea synthesis is

$$CO_2 + 2NH_3 \rightarrow CO(NH_2)_2 + H_2O \qquad (16.6.4)$$

involving a rather complicated process in which ammonium carbamate, $NH_2CO_2NH_4$, is an intermediate.

Other compounds used as nitrogen fertilizers include sodium nitrate (obtained largely from Chilean deposits, see Section 15.2), calcium nitrate, potassium nitrate, and ammonium phosphates. Ammonium sulfate, a byproduct of coke ovens, used to be widely applied as fertilizer. The alkali metal nitrate tends to make soil alkaline, whereas ammonium sulfate leaves an acidic residue.

Phosphate minerals are found in several states, including Idaho, Montana, Utah, Wyoming, North Carolina, South Carolina, Tennessee, and Florida. The principal mineral is fluorapatite, $Ca_5(PO_4)_3F$. The phosphate from fluorapatite is relatively unavailable to plants and is frequently treated with phosphoric or sulfuric acids to produce superphosphates:

$$2Ca(PO_4)_3F(s) + 14H_3PO_4 + 10H_2O \rightarrow 2HF(g)$$
$$+ \; 10CaH_4(PO_4)_2 \cdot H_2O \qquad (16.6.5)$$

$$2Ca(PO_4)_3F(s) + 7H_2SO_4 + 3H_2O \rightarrow 2HF(g)$$
$$+ \; 3CaH_4(PO_4)_2 \cdot H_2O + 7CaSO_4 \qquad (16.6.6)$$

The superphosphate products are much more soluble than the parent phosphate minerals. The HF produced as a byproduct of superphosphate production can create air pollution problems.

Phosphate minerals are rich in trace elements required for plant growth, such as boron, copper, manganese, molybdenum, and zinc. Ironically, these elements are lost in processing phosphate for fertilizers and are sometimes added for fertilizers later.

Ammonium phosphates are excellent, highly soluble phosphate fertilizers. Liquid ammonium polyphosphate fertilizers consisting of ammonium salts of pyrophosphate, triphosphate, and small quantities of higher polymeric phosphate anions in aqueous solution are becoming very popular as phosphate fertilizers. The polyphosphates are believed to have the additional advantage of chelating iron and other micronutrient metal ions, thus making the metals more available to plants.

Potassium fertilizer components consist of potassium salts, generally KCl. Such salts are found as deposits in the ground or may be obtained from some brines. Very large deposits are found in Saskatchewan, Canada. These salts are all quite soluble in water. One problem encountered with potassium fertilizers is the luxury uptake of potassium by some crops, which absorb more potassium than is really needed for their maximum growth. In a crop where only the grain is harvested, leaving the rest of the plant in the field, luxury uptake does not create much of a problem because most of the potassium is returned to the soil with the dead plant. However, when hay or forage is harvested, potassium contained in the plant as a consequence of luxury uptake is lost from the soil.

16.7. WASTES AND POLLUTANTS IN SOIL

Soil receives large quantities of waste products. Much of the sulfur dioxide emitted in the burning of sulfur-containing fuels ends up on soil as sulfates. Atmospheric nitrogen oxides are converted to nitrates in the atmosphere, and the nitrates eventually are deposited on soil. Soil sorbs NO and NO_2 readily, and these gases are oxidized to nitrate in the soil. Carbon monoxide is converted to CO_2 and possibly to biomass by soil bacteria and fungi (see Chapter 11). Particulate lead from automobile exhausts is found at elevated levels in soil along heavily traveled highways. Elevated levels of lead from lead mines and smelters are found on soil near such facilities.

Soil also receives enormous quantities of pesticides as an inevitable result of their application to crops. The degradation and eventual fate of these pesticides on soil largely determines the ultimate environmental effects of the pesticides, and detailed knowledge of these effects are now required for licensing of a new pesticide (in the U.S. under the Federal Insecticide, Rodenticide, and Fungicide act, FIFRA). Among the factors to be considered are the sorption of the pesticide by soil; leaching of the pesticide into water, as related to its potential for water pollution; effects of the pesticide on microorganisms and animal life is the soil; and possible production of relatively more toxic degradation products.

Adsorption by soil is a key step in the degradation of a pesticide. The degree of adsorption and the speed and extent of ultimate degradation are influenced by a number of factors. Some of these, including solubility, volatility, charge, polarity, and molecular structure and size, are properties of the medium. Adsorption of a pesticide by soil components may have several effects. Under some circumstances, it retards degradation by separating the pesticide from the microbial enzymes that degrade it, whereas under other circumstances the reverse is true. Purely chemical degradation reactions may be catalyzed by adsorption. Loss of the pesticide by volatilization or leaching is diminished. The toxicity of a herbicide to plants may be strongly affected by soil sorption.

The forces holding a pesticide to soil particles may be of several types. Physical adsorption involves van dar Waals forces arising from dipole-dipole interactions between the pesticide molecule and charged soil particles. Ion exchange is especially effective in holding cationic organic compounds, such as the herbicide paraquat,

$$H_3C-N\bigcirc-\bigcirc N^+-CH_3 \cdot 2Cl^-$$

to anionic soil particles. Some neutral pesticides become cationic by protonation and are bound as the protonated positive form. Hydrogen bonding is another mechanism by which some pesticides are held to soil. In some cases, a pesticide may act as a ligand coordinating to metals in soil mineral matter.

The three primary ways in which pesticides are degraded in or on soil are *biodegradation*, *chemical degradation*, and *photochemical reactions*. Various combinations of these processes may operate in the degradation of a pesticide.

Although insects, earthworms, and plants may play roles in the **biodegradation** of pesticides, microorganisms have the most important role. Several examples of microorganism-mediated degradation of pesticides are given in Chapter 6.

Chemical degradation of pesticides has been observed experimentally in soils and clays sterilized to remove all microbial activity. For example, clays have been shown to catalyze the hydrolysis of o,o-Dimethyl-o-2,4,5-trichlorophenyl thiophosphate (also called Trolene, ronnel, Etrolene, or trichlorometafos), and effect attributed to -OH groups on the mineral surface:

$$(CH_3O)_2\overset{\overset{S}{\|}}{P}-O-\underset{Cl}{\overset{Cl}{\bigcirc}}-Cl \xrightarrow[\text{surfaces}]{\underset{\text{Mineral}}{H_2O}}$$

$$HO-\underset{Cl}{\overset{Cl}{\bigcirc}}-Cl + \overset{\overset{S}{\|}}{P}(OH)_3 + 2CH_3OH \qquad (16.9.1) \qquad (16.7.1)$$

Many other purely chemical hydrolytic reactions of pesticides occur in soil.

A number of pesticides have been shown to undergo **photochemical reactions**, that is, chemical reactions brought about by the absorption of light (see Chapter 9). Frequently, isomers of the pesticides are produced as products. Many of the studies reported apply to pesticides in water or on thin films, and the photochemical reactions of pesticides on soil and plant surfaces remain largely a matter of speculation.

Soil is the receptor of many hazardous wastes from landfill leachate, lagoons, and other sources (see Section 18.6). In some cases, land farming of degradable hazardous organic wastes is practiced as a means of disposal and degradation. The degradable material is worked into the soil, and soil microbial processes bring about its degradation. As discussed in Chapter 8, sewage and fertilizer-rich sewage sludge may be applied to soil.

16.8. SOIL EROSION

Soil erosion can occur by the action of both water and wind, although water is the primary source of erosion. To provide an idea of the magnitude of the problem, U.S. Department of Agriculture officials estimate that 15 million tons per minute of topsoil are swept from the mouth of the Mississippi. About one-third of U.S. topsoil has been lost since cultivation began on the continent. At the present time approximately one-third of U.S. cultivated land is eroding at a rate sufficient to reduce soil productivity. It is estimated that 48 million acres of land, somewhat more than 10 percent of that under cultivation, is eroding at unacceptable levels, taken to mean a loss of more than 14 tons of topsoil per acre each year. Specific areas in which the greatest erosion is occurring include northern Missouri, southern Iowa, west Texas, western Tennessee, and the Mississippi Basin. Figure 16.3 shows the pattern of soil erosion in the continental U.S. in 1977.

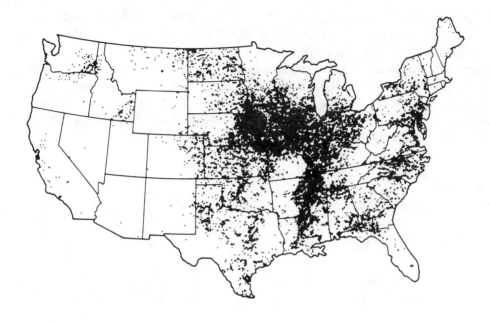

Figure 16.3. Pattern of soil erosion in the continental U.S. as of 1977. The dark areas indicate locations where the greatest erosion is occurring.

Problems involving soil erosion were aggravated in the 1970s and early 1980s when high prices for farmland resulted in the intensive cultivation of high-income crops, particularly corn and soybeans. These crops grow in rows with bare soil in between, which tends to wash away with each rainfall. Furthermore, the practice became widespread of planting corn and soybeans year after year, without intervening plantings of soil-restoring clover or grass. The problem of decreased productivity due to soil erosion has been masked somewhat by increased use of chemical fertilizers.

Wind erosion, such as occurs on the generally dry, high plains soils of eastern Colorado, poses another threat. After the Dust Bowl days of the 1930s, much of this land was allowed to revert to grassland, and the topsoil was held in place by the strong root systems of the grass cover. However, in an effort to grow more wheat and improve the sale value of the land, much of it has been cultivated in recent years. For example, from 1979 through 1982, more than 450,000 acres of Colorado grasslands were plowed. Much of this was done by speculators who purchased grassland at a low price of $100-$200 per acre, broke it up and sold it as cultivated land at more than double the original purchase price. Although freshly cultivated grassland may yield well for one or two years, the nutrients and soil moisture are rapidly exhausted, and the land becomes very susceptible to wind erosion.

There are a number of solutions to the soil erosion problem. Some are old, well-known agricultural practices, such as terracing, contour plowing, and periodically planting fields with cover crops, such as clover. For some crops **no-till agriculture** greatly reduces erosion. This practice consists of planting a crop among the residue of the previous year's crop, without plowing. Weeds are killed in the newly planted crop row by application of a herbicide prior to planting. The surface residue of plant material left on top of the soil prevents erosion.

Another, more experimental, solution to the soil erosion problem is the cultivation of perennial plants, which develop a large root system and come up each spring after being harvested the previous fall. For example, a perennial corn plant has been developed by crossing corn with a distant, wild relative, teosinte, which grows in Central America. Unfortunately, the resulting plant does not give outstanding grain yields. It should be noted that an annual plant's ability to propagate depends upon producing large quantities of seeds, whereas a perennial plant must develop a strong root system with bulbous growths called rhizomes, which store food for the coming year. However, it is possible that the application of genetic engineering (see Section 16.9) may result in the development of perennial crops with good seed yields. The cultivation of such a crop would cut down on a great deal of soil erosion.

The best known perennial plant is the tree, which is very effective in stopping soil erosion. Wood from trees can be used as biomass fuel, as a source of raw materials, and as food (see below). There is a tremendous unrealized potential for an increase in the production of biomass from trees. For example, the production of biomass from natural forests of loblolly pine trees in South Carolina has been about 3 dry tons per hectare per year. This has now been increased to 11 tons through selection of superior trees, and 30 tons may eventually be possible. In Brazil, experiments have been conducted with a species of Eucalyptus, which has a 7-year growth cycle. With improved selection of trees, the annual yields for three successive cycles of these trees in dry tons per hectare per year has been 23, 33, and 40.

The most important use for wood is, of course, as lumber for construction. This use will remain important as higher energy costs increase the costs of other construction materials, such as steel, aluminum, and cement. Wood is about 50 percent cellulose, which can be hydrolyzed by rapidly improving enzyme processes to yield glucose sugar. The glucose can be used directly as food, fermented to ethyl alcohol for fuel (gasohol), or employed as a carbon and energy source for protein-producing yeasts. Given these and other potential uses, the future of trees as an environmentally desirable and profitable crop is very bright.

16.9. GENETIC ENGINEERING AND AGRICULTURE

The nuclei of living cells contain the genetic instruction for cell reproduction. These instructions are in the form of a special material called deoxyribonucleic acid, DNA. In combination with proteins, DNA makes up the cell chromosomes. During the 1970s the ability to manipulate DNA through genetic engineering became a reality and during the 1980s became the basis of a major industry. Such manipulation falls into the category of recombinant DNA technology. Recombinant DNA gets its name from the fact that it contains DNA from two different organisms, recombined together. This technology promises some exciting developments in agriculture.

The "green revolution" of the mid-1960s used conventional plant-breeding techniques of selective breeding, hybridization, cross-pollination, and back-crossing to develop new strains of rice, wheat, and corn, which, when combined with chemical fertilizers, yielded spectacularly increased crop yields. For example, India's output of grain increased 50 percent. By working at the cellular level, however, it is now possible to greatly accelerate the process of plant breeding. Thus, plants may be developed that resist particular diseases, grow in seawater, or have much higher productivity. The possibility exists for developing entirely new kinds of plants.

One exciting possibility with genetic engineering is the development of plants other than legumes which fix their own nitrogen. For example, if nitrogen-fixing corn could be developed, the savings in fertilizer would be enormous. Furthermore, since the nitrogen is fixed in an organic form in plant root structures, there would be no pollutant runoff of chemical fertilizers.

Another promising possibility with genetic engineering is increased efficiency of photosynthesis. Plants utilize only about 1 percent of the sunlight striking their leaves, so there is appreciable room for improvement in that area.

Cell-culture techniques can be applied in which billions of cells are allowed to grow in a medium and develop mutants which, for example, might be resistant to particular viruses or herbicides or have other desirable qualities. If the cells with the desired qualities can be regenerated into whole plants, results can be obtained that might have taken decades using conventional plant-breeding techniques.

Despite the enormous potential of the "green revolution," genetic engineering, and more intensive cultivation of land to produce food and fiber, these technologies cannot be relied upon to support an uncontrolled increase in world population and may even simply postpone an inevitable day of reckoning with the consequences of population growth. Changes in climate resulting from global warming (greenhouse effect, Section 14.2), ozone depletion (by chlorofluorcarbons, Section 14.4), or natural disasters, such as massive volcanic eruptions or collisions with large meteorites can, and almost certainly will, result in worldwide famine conditions in the future that no agricultural technology will be able to alleviate.

16.10. AGRICULTURE AND HEALTH

Some authorities hold that soil has an appreciable effect upon health. An obvious way in which such an effect might be manifested is the incorporation into food of micronutrient elements essential for human health. One such nutrient (which is toxic at overdose levels) is selenium. It is definitely known that the health of animals is adversely affected in selenium-deficient areas, as it is in areas of selenium excess. Human health might be similarly affected.

There are some striking geographic correlations with the occurrence of cancer. Some of these correlations may be due to soil type. A high incidence of stomach cancer has been shown to occur in areas with certain types of soil in the Netherlands, the United States, France, Wales, and Scandinavia. These soils are high in organic matter content, are acidic, and frequently are waterlogged. A "stomach cancer-prone life style" has been described[6], which includes consumption of home-grown food, consumption of water from one's own well, and reliance on native and uncommon foodstuffs.

One possible reason for the existence of "stomach cancer-producing soils" is the production of cancer-causing secondary metabolites by plants and microorganisms. Secondary metabolites are biochemical compounds that are of no apparent use to the organism producing them. It is believed that they are formed from the precursors of primary metabolites when the primary metabolites accumulate to excessive levels.

The role of soil in environmental health is not well known, nor has it been extensively studied. The amount of research on the influence of soil in producing foods that are more nutritious and lower in content of naturally occurring toxic substances is quite small compared to research on higher soil productivity. It is to be hoped that the environmental health aspects of soil and its products will receive much greater emphasis in the future.

Chemical Contamination

Sometimes human activities contaminate food grown on soil. Most often this occurs through contamination by pesticides. An interesting example of such contamination occurred in Hawaii in early 1982[7]. It was found that milk from several sources on Oahu contained very high levels of heptachlor (see Table 7.5). This pesticide causes cancer and liver disorders in mice; therefore, it is a suspected human carcinogen. Remarkably, in this case it was not until 57 days after the initial discovery that the public was informed of the contamination by the Department of Health. The source of heptachlor was traced to contaminated "green chop," chopped-up pineapple leaves fed to cattle. Although heptachlor was banned for most applications, Hawaiian pineapple growers had obtained special Federal permission to use it to control mealybug wilt. Although it was specified that green chop could not be collected within 1 year of the last application of the pesticide, apparently this regulation was violated, and the result was distribution of contaminated milk to consumers.

In the late 1980s Alar residues on food caused considerable controversy in the marketplace. **Alar**, daminozide, is a growth regulator that was widely used on apples to bring about uniform ripening of the fruit and to improve firmness and color of the apples. It was discontinued for this purpose after 1988 because of concerns that it might cause cancer, particularly in those children who consume relatively large amounts of apples, apple juice, and other apple products. Dire predictions were made in the industry of crop losses and financial devastation. However, the 1989 apple crop, which was the first without Alar in the U.S. had a value of $1.0 billion, only $0.1 billion less than that of the 1988 crop, and 1990 was predicted to be a good year for the sale of apples.[8]

LITERATURE CITED

1. Sposito, Garrison, *The Chemistry of Soils*, Oxford University Press, New York, 1989.

2. Elprince, A. M., *Chemistry of Soil Solutions*, Van Nostrand Reinhold, New York, 1986.

3. Page, A. L., R. H. Miller, and Dr. R. Keeney, *Methods of Soil Analysis, Part 2, Chemical and Microbiological Properties*, American Society of Agronomy, Madison, Wisconsin, 1982.

4. Malaise, F., "*Aeloanthus Biformifolius* De Wild: A Hyperaccumulator of Copper from Zaire," *Science* **199**, 887-8 (1978).

5. Hevesi, Dennis, "Coast Guard Orders Draining of Hazardous Tank," *New York Times*, April 16, 1989, p. 23.

6. Adams, R. S., Jr., "Soil Variability and Cancer," *Chemical and Engineering News*, June 12, 1978, p. 84.

7. Smith, R. J., "Hawaiian Milk Contamination Creates Alarm," *Science*, **217**, 137 40 (1982).

8. "Apple Sales Strong Despite Scare in '89 About Chemical Use," *New York Times*, November 13, 1990, p. 1.

SUPPLEMENTARY REFERENCES

Conway, Gordon R., and Edward B. Barbier, *After the Green Revolution. Sustainable Agriculture for Development*, Earthscan Publications, London, 1990.

Bollag, Jean-Marc, and G. Stotzky, Eds., *Soil Biochemistry*, Vol. 6, Marcel-Dekker, New York, 1990.

Heling Ed., *Sediments and Environmental Geochemistry. Selected Aspects and Case Histories*, Springer-Verlag, New York, 1990.

Lal, R., and B. A. Stewart, Eds., *Soil Degradation*, Springer-Verlag, New York, 1990.

Roberts, Willard Lincoln, Thomas J. Campbell, and George Robert Rapp, Jr., *Encyclopedia of Minerals*, 2nd ed., Van Nostrand Reinhold, New York, 1989.

Simkiss, Kenneth, and Karl M. Wilbur, *Biomineralization. Cell Biology and Mineral Deposition*, Academic Press, San Diego, CA, 1989.

Sparks, Donald L., *Kinetics of Soil Chemical Processes*, Academic Press, San Diego, CA, 1989.

Paul, E. A., and F. E. Clark, *Soil Microbiology and Biochemistry*, Academic Press, San Diego, CA, 1988.

Rump, H. H., and H. Krist, *Laboratory Manual for the Examination of Water, Waste Water, and Soil.*, VCH, New York, 1989.

Gieseking, J. E., Ed., *Soil Components, Vol. 1, Organic Components*, Springer-Verlag, Inc., New York.

Wolf, K., W. J. Van Den Brink, and F. J. Colon, *Contaminated Soil '88*, Vols 1 and 2, Kluwer Academic Publishers, Norwell, MA, 1988.

Lindsay, W. L., *Chemical Equilibria in Soils*, Wiley-Interscience, New York, 1979.

Morrill, L. G., B. C. Mahilum, and S. H. Mohiuddin, *Organic Compounds in Soils: Sorption, Degradation, and Persistence*, Ann Arbor Science Publishers, Inc., Ann Arbor, Michigan, 1982.

Olson, G. W., *Soils and the Environment*, Methuen, Inc., New York, 1981.

Paton, T. R., *The Formation of Soil Material*, Allen and Unwin, Inc., Winchester, Mass., 1979.

Stevenson, F. J., *Humus Chemistry: Genesis, Composition, Reactions*, Wiley-Interscience, New York, 1982.

QUESTIONS AND PROBLEMS

(Some answers may require reference to Chapter 15.)

1. Give two examples of reactions involving manganese and iron compounds that may occur in waterlogged soil.

2. What temperature and moisture conditions favor the buildup of organic matter in soil?

3. "Cat clays" are soils containing a high level of iron pyrite, FeS_2. Hydrogen peroxide, H_2O_2, is added to such a soil, producing sulfate, as a test for cat clays. Suggest the chemical reaction involved in this test.

4. What effect upon soil acidity would result from heavy fertilization with ammonium nitrate accompanied by exposure of the soil to air and the action of aerobic bacteria?

5. How many moles of H^+ ion are consumed when 200 kilograms of $NaNO_3$ undergo denitrification in soil?

6. What is the primary mechanism by which organic material in soil exchanges cations?

7. Prolonged waterlogging of soil does **not** (a) increase NO_3^- production, (b) increase Mn^{2+} concentration, (c) increase Fe^{2+} concentration, (d) have harmful effects upon most plants, (e) increase production of NH_4^+ from NO_3^-.

8. Of the following phenomena, the one that eventually makes soil more basic is (a) removal of metal cations by roots, (b) leaching of soil with CO_2-saturated water, (c) oxidation of soil pyrite, (d) fertilization with $(NH_4)_2SO_4$, (e) fertilization with KNO_3.

9. How many metric tons of farm manure are equivalent 100 kg of 10-5-10 fertilizer?

10. How are the chelating agents that are produced from soil microorganisms involved in soil formation?

11. What specific compound is both a particular animal waste product and a major fertilizer?

12. What happens to the nitrogen/carbon ratio as organic matter degrades in soil?

13. To prepare a rich potting soil, a greenhouse operator mixed 75% "normal" soil with 25% peat. Estimate the cation-exchange capacity in milliequivalents/100 g of the product.

14. Explain why plants grown on either excessively acidic or excessively basic soils may suffer from calcium deficiency.

15. What are two mechanisms by which anions may be held by soil mineral matter?

16. What are the three major ways in which pesticides are degraded in or on soil?

17. Lime from lead mine tailings containing 0.5% lead was applied at a rate of 10 metric tons per acre of soil and worked in to a depth of 20 cm. The soil density was 2.0 g/cm. To what extent did this add to the burden of lead in the soil?

18. Match the soil or soil-solution constituent in the left column with the soil condition described on the right, below:

    (1) High Mn^{2+} content in soil solution
    (2) Excess H^+
    (3) High H^+ and SO_4^{2-} content
    (4) High organic content

    (a) "Cat clays" containing initially high levels of pyrite, FeS_2.
    (b) Soil in which biodegradation has not occurred to a great extent
    (c) Waterlogged soil
    (d) Soil whose fertility can be improved by adding limestone

19. What are the processes occurring in soil that operate to reduce the harmful effects of pollutants?

Nature and Sources of Hazardous Waste

17.1. INTRODUCTION

A **hazardous substance** is a material that may pose a danger to living organisms, materials, structures, or the environment by explosion or fire hazards, corrosion, toxicity to organisms, or other detrimental effects. What, then is a hazardous waste? Although it has has been stated that,[1] "The discussion on this question is as long as it is fruitless," a simple definition of a **hazardous waste** is that it is a hazardous substance that has been discarded, abandoned, neglected, released or designated as a waste material, or one that may interact with other substances to be hazardous. The definition of hazardous waste is addressed in greater detail in Section 1.2, but in a simple sense it is a material that has been left where it should not be and that may cause harm to you if you encounter it![2]

History of Hazardous Substances

Humans have always been exposed to hazardous substances going back to prehistoric times when they inhaled noxious volcanic gases or succumbed to carbon monoxide from inadequately vented fires in cave dwellings sealed too well against Ice-Age cold. Slaves in Ancient Greece developed lung disease from weaving mineral asbestos fibers into cloth to make it more degradation-resistant. Some archaeological and historical studies have concluded that lead wine containers were a leading cause of lead poisoning in the more affluent ruling class of the Roman Empire leading to erratic behavior such as fixation on spectacular sporting events, chronic unmanageable budget deficits, poorly regulated financial institutions, and ill-conceived, overly ambitious military ventures in foreign lands. Alchemists who worked during the Middle Ages often suffered debilitating injuries and illnesses resulting from the hazards of their explosive and toxic chemicals. During the 1700s runoff from mine spoils piles began to create serious contamination problems in Europe. As the production of dyes and other organic chemicals developed from the coal tar industry in Germany during the 1800s, pollution and poisoning from coal tar byproducts was observed. By around 1900 the quantity and variety of chemical wastes produced each year was increasing sharply with the addition of wastes such as spent steel and iron pickling liquor, lead battery wastes, chromic wastes, petroleum refinery wastes, radium wastes, and fluoride wastes from aluminum ore refining. As the century progressed into the World War II era, the wastes and hazardous byproducts of manufacturing increased markedly from sources such as chlorinated solvents manufacture, pesticides synthesis, polymers manufacture, plastics, paints, and wood preservatives.

The Love Canal affair of the 1970s and 1980s brought hazardous wastes to the public attention as a major political issue in the U.S. Starting around 1940, this site in Niagara Falls, New York, had received about 20,000 metric tons of chemical wastes containing at least 80 different chemicals. By 1989 state and Federal governments had spent $140 million to clean up the site and relocate residents.[3]

Other areas containing hazardous wastes that received attention included an industrial site in Woburn, Massachusetts, that had been contaminated by wastes from tanneries, glue-making factories, and chemical companies dating back to about 1850; the Stringfellow Acid Pits near Riverside, California; the Valley of the Drums in Kentucky; and Times Beach, Missouri, an entire town that was abandoned because of contamination by TCDD (dioxin).

Legislation

Governments in a number of nations have passed legislation to deal with hazardous substances and wastes. In the U.S. such legislation has included the following:

- Toxic Substances Control Act of 1976
- Resource, Recovery and Conservation Act (RCRA) of 1976 (amended and strengthened by the Hazardous and Solid Wastes Amendments (HSWA) of 1984)
- Comprehensive Environmental Response, Compensation, and Liability Act (CERCLA) of 1980.

RCRA legislation charged the U.S. Environmental Protection Agency (EPA) with protecting human health and the environment from improper management and disposal of hazardous wastes by issuing and enforcing regulations pertaining to such wastes. RCRA requires that hazardous wastes and their characteristics be listed and controlled from the time of their origin until their proper disposal or destruction. Regulations pertaining to firms generating and transporting hazardous wastes require that they keep detailed records, including reports on their activities and manifests to ensure proper tracking of hazardous wastes through transportation systems. Approved containers and labels must be used, and wastes can only be delivered to facilities approved for treatment, storage, and disposal. There are about 290 million tons of wastes regulated by RCRA. In the U.S. about 3,000 facilities are involved in the treatment, storage, or disposal of RCRA wastes.

CERCLA (Superfund) legislation deals with actual or potential releases of hazardous materials that have the potential to endanger people or the surrounding environment at uncontrolled or abandoned hazardous waste sites in the U.S.[4] The act requires responsible parties or the government to clean up waste sites. Among CERCLA's major purposes are the following:

- Site identification
- Evaluation of danger from waste sites
- Evaluation of damages to natural resources
- Monitoring of release of hazardous substances from sites
- Removal or cleanup of wastes by responsible parties or government

CERCLA was extended for 5 years by the passage of the Superfund Amendments and Reauthorization Act (SARA) of 1986, legislation with greatly increased scope and $8.5 billion in funding. Actually longer than CERCLA, SARA has the following important objectives and provisions:

- Five-fold increase in funding to $8.5 billion for five years
- Alternatives to land disposal that favor permanent solutions reducing volume, mobility, and toxicity of wastes
- Increased emphasis upon public health, research, training, and state and citizen involvement
- Codification of regulations that had been policy under CERCLA
- Mandatory schedules and goals over the lifetime of the legislation
- New procedures and authorities for enforcement
- A new program for leaking underground (petroleum) storage tanks

17.2 CLASSIFICATION OF HAZARDOUS SUBSTANCES AND WASTES

Many specific chemicals in widespread use are hazardous because of their chemical reactivities, fire hazards, toxicities, and other properties.[5,6] There are numerous kinds of hazardous substances, usually consisting of mixtures of specific chemicals. These include the following:

- **Explosives**, such as dynamite, or ammunition
- **Compressed gases** such as hydrogen and sulfur dioxide
- **Flammable liquids**, such as gasoline and aluminum alkyls
- **Flammable solids** such as magnesium metal, sodium hydride, and calcium carbide that burn readily, are water-reactive, or spontaneously combustible
- **Oxidizing materials**, such as lithium peroxide, that supply oxygen for the combustion of normally nonflammable materials
- **Corrosive materials**, including oleum, sulfuric acid and caustic soda, which may wound exposed flesh or cause disintegration of metal containers
- **Poisonous materials**, such as hydrocyanic acid or aniline
- **Etiologic agents**, including causative agents of anthrax, botulism, or tetanus
- **Radioactive materials**, including plutonium, cobalt-60, and uranium hexafluoride.

Characteristics and Listed Wastes

For regulatory and legal purposes in the U.S. hazardous substances are listed specifically and are defined according to general characteristics. Under the authority of the Resource Conservation and Recovery Act (RCRA) the United States Environmental Protection Agency (EPA) defines hazardous substances in terms of the following **characteristics:**[7]

- **Ignitability,** characteristic of substances that are liquids whose vapors are likely to ignite in the presence of ignition sources, nonliquids that may catch fire from friction or contact with water and which burn vigorously or persistently, ignitable compressed gases, and oxidizers.

- **Corrosivity** characteristic of substances that exhibit extremes of acidity or basicity or a tendency to corrode steel.
- **Reactivity** characteristic of substances that have a tendency to undergo violent chemical change (an explosive substance is an obvious example).
- **Toxicity** defined in terms of a standard extraction procedure followed by chemical analysis for specific substances.

In addition to classification by characteristics, EPA designates more than 450 **listed wastes** which are specific substances or classes of substances known to be hazardous. Each such substance is assigned an EPA **hazardous waste number** in the format of a letter followed by 3 numerals, where a different letter is assigned to substances from each of the four following lists:

- **F-type wastes from non-specific sources:** For example, quenching waste water treatment sludges from metal heat treating operations where cyanides are used in the process (F012).
- **K-type wastes from specific sources:** For example, heavy ends from the distillation of ethylene dichloride in ethylene dichloride production (K019).
- **P-type acute hazardous wastes:** These are mostly specific chemical species such as fluorine (P056) or 3-chloropropane nitrile (P027).
- **U-Type generally hazardous wastes:** These are predominantly specific compounds such as calcium chromate (U032) or phthalic anhydride (U190).

Compared to RCRA, CERCLA gives a rather broad definition of hazardous substances that includes the following:

- Any element, compound, mixture, solution, or substance, release of which may substantially endanger public health, public welfare, or the environment.
- Any element, compound, mixture, solution, or substance in reportable quantities designated by CERCLA Section 102.
- Certain substances or toxic pollutants designated by the Federal Water Pollution Control Act.
- Any hazardous air pollutant listed under Section 112 of the Clean Air Act.
- Any imminently hazardous chemical substance or mixture that has been the subject of government action under Section 7 of the Toxic Substances Control Act (TSCA).
- With the exception of those suspended by Congress under the Solid Waste Disposal Act, any hazardous waste listed or having characteristics identified by RCRA § 3001.

Hazardous Wastes

Having defined *hazardous substance* in some detail above, it is now appropriate to go into greater detail regarding the meaning of *hazardous waste*. Three basic approaches to defining hazardous wastes are (1) a qualitative description by origin, type, and constituents, (2) classification by characteristics largely based upon testing procedures; and (3) by means of concentrations of specific hazardous substances. Wastes may be classified by general type such as "spent halogenated solvents" or by industrial sources such as "pickling liquor from steel manufacturing."

Various countries have different definitions of hazardous waste.[8] For example, The Federal Republic of Germany Federal Act on Disposal of Waste (1972, as amended, 1976) mentions special wastes that are, ".....especially hazardous to human health, air, or water, or which are explosive, flammable, or may cause diseases." The United Kingdom's Deposit of Poisonous Waste Act (1972) refers to waste, "....of a kind which is poisonous, noxious, or polluting and whose presence on the land is liable to give rise to an environmental hazard." The Ontario Waste Management Corporation, a provincial crown agency created by the Ontario, Canada, Legislature, defines **special waste** as liquid industrial and hazardous wastes that is unsuitable for treatment or disposal in municipal wastewater treatment systems, incinerators or landfills and that therefore requires special treatment.[9]

Radioactive wastes are a problem for any countries with a significant nuclear power industry. In the U.S., such wastes are regulated under the Nuclear Regulatory Commission (NRC) and Department of Energy (DOE). Special problems are posed by **mixed waste** containing both radioactive and chemical wastes.[10]

Hazardous Wastes and Air and Water Pollution Control

Somewhat paradoxically, measures taken to reduce air and water pollution (Figure 17.1) have had a tendency to increase production of hazardous wastes. Most

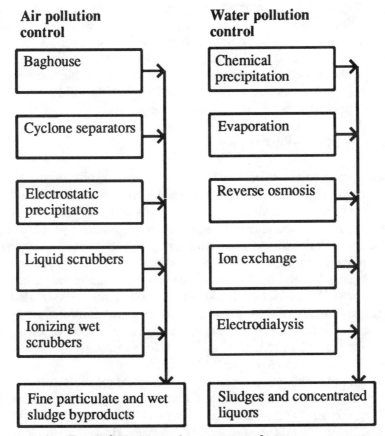

Figure 17.1. Potential contributions of air and water pollution control measures to hazardous wastes production.

water treatment processes yield sludges or concentrated liquors that require stabilization and disposal. Air scrubbing processes likewise produce sludges. Baghouses and precipitators used to control air pollution all yield significant quantities of solids, some of which are hazardous.

17.3. ORIGIN AND AMOUNTS OF WASTES

This section deals with the kinds and amounts of both hazardous and nonhazardous wastes. In a nonregulatory sense there is no sharp demarcation between hazardous and nonhazardous wastes. Some wastes such as soluble toxic heavy metal salt wastes are obviously hazardous. By comparison, discarded leaves and tree trimmings, would be regarded as posing no danger. But, if properly treated and immobilized, the heavy metal wastes are of little danger, whereas discarded tree limbs pose a fire hazard under certain circumstances. Materials that by themselves are nonhazardous may interact with hazardous substances to increase the dangers from the latter. For example, soluble humic substances from the decay of tree leaves may solubilize and transport heavy metal ions.[11]

Staggering amounts of wastes of all kinds are produced by human activities. Such wastes include municipal refuse, sewage sludge, agricultural residues, and toxic, chemically reactive byproducts of manufacturing processes.

An idea of quantities of solid wastes generated can be obtained by considering mining and milling wastes.[12] The quantities of such wastes are enormous because large quantities of rock must be removed to get to the ore and because the metal or other economically valuable constituent is usually a small percentage of the ore. Therefore, byproducts such as overburden and beneficiation wastes accumulate in vast amounts. Mining wastes make up somewhat less than half of solid wastes generated in the U.S. in quantities of about 2 billion metric tons per year.

Nonhazardous Solid Wastes

It is appropriate to consider "nonhazardous" waste (solid waste, the municipal refuse and garbage produced by human activities) along with hazardous waste because it may not be nonhazardous in all cases and situations and may interact with hazardous wastes. Furthermore, the amounts of solid waste produced each year are so enormous that our capacity to deal with the problem is under severe strain.[13] Disposal of about 92 percent of municipal refuse in the U.S. is in landfills. However, as total quantities of solid waste have increased, the landfill capacity to handle waste has decreased. When the original RCRA act was passed in 1976 about 30,000 landfills were operating (although many of these were little more than "dumping grounds"). As of 1988 the number of municipal landfills in operation had dropped to about 6,500. As a result, some cities have had to haul municipal refuse as far as 70 miles or more and disposal costs in some areas have reached $75/ton.

The potential of incineration for handling municipal refuse is very high because it can reduce waste mass by 75 percent and volume by 90 percent. However, environmental concern about organic pollutants (particularly dioxins) in stack emissions and heavy metals in incinerator ash have slowed municipal incinerator development in the U.S.

Recycling can certainly reduce quantities of solid waste, perhaps as much as 50 percent, but it is not the panacea claimed by its most avid advocates. The overall solution to the solid waste problem must involve several kinds of measures, particularly (1) reduction of wastes at the source, (2) recycling as much waste as is practical, (3) reducing the volume of remaining wastes by measures such as incineration, (4) treating residual material as much as possible to render it nonleachable and innocuous, and (5) placing the residual material in landfills, properly protected from leaching or release by other pathways.

Origin and Amounts of Hazardous Wastes

Quantities of hazardous wastes produced each year are not known with certainty and depend upon the definitions used for such materials. In 1988 the figure for RCRA-regulated wastes in the U.S. was placed at 290 million tons. However, most of this material is water, with only a few million tons consisting of solids. Some high-water-content wastes are generated directly by processes that require large quantities of water in waste treatment and other aqueous wastes are produced by mixing hazardous wastes with wastewater.

Some wastes that might exhibit a degree of hazard are exempt from RCRA regulation by legislation. These exempt wastes include the following:

- Fuel ash and scrubber sludge from power generation by utilities
- Oil and gas drilling muds
- Byproduct brine from petroleum production
- Cement kiln dust
- Waste and sludge from phosphate mining and beneficiation
- Mining wastes from uranium and other minerals

Eventual reclassification of these kinds of low-hazard wastes could increase the quantities of RCRA-regulated wastes several fold. One problem in dealing with hazardous wastes is the lack of information about these materials. There is, in fact, a shortage of hard data that effectively quantifies the extent of the hazardous waste problem or that documents what actually happens to a large fraction of hazardous wastes.

Types of Hazardous Wastes

In terms of quantity by weight, more wastes than all others combined are those from categories designated by hazardous waste numbers preceded by F and K, respectively. The former are those from nonspecific sources and include the following examples:

- F001 The spent halogenated solvents used in degreasing: tetrachloroethylene, trichloroethylene, methylene chloride, 1,1,1,-trichloroethane, carbon tetra-chloride, and the chlorinated fluorocarbons; and sludges from the recovery of these solvents in degreasing operations
- F004 The spent non-halogenated solvents: Cresols, cresylic acid, and nitro benzene; and still bottoms from the recovery of these solvents
- F007 Spent plating-bath solutions from electroplating operations
- F010 Quenching-bath sludge from oil baths from metal heat treating operations

The "K-type" hazardous wastes are those from specific sources produced by industries such as the manufacture of inorganic pigments, organic chemicals, pesticides, explosives, iron and steel, and nonferrous metals, and from processes such as petroleum refining or wood preservation; some examples are given below:

- K001 Bottoms sediment sludge from the treatment of wastewaters from wood-preserving processes that use creosote and/or pentachlorophenol
- K002 Wastewater treatment sludge from the production of chrome yellow and orange pigments
- K020 Heavy ends (residue) from the distillation of vinyl chloride in vinyl chloride monomer production
- K043 2,6-Dichlorophenol waste from the production of 2,4-D
- K047 Pink/red water from TNT operations
- K049 Slop oil emulsion solids from the petroleum refining industry
- K060 Ammonia lime still sludge from coking operations
- K067 Electrolytic anode slimes/sludges from primary zinc production

The second largest category of wastes generated are reactive wastes, followed by corrosive wastes and toxic wastes. About 1% of wastes are designated as ignitable and another 1% are type "P" wastes (discarded commercial chemical products, off-specification species, containers, and spill residues) or "U" wastes (chemical intermediates from the manufacture of commercial chemicals or chemical formulations). Several percent of wastes are unspecified types.

Hazardous Waste Generators

About 650,000 companies generate hazardous wastes in the U. S., but most of these generators produce only small quantities. About 99% of hazardous wastes come from only about 2% of the generators. Hazardous waste generators are unevenly distributed geographically across the continental U.S., with a relatively large number located in the industrialized upper midwest, including the states of Illinois, Indiana, Ohio, Michigan, and Wisconsin.

Distribution of Quantities of Hazardous Wastes

Industry types of hazardous waste generators can be divided among the 7 following major categories, each containing of the order of 10-20 percent of hazardous waste generators: chemicals and allied products manufacture, petroleum-related industries, fabricated metals, metal-related products, electrical equipment manufacture constitutes another class of generators, "all other manufacturing," and nonmanufacturing and nonspecified generators. About 10% of the generators produce more than 95% of all hazardous wastes. Whereas, as noted above, the number of hazardous waste generators is distributed relatively evenly among several major types of industries, 70-85% of the <u>quantities</u> of hazardous wastes are generated by the chemical and petroleum industries. Of the remainder, about 3/4 comes from metal-related industries and about 1/4 from all other industries.

17.4. FLAMMABLE AND COMBUSTIBLE SUBSTANCES

In a broad sense a **flammable substance** is something that will burn readily, whereas a **combustible substance** requires relatively more persuasion to burn. Before trying to sort out these definitions it is necessary to define several other terms. Most chemicals that are likely to burn accidentally are liquids. Liquids form **vapors** which are usually more dense than air, and thus tend to settle. The tendency of a liquid to ignite is measured by a test in which the liquid is heated and periodically exposed to a flame until the mixture of vapor and air ignites at the liquid's surface. The temperature at which this occurs is called the **flash point.**

With these definitions in mind it is possible to divide ignitable materials into four major classes. A **flammable solid** is one that can ignite from friction or from heat remaining from its manufacture, or which may cause a serious hazard if ignited. Explosive materials are not included in this classification. A **flammable liquid** is one having a flash point below 37.8°C (100°F). A **combustible liquid** has a flash point in excess of 37.8°C, but below 93.3°C. Gases are substances that exist entirely in the gaseous phase at 0°C and 1 atm pressure. A **flammable compressed gas** meets specified criteria for lower flammability limit, flammability range (see below), and flame projection. Especially in the case of liquids, there are several subclassifications of flammability and combustibility such as those of the U.S. Department of Transportation and the National Fire Protection Association.[14]

In considering the ignition of vapors, two important concepts are those of flammability limit and flammability range. Values of the vapor/air ratio below which ignition cannot occur because of insufficient fuel define the lower **flammability limit.** Similarly, values of the vapor/air ratio above which ignition cannot occur because of insufficient air define the upper flammability limit. The difference between upper and lower flammability limits at a specified temperature is the **flammability range.** Table 17.1 gives some examples of these values for common liquid chemicals.

Table 17.1. Flammabilities of Some Common Organic Liquids

		Volume percent in air	
Liquid	Flash point (°C)[a]	LFL[b]	UFL[b]
Diethyl ether	-43	1.9	36
Pentane	-40	1.5	7.8
Acetone	-20	2.6	13
Toluene	4	1.27	7.1
Methanol	12	6.0	37
Gasoline (2,2,4-tri-methylpentane)	---	1.4	7.6
Naphthalene	157	0.9	5.9

[a] Closed-cup flash point test

[b] LFL, lower flammability limit; UFL, upper flammability limit at 25°C.

The percentage of flammable substance for best combustion (most explosive mixture) is labelled "optimal."[15] In the case of acetone, for example, the optimal flammable mixture is 5.0% acetone.

One of the more disastrous problems that can occur with flammable liquids is a boiling liquid expanding vapor explosion, BLEVE. These are caused by rapid pressure buildup in closed containers of flammable liquids heated by an external source. The explosion occurs when the pressure buildup is sufficient to break the container walls.

Combustion of Finely Divided Particles

Finely divided particles of combustible materials are somewhat analogous to vapors in respect to flammability. One such example is a spray or mist of hydrocarbon liquid in which oxygen has the opportunity for intimate contact with the liquid particles. In this case the liquid may ignite at a temperature below its flash point.

Dust explosions can occur with a large variety of solids that have been ground to a finely divided state. Many metal dusts, particularly those of magnesium and its alloys, zirconium, titanium, and aluminum can burn explosively in air. In the case of aluminum, for example, the reaction is the following:

$$4Al(powder) + 3O_2(from\ air) \rightarrow 2Al_2O_3 \qquad (17.4.1)$$

Coal dust and grain dusts have caused many fatal fires and explosions in coal mines and grain elevators, respectively. Dusts of polymers such as cellulose acetate, polyethylene, and polystyrene can also be explosive.

Oxidizers

Combustible substances are reducing agents that react with **oxidizers** (oxidizing agents or oxidants) to produce heat. Diatomic oxygen, O_2, from air is the most common oxidizer. Many oxidizers are chemical compounds that contain oxygen in their formulas. The halogens (periodic table group 7A) and many of their compounds are oxidizers. Some examples of oxidizers are given in Table 17.2.

Table 17.2. Examples of Some Oxidizers

Name	Formula	State of matter
Ammonium nitrate	NH_4NO_3	Solid
Ammonium perchlorate	NH_4ClO_4	Solid
Bromine	Br_2	Liquid
Chlorine	Cl_2	Gas (stored as liquid)
Fluorine	F_2	Gas
Hydrogen peroxide	H_2O_2	Solution in water
Nitric acid	HNO_3	Concentrated solution
Nitrous oxide	N_2O	Gas (stored as liquid)
Ozone	O_3	Gas
Perchloric acid	$HClO_4$	Concentrated solution
Potassium permanganate	$KMnO_4$	Solid
Sodium dichromate	$Na_2Cr_2O_7$	Solid

An example of a reaction of an oxidizer is that of concentrated HNO_3 with copper metal, which gives toxic NO_2 gas as a product:

$$4HNO_3 + Cu \rightarrow Cu(NO_3)_2 + 2H_2O + 2NO_2 \qquad (17.4.2)$$

The toxic effects of some oxidizers are due to their ability to oxidize biomolecules in living systems.

Whether or not a substance acts as an oxidizer depends upon the reducing strength of the material that it contacts. For example, carbon dioxide is a common fire extinguishing material that can be sprayed onto a burning substance to keep air away. However, aluminum is such a strong reducing agent that carbon dioxide in contact with hot, burning aluminum reacts as an oxidizing agent to give off toxic combustible carbon monoxide gas:

$$2Al + 3CO_2 \rightarrow Al_2O_3 + 3CO \qquad (17.4.3)$$

Oxidizers can contribute strongly to fire hazards because fuels may burn explosively in contact with an oxidizer.

Spontaneous Ignition

Substances that catch fire spontaneously in air without an ignition source are called **pyrophoric**. These include several elements—white phosphorus, the alkali metals (group 1A), and powdered forms of magnesium, calcium, cobalt, manganese, iron, zirconium and aluminum. Also included are some organometallic compounds, such as lithium ethyl (LiC_2H_4) and lithium phenyl (LiC_6H_5), and some metal carbonyl compounds, such as iron pentacarbonyl, $Fe(CO)_5$. Another major class of pyrophoric compounds consists of metal and metalloid hydrides, including lithium hydride, LiH; pentaborane, B_5H_9; and arsine, AsH_3. Moisture in air is often a factor in spontaneous ignition. For example, lithium hydride undergoes the following reaction with water from moist air:

$$2LiH + H_2O \rightarrow 2LiOH + H_2 + heat \qquad (17.4.4)$$

The heat generated from this reaction can be sufficient to ignite the hydride so that it burns in air:

$$2LiH + O_2 \rightarrow Li_2O + H_2O \qquad (17.4.5)$$

Some compounds with organometallic character are also pyrophoric. An example of such a compound is diethylethoxyaluminum:

$$\begin{array}{c} H_5C_2 \diagdown \\ \qquad\quad Al-OC_2H_5 \quad \text{Diethylethoxyaluminum} \\ H_5C_2 \diagup \end{array}$$

Many mixtures of oxidizers and oxidizable chemicals catch fire spontaneously and are called **hypergolic mixtures**. Nitric acid and phenol form such a mixture.

Toxic Products of Combustion

Some of the greater dangers of fires are from toxic products and byproducts of combustion. The most obvious of these is carbon monoxide, CO, which can cause serious illness or death because it forms carboxyhemoglobin with hemoglobin in the blood so that the blood no longer carries oxygen to body tissues. Toxic SO_2, P_4O_{10}, and HCl are formed by the combustion of sulfur, phosphorus, and organochloride compounds, respectively. A large number of noxious organic compounds such as aldehydes are generated as byproducts of combustion. In addition to forming carbon monoxide, combustion under oxygen-deficient conditions produces polycyclic aromatic hydrocarbons consisting of fused ring structures. Some of these compounds, such as benzo(a)pyrene, below, are precarcinogens that are acted upon by enzymes in the body to yield cancer-producing metabolites.

Benzo(a)pyrene

17.5. REACTIVE SUBSTANCES

Reactive substances are those that tend to undergo rapid or violent reactions under certain conditions. Such substances include those that react violently or form potentially explosive mixtures with water. An example is sodium metal which reacts strongly with water as follows:

$$2Na \ + \ 2H_2O \ \rightarrow \ 2NaOH \ + \ H_2 \ + \ heat \qquad\qquad (17.5.1)$$

This reaction usually generates enough heat to ignite the sodium. Explosives constitute another class of reactive substances. For regulatory purposes substances are also classified as reactive that react with water, acid, or base to produce toxic fumes, particularly those of hydrogen sulfide or hydrogen cyanide.

Heat and temperature are usually very important factors in reactivity. Many reactions require energy of activation to get them started. The rates of most reactions tend to increase sharply with increasing temperature and most chemical reactions give off heat. Therefore, once a reaction is started in a reactive mixture lacking an effective means of heat dissipation, the rate may increase exponentially with time, leading to an uncontrollable event. Other factors that may affect reaction rate include physical form of reactants (for example, a finely divided metal powder that reacts explosively with oxygen, whereas a single mass of metal barely reacts), rate and degree of mixing of reactants, degree of dilution with nonreactive media (solvent), presence of a catalyst, and pressure.

Some chemical compounds are self-reactive, in that they contain oxidant and reductant in the same compound. Nitroglycerin, a strong explosive with the formula $C_3H_5(ONO_2)_3$ decomposes spontaneously to CO_2, H_2O, O_2, and N_2 with a rapid release of a very high amount of energy. Pure nitroglycerin has such a high inherent instability that only a slight blow may be sufficient to detonate it. Trinitrotoluene (TNT) is also an explosive with a high degree of reactivity. However, it is inherently relatively stable in that some sort of detonating device is required to cause it to explode.

Chemical Structure and Reactivity

As shown in Table 17.3, some chemical structures are associated with high reactivity.[16] High reactivity in some organic compounds

Table 17.3. Examples of Reactive Compounds and Structures

Name	Structure or formula
Organic	
Allenes	C=C=C
Dienes	C=C–C=C
Azo compounds	C=N–N=C
Triazenes	C–N=N–N
Hydroperoxides	R–OOH
Peroxides	R–OO–R'
Alkyl nitrates	R–O–NO$_2$
Nitro compounds	R–NO$_2$
Inorganic	
Nitrous oxide	N$_2$O
Nitrogen halides	NCl$_3$, NI$_3$
Interhalogen compounds	BrCl
Halogen oxides	ClO$_2$
Halogen azides	ClN$_3$
Hypohalites	NaClO

results from unsaturated bonds in the carbon skeleton, particularly where multiple bonds are adjacent (allenes, C=C=C) or separated by only one carbon-carbon single bond (dienes, C=C-C=C). Some organic structures involving oxygen are very reactive. Examples are oxiranes, such as ethylene oxide,

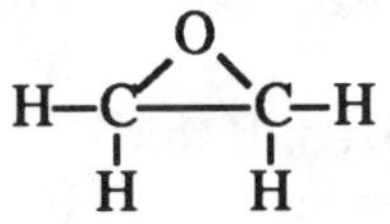

$$H-\underset{\underset{H}{|}}{C}\overset{O}{\diagup\diagdown}\underset{\underset{H}{|}}{C}-H \quad \text{Ethylene oxide}$$

hydroperoxides (ROOH), and peroxides (ROOR'), where R and R' stand for hydrocarbon moieties such as the methyl group, -CH$_3$. Many organic compounds containing nitrogen along with carbon and hydrogen are very reactive. Included are triazenes (R-N=N-N), some azo compounds (R-N=N-R'), and some nitriles:

$$R-C≡N \quad \text{Nitrile}$$

Functional groups containing both oxygen and nitrogen tend to impart reactivity to an organic compound. Examples are alkyl nitrates ($R-O-NO_2$), alkyl nitrites ($R-O-N=O$), nitroso compounds ($R-N=O$), and nitro compounds ($R-NO_2$).

Many different classes of inorganic compounds are reactive. These include some of the halogen compounds of nitrogen (shock-sensitive nitrogen triiodide, NI_3, is an outstanding example), compounds with metal-nitrogen bonds, halogen oxides (ClO_2), and compounds with oxyanions of the halogens. An example of the last group of compounds is ammonium perchlorate, NH_4ClO_4, which was involved in a series of massive explosions that destroyed 8 million lb of the compound and demolished a 40 million lb/year U.S. rocket fuel plant near Henderson, Nevada, in 1988. (By late 1989 a new $92 million plant for the manufacture of ammonium perchlorate had been constructed near Cedar City in a remote region of southwest Utah.[17] Prudently, the buildings at the new plant have been placed at large distances from each other!)

Explosives such as nitroglycerin or TNT that are single compounds containing both oxidizing and reducing functions in the same molecule are called **redox compounds**. Some redox compounds have more oxygen than is needed for a complete reaction and are said to have a positive balance of oxygen, some have exactly the stoichiometric quantity of oxygen required (zero balance, maximum energy release), and others have a negative balance and require oxygen from outside sources to completely oxidize all components. Trinitrotoluene has a substantial negative balance; ammonium dichromate (($NH_4)_2Cr_2O_7$) has a zero balance, reacting with exact stoichiometry to H_2O, N_2, and Cr_2O_3; and nitroglycerin has a positive balance as shown by the following reaction:

$$4C_3H_5N_3O_9 \rightarrow 12CO_2 + 10H_2O + 6N_2 + O_2 \qquad (17.5.2)$$

17.6. CORROSIVE SUBSTANCES

Conventionally, **corrosive substances** are regarded as those that dissolve metals or cause oxidized material to form on the surface of metals—rusted iron is a prime example. In a broader sense corrosives cause deterioration of materials, including living tissue, that they contact.[18] Most corrosives belong to at least one of the four following chemical classes:[19]   (1) strong acids, (2) strong bases, (3) oxidants, (4) dehydrating agents. Table 17.4 lists some of the major corrosive substances and their effects.

Sulfuric Acid

Sulfuric acid is a prime example of a corrosive substance. As well as being a strong acid, concentrated sulfuric acid is also a dehydrating agent and oxidant. The tremendous affinity of H_2SO_4 for water is illustrated by the heat generated when water and concentrated sulfuric acid are mixed. If this is done incorrectly by adding water to the acid, localized boiling and spattering can occur that result in personal injury. The major destructive effect of sulfuric acid on skin tissue is removal of water with accompanying release of heat. Sulfuric acid decomposes carbohydrates by removal of water. In contact with sugar, for example, concentrated sulfuric acid reacts to leave a charred mass. The reaction is

$$C_{12}H_{22}O_{11} \xrightarrow{H_2SO_4} 11H_2O(H_2SO_4) + 12C + heat \qquad (17.6.1)$$
Dextrose sugar

Table 17.4. Examples of Some Corrosive Substances

Name and formula	Properties and effects
Nitric acid, HNO_3	Strong acid and strong oxidizer, corrodes metal, reacts with protein in tissue to form yellow xanthoproteic acid, lesions are slow to heal
Hydrochloric acid, HCl	Strong acid, corrodes metals, gives off HCl gas vapor, which can damage respiratory tract tissue
Hydrofluoric acid, HF	Corrodes metals, dissolves glass, causes particularly bad burns to flesh
Alkali metal hydroxides,	Strong bases, corrode zinc, lead, and NaOH and KOH aluminum, caustic substances that dissolve tissue and cause severe burns
Hydrogen peroxide,	Oxidizer, all but very dilute solutions H_2O_2 cause severe burns
Interhalogen compounds such as ClF, BrF_3	Powerful corrosive irritants that acidify, oxidize, and dehydrate tissue
Halogen oxides such as OF_2, Cl_2O, Cl_2O_7	Powerful corrosive irritants that acidify, oxidize, and dehydrate tissue
Elemental fluorine, chlorine, bromine (F_2, Cl_2, Br_2,)	Very corrosive to mucous membranes and moist tissue, strong irritants

Some dehydration reactions of sulfuric acid can be very vigorous. For example, the reaction with perchloric acid produces unstable Cl_2O_7, and a violent explosion can result. Concentrated sulfuric acid produces dangerous or toxic products with a number of other substances, such as toxic carbon monoxide (CO) from reaction with oxalic acid, $H_2C_2O_4$; toxic bromine and sulfur dioxide (Br_2, SO_2) from reaction with sodium bromide, NaBr; and toxic, unstable chlorine dioxide (ClO_2) from reaction with sodium chlorate, $NaClO_3$.

Contact of sulfuric acid with tissue results in tissue destruction at the point of contact. A severe burn results, which may be difficult to heal. Inhalation of sulfuric acid fumes or mists damages tissues in the upper respiratory tract and eyes. Long term exposure to sulfuric acid fumes or mists has caused erosion of teeth!

17.7. TOXIC SUBSTANCES

Toxicity is of the utmost concern in dealing with hazardous substances. This includes both long term chronic effects from continual or periodic exposures to low levels of toxicants and acute effects from a single large exposure. Toxic substances are covered in greater detail in Chapter 20.

Toxicity Characteristic Leaching Procedure

For regulatory and remediation purposes a standard test is needed to measure the likleihood of toxic substances getting into the environment and causing harm to organisms. The test required by the U.S. EPA is the **Toxicity Characteristic Leaching Procedure** (TCLP) designed to determine the mobility of both organic and inorganic contaminants present in liquid, solid, and multiphasic wastes. For analysis of toxic species a solution is leached from the waste and is designated as the TCLP extract. If no significant solid material is present, the waste is filtered through a 0.6-0.8 µm glass fiber filter and designated as the TCLP extract. In mixed liquid-solid wastes the liquid is separated and analyzed separately. Solid wastes to be extracted are required to have a surface area per gram of material equal to or greater than 3.1 cm^2 or to consist of particles smaller than 1 cm in their most narrow dimension. The kind of extraction fluid used on the solids is determined from the pH of a mixture of 5 g of the solids (reduced to approximately 1 mm in size if necessary) shaken vigorously with 96.5 mL of water. If the pH of the water after mixing is less than 5.0, the extraction fluid used is an acetic acid/sodium acetate buffer of pH 4.93±0.05 and if the pH is greater than 5.0, the extraction fluid is a dilute acetic acid solution with a pH of 2.88±0.05. After the fluid to be used is determined, an amount of extraction fluid equal to 20 times the mass of the solid is used for the extraction which is carried out for 18 hours in a sealed container held on a device that rotates it end-over-end for 18 hours. After the TCLP extract is separated from the solids it is analyzed for 39 specified volatile organic compounds, semivolatile organic compounds, and metals to determine if the waste exceeds specified levels of these contaminants.[20]

17.8. CHEMICAL CLASSES OF HAZARDOUS SUBSTANCES

Another way of viewing hazardous substances in the context of their chemical properties is to divide them into classes of chemicals. That is done briefly in this section.

A number of **elements** are used industrially in their elemental forms, in many cases for chemical synthesis. Some of these elements pose hazards of flammability, corrosivity, reactivity, or toxicity. Elemental hydrogen, H_2, is extremely flammable and forms explosive mixtures with air. Three of the halogens, fluorine, chlorine, and bromine, are widely produced as elemental F_2, Cl_2, and Br_2, respectively. Fluorine is the strongest elemental oxidant and extremely reactive. It is very corrosive to the skin and inhalation of F_2 can cause severe lung damage. Chlorine, one of the most widely produced industrial chemicals, is a reactive oxidant that forms acid in water and is a corrosive poison to tissue, especially in the respiratory tract. Bromine is a volatile brown liquid which is corrosive to skin in both the liquid and vapor form. Elemental white phosphorus is a reactive substance that may catch fire spontaneously in air. It is a systemic poison. Elemental lithium, sodium, and potassium react with a large number of chemicals and burn readily to give off caustic oxide and hydroxide fumes. Elemental mercury vapor is especially toxic by inhalation. Some metals, commonly known as heavy metals, are particularly toxic in their chemically combined forms. These include lead, cadmium, mercury, beryllium, and arsenic.

Many **inorganic compounds** are hazardous because of reactivity (NH_4ClO_4), corrosivity (HNO_3), and toxicity (KCN). Many **organometallic compounds**, which have a metal atom or metalloid atom (such as silicon or arsenic) bonded directly to carbon in a hydrocarbon group or in carbon monoxide, CO, are volatile, reactive, and toxic.

Organic Compounds

There are millions of known **organic compounds**, most of which can be hazardous in some way and to some degree. Most organic compounds can be divided among hydrocarbons, oxygen-containing compounds, nitrogen-containing compounds, organohalides, sulfur-containing compounds, phosphorus-containing compounds, or combinations thereof.

17.9. PHYSICAL FORMS AND SEGREGATION OF WASTES

Three major categories of wastes based upon their physical forms are **organic materials, aqueous wastes,** and **sludges**. These forms largely determine the course of action taken in treating and disposing of the wastes. The **level of segregation,** a concept illustrated in Figure 17.2, is very important in treating, storing, and disposing

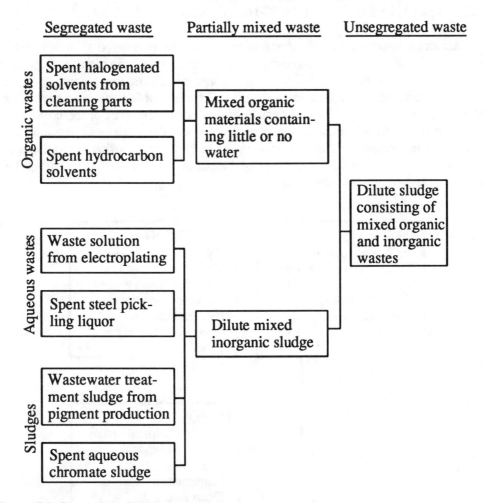

Figure 17.2. Illustration of waste segregation.

of different kinds of wastes. It is relatively easy to deal with wastes that are not mixed with other kinds of wastes, that is, those that are highly segregated. For example, spent hydrocarbon solvents can be used as fuel in boilers. However, if these solvents are mixed with spent organochloride solvents, the production of contaminant hydrogen chloride during combustion may prevent fuel use and require disposal in special hazardous waste incinerators. Further mixing with inorganic sludges adds mineral matter and water. These impurities complicate the treatment processes required by producing mineral ash in incineration or lowering the heating value of the material incinerated because of the presence of water. Among the most difficult types of wastes to handle and treat are those with the least segregation, of which a "worst case scenario" would be "dilute sludge consisting of mixed organic and inorganic wastes," as shown in Figure 17.2.

Concentration of wastes is an important factor in their management. A waste that has been concentrated or preferably never diluted is generally much easier and more economical to handle than one that is dispersed in a large quantity of water or soil. Dealing with hazardous wastes is greatly facilitated when the original quantities of wastes are minimized and the wastes remain separated and concentrated insofar as possible.

17.10. GENERATION, TREATMENT, AND DISPOSAL

Hazardous waste **management** refers to a carefully organized system in which wastes go through appropriate pathways to their ultimate elimination or disposal in ways that protect human health and the environment. The management of hazards posed by hazardous substances and wastes is a crucial part of the operation of any modern chemical industry.[21] It is a significant and increasing part of the cost of any business dealing with chemical products and processes. Personnel working with such products and processes must have a good understanding of hazardous substances and hazardous wastes. Three main aspects of hazardous waste management involve **generation**, **treatment**, and **disposal**[22] as illustrated in Figure 17.3.

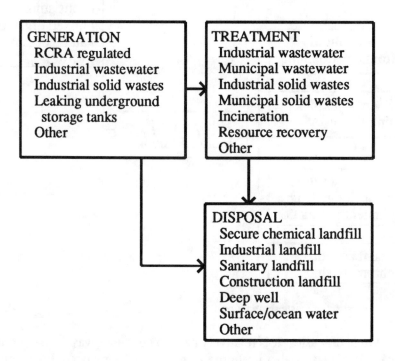

Figure 17.3. System of generation, treatment, and disposal of hazardous wastes.

The **effectiveness** of a hazardous waste system is a measure of how well it reduces the quantities and hazards of wastes, ideally approaching zero for both.[23] In decreasing order of effectiveness the options for handling hazardous wastes are the following:

• Measures that prevent generation of wastes
• Recovery and recycle of waste constituents
• Destruction and treatment, conversion to non-hazardous waste forms
• Disposal (storage, landfill)

Treatment, Storage, and Disposal Facilities

A crucial part of the regulation of hazardous wastes in the United States pertains to **treatment, storage, and disposal facilities** (TSDF). Treatment alters the physical, chemical, or biological character or composition of a waste to make it safer. Storage refers to the holding of hazardous wastes for a temporary period pending treatment or disposal. Disposal refers to the ultimate fate of hazardous substances or their treatment products.

Waste Reduction and Waste Minimization

Many hazardous waste problems can be avoided at early stages by **waste reduction**[24] and **waste minimization**. As these terms are most commonly used, waste reduction refers to source reduction—less waste-producing materials in, less waste out. Waste minimization can include treatment processes, such as incineration, which reduce the quantities of wastes for which ultimate disposal is required. Waste reduction and minimization are further addressed in Chapter 19.

Waste Treatment

Waste treatment is the major topic of Chapter 19 and is addressed briefly here. An overall scheme for the treatment of hazardous wastes is shown in Figure 17.4.

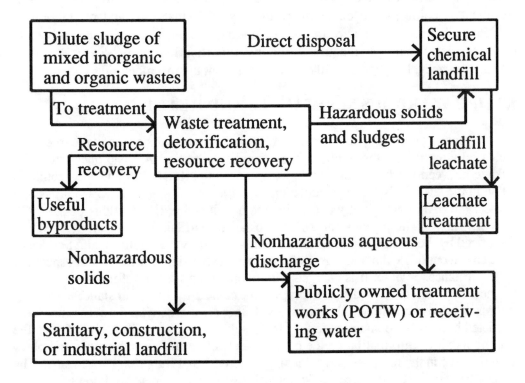

Figure 17.4. Treatment options for mixed hazardous wastes.

Under the category of treatment it is necessary to consider both municipal wastewater and municipal solid wastes along with hazardous wastes. The goal of many industrial wastewater and sludge treatment processes is to produce an effluent that meets standards for release to a municipal wastewater treatment plant (publicly owned treatment works, POTW, typically an activated sludge sewage treatment plant, see Section 8.4) and in some cases to produce solids that can be codisposed with municipal solid wastes. Incineration of muncipal solid wastes may produce some solids, particularly fly ash, that have to be treated as hazardous.

The scheme outlined in Figure 17.4 may serve as a frame of reference for subsequent discussions of waste treatment.The many options for the treatment of hazardous wastes are discussed in more detail later in Chapter 19. The ideal treatment process reduces the quantity of hazardous waste material to a small fraction of the original amount and converts it to a non-hazardous form. However, most treatment processes yield material, such as sludge from wastewater treatment or incinerator ash, which requires disposal and which may be hazardous to some extent.

Direct disposal of minimally treated hazardous wastes is becoming more severely limited with new regulations coming from the Hazardous and Solid Waste Amendments of 1984 (HSWA). Under its "land-ban" rules, this act prohibits the land disposal of more than 400 chemicals and waste streams unless they are treated or can be shown not to migrate during the time that they remain hazardous. The ultimate objective of these rules is to reduce the amounts of hazardous wastes generated, although quantities are expected to increase during the next decade. More emphasis in treatment is being placed on recovery of recyclable materials and production of innocuous byproducts. There are strong regulatory and economic incentives to generate less wastes in manufacturing by modification of processes, product substitution, recycling, and careful control throughout the manufacturing system.

1.7.11. HAZARDOUS SUBSTANCES AND HEALTH

In recent years the health aspects of hazardous substances have received increased attention by the public and by legislative bodies. A basic question is the linkage between the health of people living near Superfund sites and the chemicals found in the sites. This concern gained increased recognition in the U.S. with passage of the 1986 SARA act, greatly expanding the health authorities sections of the 1980 CERCLA act.[25] The Agency for Toxic Substances and Disease Registry (ATSDR), authorized by the 1980 CERCLA act and administered by the Public Health Service of the Department of Health and Human Services is responsible for the health aspects of toxic substances release. It is charged with maintaining files of information and data on the health effects and diseases potentially caused by toxic substances, keeping records of exposure to toxic substances, and listing areas where public access has been restricted because of contamination by toxic substances. In addition, ATSDR is the major conduit of information on the health effects of hazardous substances and plays an active role in the response and remediation activities at Superfund waste sites. The agency has prepared extensive Toxicological Profile documents pertaining to specific hazardous substances encountered at Superfund sites. The materials that are the subjects of these profiles are those both commonly encountered at hazardous waste sites and likely to pose substantial health hazards.

LITERATURE CITED

1. Wolbeck, Bernd, "Political Dimensions and Implications of Hazardous Waste Disposal," in *Hazardous Waste Disposal*, Lehman, John P., Ed., Plenum Press, New York, 1982, pp. 7-18.

2. Manahan, Stanley E., "Hazardous Substances and Hazardous Wastes," Chapter 1 in *Hazardous Waste Chemistry, Toxicology and Treatment*, Lewis Publishers, Inc, Chelsea, Michigan, pp. 1-26.

3. Ember, Lois, "Occidental Agrees to Store, Treat Love Canal Wastes," *Chemical and Engineering News*, June 19, 1989, pp. 20-21.

4. Nott, Sam, Caren Arnstein, Stephen Ramsey, and Maureen Crough, *Superfund Handbook*, 2nd ed., Sidley and Austin, Chicago, Illinois, 1987.

5. Weiss, G., Ed., *Hazardous Chemicals Data Book*, 2nd ed., Noyes Data Corporation, Park Ridge, New Jersey, 1986.

6. U. S. Environmental Protection Agency, *Extremely Hazardous Substances*, Noyes Data Corporation, Park Ridge, New Jersey, 1988.

7. "Identification and Listing of Hazardous Waste," *Code of Federal Regulations*, **40**, July 1, 1986, Part 261, U.S. Government Printing Office, Washington, DC, pp. 359-408.

8. Lehman, John P., "Hazardous Waste Definition and Recommended Procedures," in *Hazardous Waste Disposal*, Lehman, John P., Ed., Plenum Press, New York, 1982, pp. 45-68.

9. Monenco Limited, "OWMC Treatment Facility: Status Report, Facility Design," Ontario Waste Management Corporation, Toronto, Ontario, Canada, 1985.

10. "Summary Report of ASME Mixed Waste Workshop," The 1989 Incineration Conference, Knoxville, Tennessee, May 1, 1989.

11. Manahan, Stanley E., "Humic Substances and the Fates of Hazardous Waste Chemicals," Chapter 6 in *Influence of Aquatic Humic Substances on Fate and Treatment of Pollutants*, Advances in Chemistry Series No. 219, American Chemical Society, Washington, DC, 1989, pp. 83-92.

12. Hoye, Robert L., and S. Jackson Hubbard, "Mining Waste," Section 4.5 in *Standard Handbook of Hazardous Waste Treatment and Disposal*, Harry M. Freeman, Ed., McGraw-Hill, New York, 1989, pp. 4.47-4.51.

13. Dorian, Gerry, "Household Hazardous Waste Management," Section 4.6 in *Standard Handbook of Hazardous Waste Treatment and Disposal*, Freeman, Harry M., Ed., McGraw-Hill Book Company, New York, 1989, pp. 4.53-4.60.

14. Gerlach, Rudolph, "Flammability, Combustibility," Chapter 6 in *Improving Safety in the Chemical Laboratory: A Practical Guide*, Young, Jay A., Ed., John Wiley and Sons, New York, 1987, pp. 59-91.

15. Wray, Tom, "Explosive Limits," *Hazmat World*, June, 1989, p. 52.

16. Bretherick, Leslie, "Chemical Reactivity: Instability and Incompatible Combinations," Chapter 7 in *Improving Safety in the Chemical Laboratory: A Practical Guide*, Young, Jay A., Ed., John Wiley and Sons, New York, 1987, pp. 93-113.

17. Seltzer, Richard, "New Plant Ends Rocket Oxidizer Shortage," *Chemical and Engineering News*, October 8, 1989, p. 5.

18. Weiss, G., Ed., *Hazardous Chemicals Data Book*, 2nd ed., Noyes Data Corporation, Park Ridge, New Jersey, 1986. (See also U. S. Environmental Protection Agency, *Extremely Hazardous Substances*, Noyes Data Corporation, Park Ridge, New Jersey, 1988.)

19. Manahan, Stanley E., *Toxicological Chemistry*, Lewis Publishers, Chelsea, Michigan, 1989.

20. "Identification and Listing of Hazardous Waste," *Code of Federal Regulations*, **40**, July 1, 1986, Part 261, U.S. Government Printing Office, Washington, DC, pp. 359-408.

21. Kasperson, Roger E., Jeanne X. Kasperson, Christoph Hohenemser, and Robert W. Kates, *Corporate Management of Health and Safety Hazards: A Comparison of Current Practice*, Westview Press, Boulder, Colorado, 1988.

22. "The Hazardous Waste System," U.S. Environmental Protection Agency Office of Solid Waste and Emergency Response, Washington, D.C., 1987.

23. Andrews, Richard N. L., and Francis M. Lynn, "Siting of Hazardous Waste Facilities," Section 3.1 in *Standard Handbook of Hazardous Waste Treatment and Disposal*, Harry M. Freeman, Ed., McGraw Hill, New York, 1989, pp. 3.3 3.16.

24. Bishop, Jim, "Waste Reduction," *Hazmat World*, October, 1988, pp. 56-61.

25. Siegel, Martin R., "Agency for Toxic Substances and Disease Registry Health Related Activities," in *Hazardous Wastes, Superfund, and Toxic Substances*, American Law Institute American Bar Association Committee on Continuing Professional Education, Philadelphia, PA, 1987, pp. 19-27.24.

SUPPLEMENTAL REFERENCES

Kharbanda, O. P., and E. A. Stallworthy, *Waste Management. Towards a Sustainable Society*, Auburn House, New York, 1990.

Existing Chemicals of Environmental Relevance. Criteria and List of Chemicals. GDCH-Advisory Committee on Existing Chemicals of Environmental Relevance. VCH, New York, 1989.

Higgins, Thomas E., *Hazardous Waste Minimization Handbook*, Lewis Publishers, Inc., Chelsea, MI, 1989.

Evans, Jeffrey C., Ed.,*Toxic and Hazardous Wastes: Proceedings of the Nineteenth Mid-Atlantic Industrial Waste Conference*, Technomic Publishing, Lancaster, PA, 1987.

Extremely Hazardous Substances: Superfund Chemical Profiles, Vols. 1 & 2, U.S. Environmental Protection Agency, Noyes Publications, Park Ridge, NJ, 1989.

Bell, John M., Ed.,*Proceedings of the 42nd Industrial Waste Conference, May 12, 13, 14, 1987*, Lewis Publishers, Chelsea, MI, 1988.

Wolf, K., W. J. Van Den Brink, and F. J. Colon, *Contaminated Soil '88*. Vols 1 & 2, Kluwer Academic Publishers, Norwell, MA, 1988.

Techniques for Assessing Industrial Hazards: A Manual, Technica Ltd., World Bank, Washington, DC, 1988.

Cheremisinoff, Paul N., *Hazardous Materials Emergency Response Pocket Handbook*, Technomic Publishing, Lancaster, PA, 1989.

Harris, Christopher, and Donavee A. Berger, *SARA Title III: A Guide to Emergency Preparedness and Community Right to Know*, Executive Enterprises Publications Co., New York, NY, 1988.

Howard, Philip H., *Handbook of Environmental Fate and Exposure Data for Organic Chemicals*, Vol. 1: *Large Production and Priority Pollutants*, Lewis Publishers, Chelsea, MI, 1989.

QUESTIONS AND PROBLEMS

1. Match the following kinds of hazardous substances on the left with a specific example of each from the right, below:

1. Explosives	(a) Oleum, sulfuric acid, caustic soda
2. Compressed gases	(b) Magnesium metal, sodium hydride
3. Radioactive materials	(c) Lithium peroxide
4. Flammable solids	(d) Hydrogen, sulfur dioxide
5. Oxidizing materials	(e) Dynamite, ammunition
6. Corrosive materials	(f) Plutonium, cobalt-60

2. Of the following, the property that is **not** a member of the same group as the other properties listed is (a) substances that are liquids whose vapors are likely to ignite in the presence of ignition sources, (b) nonliquids that may catch fire from friction or contact with water and which burn vigorously or persistently, (c) ignitable compressed gases, (d) oxidizers, (e) substances that exhibit extremes of acidity or basicity.

3. In what respects may it be said that measures taken to alleviate air and water pollution tend to aggravate hazardous waste problems?

4. Place the following in descending order of desirability for dealing with wastes and discuss your rationale for doing so: (a) reducing the volume of remaining wastes by measures such as incineration,(b) placing the residual material in landfills, properly protected from leaching or release by other pathways, (c) treating residual material as much as possible to render it nonleachable and innocuous, (d) reduction of wastes at the source, (e) recycling as much waste as is practical.

5. Discuss the significance of LFL, UFL, and flammability range in determining the flammability hazards of organic liquids.

6. Concentrated HNO_3 and its reaction products pose several kinds of hazards. What are these?

7. What are substances called that catch fire spontaneously in air without an ignition source?

8. Name four or five hazardous products of combustion and specify the hazards posed by these materials.

9. What kind of property tends to be imparted to a functional group of an organic compound containing both oxygen and nitrogen?

10. Match the corrosive substance from the column on the left, below, with one of its major properties from the right column:

    1. Alkali metal hydroxides   (a) Reacts with protein in tissue to form yellow
    2. Hydrogen peroxide   xanthoproteic acid
    3. Hydrofluoric acid, HF   (b) Dissolves glass
    4. Nitric acid, HNO_3 (c) Strong bases
                                      (d) Oxidizer

11. Rank the following wastes in increasing order of segregation (a) mixed halogenated and hydrocarbon solvents containing little water, (b) spent steel pickling liquor (c) dilute sludge consisting of mixed organic and inorganic wastes, (d) spent hydrocarbon solvents free of halogenated materials, (e) dilute mixed inorganic sludge.

12. "A carefully organized system in which wastes go through appropriate pathways to their ultimate elimination or disposal in ways that protect human health and the environment," defines _____.

13. What is the role of a POTW in the treatment of hazardous wastes?

14. The Toxicity Characteristic Leaching Procedure was originally devised to mimic a "mismanagement scenario" in which hazardous wastes were disposed along with biodegradable organic municipal refuse. Discuss how this procedure reflects the conditions that might arise from circumstances in which hazardous wastes and actively decaying municipal refuse were disposed together.

Environmental Chemistry of Hazardous Wastes

18.1. INTRODUCTION

Having outlined the nature and sources of hazardous substances and hazardous wastes in Chapter 17, it is now possible to discuss their environmental chemistry. The treatment and minimization of hazardous wastes are covered in Chapter 19.

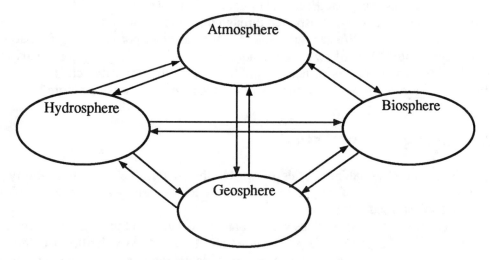

Figure 18.1. Scheme of interactions of hazardous wastes in the environment.

The environmental chemistry of hazardous waste materials in the environment may be considered on the basis of the definition of environmental chemistry (Section 1.2) according to the following factors:

- Origin
- Transport
- Reactions
- Effects
- Ultimate fate

In addition, consideration must be given to the distribution of hazardous wastes among the geosphere, hydrosphere, atmosphere, and biosphere, as shown in Figure 18.1.

18.2. ORIGIN OF HAZARDOUS WASTES

Chapter 17 discusses sources of hazardous wastes. For purposes of discussion in this chapter, *origin* of hazardous wastes refers to their points of entry into the environment. These may consist of the following:

- Deliberate addition to soil, water, or air by humans
- Evaporation or wind erosion from waste dumps into the atmosphere
- Leaching from waste dumps into groundwater, streams, and bodies of water
- Leakage, such as from underground storage tanks or pipelines
- Evolution and subsequent deposition by accidents, such as fire or explosion
- Release from improperly operated waste treatment or storage facilities

18.3. TRANSPORT OF HAZARDOUS WASTES

The transport of hazardous wastes is largely a function of their physical properties, the physical properties of their surrounding matrix, the physical conditions to which they are subjected, and chemical factors. Highly volatile wastes are obviously more likely to be transported through the atmosphere and more soluble ones to be carried by water. Wastes will move farther, faster in porous sandy formations than in denser soils. Volatile wastes are more mobile under hot, windy conditions and soluble ones during periods of heavy rainfall. Wastes that are more chemically and biochemically reactive will not move so far as less reactive wastes before breaking down.

Physical Properties of Wastes

The major physical properties of wastes that determine their amenability to transport are volatility, solubility, and the degree to which they are sorbed to solids, including soil and sediments.

The distribution of hazardous waste compounds between the atmosphere and the geosphere or hydrosphere is largely a function of compound volatility. Compound volatilities are usually measured by vapor pressures, which vary over a wide range. A parameter called **evaporation rate** is used on Material Safety Data Sheets (MSDSs) to express the likelihood of a compound's going into the vapor state. Evaporation rate is based upon the vapor pressure at 20°C of butyl acetate, a solvent that is widely used in making lacquers, plastics, and safety glass.[1] The value of the vapor pressure of butyl acetate under these conditions is 10 mm Hg and the evaporation rate of a compound is given as,

$$\text{Evaporation rate} = \frac{\text{Vapor pressure of compound}}{10 \text{ mm}} \qquad (18.3.1)$$

where the vapor pressure of the compound is given in mm Hg. Examples of readily vaporizable hazardous waste compounds are acetone (evaporation rate 22), ethyl ether (evaporation rate 44), and normal pentane (evaporation rate 42.6). By way of contrast, the evaporation rate of the PCB Arochlor 1254 is only 6×10^{-5}.

Usually, in the hydrosphere, and often in soil, hazardous waste compounds are dissolved in water; therefore, the tendency of water to hold the compound is a factor in its mobility. For example, although ethyl alcohol has a higher evaporation rate and lower boiling temperature (4.3 and 77.8°C, respectively) than toluene (2.2 and 110.6 °C), vapor of the latter compound is more readily evolved from soil because of its limited solubility in water compared to ethanol, which is totally miscible with water.

Chemical Factors

As an illustration of chemical factors involved in transport of wastes, consider largely cationic inorganic species. Inorganic species can be divided into three groups based upon their retention by clay minerals. Elements that tend to be highly retained by clay include cadmium mercury, lead, and zinc. Potassium, magnesium, iron, silicon, and NH_4^+ are moderately retained by clay, whereas sodium, chloride, calcium, manganese, and boron are poorly retained. The retention of the last three elements is probably biased in that they are leached from clay, so that negative retention (elution) is often observed. It should be noted, however, that the retention of iron and manganese is a strong function of oxidation state in that the reduced forms of Mn and Fe are relatively poorly retained, whereas the oxidized forms of $Fe_2O_3 \cdot xH_4O$ and MnO_2 are very insoluble and stay on soil as solids.

18.4. EFFECTS OF HAZARDOUS WASTES

The effects of hazardous wastes in the environment may be divided among effects on organisms, effects on materials, and effects on the environment. These are addressed briefly here and in greater detail in later sections.

The ultimate concern with wastes has to do with their toxic effects to animals, plants, and microbes. Virtually all hazardous waste substances are poisonous to a degree, some extremely so. The toxicity of a waste is a function of many factors, including the chemical nature of the waste, the matrix in which it is contained, circumstances of exposure, the species exposed, manner of exposure, degree of exposure, and time of exposure. The toxicities of hazardous wastes are discussed in more detail in Chapter 20, "Toxicological Chemistry."

As defined in Section 17.6, many hazardous wastes are *corrosive* to materials, usually because of extremes of pH or because of dissolved salt content. Oxidant wastes can cause combustible substances to burn uncontrollably. Highly reactive wastes can explode, causing damage to materials and structures. Contamination by wastes, such as by toxic pesticides in grain, can result in substances becoming unfit for use.

In addition to their toxic effects in the biosphere, hazardous wastes can damage air, water, and soil. Wastes that get into air can cause deterioration of air quality, either directly, or by the formation of secondary pollutants (see Section 18.8). Hazardous waste compounds dissolved in, suspended in, or floating as surface films on, the surface of water can render it unfit for use and for sustenance of aquatic organisms.

Soil exposed to hazardous wastes can be severely damaged by alteration of its physical and chemical properties and ability to support plants. For example, soil exposed to concentrated brines from petroleum production may become unable to support plant growth so that the soil becomes extremely susceptible to erosion.

18.5. FATES OF HAZARDOUS WASTES

The fates of hazardous waste substances are addressed in more detail in subsequent sections. As with all environmental pollutants, such substances eventually reach a state of physical and chemical stability, although that may take many centuries to occur. In some cases, the fate of a hazardous waste material is a simple function of its physical properties and surroundings.

The fate of a hazardous waste substance in water is a function of the substance's solubility, density, biodegradability, and chemical reactivity. Dense, water-immiscible liquids may simply sink to the bottoms of bodies of water or aquifers and accumulate there as "blobs" of liquid. This has happened, for example, with hundreds of tons of PCB wastes that have accumulated in sediments in the Hudson River in New York State. Biodegradable substances are broken down by bacteria, a process for which the availability of oxygen is an important variable. Substances that readily undergo bioaccumulation are taken up by organisms, exchangeable cationic materials become bound to sediments, and organophilic materials may be sorbed by organic matter in sediments.

The fates of hazardous waste substances in the atmosphere are often determined by photochemical reactions. Ultimately such substances may be converted to nonvolatile, insoluble matter and precipitate from the atmosphere onto soil or plants.

18.6. HAZARDOUS WASTES IN THE GEOSPHERE

The sources, transport, interactions, and fates of contaminant hazardous wastes in the geosphere involve a complex scheme, some aspects of which are illustrated in Figure 18.2. The primary environmental concern regarding hazardous wastes in the

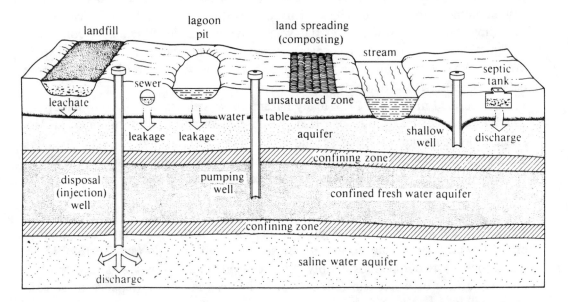

Figure 18.2. Sources, disposal, and movement of hazardous wastes in the geosphere.

geosphere is the possible contamination of groundwater aquifers by waste leachates and leakage from wastes. As the figure shows, there are a number of possible contamination sources. The most obvious one is leachate from landfills containing hazardous wastes. In some cases, liquid hazardous materials are placed in lagoons, which can leak into aquifers. Leaking sewers can also result in contamination, as can the discharge from septic tanks. Hazardous wastes spread on land can result in aquifer contamination by leachate. Hazardous chemicals are sometimes deliberately disposed of underground in waste disposal wells. This means of disposal can result in interchange of contaminated water between surface water and ground water at discharge and recharge points.

The transport of contaminants in the geosphere depends largely upon the hydrologic factors governing the movement of water underground and the interactions of hazardous waste constituents with geological strata, particularly unconsolidated earth materials. As shown in Figure 18.3, groundwater contaminated with hazardous wastes tends to flow as a relatively undiluted plug or plume along with the groundwater in an aquifer. The groundwater flow rate depends upon the water gradient and aquifer characteristics, such as permeability and cross-section area. The rate of flow is generally relatively slow; 1 meter per day would be considered fast. As discussed in Section 18.7, contaminated groundwater can result in contamination of a surface water source. This can occur at a discharge area where the groundwater flows into a lake or stream.

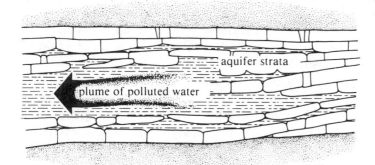

Figure 18.3. Plug-flow of hazardous wastes in groundwater.

As discussed in the preceding section, hazardous waste dissolved in groundwater can be attenuated by soil or rock by means of various sorption mechanisms. Mathematically the distribution of a solute between groundwater or leachate water and soil is expressed by a **distribution coefficient**, K_d,

$$K_d = \frac{C_S}{C_W} \tag{18.6.1}$$

where C_S is the the concentration of the species in the solid phase and C_W is its concentration in water. This equation assumes that the relative degree of sorption is independent of C_W, that is, it assumes a linear sorption isotherm. For the more common case of nonlinear sorption isotherms, C_S is expressed as a function of the equilibrium concentration of sorbate in water, C_{eq}, by the Freundlich equation,

$$C_S = K_F C_{eq}^{1/n} \tag{18.6.2}$$

where and K_F and $1/n$ are empirical constants.

The degree of attenuation depends upon the surface properties of the solid, particularly its surface area. The chemical nature of the attenuating solid is also important because attenuation is a function of the organic matter (humus) content, presence of hydrous metal oxides, and the content and types of clays present. The chemical characteristics of the leachate also affect attenuation greatly. For example, attenuation of metals is very poor in acidic leachate because precipitation reactions, such as,

$$M^{2+} + 2OH^- \rightarrow M(OH)_2(s) \tag{18.6.3}$$

are reversed in acid:

$$M(OH)_2(s) + 2H^+ \rightarrow M^{2+} + 2H_2O \tag{18.6.4}$$

Organic solvents in leachates tend to prevent attenuation of organic hazardous waste constituents.

Sorption of nonionic organic matter by soil depends upon the organic content of the soil. In a study of the sorption of trichloromethane, 1,1,1-trichloroethane, trichloroethylene, and perchloroethylene on tertiary shale, jurassic shale, peat, lignite, bituminous coal, and anthracite coal, it has been shown that the type of organic matter in soil is also quite important[2] According to the study, organic matter with a low oxygen and high hydrogen content, such as that in unweathered shales, is about an order of magnitude more effective in sorbing organic solutes than is more oxidized organic matter, such as that in weathered shales.

The degree of attenuation of a pollutant by soil depends upon the water content of the soil. As shown in Figure 15.4, above the water table there is an unsaturated zone of soil in which attenuation is more highly favored. Normally soil has a greater surface area at liquid-solid interfaces in this zone so that absorption and ion-exchange processes are favored. Aerobic degradation (see Chapter 6) is possible in the unsaturated zone, enabling more rapid and complete degradation of biodegradable hazardous wastes.

Heavy metals are particularly damaging to groundwater and their movement through the geosphere is of considerable concern. Heavy-metal ions may be sorbed by the soil, held by ion-exchange processes, interact with organic matter in soil, undergo oxidation-reduction processes leading to mobilization or immobilization, or even be volatilized as organometallic compounds formed by methylating bacteria. A large number of factors affect heavy-metal mobility and attenuation in soil. These include pH, pE (see Chapter 4), temperature, cation-exchange capacity (see Chapter 5), nature of soil mineral matter, and kinds of soil organic matter present.

Normally, the mobility of heavy metals in soil and mineral matter is relatively low. A study of relative mobilities in clay mineral columns[3] showed that Pb, Zn, Cd, and Hg were strongly attenuated by the clay, primarily by precipitation and exchange processes. Iron was only moderately attenuated, which had to be due to reduction of highly insoluble iron(III) to iron(II):

$$Fe_2O_3 \cdot xH_2O(s) + 2e^- + 6H^+ \rightarrow 2Fe^{2+} + (3 + x)H_2O \tag{18.6.5}$$

Manganese was actually eluted from clay, probably because of reduction to soluble manganese(II) of insoluble higher-oxidation-state manganese originally bound to the clay.

Clays vary in their abilities to remove hazardous waste constituents from water. Montmorillonite tends to be more effective than illite, which is followed by kaolinite.

As illustrated by a study of the mobilization of radionuclides,[4] codisposal of chelating agents with heavy metals can have a strong effect upon the mobility of metal ions in soil. This was observed in a study of the effects of codsiposal of intermediate-level nuclear wastes with chelating agents during the period 1951-1965 at Oak Ridge National Laboratory. The presence of chelating agents resulted from the use of salts of ethylenediaminetetraacetic acid (EDTA, see Section 3.8) in decontaminating facilities exposed to nuclear wastes. Whereas metal cations are readily held by ion exchange processes and precipitation on soil,

$$2\text{Soil}\}^-\text{H}^+ + \text{Co}^{2+} \rightarrow (\text{Soil}\}^-)_2\text{Co}^{2+} + 2\text{H}^+ \tag{18.6.6}$$

$$\text{Co}^{2+} + 2\text{OH}^- \rightarrow \text{Co(OH)}_2(s) \tag{18.6.7}$$

chelated anionic species, such as CoY^{2-} (where Y^{4-} is the chelating EDTA anion), are not strongly retained by the negatively charged functional groups in soil.

Radionuclides have been buried in shallow trenches on the grounds of Oak Ridge National Laboratory since 1944, so ample time has elapsed to observe the effects of this means of radioactive waste disposal. The predominant geological formation at these burial sites is Conasauga shale. This bedrock material has a very high sorptive capacity for most of the radionuclides produced as byproducts of nuclear fission, particularly those that are cationic. Despite this, some migration of radionuclides has been observed from sites used to dispose of solid and liquid wastes. Some of this migration has been attibuted to the high rainfall in the area, shallow groundwater levels, fractures in the underlying rock that allow for rapid infiltration of dissolved wastes, and other physical factors.

In addition to the factors listed above as contributing to the migration of radionuclides from waste disposal trenches, it has been found that chelating agents used for decontamination, as well as naturally occurring humic substance chelators, are responsible for migration in excess of that expected. Most notably, ^{60}Co has been found outside the disposal trenches. Levels of radioactive contamination from this isotope adjacent to the disposal trenches have been observed as high as 1×10^5 disintegrations per minute (dpm) per gram (45,000 picoCuries/g) in soil and as high as 1×10^3 dpm/mL in soil water. In addition, traces of various isotopes of the alpha emitters, uranium, plutonium, radium, thorium, and californium, have been found outside the disposal area. Experiments have been conducted to determine the values of K_d for the distribution of ^{60}Co between water sampled from wells at the disposal sites and the shale at the sites. (The distribution coefficient is a measure of the affinity of a solute for a solid phase; the higher its value, the greater the tendency of the solute to be sorbed by the solid.) For well water samples ranging in pH from 6.0 to 8.5, values of K_d were measured as 7 to 70 with an average of about 35. This is in marked contrast to the value of 7.0×10^4 for K_d obtained with a standard ^{60}Co solution prepared from inorganic cobalt in the absence of chelating agent, indicative of a tremendous affinity of inorganic cobalt for shale. Similar solutions prepared containing 1×10^{-5} M EDTA and cobalt at the same pH gave K_d values of only 2.9. The actual EDTA concentration found in the well samples was 3.4×10^{-7} M, thus explaining why the distribution coefficient in the well water samples was somewhat higher than that observed in the experimental samples containing 1×10^{-5} M EDTA.

Species other than EDTA possess the potential to mobilize radionuclides or heavy metals. Of these, palmitic and and phthalic acid

Phthalic acid   Palmitic acid

were found in leachate from the disposal trenches.[5] Other species that might be codisposed with radionuclides that could also increase radionuclide mobilities are citrate, fluoride, oxalate, and gluconate salts.

In another study of radioactive waste disposal sites,[6] it was observed that organic chelating agents, particularly EDTA, were present and dramatically increase the migration of radionuclides from the site. According to this study, water samples at the Maxey flats disposal site in Kentucky, where EDTA-containing plutonium wastes were disposed, showed levels of 300,000 picoCuries/liter, far in excess of those found in the absence of the chelating agent.

The evidence just cited suggests that strong chelating agents would have a tendency to transport heavy metal ions from disposal sites. Codisposal of EDTA with radionuclides and heavy metals should be avoided.

18.7. HAZARDOUS WASTES IN THE HYDROSPHERE

Figure 18.4 illustrates a typical pathway for the entry of hazardous waste materials into the hydrosphere. Other sources consist of precipitation from the atmosphere with rainfall, deliberate release to streams and bodies of water, runoff from soil, and mobilization from sediments. Once into an aquatic system, hazardous waste species are subject to a number of chemical and biochemical processes, including acid-base, oxidation-reduction, precipitation-dissolution, and hydrolysis reactions, as well as biodegradation.

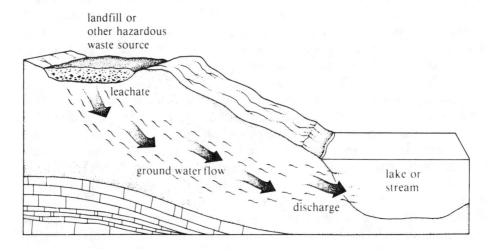

Figure 18.4. Discharge of groundwater contaminated from hazardous waste landfill into a body of water.

The presence of organic matter in water has a tendency to increase the solubility of hazardous organic substances. Typically, the solubility of hexachlorobenzene is 1.8 µg/L in pure water at 25°C whereas it is 2.3 µg/L in creek water containing organic solutes and 4-4.5 µg/L in landfill leachate.

In considering the processes that hazardous wastes undergo in water, it is important to recall the nature of aquatic systems and the unique properties of water discussed in detail in Chapter 2. Water in the environment is far from pure. Just as the atmosphere is a constantly changing mass of bodies of moving air with different temperatures, pressures, and humidities, bodies of water are highly dynamic systems. Rivers, impoundments, and groundwater aquifers are subject to the input and loss of a variety of materials from both natural and anthropogenic sources. These materials may be gases, liquids, or solids. They interact chemically with each other and with living organisms — particularly bacteria — in the water. They are subject to dispersion and transport by stream flow, convection currents, and other physical phenomena. Hazardous substances or their byproducts in water may undergo bioaccumulation through food chains involving aquatic organisms.

Several physical, chemical, and biochemical processes are particularly important in determining the transformations and ultimate fates of hazardous chemical species in the hydrosphere. These include **hydrolysis reactions**, through which a molecule is cleaved with the addition of H_2O; **precipitation reactions**, generally accompanied by **aggregation** of colloidal particles suspended in water; **oxidation-reduction reactions**, generally mediated by microorganisms (see Chapters 4 and 6); **sorption** of hazardous solutes by sediments and by suspended mineral and organic matter; **biochemical processes**, often involving hydrolysis and oxidation-reduction reactions; **photolysis reactions**; and miscellaneous chemical phenomena.

The hydrolysis of two hazardous waste compounds, an acid anhydride and an ester, is illustrated by the following reactions:

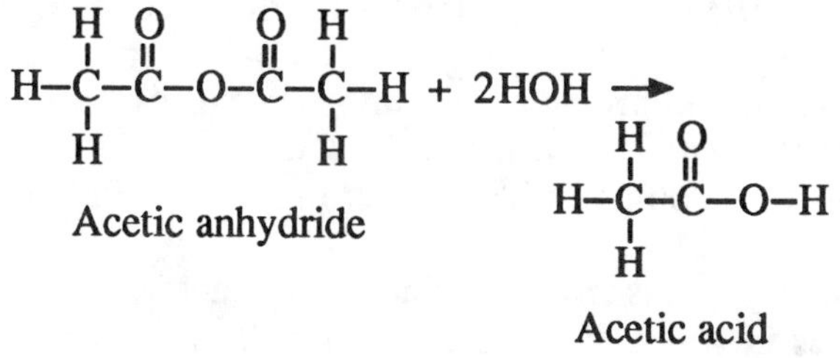

Acetic anhydride

Acetic acid   (18.7.1)

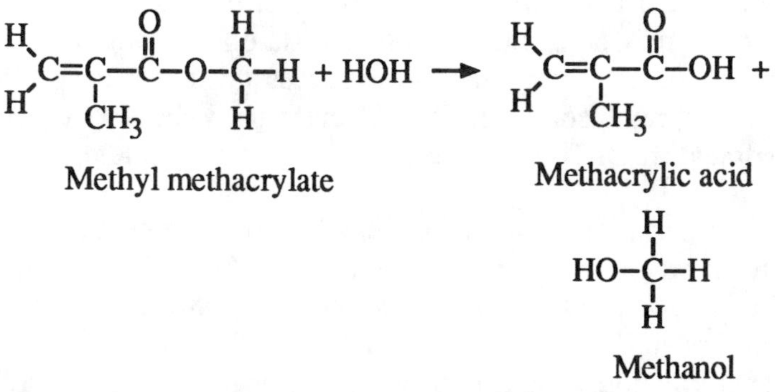

Methyl methacrylate   Methacrylic acid

Methanol   (18.7.2)

The rates at which compounds hydrolyze in water vary widely. Acetic anhydride hydrolyzes very rapidly. In fact, the great affinity of this compound for water (including water in skin) is one of the reasons that it is hazardous. Once in the aquatic environment, though, acetic anhydride is converted very rapidly to essentially harmless acetic acid. Many ethers, esters, and other compounds formed originally by the joining together of two or more molecules with the loss of water hydrolyze very slowly, although the rate may be greatly increased by the action of enzymes in micoorganisms (biochemical processes). Hydrolysis of some compounds results in the loss of halogen atoms. For example, bis(chloromethyl)ether hydrolyzes rapidly to produce HCl and formaldehyde:

$$\underset{\underset{H}{|}}{\overset{\overset{H}{|}}{Cl-C}}-O-\underset{\underset{H}{|}}{\overset{\overset{H}{|}}{C}}-Cl + H_2O \longrightarrow 2H-\overset{\overset{O}{\|}}{C}-H + 2HCl \qquad (18.7.3)$$

Hydrolysis of a large quantity of this chloro ether in the aquatic environment could produce harmful amounts of corrosive HCl and formaldehyde.

As discussed in Section 19.5, the formation of precipitates in the form of sludges is one of the most common means of isolating hazardous components from an unsegregated waste. Although solid inorganic ionic compounds are often discussed in terms of very simple formulas, such as $PbCO_3$ for lead carbonate, much more complicated species (for example, $2PbCO_3 \cdot Pb(OH)_2$) generally result when precipitates are formed in the aquatic environment. For example, a hazardous heavy metal ion in the hydrosphere may be precipitated as a relatively complicated compound, coprecipitated as a minor constituent of some other compound, or be sorbed by the surface of another solid.

The major anions present in natural waters and wastewaters are OH^-, CO_3^{2-}, and SO_4^{2-} Since these anions are all capable of forming precipitates with cationic impurities, such pollutants tend to precipitate as hydroxides, carbonates, and sulfates. Sometimes a distinction can be made between hydroxides and hydrated oxides with similar, or identical, empirical formulas. For example, iron(III) hydroxide, $Fe(OH)_3$, is a relatively uncommon species; iron(III) usually is precipitated from water as hydrated iron(III) oxides, such as β ferric oxide monoydrate, $Fe_2O_3 \cdot H_2O$. Basic salts containing OH^- ion along with some other anion are very common in solids formed by precipitation from water. A typical example is azurite, $2CuCO_3 \cdot Cu(OH)_2$. Two or more metal ions may be present in a compound, as is the case with chalcopyrite, $CuFeS_2$.

Two aspects of the precipitation process are particularly important in determining the fate of hazardous ionic solutes in water. If precipitation occurs very rapidly and with a high degree of supersaturation, the solid tends to form as a large number of small colloidal particles that may persist in the colloidal state for a long time. In this form, hazardous substances are much more mobile and accessible to organisms than as precipitates. A second important consideration is that many heavy metals are coprecipitated with hydrated iron(III) oxide ($Fe_2O_3 \cdot xH_2O$) or manganese (IV) oxide ($MnO_2 \cdot xH_2O$).

Sorption processes are particularly common methods for the removal of low level hazardous materials from water. As discussed in Chapter 5, freshly precipitated $MnO_2 \cdot xH_2O$ very effectively scavenges other metal ions, such as Ba^{2+}, from water. As shown by the examples in Table 18.1, oxidation-reduction reactions are very important means of transformation of hazardous wastes in water. The degradation of most organic wastes proceeds by way of oxidation, as discussed in Chapters 4 and 6.

Table 18.1. Oxidation-Reduction of Wastes in Water

Reaction	Significance
Oxidation half-reactions	
$SO_2(aq) + 2 H_2O \rightarrow 4 H^- + SO_4^{2-} + 2e^-$	Conversion of dissolved SO_2 gas to sulfuric acid
$\underset{\substack{\|\\O}}{CH_3C}-H + H_2O \rightarrow \underset{\substack{\|\\O}}{CH_3C}-OH + 2 H^+ + 2 e^-$	Conversion of acetaldehyde to acetic acid
$\{CH_2O\} + H_2O \rightarrow CO_2 + 4 H^- + 4 e^-$	Degradation of biomass
$C_nH_{2n+2} + 2n H_2O \rightarrow n CO_2 + (6n + 2) H^+ + (6n + 2) e^-$	Degradation of hydrocarbons
Reduction half-reactions	
$O_2(aq) + 4 H^+ + 4 e^- \rightarrow 2 H_2O$	Removal of O_2 from water; O_2 is an electron receptor (source of O_2) for oxidation half-reactions above*
$Fe_2O_3 \cdot x H_2O(s) + 6 H^+ + 2 e^- \rightarrow 2 Fe^{2+} + (3 + x) H_2O$	Formation of soluble Fe^{2+}
$MnO_2(s) + 4 H^+ + 2 e^- \rightarrow Mn^{2+} + 2 H_2O$	Production of soluble Mn^{2+}

*For example, the overall reaction for the oxidation of dissolved SO_2 by O_2 is obtained as follows:

$$2 \{SO_2(aq) + 2 H_2O \rightarrow 4 H^+ + SO_4^{2-} + 2 e^-\}$$
$$O_2(aq) + 4 H^+ + 4 e^- \rightarrow 2 H_2O$$
$$\overline{2 SO_2(aq) + O_2(aq) + 2 H_2O \rightarrow 4 H^+ + 2 SO_4^{2-}}$$

Under many circumstances, biochemical processes largely determine the fates of hazardous chemical species in the hydrosphere. The most important such processes are those mediated by microorganisms, as discussed in Chapter 6. In particular, the oxidation of biodegradable hazardous organic wastes in water generally occurs by means of microorganism-mediated biochemical reactions. Bacteria produce organic acids and chelating agents, such as citrate, which have the effect of solubilizing hazardous heavy metal ions. Some mobile methylated forms, such as compounds of methylated arsenic and mercury, are produced by bacterial action.

As discussed in Chaptes 9-14, photolysis reactions are those initiated by the absorption of light. The effect of photolytic processes on the destruction of hazardous wastes in the hydrosphere is minimal, although some photochemical reactions of hazardous waste compounds can occur when the compounds are present as surface films on water exposed to sunlight.

Groundwater is the part of the hydrosphere most vulnerable to damage from hazardous wastes. Although surface water supplies are subject to contamination, groundwater can become almost irreversibly contaminated by the improper land disposal of hazardous chemicals. This matter has been discussed in some detail in Section 18.6.

18.8. HAZARDOUS WASTES IN THE ATMOSPHERE

Some chemicals found in hazardous waste sites may enter the atmosphere by evaporation or even as windblown particles. Typical of such chemicals are acrylonitrile, ethylene dichloride, perchloroethylene, vinylidene chloride, and benzo-(a)pyrene, the structural formulas of which are given on the following page:

$$\underset{\text{Acrylonitrile}}{\overset{\displaystyle \underset{H}{\overset{H}{\diagup}}\overset{\displaystyle H}{\underset{H}{\diagdown}}C=C-C\equiv N}{}}
\qquad
\underset{\substack{\text{Ethylene}\\\text{dichloride}}}{\overset{\displaystyle H-\overset{H}{\underset{Cl}{\overset{|}{\underset{|}{C}}}}-\overset{H}{\underset{Cl}{\overset{|}{\underset{|}{C}}}}-H}{}}
\qquad
\underset{\text{Perchloroethylene}}{\overset{\displaystyle \underset{Cl}{\overset{Cl}{\diagup}}C=C\underset{Cl}{\overset{Cl}{\diagdown}}}{}}$$

$$\underset{\text{Vinylidene chloride}}{\overset{\displaystyle \underset{Cl}{\overset{Cl}{\diagup}}C=C\underset{H}{\overset{H}{\diagdown}}}{}}
\qquad
\text{Benzo(a)pyrene}$$

Three major areas of interest in respect to hazardous waste compounds in the atmosphere are their **pollution potential, atmospheric fate,** and **residence time.** These strongly interrelated factors are discussed in this section.

Air Pollution Potential of Hazardous Waste Compounds

The pollution potential of hazardous wastes in the atmosphere depends upon whether they are *primary pollutants* that have a direct effect or *secondary pollutants* that are converted to harmful substances by atmospheric chemical processes. Hazardous waste sites do not usually evolve sufficient quantities of pollutants to give significant amounts of secondary pollutants, so primary air pollutants are the greater concern. Examples of primary air pollutants include toxic organic vapors (vinyl chloride), corrosive acid gases (HCl), and toxic inorganic gases, such as H_2S released by the accidental mixing of waste acid (HCl from waste steel pickling liquor) and waste metal sulfides:

$$2HCl + FeS \rightarrow FeCl_2 + H_2S(g) \tag{18.8.1}$$

Primary air pollutants are most dangerous in the immediate vicinity of a site, usually to workers involved in disposal or cleanup or people living adjacent to the site. Quantities are rarely sufficient to pose any kind of regional air pollution hazard.

The two major kinds of secondary air pollutants from hazardous wastes are those that are oxidized in the atmosphere to corrosive substances and organic materials that undergo photochemical oxidation. Plausible examples of the former are sulfur dioxide released from the action of waste strong acids on sulfites and subsequently oxidized in the atmosphere to corrosive sulfuric acid,

$$SO_2 + \tfrac{1}{2}O_2 + H_2O \rightarrow H_2SO_4(aerosol) \tag{18.8.2}$$

and nitrogen dioxide (itself a toxic primary air pollutant) produced by the reaction of waste nitric acid with reducing agents such as metals and oxidized to corrosive nitric acid or converted to corrosive nitrate salts:

$$4HNO_3 + Cu \rightarrow Cu(NO_3)_2 + 2NO_2(g) + 2H_2O \tag{18.8.3}$$

$$2NO_2(g) + \tfrac{1}{2}O_2 + H_2O \rightarrow 2HNO_3(aerosol) \tag{18.8.4}$$

$$HNO_3(aerosol) + NH_3(g) \rightarrow NH_4NO_3(aerosol) \tag{18.8.5}$$

Organic species that produce secondary air pollutants are those that form photochemical smog (see Chapter 13). The more reactive of these are unsaturated compounds that react with atomic oxygen or hydroxyl radical in air,

$$R\text{-}CH{=}CH_2 + HO^\bullet \rightarrow RCH_2CH_2O^\bullet \tag{18.8.6}$$

to yield reactive radicals that participate in chain reactions to eventually yield ozone, organic oxidants, noxious aldehydes, and other products characteristic of photochemical smog.

Fate and Residence Times of Hazardous Waste Compounds in the Atmosphere

An obvious means by which hazardous waste species may be removed from the atmosphere is by **dissolution** in water in the form of cloud or rain droplets. Inorganic acid, base, and salt compounds, such as H_2SO_4, HNO_3, and NH_4NO_3 mentioned above, are readily removed from the atmosphere by dissolution. For vapors of compounds such as ethylene chloride, tetrachloroethylene, and vinylidene chloride, which are not highly soluble in water, Henry's law (Chapter 5) combined with information about rainfall amounts and mixing in the atmosphere can be used to estimate the atmospheric half-life, $\tau_{1/2}$, of the species. Solubility rates may be used to estimate half-lives for substances that are more miscible in water. For poorly water-soluble compounds, such calculations tend to drastically underestimate lifetimes, which indicates that other removal mechanisms must predominate.

The lifetimes of vaporized hazardous waste species removed from the atmosphere through **adsorption by aerosol particles** is limited to that of the sorbing aerosol particles (typically about 7 days) plus the time spent in the vapor phase before adsorption. This mechanism appears to be viable only for highly nonvolatile constituents such as benzo(a)pyrene.

Sorptive removal by soil, water, or plants on the Earth's surface, called **dry deposition**, is another means for physical removal of hazardous substances from the atmosphere. Predictions of dry deposition rate vary greatly with type of compound, type of surface, and weather conditions. For highly volatile organic compounds, such as low molecular mass organohalide compounds, predicted rates of dry deposition give atmospheric lifetimes many-fold higher than those actually observed so, for such compounds, dry deposition is probably not a common removal mechanism.

Predicted rates of physical removal of a number of volatile organic compounds that are not very soluble in water are far too slow to account for the loss of such compounds from the atmosphere, so chemical processes must predominate. As discussed in Chapters 12 and 13, the most important of these processes is reaction with hydroxyl radical, $HO^\bullet$, in the troposphere. Ozone can react with compounds having a double bond, such as acrylonitrile or vinylidene chloride. Other oxidant species that might react with hazardous waste compounds in the troposphere and stratosphere are atomic oxygen (O), peroxyl radicals ($HOO^\bullet$), alkylperoxyl radicals ($ROO^\bullet$), and NO_3.

Despite the fact that its concentration in the troposphere is relatively low, $HO^\bullet$ is so reactive that it tends to initiate most of the reactions leading to the chemical removal of most refractory organic compounds from the atmosphere. As noted in Section 13.3, hydroxyl radical undergoes *abstraction reactions* to remove H atoms from organic compounds containing R-H,

$$R\text{-}H + HO^{\bullet} \rightarrow R^{\bullet} + H_2O \tag{18.8.7}$$

and may react with those containing unsaturated bonds by addition as illustrated in Reaction 18.8.6. In both cases reactive free radicals are formed that undergo further reactions, leading to nonvolatile and/or water-soluble species, which are scavenged from the atmosphere by physical means. These scavengeable species tend to be aldehydes, ketones, or acids. Halogenated organic compounds may lose halogen atoms in the form of halo-oxy radicals and undergo further reactions to form scavengeable species.

Reaction with ozone may be a significant pathway for the removal of unsaturated compounds from the amosphere. Ozone adds across double bonds, leading to the formation of reactive species that undergo further reactions to form products that precipitate from the atmosphere (see Chapter 13). Rate constants for the reactions of unsaturated compounds with O_3 are typically only about 1 millionth as fast as those for reactions with $HO^{\bullet}$. Despite this, ozone's concentration in the troposphere is normally so much higher than that of $HO^{\bullet}$ that ozonolysis mechanisms are competitive with those involving $HO^{\bullet}$ for the removal of alkenyl compounds from the atmosphere. Although aromatic compounds also react with O_3, the rate is so slow compared to the rate of reaction with $HO^{\bullet}$ that reaction with ozone is not a significant removal mechanism for these substances.

In general, reactions with species other than $HO^{\bullet}$ or O_3 are not considered significant in the removal of hazardous organic waste compounds from the troposphere. Perhaps in some cases such reactions do contribute to a very slow removal of such contaminants compounds.

Photolytic transformations involve direct cleavage (photodissociation of compounds by reactions with light and ultraviolet radiation:

$$R\text{-}X + h\nu \rightarrow R^{\bullet} + X^{\bullet} \tag{18.8.8}$$

The extent of these reactions varies greatly with light intensity, quantum yields (chemical reactions per quantum absorbed) and other factors. In order for photolysis to be an important process for its removal from the atmosphere, a molecule must have a **chromophore** (light-absorbing group) that absorbs light in a wavelength region of significant intensity in the impinging light spectrum. This requirement limits the importance of photolysis as a removal mechanism to only a few classes of compounds, including conjugated alkenes, carbonyl compounds, some halides, and some nitrogen compounds, particularly nitro compounds. However, these do include a number of the more important hazardous waste compounds.

Table 18.1 summarizes possible fates and estimated residence times of hazardous waste chemicals in the atmosphere. Extremely volatile and refractory hazardous waste compounds that escape destruction and removal in the troposphere enter the stratosphere and are destroyed by chain reactions that begin with direct photolytic reactions. This is possible because the flux of highly energetic short wavelength ultraviolet radiation is very intense. Direct photolytic destruction is most clearly illustrated by chlorofluorocarbon (Freon) compounds, which do not undergo photodissociation until they are subjected to ultraviolet radiation in the stratosphere. Here, ultraviolet photolysis can result in the removal of halide atoms from compounds as illustrated by the reaction,

$$F_2CCl_2 + h\nu \rightarrow F_2CCl\cdot + Cl\cdot \qquad (18.8.9)$$

after which the chlorine atom may catalyze ozone destruction (see Section 14.4) and the highly reactive radical products undergo further reactions to eventually form species that are scavenged from the atmosphere.

18.9. HAZARDOUS WASTES IN THE BIOSPHERE

One of the most crucial aspects of fate and toxic effects of environmental chemicals is their accumulation by organisms from their surroundings. The factors involved in bioaccumulation and biomagnification have been summarized.[7]

The biotransformations of environmental chemicals, including pesticides and industrial chemicals, in vertebrates (birds, mammals, amphibians, fish, and reptiles) have been summarized.[8]

Biodegradation of wastes is their conversion by biological processes to simple inorganic molecules and, to a certain extent, to biological materials. The complete bioconversion of a substance to inorganic species such as CO_2, NH_3, and phosphate is called **mineralization. Detoxification** refers to the biological conversion of a toxic substance to a less toxic species, which may still be a relatively complex, or biological conversion to an even more complex material. An example of detoxification is illustrated below for the enzymatic conversion of paraoxon (a highly toxic organophosphate insecticide) to p-nitrophenol, which has only about 1/200 the toxicity of the parent compound:

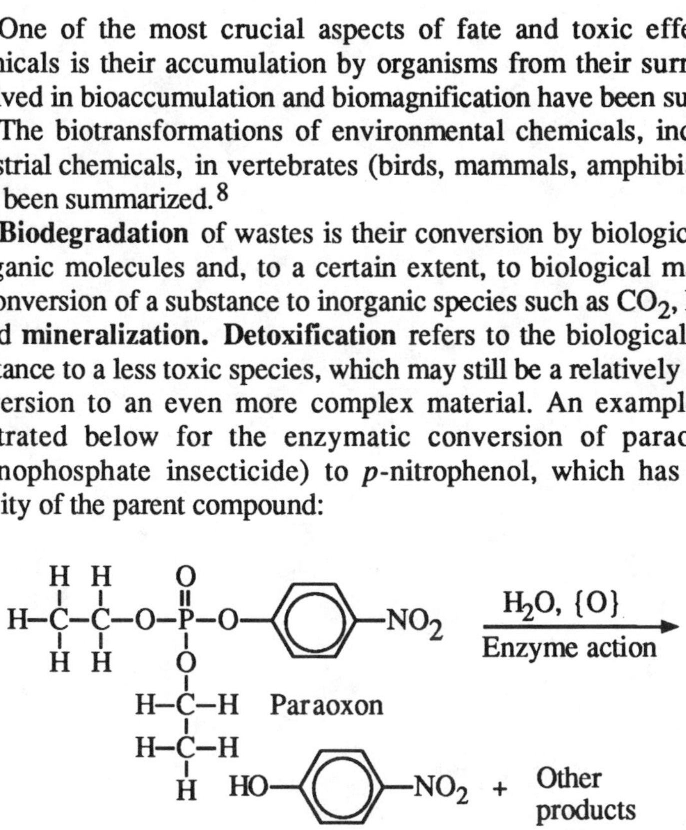

$$(18.9.1)$$

Usually the products of biodegradation are molecular forms that tend to occur in nature. Because the organisms that carry out biodegradation do so as a means of extracting free energy for their metabolic and growth needs (see Chapter 6) the form products that are in greater thermodynamic equilibrium with their surroundings. The definition of biodegradation is illustrated by an example in Figure 18.5. Biodegradation is usually carried out by the action of microorganisms, particularly bacteria and fungi.

Biodegradation Processes

Biotransformation is what happens to any substance that is **metabolized** and thereby altered by biochemical processes in an organism. **Metabolism** is divided into the two general categories of **catabolism**, which is the breaking down of more complex molecules, and **anabolism**, which is the building up of life molecules from

Table 18.2. Estimated Residence Times and Possible Fates of Hazardous Waste Chemicals in the Atmosphere*

Compound	$k_{OH} \times 10^{12}$**	$k_{O_3} \times 10^{18}$**	Photolysis probability	Physical removal probability	Residence time, days***	Possible reaction products
Acetaldehyde	16	—	Probable	Unlikely	0.03–0.7	H_2CO, CO_2
Acrolein	44	4	Probable	Unlikely	0.2	OCH—CHO, H_2CO, HCO_2H, CO_2
Acrylonitrile	2	≤0.05	—	Unlikely	5.6	H_2CO, HC(O)CN, HCO_2H, CN·
Allyl chloride	28	18.3	Possible	Unlikely	0.3	HCO_2H, H_2CO, $ClCH_2CHO$, chlorinated hydroxyl carbonyls, $ClCH_2CO_2H$
Benzo(a)pyrene	—	—	Possible	Probable	8	1-6-Quinone of benzo(a)pyrene
Benzyl chloride	3	0.004	Possible	Unlikely	3.9	φCHO, Cl·, chloromethylphenols, ring-cleavage products
Bis(chloromethyl)ether	4	—	Possible	Probable	0.02–2.9	HCl, H_2CO, chloromethylformate, ClHCO
Carbon tetrachloride	<0.001	—	—	Unlikely	>11,000	Cl_2CO, Cl·
Chlorobenzene	0.4	<5 × 10^{-5}	Possible	Unlikely	28	Chlorophenols, ring-cleavage products
Chloroform	0.1	—	—	Unlikely	120	Cl_2CO, Cl·
Chloromethyl methyl ether	3	—	Possible	Probable	0.004–3.9	Decomposition products, chloromethyl and methyl formate, ClHCO

Compound						Products
Chloroprene	46	8	Probable	Unlikely	0.3	H_2CO, $H_2C=CClCHO$, OHCCHO, ClCOCHO, $H_2CCHCClO$, chlorohydroxyl acids, aldehydes
o-, m-, p-cresol	55	0.6	—	Unlikely	0.2	Hydroxynitrotoluenes, ring-cleavage products
Dichlorobenzene	2.03	$\leq 5 \times 10^-$	Possible	Unlikely	39	Chlorinated phenols, ring-cleavage products, nitro compounds
Dioxane	3	—	—	Unlikely	3.9	$OHCOCH_2CH_2OCHO$, OHCOCHO, oxygenated formates
Dioxin	—	—	Probable	—	—	—
Epichlorhydrin (chloropropylene oxide)	2	—	Possible	Unlikely	5.8	H_2CO, OHCOCHO, $ClCH_2C(O)OHCO$
Ethylene dibromide	0.25	—	Possible	Unlikely	45	Br·, $BrCH_2CHO$, H_2CO, BrHCO
Ethylene dichloride	0.22	—	Possible	Unlikely	53	ClHCHO, $H_2CClCOCl$, H_2CO, $H_2CClCHO$
Ethylene oxide	2	—	—	Unlikely	5.8	OHCOCHO
Formaldehyde	10	$<2 \times 10^{-5}$	Probable	Unlikely	0.1–1.2	CO, CO_2
Hexachlorocyclopentadiene	59	8	Probable	—	0.2	Cl_2CO, diacylchlorides, ketones, Cl·

Table 18.2. Estimated Residence Times and Possible Fates of Hazardous Waste Chemicals in the Atmosphere (*Cont.*)

Compound	$k_{OH} \times 10^{12}$**	k_{O}, $\times 10^{18}$**	Photolysis probability	Physical removal probability	Residence time, days***	Possible reaction products
Maleic anhydride	60	160	Possible	Possible	0.1	CO_2, CO, acids, aldehydes and esters that should photolyze
Methyl chloride	0.14	—	Possible	Unlikely	83	Cl_2CO, CO, ClHCO, Cl·
Methyl chloroform	0.012	—	Possible	Unlikely	970	H_2CO, Cl_2CO, Cl·
Methyl iodide	0.004	—	Possible	Unlikely	2,900	H_2CO, I·, IHCO, CO
Nitrobenzene	0.06	$<5 \times 10^{-5}$	Possible	Unlikely	190	Nitrophenols, ring-cleavage products
2-Nitropropane	55	—	Possible	Unlikely	0.2	H_2CO, CH_3CHO
N-Nitrosodiethylamine	26	—	Probable	—	≤ 0.4	Photolysis products, aldehydes, nitramines
Nitrosoethylurea	13	—	Possible	—	≤ 0.9	Photolysis products, aldehydes, nitramines
Nitrosomethylurea	20	—	Possible	—	≤ 0.6	Photolysis products, aldehydes; nitramines
Nitrosomorpholine	28	—	Possible	—	≤ 0.4	Photolysis products, aldehydic ethers
Perchloroethylene	0.17	0.002	Possible	Unlikely	67	Cl_2CO, $Cl_2C(OH)COCl$, Cl·
Phenols	17	1	—	Possible	0.6	Dihydroxybenzenes, nitro phenols, ring-cleavage products
Phosgene	0	—	—	Possible	—	CO_2, Cl·, HCl
Polychlorinated biphenyls (PCB's)	< 1	5×10^{-5}	Possible	Unlikely	>11	Hydroxy PCB's, ring-cleavage products

					Products
Propylene oxide	1.3	—	—	Unlikely	> 8.9	$CH_3C(O)OCHO$, $CH_3C(O)OCHO$, H_2CO, $HC(O)OCHO$
Toluene	6	0.0001	—	Unlikely	1.9	Benzaldehyde, cresols, ring-cleavage products, nitro compounds
Trichloroethylene	2.2	0.006	Possible	Unlikely	5.2	Cl_2CO, $ClHCO$, CO, $Cl\cdot$
Vinylidene chloride	4	0.04	Possible	Unlikely	2.9	H_2CO, Cl_2CO, HCO_2H
o-, m-, p-Xylene	~16	~0.001	—	Unlikely	0.7	—

*From Reference 15.

**In units of cm³ per molecule per second; most rate constants calculated theoretically. The term k_{OH} represents the rate constant for reaction with hydroxyl radical; k_{O_3} represents the rate constant for reaction with ozone.

***The shorter residence time (if a range is indicated) includes removal by photolysis.

simpler materials. The substances subjected to biotransformation may be naturally occurring or *anthropogenic* (made by human activities). They may consist of *xenobiotic* molecules that are foreign to living systems.

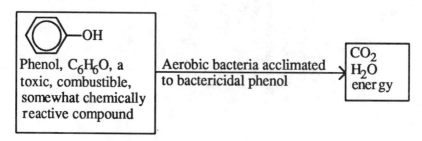

Figure 18.5.   Illustration of biological action on a hazardous waste constituent.

An important biochemical process that occurs in the biodegradation of many synthetic and hazardous waste materials is **cometabolism**. Cometabolism does not serve a useful purpose to an organism in terms of providing energy or raw material to build biomass, but occurs concurrently with normal metabolic processes. An example of cometabolism of hazardous wastes is provided by the white rot fungus, *Phanerochaete chrysosporium*, which degrades a number of kinds of organochlorine compounds — including DDT, PCBs, and chlorodioxins — under the appropriate conditions. The enzyme system responsible for this degradation is one that the fungus uses to break down lignin in plant material under normal conditions.

Enzymes in Waste Degradation

Enzyme systems hold the key to biodegradation of hazardous wastes. For most biological treatment processes currently in use, enzymes are present in living organisms in contact with the wastes. However, in some cases it is possible to use cell-free extracts of enzymes removed from bacterial or fungal cells to treat hazardous wastes. For this application the enzymes may be present in solution or, more commonly, immobilized in biochemical reactors.[9]

Biodegradation of municipal wastewater and solid wastes in landfills occurs by design. Biodegradation of any kind of waste that can be metabolized takes place whenever the wastes are subjected to conditions conducive to biological processes. The most common type of biodegradation is that of organic compounds in the presence of air, that is, **aerobic processes**. However, in the absence of air, **anaerobic biodegradation** may also take place. Furthermore, inorganic species are subject to both aerobic and anaerobic biological processes.

Although biological treatment of wastes is normally regarded as degradation to simple inorganic species such as carbon dioxide, water, sulfates, and phosphates, the possibility must always be considered of forming more complex or more hazardous chemical species. An example of the latter is the production of volatile, soluble, toxic methylated forms of arsenic and mercury from inorganic species of these elements by bacteria under anaerobic conditions.

For the most part, anthropogenic compounds resist biodegradation much more strongly than do naturally occurring compounds. This is generally due to the absence of enzymes that can bring about an initial attack on the compound (see Phase I reactions, Section 20.3). A number of physical and chemical characteristics of a compound are involved in its amenability to biodegradation. Such characteristics include hydrophobicity, solubility, volatility, and affinity for lipids. Some organic structural groups impart particular resistance to biodegradation. These include branched carbon chains, ether linkages, meta-substituted benzene rings, chlorine, amines, methoxy groups, sulfonates, and nitro groups.

Several groups of microorganisms are capable of partial or complete degradation of hazardous organic compounds. Among the aerobic bacteria, those of the *Pseudomonas* family are the most widespread and most adaptable to the degradation of synthetic compounds. These bacteria degrade biphenyl, naphthalene, DDT, and many other compounds. Anaerobic bacteria are very fastidious and difficult to study in the laboratory because they require oxygen-free (anoxic) conditions and pE values of less than -3.4 in order to survive. These bacteria catabolize biomass through hydrolytic processes, breaking down proteins, lipids, and saccharides. They are also known to reduce nitro compounds to amines, degrade nitrosamines, promote reductive dechlorination, reduce epoxide groups to alkenes, and break down aromatic structures. **Actinomycetes** are microorganisms that are morphologically similar to both bacteria and fungi. They are involved in the degradation of a variety of organic compounds, including degradation-resistant alkanes, and lignocellulose. Other compounds attacked include pyridines, phenols, nonchlorinated aromatics, and chlorinated aromatics. Fungi are particularly noted for their ability to attack long-chain and complex hydrocarbons and are more successful than bacteria in the initial attack on PCB compounds. Phototrophic microorganisms, which include algae, photosynthetic bacteria, and cyanobacteria (blue-green algae) tend to concentrate organophilic compounds in their lipid stores and induce photochemical degradation of the stored compounds. For example, *Oscillatoria* can initiate the biodegradation of naphthalene by the attachment of −OH groups.

Practically all classes of synthetic organic compounds can be at least partially degraded by various microorganisms. These classes include nonhalogenated alkanes, halogenated alkanes (trichloroethane, dichloromethane) nonhalogenated aromatic compounds (benzene, naphthalene, benzo(a)pyrene), halogenated aromatic compounds (hexachlorobenzene, pentachlorophenol), phenols (phenol, cresols), polychlorinated biphenyls, phthalate esters , and pesticides (chlordane, parathion).

LITERATURE CITED

1. Wray, Thomas K., "Evaporation Rate," *Hazmat World*, November, 1990, p. 32.

2. Grathwohl, Peter, "Influence of Organic Matter from Soils and Sediments from Various Origins on the Sorption of Some Chlorinated Aliphatic Hydrocarbons: Implications on K_{OC} Correlations," *Environmental Science and Technology* **24**, 1687-1693 (1990).

3. Griffin, R. A., and N. F. Shimp, *Attenuation of Pollutants in Municipal Landfill Leachate by Clay Minerals*, U. S. Environmental Protection Agency, Washington, D. C., 1978.

4. Means, J. L., D. A. Crear, and J. O. Duguid, "Migration of Radioactive Wastes: Radionuclide Mobilization by Complexing Agents," *Science* **200**, 1477-81 (1978).

5. Means, J. L., D. A. Crear, and J. O. Duguid, *Oak Ridge National Laboratory Report* ORNL/TM-5438, Oak Ridge, Tenn., 1976.

6. "Organic Agents Increase Plutonium Leaching," *Chemical and Engineering News*, June 8, 1981, p. 7.

7. Connell, Des W., *Bioaccumulation of Xenobiotic Compounds*, CRC Press, Inc, Boca Raton, Florida, 1989.

8. Hawkins, David R., Ed., *Biotransformations: A Survey of the Biotransformations of Drugs and Chemicals in Animals*, Vols. 1 and 2, Royal Society of Chemistry, London, 1989 (updated annually).

9. Glaser, John A., "Enzyme Systems and Related Reactive Species," Section 9.2 in *Standard Handbook of Hazardous Waste Treatment and Disposal*, Freeman, Harry M., Ed., McGraw-Hill Book Company, New York, 1989, pp. 9.61-9.73.

SUPPLEMENTARY REFERENCES

Racke, Kenneth D., and Joel R. Coats, Eds., *Enhanced Biodegradation of Pesticides in the Environment*, American Chemical Society, Washington, DC, 1990.

Environmental Restoration and Waste Management. Five-Year Plan Fiscal Years 1992-1996, U.S. Department of Energy, Washington, DC, 1990.

Berlin, Robert E., and Catherine C. Stanton, *Radioactive Waste Management*, Wiley-Interscience, New York, 1989.

Higgins, Thomas E., *Hazardous Waste Minimization Handbook*, Lewis Publishers, Chelsea, MI, 1989.

Howard, Philip H., *Handbook of Environmental Fate and Exposure Data for Organic Chemicals.Vol. 1, Large Production and Priority Pollutants*, Lewis Publishers, Chelsea, MI, 1989.

Rush to Burn. Solving America's Garbage Crisis?, Island Press, Washington, DC, 1989.

Grosse, Douglas, *Managing Hazardous Wastes Containing Heavy Metals*, SciTech Publishers, Matawan, NJ, 1990.

Allegri, Theodore C., Sr., *Handling and Management of Hazardous Materials and Wastes*, Chapman and Hall, New York, NY, 1986.

Evans, Jeffrey C., Ed., *Toxic and Hazardous Wastes: Proceedings of the Nineteenth Mid-Atlantic Industrial Waste Conference*, Technomic Publishing, Lancaster, PA, 1987.

Suffet, I. H., Patrick MacCarthy, Eds., *Aquatic Humic Substances: Influence on Fate and Treatment of Pollutants*, Advances in Chemistry Series **219**, American Chemical Society, Washington, DC, 1989.

Knox, Peter J., Ed., *Resource Recovery of Municipal Solid Wastes*, American Institute of Chemical Engineers, New York, NY, 1988.

Goldman, Benjamin A., James A. Hulme, and Cameron Johnson, *Hazardous Waste Management: Reducing the Risk*, Council on Economic Priorities, New York, 1986.

Freeman, Harry M., Ed., *Standard Handbook of Hazardous Waste Treatment and Disposal*, McGraw-Hill, New York, NY, 1988.

U.S. Environmental Protection Agency, *Extremely Hazardous Substances: Superfund Chemical Profiles*, Vols. 1 and 2, Noyes Publications, Park Ridge, NJ, 1989.

Palmer, S. A. K., *Metal/Cyanide Containing Wastes: Treatment Technologies*, Noyes Data Corp., Park Ridge, NJ, 1988.

Bell, John M., Ed., *Proceedings of the 42nd Industrial Waste Conference, May 12, 13, 14, 1987*, Lewis Publishers, Chelsea, MI, 1988.

Scholze, R. J., *Biotechnology for Degradation of Toxic Chemicals in Hazardous Wastes*, Noyes Publications, Park Ridge, NJ, 1988.

Wolf, K., W. J. Van Den Brink, and F. J. Colon, *Contaminated Soil '88*, Vols 1 & 2. Kluwer Academic Publishers, Norwell, MA, 1988.

QUESTIONS AND PROBLEMS

1. What are three major properties of wastes that determine their amenability to transport?

2. What is the role played by butyl acetate in expressing the parameter known as evaporation rate?

3. What is the influence of organic solvents in leachates upon attenuation of organic hazardous waste constituents?

4. Match the following physical, chemical, and biochemical processes dealing with the transformations and ultimate fates of hazardous chemical species in the hydrosphere on the left with the description of the process on the right, below:

1. Precipitation reactions	(a) Molecule is cleaved with the addition of H_2O
2. Biochemical processes	(b) Generally accompanied by aggregation of colloidal particles suspended in water
3. Hydrolysis reactions	(c) Generally mediated by microorganisms
4. Sorption	(d) By sediments and by suspended matter
5. Oxidation-reduction	(e) Often involve hydrolysis and oxidation-reduction

5. In general, what is the influence of soil water upon attenuation of organic hazardous waste constituents?

6. Describe the particular danger posed by codisposal of strong chelating agents with radionuclide wastes. What may be said about the chemical nature of the latter in regard to this danger?

7. Describe a beneficial effect that might result from the precipitation of either $Fe_2O_3 \cdot xH_2O$ or $MnO_2 \cdot xH_2O$ from hazardous wastes in water.

8. Why are secondary air pollutants from hazardous waste sites usually of only limited concern as compared to primary air pollutants? What is the distinction between the two?

9. What are the major means by which hazardous waste species may be removed from the atmosphere? What is meant by $\tau_{1/2}$?

10. What may be said about the relative rates of reaction of hazardous waste compounds in the atmosphere with $HO \cdot$ and O_3?

11. As applied to hazardous wastes in the biosphere distinguish among biodegradation, biotransformation, detoxification, and mineralization.

12. What is the potential role of *Phanerochaete chrysosporium* in treatment of hazardous waste compounds? For which kinds of compounds might it be most useful?

13. What is a specific example of the formation of relatively more hazardous materials by the action of biological processes on hazardous wastes?

14. Several physical and chemical characteristics are involved in determining the amenability of a hazardous waste compound to biodegradation. These include hydrophobicity, solubility, volatility, and affinity for lipids. Suggest and discuss ways in which each one of these factors might affect biodegradability.

15. List and discuss some of the important processes determining the transformations and ultimate fates of hazardous chemical species in the hydrosphere.

16. Which part of the hydrosphere is most subject to long-term, largely irreversible contamination from the improper disposal of hazardous wastes in the environment?

17. What features or characteristics should a compound possess in order for direct photolyis to be a significant factor in its removal from the atmosphere?

18. List and discuss the significance of major sources for the origin of hazardous wastes, that is their main modes of entry into the environment. What are the relative dangers posed by each of these? Which part of the environment would each be most likely to contaminate?

19. Inorganic species may be divided into three major groups based upon their retention by clays. What are the elements commonly listed in these groups? What is the chemical basis for this division? How might anions (Cl^-, NO_3^-) be classified?

20. In what form would a large quantity of hazardous waste PCB likely be found in the hydrosphere?

21. Why is attenuation of metals likely to be very poor in acidic leachate?

Reduction, Treatment, and Disposal of Hazardous Waste

19.1. INTRODUCTION

Chapters 17 and 18 have addressed the nature and sources of hazardous wastes and their environmental chemistry. These chapters have pointed out some of the major problems associated with such wastes. Since the 1970s, efforts to reduce and clean up hazardous wastes have been characterized by

- Legislation
- Regulation
- Litigation
- Procrastination
- Modelling
- Analysis
- Cleanup of a few select sites

It is perhaps fair to say that in proportion to the magnitude of the problems and the amount of money devoted to them so far, insufficient progress has been made in coping with hazardous wastes. In the U.S., huge amounts of time have been devoted to promulgating hazardous waste regulations, instrument manufacturers have prospered as more and more chemical analyses have been required, and computationists, some of whom would be offended at the sight of any chemical, much less a hazardous one, have consumed thousands of hours of computer time to model hazardous waste systems. This is to say nothing of the vast expense of litigation that has gone into lawsuits dealing with hazardous waste sites. In the future, a higher percentage of the effort and resources devoted to hazardous wastes needs to be placed on remediation of existing problems and preventive action to avoid future problems. This chapter discusses how environmental chemistry can be applied to hazardous waste management to develop measures by which chemical wastes can be minimized, recycled, treated, and disposed.[1] In descending order of desirability, hazardous waste management attempts to accomplish the following:

- Do not produce it
- If making it cannot be avoided, produce only minimum quantities
- Recycle it
- If it is produced and cannot be recycled, treat it, preferably in a way that makes it nonhazardous
- If it cannot be rendered nonhazardous, dispose of it in a safe manner
- Once it is disposed, monitor it for leaching and other adverse effects

19.2. WASTE REDUCTION AND MINIMIZATION

Many hazardous waste problems can be avoided at early stages by **waste reduction**[2] (cutting down quantities of wastes from their sources) and **waste minimization** (utilization of treatment processes which reduce the quantities of wastes requiring ultimate disposal). This section outlines basic approaches to waste minimization and reduction. For more details, the reader is referred to a reference work dealing with these topics.[3]

There are several ways in which quantities of wastes can be reduced, including source reduction, waste separation and concentration, resource recovery, and waste recycling.[4] The most effective approaches to minimizing wastes center around careful control of manufacturing processes,[5] taking into consideration discharges and the potential for waste minimization at every step of manufacturing. Viewing the process as a whole (as outlined for a generalized chemical manufacturing process in Figure 19.1) often enables crucial identification of the source of a waste, such as a raw material impurity, catalyst, or process solvent. Once a source is identified, it is much easier to take measures to eliminate or reduce the waste.

Modifications of the manufacturing process can yield substantial waste reduction. Some such modifications are of a chemical nature. Changes in chemical reaction conditions can minimize production of byproduct hazardous substances. In some cases potentially hazardous catalysts, such as those formulated from toxic substances, can be replaced by catalysts that are non-hazardous or that can be recycled rather than discarded. Wastes can be minimized by volume reduction, for example, through dewatering and drying sludge.

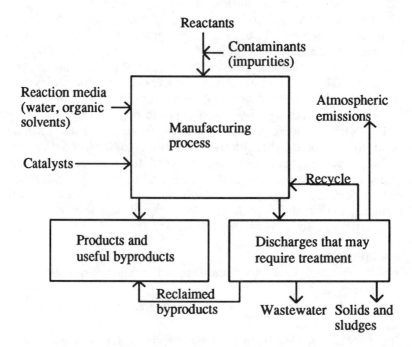

Figure 19.1. Chemical manufacturing process from the viewpoint of discharges and waste minimization.

19.3. RECYCLING

Wherever possible, recycling and reuse should be accomplished on-site because it avoids having to move wastes and because a process that produces recyclable materials is often the most likely to have use for them. The four broad areas in which something of value may be obtained from wastes are the following:

- Direct recycle as raw material to the generator, as with the return to feedstock of raw materials not completely consumed in a synthesis process
- Transfer as a raw material to another process; a substance that is a waste product from one process may serve as a raw material for another, sometimes in an entirely different industry
- Utilization for pollution control or waste treatment, such as use of waste alkali to neutralize waste acid
- Recovery of energy, for example, from the incineration of combustible hazardous wastes

Examples of Recycling

Recycling of scrap industrial impurities and products occurs on a large scale with a number of different materials. Most of these materials are not hazardous, but, as with most large-scale industrial operations, their recycle may involve the use or production of hazardous substances. Some of the more important examples are the following:

- **Ferrous metals** composed primarily of iron and used largely as feedstock for electric-arc furnaces
- **Nonferrous metals**, including aluminum (which ranks next to iron in terms of quantities recycled), copper and copper alloys, zinc, lead, cadmium, tin, silver, and mercury
- **Metal compounds**, such as metal salts
- **Inorganic substances** including alkaline compounds (such as sodium hydroxide used to remove sulfur compounds from petroleum products), acids (steel pickling liquor where impurities permit reuse), and salts (for example, ammonium sulfate from coal coking used as fertilizer)
- **Glass**, which makes up about 10 percent of municipal refuse
- **Paper**, commonly recycled from municipal refuse
- **Plastic**, consisting of a variety of moldable polymeric materials and composing a major constituent of municipal wastes
- **Rubber**
- **Organic substances**, especially solvents and oils, such as hydraulic and lubricating oils
- **Catalysts** from chemical synthesis or petroleum processing
- Materials with **agricultural uses**, such as waste lime or phosphate-containing sludges used to treat and fertilize acidic soils.

Waste Oil Utilization and Recovery

Waste oil generated from lubricants and hydraulic fluids is one of the more commonly recycled materials.[6] Annual production of waste oil in the U.S. is of the order of 4 billion liters per year. Around half of this amount is burned as fuel and lesser quantities are recycled or disposed as waste. The collection, recycling, treatment, and disposal of waste oil are all complicated by the fact that it comes from diverse, widely dispersed sources[7] and contains several classes of potentially hazardous contaminants. These are divided between organic constituents (polycyclic aromatic hydrocarbons, chlorinated hydrocarbons) and inorganic constituents (aluminum, chromium, and iron from wear of metal parts; barium and zinc from oil additives; lead from leaded gasoline).

Recycling Waste Oil

The processes used to convert waste oil to a feedstock hydrocarbon liquid for lubricant formulation are illustrated in Figure 19.2. The first of these uses distillation to remove water and light ends that have come from condensation and contaminant fuel. The second, or processing, step may be a vacuum distillation in which the three products are oil for further processing, a fuel oil cut, and a heavy residue. The processing step may also employ treatment with a mixture of solvents including isopropyl and butyl alcohols and methylethyl ketone to dissolve the oil and leave contaminants as a sludge; or contact with sulfuric acid to remove inorganic contaminants followed by treatment with clay to take out acid and contaminants that cause odor and color. The third step shown in Figure 19.2 employs vacuum distillation to separate lubricating oil stocks from a fuel fraction and heavy residue. This phase of treatment may also involve hydrofinishing, treatment with clay, and filtration.

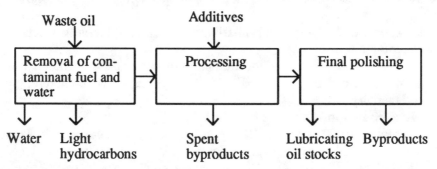

Figure 19.2. Major steps in reprocessing waste oil.

Waste Oil Fuel

For economic reasons, waste oil that is to be used for fuel is given minimal treatment of a physical nature, including settling, removal of water, and filtration. Metals in waste fuel oil become highly concentrated in its fly ash, which may be hazardous.

Waste Solvent Recovery and Recycle

The recovery and recycling of waste solvents has some similarities to the recycling of waste oil and is also an important enterprise. Among the many solvents listed as hazardous wastes and recoverable from wastes are dichloromethane, tetrachloroethylene, trichloroethylene, 1,1,1-trichloroethane, benzene, liquid alkanes, 2-nitropropane, methylisobutyl ketone, and cyclohexanone.[8] For reasons of both economics and pollution control, many industrial processes that use solvents are equipped for solvent recycle.[9] The basic scheme for solvent reclamation and reuse is shown in Figure 19.3.

A number of operations are used in solvent purification.[10] Entrained solids are removed by settling, filtration, or centrifugation. Drying agents may be used to remove water from solvents and various adsorption techniques and chemical treatment may be required to free the solvent from specific impurities. Fractional distillation, often requiring several distillation steps, is the most important operation in solvent purification and recycle. It is used to separate solvents from impurities, water, and other solvents.

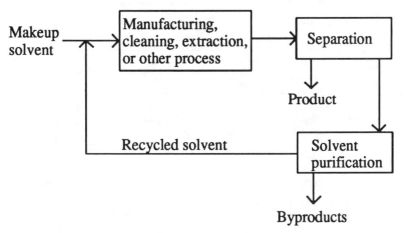

Figure 19.3. Overall process for recycling solvents.

19.4. PHYSICAL METHODS OF WASTE TREATMENT

This section addresses predominantly physical methods for waste treatment and the following section addresses methods that utilize chemical processes. It should be kept in mind that most waste treatment measures have both physical and chemical aspects. The appropriate treatment technology for hazardous wastes obviously depends upon the nature of the wastes.[11] These may consist of volatile wastes (gases, volatile solutes in water, gases or volatile liquids held by solids, such as catalysts), liquid wastes (wastewater, organic solvents), dissolved or soluble wastes (water-soluble inorganic species, water-soluble organic species, compounds soluble in organic solvents) semisolids (sludges, greases), and solids (dry solids, including granular solids with a significant water content, such as dewatered sludges, as well as solids suspended in liquids). The type of physical treatment to be applied to wastes depends strongly upon the physical properties of the material treated, including state of matter, solubility in water and organic solvents, density, volatility, boiling point, and melting point.

As shown in Figure 19.4, waste treatment may occur at three major levels — **primary**, **secondary**, and **polishing** — somewhat analogous to the treatment of wastewater (see Chapter 8). Primary treatment is generally regarded as preparation for further treatment, although it can result in the removal of byproducts and reduction of the quantity and hazard of the waste. Secondary treatment detoxifies, destroys, and removes hazardous constituents. Polishing usually refers to treatment of water that is removed from wastes so that it may be safely discharged. However, the term can be broadened to apply to the treatment of other products as well so that they may be safely discharged or recycled.

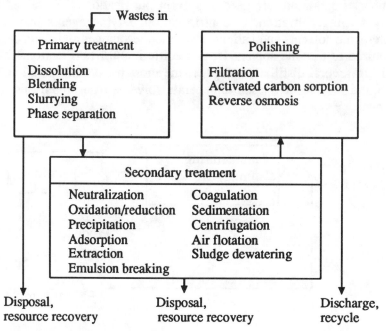

Figure 19.4. Major phases of waste treatment.

Methods of Physical Treatment

Knowledge of the physical behavior of wastes has been used to develop various unit operations for waste treatment that are based upon physical properties. These operations include the following:

- Phase separation
    Filtration
- Phase transition
    Distillation
    Evaporation
    Physical precipitation

- Phase transfer
    Extraction
    Sorption
- Membrane separations
    Reverse osmosis
    Hyper- and ultrafiltration

Phase Separations

The most straightforward means of physical treatment involves separation of components of a mixture that are already in two different phases. **Sedimentation** and **decanting** are easily accomplished with simple equipment. In many cases the separation must be aided by mechanical means, particularly **filtration** or **centrifugation**. **Flotation** is used to bring suspended organic matter or finely divided particles to the surface of a suspension. In the process of **dissolved air flotation** (DAF), air is dissolved in the suspending medium under pressure and comes out of solution when the pressure is released as minute air bubbles attached to suspended particles, which causes the particles to float to the surface.

An important and often difficult waste treatment step is **emulsion breaking** in which colloidal-sized **emulsions** are caused to aggregate and settle from suspension. Agitation, heat, acid, and the addition of **coagulants** consisting of organic polyelectrolytes or inorganic substances, such as an aluminum salt, may be used for this purpose. The chemical additive acts as a flocculating agent to cause the particles to stick together and settle out.[12]

Phase Transition

A second major class of physical separation is that of **phase transition** in which a material changes from one physical phase to another. It is best exemplified by **distillation**, which is used in treating and recycling solvents, waste oil, aqueous phenolic wastes, xylene contaminated with paraffin from histological laboratories, and mixtures of ethylbenzene and styrene. Distillation produces **distillation bottoms** (still bottoms), which are often hazardous and polluting. These consist of unevaporated solids, semisolid tars, and sludges from distillation. Specific examples with their hazardous waste numbers are distillation bottoms from the production of acetaldehyde from ethylene (hazardous waste number K009), and still bottoms from toluene reclamation distillation in the production of disulfoton (K036). The landfill disposal of these and other hazardous distillation bottoms used to be widely practiced but is now severely limited [13]:

Evaporation is usually employed to remove water from an aqueous waste to concentrate it. A special case of this technique is **thin-film evaporation** in which volatile constituents are removed by heating a thin layer of liquid or sludge waste spread on a heated surface.

Drying — removal of solvent or water from a solid or semisolid (sludge) or the removal of solvent from a liquid or suspension — is a very important operation because water is often the major constituent of waste products, such as sludges obtained from emulsion breaking. In **freeze drying**, the solvent, usually water, is sublimed from a frozen material. Hazardous waste solids and sludges are dried to reduce the quantity of waste, to remove solvent or water that might interfere with subsequent treatment processes, and to remove hazardous volatile constituents. Dewatering can often be improved with addition of a filter aid, such as diatomaceous earth, during the filtration step.

Stripping is a means of separating volatile components from less volatile ones in a liquid mixture by the partitioning of the more volatile materials to a gas phase of air or steam (steam stripping). The gas phase is introduced into the aqueous solution or suspension containing the waste in a stripping tower that is equipped with trays or packed to provide maximum turbulence and contact between the liquid and gas phases. The two major products are condensed vapor and a stripped bottoms residue. Examples of two volatile components that can be removed from water by air stripping are benzene and dichloromethane. Air stripping can also be used to remove ammonia from water that has been treated with a base to convert ammonium ion to volatile ammonia.

Physical precipitation is used here as a term to describe processes in which a solid forms from a solute in solution as a result of a physical change in the solution, as compared to chemical precipitation (see Section 19.5) in which a chemical reaction in solution produces an insoluble material. The major changes that can cause physical precipitation are cooling the solution, evaporation of solvent, or alteration of solvent composition. The most common type of physical precipitation by alteration of solvent composition occurs when a water-miscible organic solvent is added to an aqueous solution, so that the solubility of a salt is lowered below its concentration in the solution.

Phase Transfer

Phase transfer consists of the transfer of a solute in a mixture from one phase to another. An important type of phase transfer process is **solvent extraction,** a process in which a substance is transferred from solution in one solvent (usually water) to another (usually an organic solvent) without any chemical change taking place. When solvents are used to leach substances from solids or sludges, the process is called **leaching**. Solvent extraction and the major terms applicable to it are summarized in Figure 19.5. The same terms and general principles apply to leaching. The major application of solvent extraction to waste treatment has been in the removal of phenol from byproduct water produced in coal coking, petroleum refining, and chemical syntheses that involve phenol.

One of the more promising approaches to solvent extraction and leaching of hazardous wastes is the use of **supercritical fluids,** most commonly CO_2, as extraction solvents. A supercritical fluid is one that has characteristics of both liquid and gas and consists of a substance above its supercritical temperature and pressure (31.1°C and 73.8 atm, respectively, for CO_2). After a substance has been extracted from a waste into a supercritical fluid at high pressure, the pressure can be released, resulting in separation of the substance extracted. The fluid can then be compressed again and recirculated through the extraction system. Some possibilities for treatment of hazardous wastes by extraction with supercritical CO_2 include removal of organic contaminants from wastewater, extraction of organohalide pesticides from soil, extraction of oil from emulsions used in aluminum and steel processing, and regeneration of spent activated carbon. Waste oils contaminated with PCBs, metals, and water can be purified using supercritical ethane.

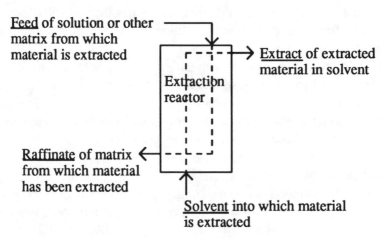

Figure 19.5. Outline of solvent extraction/leaching process with important terms underlined.

Transfer of a substance from a solution to a solid phase is called **sorption**. The most important sorbent is **activated carbon** used for several purposes in waste treatment; in some cases it is adequate for complete treatment. It can also be applied to pretreatment of waste streams going into processes such as reverse osmosis to improve treatment efficiency and reduce fouling. Effluents from other treatment processes, such as biological treatment of degradable organic solutes in water can be polished with activated carbon. Activated carbon sorption is most effective for removing from water those hazardous waste materials that are poorly water soluble and that have high molecular masses, such as xylene, naphthalene (U165), cyclohexane (U056); chlorinated hydrocarbons, phenol (U188), aniline (U012), dyes, and surfactants. Activated carbon does not work well for organic compounds that are highly water-soluble or polar.

Solids other than activated carbon can be used for sorption of contaminants from liquid wastes. These include synthetic resins composed of organic polymers and mineral substances. Of the latter, clay is employed to remove impurities from waste lubricating oils in some oil recycling processes.

Molecular Separation

A third major class of physical separation is **molecular separation**, often based upon **membrane processes** in which dissolved contaminants or solvent pass through a size-selective membrane under pressure.[14] The products are a relatively pure solvent phase (usually water) and a concentrate enriched in the solute impurities. **Hyperfiltration** allows passage of species with molecular masses of about 100 to 500, whereas **ultrafiltration**, is used for the separation of organic solutes with molecular masses of 500 to 1,000,000. With both of these techniques water and lower molecular mass solutes under pressure pass through the membrane as a stream of purified **permeate,** leaving behind a stream of **concentrate** containing impurities in solution or suspension. Ultrafiltration and hyperfiltration are especially useful for concentrating suspended oil, grease, and fine solids in water. They also serve to concentrate solutions of large organic molecules and heavy metal ion complexes.

Reverse osmosis is the most widely used of the membrane techniques. Although superficially similar to ultrafiltration and hyperfiltration, it operates on a different principle in that the membrane is selectively permeable to water and excludes ionic solutes. Reverse osmosis uses high pressures to force permeate through the membrane, producing a concentrate containing high levels of dissolved salts.

Electrodialysis, sometimes used to concentrate plating wastes, employs membranes alternately permeable to cations and to anions. The driving force for the separation is provided by electrolysis with a direct current between two electrodes. Alternate layers between the membranes contain concentrate (brine) and purified water.

19.5. CHEMICAL TREATMENT: AN OVERVIEW

The applicability of chemical treatment to wastes depends upon the chemical properties of the waste constituents, particularly acid-base, oxidation-reduction, precipitation, and complexation behavior; reactivity; flammability/combustibility; corrosivity; and compatibility with other wastes. The chemical behavior of wastes translates to various unit operations for waste treatment that are based upon chemical properties and reactions. These include the following:

- Acid/base neutralization
- Chemical extraction and leaching
- Reduction

- Chemical precipitation
- Oxidation/reduction
- Ion exchange

Some of the more sophisticated means available for treatment of wastes have been developed for pesticide disposal.[15]

Acid/Base Neutralization

Waste acids and bases are treated by **neutralization**:

$$H^+ + OH^- \rightarrow H_2O \tag{19.5.1}$$

Although simple in principle, neutralization can present some problems in practice. These include evolution of volatile contaminants, mobilization of soluble substances, excessive heat generated by the neutralization reaction, and corrosion to apparatus. By adding too much or too little of the neutralizing agent, it is possible to get a product that is too acidic or basic.

Lime, $Ca(OH)_2$, is widely used as a base for treating acidic wastes. Because of lime's limited solubility, solutions of excess lime do not reach extremely high pH values. Sulfuric acid, H_2SO_4, is a relatively inexpensive acid for treating alkaline wastes. However, addition of too much sulfuric acid can produce highly acidic products; for some applications, acetic acid, CH_3COOH, is preferable. As noted above, acetic acid is a weak acid and an excess of it does little harm. It is also a natural product and biodegradable.

Neutralization, or pH adjustment, is often required prior to the application of other waste treatment processes.[16] Processes that may require neutralization include oxidation/reduction, activated carbon sorption, wet air oxidation, stripping, and ion exchange. Microorganisms usually require a pH in the range of 6-9, so neutralization may be required prior to biochemical treatment.

Chemical Precipitation

Chemical precipitation is used in hazardous waste treatment primarily for the removal of heavy metal ions from water as shown below for the chemical precipitation of cadmium:

$$Cd^{2+}(aq) + HS^-(aq) \rightarrow CdS(s) + H^+(aq) \tag{19.5.2}$$

Precipitation of Metals

The most widely used means of precipitating metal ions is by the formation of hydroxides such as chromium(III) hydroxide:

$$Cr^{3+} + 3OH^- \rightarrow Cr(OH)_3 \tag{19.5.3}$$

The source of hydroxide ion is a base (alkali), such as lime ($Ca(OH)_2$), sodium hydroxide (NaOH), or sodium carbonate (Na_2CO_3). Most metal ions tend to produce basic salt precipitates, such as basic copper(II) sulfate, $CuSO_4 \cdot 3Cu(OH)_2$, formed as a solid when hydroxide is added to a solution containing Cu^{2+} and SO_4^{2-} ions. The solubilities of many heavy metal hydroxides reach a minimum value, often at a pH in the range of 9-11, then increase with increasing pH values due to the formation of soluble hydroxo complexes, as illustrated by the following reaction:

$$Zn(OH)_2(s) \; + \; OH^-(aq) \; \rightarrow \; Zn(OH)_3^-(aq) \qquad (19.5.4)$$

The chemical precipitation method that is used most is precipitation of metals as hydroxides and basic salts with lime. Sodium carbonate can be used to precipitate hydroxides ($Fe(OH)_3 \cdot xH_2O$) carbonates ($CdCO_3$), or basic carbonate salts ($2PbCO_3 \cdot Pb(OH)_2$). The carbonate anion produces hydroxide by virtue of its hydrolysis reaction with water:

$$CO_3^{2-} \; + \; H_2O \; \rightarrow \; HCO_3^- \; + \; OH^- \qquad (19.5.5)$$

Carbonate, alone, does not give as high a pH as do alkali metal hydroxides, which may have to be used to precipitate metals that form hydroxides only at relatively high pH values.

The solubilities of some heavy metal sulfides are extremely low, so precipitation by H_2S or other sulfides (see Reaction 19.5.2) can be a very effective means of treatment. Hydrogen sulfide is a toxic gas that is itself considered to be a hazardous waste (U135). Iron(II) sulfide (ferrous sulfide) can be used as a safe source of sulfide ion to produce sulfide precipitates with other metals that are less soluble than FeS. However, toxic H_2S can be produced when metal sulfide wastes contact acid:

$$MS \; + \; 2H^+ \; \rightarrow \; M^{2+} \; + \; H_2S \qquad (19.5.6)$$

Some metals can be precipitated from solution in the elemental metal form by the action of a reducing agent, such as sodium borohydride,

$$8Cu^{2+} \; + \; NaBH_4 \; + \; 2H_2O \; \rightarrow \; 8Cu \; + \; NaBO_2 \; + \; 8H^+ \qquad (19.5.7)$$

or with more active metals in a process called **cementation**:

$$Cd^{2+} \; + \; Zn \; \rightarrow \; Cd \; + \; Zn^{2+} \qquad (19.5.8)$$

Oxidation/Reduction

As shown by the reactions in Table 19.1, **oxidation** and **reduction** can be used for the treatment and removal of a variety of inorganic and organic wastes.[17] Some waste oxidants can be used to treat oxidizable wastes in water and cyanides.

Ozone, O_3, is a strong oxidant that can be generated on-site by an electrical discharge through dry air or oxygen. Ozone employed as an oxidant gas at levels of 1-2 wt% in air and 2-5 wt% in oxygen has been used to treat a large variety of oxidizable contaminants, effluents, and wastes including wastewater and sludges containing oxidizable constituents.

Electrolysis

As shown in Figure 19.6, **electrolysis** is a process in which one species in solution (usually a metal ion) is reduced by electrons at the **cathode** and another gives

Table 19.1. Oxidation/Reduction Reactions Used to Treat Wastes

Waste Substance	Reaction with Oxidant or Reductant

Oxidation of Organics

Organic matter, {CH$_2$O} $\{CH_2O\}$ + $\{O\}$ $\rightarrow$ CO_2 + H_2O

Aldehyde CH_3CH_2O + $\{O\}$ $\rightarrow$ CH_3COOH (acid)

Oxidation of Inorganics

Cyanide $2CN^-$ + $5OCl^-$ + H_2O $\rightarrow$ N_2 +
$$2HCO_3^- + 5Cl^-$$

Iron(II) $4Fe^{2+}$ + O_2 + $10H_2O$ $\rightarrow$
$$4Fe(OH)_3 + 8H^+$$

Sulfur dioxide $2SO_2$ + $2O_2$ + H_2O $\rightarrow$ $2H_2SO_4$

Reduction of Inorganics

Chromate $2CrO_4^-$ + $3SO_2$ + $4H^+$ $\rightarrow$
$$Cr_2(SO_4)_3 + 2H_2O$$

Permanganate MnO_4^- + $3Fe^{2+}$ + $7H_2O$ $\rightarrow$
$$MnO_2(s) + 3Fe(OH)_3(s\ ) + 5H^+$$

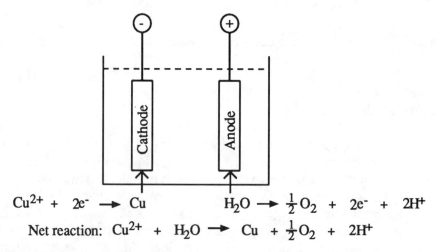

Cu^{2+} + $2e^-$ $\rightarrow$ Cu $\qquad$ H_2O $\rightarrow$ $\frac{1}{2}O_2$ + $2e^-$ + $2H^+$

Net reaction: Cu^{2+} + H_2O $\rightarrow$ Cu + $\frac{1}{2}O_2$ + $2H^+$

Figure 19.6. Electrolysis of copper solution.

up electrons to the **anode** and is oxidized there. In hazardous waste applications electrolysis is most widely used in the recovery of cadmium, copper, gold, lead, silver, and zinc. Metal recovery by electrolysis is made more difficult by the presence of cyanide ion, which stabilizes metals in solution as the cyanide complexes, such as $Ni(CN)_4^{2-}$.

Hydrolysis

One of the ways to dispose of chemicals that are reactive with water is to allow them to react with water under controlled conditions, a process called **hydrolysis**. Inorganic chemicals that can be treated by hydrolysis include metals that react with water; metal carbides, hydrides, amides, alkoxides, and halides; and nonmetal oxyhalides and sulfides. Examples of the treatment of these classes of inorganic species are given in Table 19.2.

Organic chemicals may also be treated by hydrolysis. For example, toxic acetic anhydride is hydrolyzed to relatively safe acetic acid:

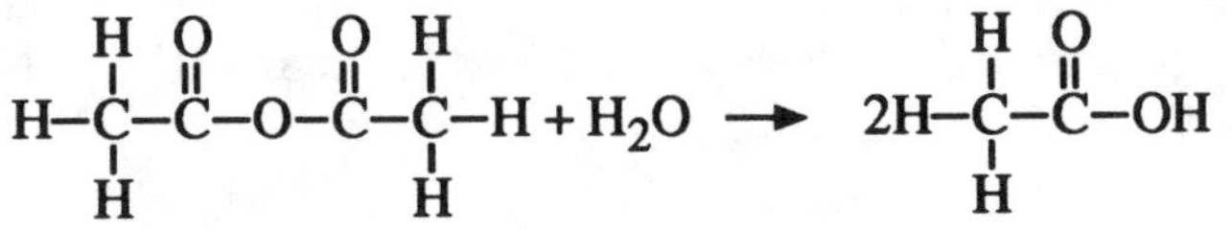

$$H-\underset{\underset{H}{|}}{\overset{\overset{H}{|}}{C}}-\overset{\overset{O}{\|}}{C}-O-\overset{\overset{O}{\|}}{C}-\underset{\underset{H}{|}}{\overset{\overset{H}{|}}{C}}-H + H_2O \longrightarrow 2H-\underset{\underset{H}{|}}{\overset{\overset{H}{|}}{C}}-\overset{\overset{O}{\|}}{C}-OH \quad (19.5.9) \qquad (19.5.9)$$

Acetic anhydride (an
acid anhydride)

Table 19.2 Inorganic Chemicals That May be Treated by Hydrolysis

Class of Chemical	Reaction with Water
Active metals (calcium)	$Ca + 2H_2O \longrightarrow H_2 + Ca(OH)_2$
Hydrides (sodium aluminum hydride)	$NaAlH_4 + 4H_2O \longrightarrow 4H_2 + NaOH + Al(OH)_3$
Carbides (calcium carbide)	$CaC_2 + 2H_2O \longrightarrow Ca(OH)_2 + C_2H_2$
Amides (sodium amide	$NaNH_2 + H_2O \longrightarrow NaOH + NH_3$
Halides (silicon tetrachloride)	$SiCl_4 + 2H_2O \longrightarrow SiO_2 + 4HCl$
Alkoxides (sodium ethoxide)	$NaOC_2H_5 + H_2O \longrightarrow NaOH + C_2H_5OH$

Chemical Extraction and Leaching

Chemical extraction or **leaching** in hazardous waste treatment is the removal of a hazardous constituent by chemical reaction with an extractant in solution. Poorly soluble heavy metal salts can be extracted by reaction of the salt anions with H^+ as illustrated by the following:

$$PbCO_3 + H^+ \rightarrow Pb^{2+} + HCO_3^- \qquad (19.5.10)$$

Acids also dissolve basic organic compounds such as amines and aniline. Extraction with acids should be avoided if cyanides or sulfides are present to prevent formation of toxic hydrogen cyanide or hydrogen sulfide. Nontoxic weak acids are usually the safest to use. These include acetic acid, CH_3COOH, and the acid salt, NaH_2PO_4.

Chelating agents, such as dissolved ethylenedinitrilotetraacetate (EDTA, HY^{2-}), dissolve insoluble metal salts by forming soluble species with metal ions:

$$FeS + HY^{3-} \rightarrow FeY^{2-} + HS^- \qquad (19.5.11)$$

Heavy metal ions in soil contaminated by hazardous wastes may be present in a coprecipitated form with insoluble iron(III) and manganese(IV) oxides, Fe_2O_3 and MnO_2, respectively. These oxides can be dissolved by reducing agents, such as solutions of sodium dithionate/citrate or hydroxylamine. This results in the production of soluble Fe^{2+} and Mn^{2+} and the release of heavy metal ions, such as Cd^{2+} or Ni^{2+}, which are removed with the water.

Ion Exchange

Ion exchange is a means of removing cations or anions from solution onto a solid resin, which can be regenerated by treatment with acids, bases or salts. The greatest use of ion exchange in hazardous waste treatment is for the removal of low levels of heavy metal ions from wastewater:

$$2H^{+-}\{CatExchr\} + Cd^{2+} \rightarrow Cd^{2+-}\{CatExchr\}_2 + 2H^+ \qquad (19.5.12)$$

Ion exchange is employed in the metal plating industry to purify rinsewater and spent plating bath solutions. Cation exchangers are used to remove cationic metal species, such as Cu^{2+}, from such solutions. Anion exchangers remove anionic cyanide metal complexes (for example, $Ni(CN)_4^{2-}$) and chromium(VI) species, such as CrO_4^{2-}. Radionuclides may be removed from radioactive wastes and mixed waste by ion exchange resins.

19.6. Photolytic Reactions

Photolytic reactions were discussed in Chapter 9. *Photolysis* can be used to destroy a number of kinds of hazardous wastes. In such applications it is most useful in breaking chemical bonds in refractory organic compounds. TCDD (see Section 7.10), one of the most troublesome and refractory of wastes,[18] can be treated by ultraviolet light in the presence of hydrogen atom donors {H} resulting in reactions such as the following:

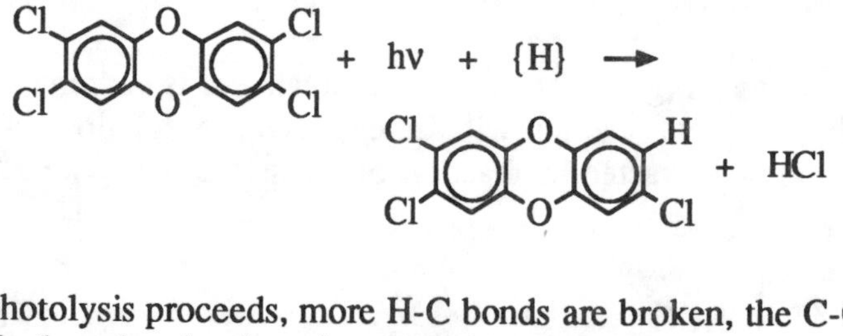

As photolysis proceeds, more H-C bonds are broken, the C-O bonds are broken, and the final product is a harmless organic polymer.

An initial photolysis reaction can result in the generation of reactive intermediates that participate in **chain reactions** that lead to the destruction of a compound. One of the most important reactive intermediates is free radical HO$^\bullet$. In some cases **sensitizers** are added to the reaction mixture to absorb radiation and generate reactive species that destroy wastes.

Hazardous waste substances other than TCDD that have been destroyed by photolysis are herbicides (atrazine), 2,4,6-trinitrotoluene (TNT), and polychlorinated biphenyls (PCBs). The addition of a chemical oxidant, such as potassium peroxidisulfate, $K_2S_2O_8$, enhances destruction by oxidizing active photolytic products.

19.7. THERMAL TREATMENT METHODS

Thermal treatment of hazardous wastes can be used to accomplish most of the common objectives of waste treatment — volume reduction; removal of volatile, combustible, mobile organic matter; and destruction of toxic and pathogenic materials. The most widely applied means of thermal treatment of hazardous wastes is **incineration**. Incineration utilizes high temperatures, an oxidizing atmosphere, and often turbulent combustion conditions to destroy wastes. As of 1989, there were about 100 hazardous waste incinerators operating in the U. S. and another 100 were going through the permitting process.[19] Methods other than incineration that make use of high temperatures to destroy or neutralize hazardous wastes are discussed briefly at the end of this section.

Incineration

Hazardous waste incineration will be defined here as a process that involves exposure of the waste materials to oxidizing conditions at a high temperature, usually in excess of 900°C. Normally the heat required for incineration comes from the oxidation of organically bound carbon and hydrogen contained in the waste material or in supplemental fuel:

$$C(organic) + O_2 \rightarrow CO_2 + heat \tag{19.7.1}$$

$$4H(organic) + O_2 \rightarrow 2H_2O + heat \tag{19.7.2}$$

These reactions destroy organic matter and generate heat required for endothermic reactions, such as the breaking of C-Cl bonds in organochlorine compounds.

Incinerable Wastes

Ideally, incinerable wastes are predominantly organic materials that will burn with a heating value of at least 5,000 Btu/lb and preferably over 8,000 Btu/lb. Such heating values are readily attained with wastes having high contents of the most commonly incinerated waste organic substances, including methanol, acetonitrile, toluene, ethanol, amyl acetate, acetone, xylene, methylethyl ketone, adipic acid, and ethyl acetate. In some cases, however, it is desirable to incinerate wastes that will not burn alone and which require **supplemental fuel,** such as methane and petroleum liquids. Examples of such wastes are nonflammable organochloride wastes, some aqueous wastes, or soil in which the elimination of a particularly troublesome contaminant is worth the expense and trouble of incinerating it. Inorganic matter, water, and organic hetero element contents of liquid wastes are important in determining their incinerability.

Hazardous Waste Fuel

Many industrial wastes, including hazardous wastes, are burned as **hazardous waste fuel** for energy recovery in industrial furnaces and boilers and in incinerators for nonhazardous wastes, such as sewage sludge incinerators. This process is called **coincineration**, and more combustible wastes are utilized by it than are burned solely for the purpose of waste destruction. In addition to heat recovery from combustible wastes, it is a major advantage to use an existing on-site facility for waste disposal rather than a separate hazardous waste incinerator.

Incineration Systems

The four major components of hazardous waste incineration systems are shown in Figure 19.7.

Waste preparation for liquid wastes may require filtration, settling to remove solid material and water, blending to obtain the optimum incinerable mixture, or heating to decrease viscosity. Solids may require shredding and screening. Atomization is commonly used to feed liquid wastes. Several mechanical devices, such as rams and augers, are used to introduce solids into the incinerator.

The most common kinds of **combustion chambers** are liquid injection, fixed hearth, rotary kiln, and fluidized bed. These types are discussed in more detail later in this section.

Often the most complex part of a hazardous waste incineration system is the **air pollution control system**, which involves several operations. The most common operations in air pollution control from hazardous waste incinerators are combustion gas cooling, heat recovery, quenching, particulate matter removal, acid gas removal, and treatment and handling of byproduct solids, sludges, and liquids.

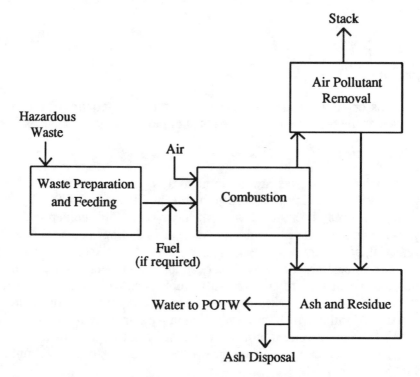

Figure 19.7. Major components of a hazardous waste incinerator system.

Hot ash is often quenched in water. Prior to disposal it may require dewatering and chemical stabilization. A major consideration with hazardous waste incinerators and the types of wastes that are incinerated is the disposal problem posed by the ash, especially in respect to potential leaching of heavy metals.

Types of Incinerators

Hazardous waste incinerators may be divided among the following, based upon type of combustion chamber:[20]

- Rotary kiln (about 40% of U.S. hazardous waste incinerator capacity) in which the primary combustion chamber is a rotating cylinder lined with refractory materials and an afterburner downstream from the kiln to complete destruction of the wastes
- Liquid injection incinerators (also about 40% of U.S. hazardous waste incinerator capacity) that burn pumpable liquid wastes dispersed as small droplets
- **Fixed-hearth incinerators** with single or multiple hearths upon which combustion of liquid or solid wastes occurs.
- **Fluidized-bed incinerators** that have a bed of granular solid (such as sand) maintained in a suspended state by injection of air[21] to remove pollutant acid gas and ash products
- **Advanced design incinerators** including **plasma incinerators** that make use of an extremely hot plasma of ionized air injected through an electrical arc; **electric reactors** that use resistance-heated incinerator walls at around 2,200°C to heat and pyrolyze wastes by radiative heat transfer; **infrared systems**, which generate intense infrared radiation by passing electricity through silicon carbide resistance heating elements; **molten salt combustion** that use a bed of molten sodium carbonate at about 900°C to destroy the wastes and retain gaseous pollutants; and **molten glass processes** that use a pool of molten glass to transfer heat to the waste and to retain products in a poorly leachable glass form

Combustion Conditions

The key to effective incineration of hazardous wastes lies in the combustion conditions. These require (1) sufficient free oxygen in the combustion zone; (2) turbulence for thorough mixing of waste, oxidant, and (where used) supplemental fuel; (3) high combustion temperatures above about 900°C to ensure that thermally resistant compounds do react; and (4) sufficient residence time (at least 2 seconds) to allow reactions to occur.

Effectiveness of Incineration

EPA standards for hazardous waste incineration are based upon the effectiveness of destruction of the **principal organic hazardous constituents** (POHC). Measurement of these compounds before and after incineration gives the **destruction removal efficiency** (DRE) according to the formula,[22]

$$DRE = \frac{W_{in} - W_{out}}{W_{in}} \times 100 \qquad (19.7.3)$$

where W_{in} and W_{out} are the mass flow rates of the principal organic hazardous constituent (POHC) input and output (at the stack downstream from emission controls), respectively.

Wet Air Oxidation

Organic compounds and oxidizable inorganic species can be oxidized by oxygen in aqueous solution.[23] The source of oxygen usually is air. Rather extreme conditions of temperature and pressure are required, with a temperature range of 175-327°C and a pressure range of 300-3,000 psig (2070-20,700 kPa). The high pressures allow a high concentration of oxygen to be dissolved in the water and the high temperatures enable the reaction to occur.

Wet air oxidation has been applied to the destruction of cyanides in electroplating wastewaters.[24] The oxidation reaction for sodium cyanide is the following:

$$2Na^+ + 2CN^- + O_2 + 4H_2O \rightarrow$$
$$2Na^+ + 2HCO_3^- + 2NH_3 \qquad (19.7.4)$$

UV-Enhanced Wet Oxidation

Hydrogen peroxide (H_2O_2) can be used as an oxidant in solution assisted by ultraviolet radiation (hv).[25] For the oxidation of organic species represented in general as {CH_2O}, the overall reaction is

$$2H_2O_2 + \{CH_2O\} + hv \rightarrow CO_2 + 3H_2O \qquad (19.7.5)$$

The ultraviolet radiation breaks chemical bonds and serves to form reactive oxidant species, such as HO•.

19.8. BIODEGRADATION OF WASTES

Biodegradation of wastes is their conversion by biological processes to simple inorganic molecules (mineralization) and, to a certain extent, to biological materials. Usually the products of biodegradation are molecular forms that tend to occur in nature and that are in greater thermodynamic equilibrium with their surroundings than are the starting materials. **Detoxification** refers to the biological conversion of a toxic substance to a less toxic species. Microbial bacteria and fungi possessing enzyme systems required for biodegradation of wastes are usually best obtained from populations of indigenous microorganisms at a hazardous waste site where they have developed the ability to degrade particular kinds of molecules. Although it has some shortcomings in the degradation of complex chemical mixtures,[26] biological treatment offers a number of significant advantages and has considerable potential for the degradation of hazardous wastes, even *in situ*.[27]

Biodegradability

The **biodegradability** of a compound is influenced by its physical characteristics, such as solubility in water and vapor pressure, and by its chemical properties, including molecular mass, molecular structure, and presence of various kinds of functional groups, some of which provide a "biochemical handle" for the initiation of biodegradation. With the appropriate organisms and under the right conditions, even substances such as phenol that are considered to be biocidal to most microorganisms can undergo biodegradation.

Recalcitrant or **biorefractory** substances are those that resist biodegradation and tend to persist and accumulate in the environment. Such materials are not necessarily toxic to organisms, but simply resist their metabolic attack. However, even some compounds regarded as biorefractory may be degraded by microorganisms adapted to their biodegradation; for example DDT is degraded by properly acclimated *Pseudomonas*. Chemical pretreatment, especially by partial oxidation, can make some kinds of recalcitrant wastes much more biodegradable.

Properties of hazardous wastes and their media can be changed to increase biodegradability. This can be accomplished by adjustment of conditions to optimum temperature, pH (usually in the range of 6-9), stirring, oxygen level, and material load. Biodegradation can be aided by removal of toxic organic and inorganic substances, such as heavy metal ions.

Aerobic Treatment

Aerobic waste treatment processes utilize aerobic bacteria and fungi that require molecular oxygen, O_2. These processes are often favored by microorganisms, in part because of the high energy yield obtained when molecular oxygen reacts with organic matter. Aerobic waste treatment is well adapted to the use of an activated sludge process. It can be applied to hazardous wastes such as chemical process wastes and landfill leachates. Some systems use powdered activated carbon as an additive to absorb nonbiodegradable organic wastes.

Contaminated soils can be mixed with water and treated in a bioreactor to eliminate biodegradable contaminants in the soil. It is possible in principle to treat contaminated soils biologically in place by pumping oxygenated, nutrient-enriched water through the soil in a recirculating system.

Anaerobic Treatment

Anaerobic waste treatment in which microorganisms degrade wastes in the absence of oxygen can be practiced on a variety of organic hazardous wastes.[28] Compared to the aerated activated sludge process, anaerobic digestion requires less energy; yields less sludge byproduct; generates sulfide (H_2S), which precipitates toxic heavy metal ions; and produces methane gas, CH_4, which can be used as an energy source.

The overall process for anaerobic digestion, is a fermentation process in which organic matter is both oxidized and reduced. The simplified reaction for the anaerobic fermentation of a hypothetical organic substance, "{CH_2O}", is the following:

$$2\{CH_2O\} \rightarrow CO_2 + CH_4 \tag{19.8.1}$$

In practice, the microbial processes involved are quite complex. Most of the wastes for which anaerobic digestion is suitable consist of oxygenated compounds, such as acetaldehyde or methylethyl ketone.

19.9. LAND TREATMENT AND COMPOSTING

Land Treatment

Soil may be viewed as a natural filter for wastes. Soil has physical, chemical, and biological characteristics that can enable waste detoxification, biodegradation, chemical decomposition, and physical and chemical fixation. Therefore, **land treatment** of wastes may be accompished by mixing the wastes with soil under appropriate conditions.

Soil is a natural medium for a number of living organisms that may have an effect upon biodegradation of hazardous wastes. Of these, the most important are bacteria, including those from the genera *Agrobacterium*, *Arthrobacteri*, *Bacillus*, *Flavobacterium*, and *Pseudomonas*. *Actinomycetes* and fungi are important organisms in decay of vegetable matter and may be involved in biodegradation of wastes.

Wastes that are amenable to land treatment are biodegradable organic substances. However, in soil contaminated with hazardous wastes, bacterial cultures may develop that are effective in degrading normally recalcitrant compounds through acclimation over a long period of time. Land treatment is most used for petroleum refining wastes[29] and is applicable to the treatment of fuels and wastes from leaking underground storage tanks. It can also be applied to biodegradable organic chemical wastes, including some organohalide compounds. Land treatment is not suitable for the treatment of wastes containing acids, bases, toxic inorganic compounds, salts, heavy metals, and organic compounds that are excessively soluble, volatile, or flammable.[30]

Composting

Composting of hazardous wastes is the biodegradation of solid or solidified materials in a medium other than soil. Bulking material, such as plant residue, paper, municipal refuse, or sawdust may be added to retain water and enable air to penetrate to the waste material. Successful composting of hazardous waste depends upon a number of factors,[31] including those discussed above under land farming. The first of these is the selection of the appropriate microorganism or **inoculum**. Once a successful composting operation is underway, a good inoculum is maintained by recirculating spent compost to each new batch. Other parameters that must be controlled include oxygen supply, moisture content (which should be maintained at a minimum of about 40%), pH (usually around neutral), and temperature. The composting process generates heat so, if the mass of the compost pile is sufficiently high, it can be self-heating under most conditions. Some wastes are deficient in nutrients, such as nitrogen, which must be supplied from commercial sources or from other wastes.

19.10. PREPARATION OF WASTES FOR DISPOSAL

Immobilization, stabilization, fixation, and solidification are terms that describe techniques whereby hazardous wastes are placed in a form suitable for long term disposal. These aspects of hazardous waste management are addressed in detail in reference works on the subject.[32,33]

Immobilization

Immobilization includes physical and chemical processes that reduce surface areas of wastes to minimize leaching.[34] It isolates the wastes from their environment, especially groundwater, so that they have the least possible tendency to migrate. This is accomplished by physically isolating the waste, reducing its solubility, and decreasing its surface area. Immobilization usually improves the handling and physical characteristics of wastes.

Stabilization

Stabilization means the conversion of a waste from its original form to a physically and chemically more stable material. Stabilization may include chemical reactions that produce products that are less volatile, soluble, and reactive. Solidification, which is discussed below, is one of the most common means of stabilization. Stabilization is required for land disposal of wastes. **Fixation** is a process that binds a hazardous waste in a less mobile and less toxic form; it means much the same thing as stabilization.

Solidification

Solidification may involve chemical reaction of the waste with the solidification agent, mechanical isolation in a protective binding matrix, or a combination of chemical and physical processes. It can be accomplished by evaporation of water from aqueous wastes or sludges, sorption onto solid material, reaction with cement, reaction with silicates, encapsulation, or imbedding in polymers or thermoplastic materials.

In many solidification processes, such as reaction with Portland cement, water is an important ingredient of the hydrated solid matrix. Therefore, the solid should not be heated excessively or exposed to extremely dry conditions, which could result in diminished structural integrity from loss of water. In some cases, however, heating a solidified waste is an essential part of the overall solidification procedure. For example, an iron hydroxide matrix can be converted to highly insoluble, refractory iron oxide by heating. Organic constituents of solidified wastes may be converted to inert carbon by heating. Heating is an integral part of the process of vitrification (see below).

Sorption to a Solid Matrix Material

Hazardous waste liquids, emulsions, sludges, and free liquids in contact with sludges may be solidified and stabilized by fixing onto solid **sorbents**, including activated carbon (for organics), fly ash, kiln dust, clays, vermiculite, and various proprietary materials. Sorption may be done to convert liquids and semisolids to dry solids, improve waste handling, and reduce solubility of waste constituents. Sorption can also be used to improve waste compatibility with substances such as Portland cement used for solidification and setting. Specific sorbents may also be used to stabilize pH and pE (a measure of the tendency of a medium to be oxidizing or reducing, see Chapter 4).

The action of sorbents can include simple mechanical retention of wastes, physical sorption, and chemical reactions. It is important to match the sorbent to the waste. A substance with a strong affinity for water should be employed for wastes containing excess water and one with a strong affinity for organic materials should be used for wastes with excess organic solvents.

Thermoplastics and Organic Polymers

Thermoplastics are solids or semisolids that become liquified at elevated temperatures. Hazardous waste materials may be mixed with hot thermoplastic liquids and solidified in the cooled thermoplastic matrix, which is rigid but deformable. The thermoplastic material most used for this purpose is asphalt bitumen. Other thermoplastics, such as paraffin and polyethylene have also been used to immobilize hazardous wastes.

Among the wastes that can be immobilized with thermoplastics are those containing heavy metals, such as electroplating wastes. Organic thermoplastics repel water and reduce the tendency toward leaching in contact with groundwater. Compared to cement, thermoplastics add relatively less material to the waste.

A technique similar to that described above uses **organic polymers** produced in contact with solid wastes to imbed the wastes in a polymer matrix. Three kinds of polymers that have been used for this purpose include polybutadiene, urea-formaldehyde, and vinyl ester-styrene polymers. This procedure is more complicated than is the use of thermoplastics but, in favorable cases, yields a product in which the waste is held more strongly.

Vitrification

Vitrification or **glassification** consists of imbedding wastes in a glass material. In this application, glass may be regarded as a high-melting-temperature inorganic thermoplastic. Molten glass can be used, or glass can be synthesized in contact with the waste by mixing and heating with glass constituents—silicon dioxide (SiO_2), sodium carbonate (Na_2CO_3), and calcium oxide (CaO). Other constituents may include boric oxide, B_2O_3, which yields a borosilicate glass that is especially resistant to changes in temperature and chemical attack. In some cases glass is used in conjunction with thermal waste destruction processes, serving to immobilize hazardous waste ash consituents. Some wastes are detrimental to the quality of the glass. Aluminum oxide, for example, may prevent glass from fusing.

Vitrification is relatively complicated and expensive, the latter because of the energy consumed in fusing glass. Despite these disadvantages, it is the best immobilization technique for some special wastes and has been promoted for solidification of radionuclear wastes because glass is chemically inert and resistant to leaching. However, high levels of radioactivity can cause deterioration of glass and lower its resistance to leaching.

Solidification with Cement

Portland cement is widely used for solidification of hazardous wastes. In this application, Portland cement provides a solid matrix for isolation of the wastes, chemically binds water from sludge wastes, and may react chemically with wastes (for example, the calcium and base in Portland cement react chemically with inorganic arsenic sulfide wastes to reduce their solubilities). However, most wastes are held physically in the rigid Portland cement matrix and are subject to leaching.

As a solidification matrix, Portland cement is most applicable to inorganic sludges containing heavy metal ions that form insoluble hydroxides and carbonates in the basic carbonate medium provided by the cement. The success of solidification with Portland cement strongly depends upon whether or not the waste adversely affects the strength and stability of the concrete product. A number of substances — organic matter such as petroleum or coal; some silts and clays; sodium salts of arsenate, borate, phosphate, iodate, and sulfide; and salts of copper, lead, magnesium, tin, and zinc — are incompatible with Portland cement because they interfere with its set and cure and cause deterioration of the cement matrix with time.[35] However, a reasonably good disposal form can be obtained by absorbing organic wastes with a solid material, which in turn is set in Portland cement. This approach has been used with hydrocarbon wastes sorbed by an activated coal char matrix.[36]

Solidification with Silicate Materials

Water-insoluble **silicates**, (pozzolanic substances) containing oxyanionic silicon such as SiO_3^{2-} are used for waste solidification. These substances include fly ash, flue dust, clay, calcium silicates, and ground-up slag from blast furnaces. Soluble silicates, such as sodium silicate, may also be used. Silicate solidification usually requires a setting agent, which may be Portland cement (see above), gypsum (hydrated $CaSO_4$), lime, or compounds of aluminum, magnesium, or iron. The product may vary from a granular material to a concrete-like solid. In some cases the product is improved by additives, such as emulsifiers, surfactants, activators, calcium chloride, clays, carbon, zeolites, and various proprietary materials.

Success has been reported for the solidification of both inorganic wastes and organic wastes (including oily sludges) with silicates. The advantages and disadvantages of silicate solidification are similar to those of Portland cement discussed above. One consideration that is especially applicable to fly ash is the presence in some silicate materials of leachable hazardous substances, which may include arsenic and selenium.

Encapsulation

As the name implies, **encapsulation** is used to coat wastes with an impervious material so that they do not contact their surroundings. For example, a water-soluble waste salt encapsulated in asphalt would not dissolve, so long as the asphalt layer remains intact. A common means of encapsulation uses heated, molten thermoplastics, asphalt, and waxes that solidify when cooled. A more sophisticated approach to encapsulation is to form polymeric resins from monomeric substances in the presence of the waste.

Chemical Fixation

Chemical fixation is a process that binds a hazardous waste substance in a less mobile, less toxic form by a chemical reaction that alters the waste chemically. Physical and chemical fixation often occur together. Polymeric inorganic silicates containing some calcium and often some aluminum are the inorganic materials most widely used as a fixation matrix. Many kinds of heavy metals are chemically bound in such a matrix, as well as being held physically by it. Similarly, some organic wastes are bound by reactions with matrix constituents. For example, humic acid wastes react with calcium ion in a solidification matrix to produce insoluble calcium humates.[37]

19.10. ULTIMATE DISPOSAL OF WASTES

Regardless of the destruction, treatment, and immobilization techniques used, there will always remain from hazardous wastes some material that has to be put somewhere. This section briefly addresses the ultimate disposal of ash, salts, liquids, solidified liquids, and other residues that must be placed where their potential to do harm is minimized.

Disposal Aboveground

In some important respects disposal aboveground, essentially in a pile designed to prevent erosion and water infiltration, is the best way to store solid wastes. Perhaps its most important advantage is that it avoids infiltration by groundwater that can result in leaching and groundwater contamination common to storage in pits and landfills. In a properly designed aboveground disposal facility any leachate that is produced drains quickly by gravity to the leachate collection system, where it can be detected and treated.

Aboveground disposal can be accomplished with a storage mound deposited on a layer of compacted clay covered with impermeable membrane liners laid somewhat above the original soil surface and shaped to allow leachate flow and collection. The slopes around the edges of the storage mound should be sufficiently great to allow good drainage of precipitation, but gentle enough to deter erosion.

Landfill

Landfill historically has been the most common way of disposing of solid hazardous wastes and some liquids, although it is being severely limited in many nations by new regulations and high land costs. Landfill involves disposal that is at least partially underground in excavated cells, quarries, or natural depressions. Usually fill is continued above ground to most efficiently utilize space and provide a grade for drainage of precipitation.

The greatest environmental concern with landfill of hazardous wastes is the generation of leachate from infiltrating surface water and groundwater with resultant contamination of groundwater supplies. Modern hazardous waste landfills provide elaborate systems to contain, collect, and control such leachate.

There are several components to a modern landfill. A landfill should be placed on a compacted low-permeability medium, preferably clay, which is covered by a flexible-membrane liner consisting of water-tight impermeable material. This liner is covered with granular material in which is installed a secondary drainage system. Next is another flexible-membrane liner above which is installed a primary drainage system for the removal of leachate. This drainage system is covered with a layer of granular filter medium, upon which the wastes are placed. In the landfill, wastes of different kinds are separated by berms consisting of clay or soil covered with liner material. When the fill is complete, the waste is capped to prevent surface water infiltration and covered with compacted soil. In addition to leachate collection, provision may be made for a system to treat evolved gases, particularly when methane-generating biodegradable materials are disposed in the landfill.

The flexible-membrane liner made of rubber (including chlorosulfonated polyethylene) or plastic (including chlorinated polyethylene, high-density polyethylene, and polyvinylchloride), is a key component of state-of-the-art landfills. It controls seepage out of, and infiltration into the landfill. Obviously, liners have to meet stringent standards to serve their intended purpose. In addition to being impermeable, the liner material must be strongly resistant to biodegradation, chemical attack, and tearing.

Capping is done to cover the wastes, prevent infiltration of excessive amounts of surface water, and prevent release of wastes to overlying soil and the atmosphere. Caps come in a variety of forms and are often multilayered. Some of the problems that may occur with caps are settling, erosion, ponding, damage by rodents and penetration by plant roots.

Surface Impoundment of Liquids

Many liquid hazardous wastes, slurries, and sludges are placed in **surface impoundments**, which usually serve for treatment and often are designed to be filled in eventually as a landfill disposal site. Most liquid hazardous wastes and a significant fraction of solids are placed in surface impoundments in some stage of treatment, storage, or disposal.

A surface impoundment may consist of an excavated "pit," a structure formed with dikes, or a combination thereof. The construction is similar to that discussed above for landfills in that the bottom and walls should be impermeable to liquids and provision must be made for leachate collection. The chemical and mechanical challenges to liner materials in surface impoundments are severe so that proper geological siting and construction with floors and walls composed of low-permeability soil and clay are important in preventing pollution from these installations.

Deep-well Disposal of Liquids

Deep-well disposal of liquids consists of their injection under pressure to underground strata isolated by impermeable rock strata from aquifers. Early experience with this method was gained in the petroleum industry where disposal is required of large quantities of saline wastewater coproduced with crude oil. The method was later extended to the chemical industry for the disposal of brines, acids, heavy metal solutions, organic liquids, and other liquids.

A number of factors must be considered in deep-well disposal. Wastes are injected into a region of elevated temperature and pressure, which may cause chemical reactions to occur involving the waste constituents and the mineral strata. Oils, solids, and gases in the liquid wastes can cause problems such as clogging. Corrosion may be severe. Microorganisms may have some effects. Most problems from these causes can be mitigated by proper waste pretreatment.

The most serious consideration involving deep-well disposal is the potential contamination of groundwater. Although injection is made into permeable saltwater aquifers presumably isolated from aquifers that contain potable water, contamination may occur. Major routes of contamination include fractures, faults, and other wells. The disposal well itself can act as a route for contamination if it is not properly constructed and cased or if it is damaged.

19.12. LEACHATE AND GAS EMISSIONS

Leachate

The production of contaminated leachate is a possibility with most disposal sites. Therefore, new hazardous waste landfills require leachate collection/treatment systems and many older sites are required to have such systems retrofitted to them. Modern hazardous waste landfills typically have dual leachate collection systems, one located between the two impermeable liners required for the bottom and sides of the landfill and another just above the top liner of the double-liner system. The upper leachate collection system is called the primary leachate collection system, and the bottom is called the secondary leachate collection system. Leachate is collected in perforated pipes that are imbedded in granular drain material.

Chemical and biochemical processes have the potential to cause some problems for leachate collection systems. One such problem is clogging by insoluble manganese(IV) and iron(III) hydrated oxides upon exposure to air as described for water wells in Section 15.6.

Leachate consists of water that has become contaminated by wastes as it passes through a waste disposal site. It contains waste constituents that are soluble, not retained by soil, and not degraded chemically or biochemically. Some potentially harmful leachate constituents are products of chemical or biochemical transformations of wastes.

The best approach to leachate management is to prevent its production by limiting infiltration of water into the site. Rates of leachate production may be very low when sites are selected, designed, and constructed with minimal production of leachate as a major objective. A well-maintained, low-permeability cap over the landfill is very important for leachate minimization.

Hazardous Waste Leachate Treatment

The first step in treating leachate is to characterize it fully, particularly with a thorough chemical analysis of possible waste constituents and their chemical and metabolic products. The biodegradability of leachate constituents should also be determined.

The options available for the treatment of hazardous waste leachate are generally those that can be used for industrial wastewaters. These include biological treatment by an activated sludge, or related process, and sorption by activated carbon usually in columns of granular activated carbon. Hazardous waste leachate can be treated by a variety of chemical processes, including acid/base neutralization, precipitation, and oxidation/reduction. In some cases these treatment steps must precede biological treatment; for example, leachate exhibiting extremes of pH must be neutralized in order for microorganisms to thrive in it. Cyanide in the leachate may be oxidized with chlorine and organics with ozone, hydrogen peroxide promoted with ultraviolet radiation, or dissolved oxygen at high temperatures and pressures. Heavy metals may be precipitated with base, carbonate, or sulfide.

Leachate can be treated by a variety of physical processes. In some cases, simple density separation and sedimentation can be used to remove water-immiscible liquids and solids. Filtration is frequently required and flotation may be useful. Leachate solutes can be concentrated by evaporation, distillation, and membrane processes, including reverse osmosis, hyperfiltration, and ultrafiltration. Organic constituents can be removed by solvent extraction, air stripping, or steam stripping.

Gas Emissions

In the presence of biodegradable wastes, methane and carbon dioxide gases are produced in landfills by anaerobic degradation (see Reaction 19.8.1). Gases may also be produced by chemical processes with improperly pretreated wastes, as would occur in the hydrolysis of calcium carbide to produce acetylene:

$$CaC_2 + 2H_2O \rightarrow 2C_2H_2 + Ca(OH)_2 \tag{19.12.1}$$

Odorous and toxic hydrogen sulfide, H_2S, may be generated by the chemical reaction of sulfides with acids or by the biochemical reduction of sulfate by anaerobic bacteria (*Desulfovibrio*) in the presence of biodegradable organic matter:

$$SO_4^{2-} + 2\{CH_2O\} + 2H^+ \xrightarrow[\text{bacteria}]{\text{Anaerobic}} H_2S + 2CO + 2H_2O \tag{19.12.2}$$

Gases such as these may be toxic, they may burn, or they may explode. Furthermore, gases permeating through landfilled hazardous waste may carry along waste vapors, such as those of volatile aryl compounds and low-molecular-mass chlorinated hydrocarbons. Of these, the ones of most concern are benzene, carbon tetrachloride, chloroform, 1,2-dibromoethane, 1,2-dichloroethane, dichloromethane, tetrachloroethane, 1,1,1-trichloroethane, trichloroethylene, and vinyl chloride. Because of the hazards from these and other volatile species, it is important to minimize production of gases and, if significant amounts of gases are produced, they should be vented or treated by activated carbon sorption or flaring.

19.13. IN-SITU TREATMENT

In-situ treatment refers to waste treatment processes that can be applied to wastes in a disposal site by direct application of treatment processes and reagents to the wastes. Where possible, in-situ treatment is highly desirable as a waste site remediation option.

In-Situ Immobilization

In-situ immobilization is used to convert wastes to insoluble forms that will not leach from the disposal site.[38] Heavy metal contaminants including lead, cadmium, zinc, and mercury, can be immobilized by chemical precipitation as the sulfides by treatment with gaseous H_2S or alkaline Na_2S solution. Disadvantages include the high toxicity of H_2S and the contamination potential of soluble sulfide. Although precipitated metal sulfides should remain as solids in the anaerobic conditions of a landfill, unintentional exposure to air can result in oxidation of the sulfide and remobilization of the metals as soluble sulfate salts.

Oxidation and reduction reactions can be used to immobilize heavy metals in-situ. Oxidation of soluble Fe^{2+} and Mn^{2+} to their insoluble hydrous oxides, $Fe_2O_3 \cdot xH_2O$ and $MnO_2 \cdot xH_2O$, respectively, can precipitate these metal ions and coprecipitate other heavy metal ions. However, subsurface reducing conditions could later result in reformation of soluble reduced species. Reduction can be used in-situ to convert soluble, toxic chromate to insoluble chromium(III).

Chelation may convert metal ions to less mobile forms, although with most agents chelation has the opposite effect. A chelating agent called Tetran is supposed to form metal chelates that are strongly bound to clay minerals. The humin fraction of soil humic substances likewise immobilizes metal ions.

Solidification In-Situ

In situ solidification can be used as a remedial measure at hazardous waste sites. One approach is to inject soluble silicates followed by reagents that cause them to solidify. For example, injection of soluble sodium silicate followed by calcium chloride or lime forms solid calcium silicate.

Detoxification In-Situ

When only one, or a limited number of harmful constituents is present in a waste disposal site, it may be practical to consider detoxification in-situ. This approach is most practical for organic contaminants including pesticides (organophosphate esters and carbamates), amides, and esters. Among the chemical and biochemical processes that can detoxify such materials are chemical and enzymatic oxidation, reduction, and hydrolysis. Chemical oxidants that have been proposed for this purpose include hydrogen peroxide, ozone, and hypochlorite.

Enzyme extracts collected from microbial cultures and purified have been considered for in-situ detoxification. One cell-free enzyme that has been used for detoxification of organophosphate insecticides is parathion hydrolase. The hostile environment of a chemical waste landfill, including the presence of enzyme-inhibiting heavy metal ions, is detrimental to many biochemical approaches to in-situ treatment. Furthermore, most sites contain a mixture of hazardous constituents, which might require several different enzymes for their detoxification.

Permeable Bed Treatment

Some groundwater plumes contaminated by dissolved wastes can be treated by a permeable bed of material placed in a trench through which the groundwater must flow. For example, limestone contained in a permeable bed neutralizes acid and precipitates some kinds of heavy metals as hydroxides or carbonates. Synthetic ion exchange resins can be used in a permeable bed to retain heavy metals and even some anionic species, although competition with ionic species present naturally in the groundwater can cause some problems with the use of ion exchangers. Activated carbon in a permeable bed will remove some organics, especially less soluble, higher molecular mass organic compounds.

Permeable bed treatment requires relatively large quantities of reagent, which argues against the use of activated carbon and ion exchange resins. In such an application it is unlikely that either of these materials could be reclaimed and regenerated as is done when they are used in columns to treat wastewater. Furthermore, ions taken up by ion exchangers and organic species retained by activated carbon may be released at a later time, causing subsequent problems. Finally, a permeable bed that has been truly effective in collecting waste materials may, itself, be considered a hazardous waste requiring special treatment and disposal.

In-Situ Thermal Processes

Heating of wastes in-situ can be used to remove or destroy some kinds of hazardous substances. Both steam injection and radio frequency heating have been proposed for this purpose. Volatile wastes brought to the surface by heating can be collected and held as condensed liquids or by activated carbon.

One approach to immobilizing wastes in-situ is high temperature vitrification using electrical heating. This process involves placing conducting graphite between two electrodes poured on the surface and passing an electrical current between the electrodes. In principle, the graphite becomes very hot and "melts" into the soil leaving a glassy slag in its path. Volatile species evolved are collected and, if the operation is successful, a nonleachable slag is left in place. It is easy to imagine problems that might occur, including difficulties in getting a uniform melt, problems from groundwater infiltration, and very high consumption of electricity.

Soil Washing and Flushing

Extraction with water containing various additives can be used to cleanse soil contaminated with hazardous wastes. When the soil is left in place and the water pumped into and out of it, the process is called **flushing**; when soil is removed and contacted with liquid the process is referred to as **washing**. Here, washing is used as a term applied to both processes.

The composition of the fluid used for soil washing depends upon the contaminants to be removed. The washing medium may consist of pure water or it may contain acids (to leach out metals or neutralize alkaline soil contaminants), bases (to neutralize contaminant acids), chelating agents (to solubilize heavy metals), surfactants (to enhance the removal of organic contaminants from soil and improve the ability of the water to emulsify insoluble organic species), or reducing agents (to reduce oxidized species).[39] Soil contaminants may dissolve, form emulsions, or react chemically. Inorganic species commonly removed from soil by washing include heavy metals salts; lighter aromatic hydrocarbons, such as toluene and xylenes; lighter organohalides, such as trichloro- or tetrachloroethylene; and light-to-medium molecular mass aldehydes and ketones.

LITERATURE CITED

1. Manahan, Stanley E., *Hazardous Waste Chemistry Toxicology and Treatment*, Lewis Publishers/CRC Press, Inc., Boca Raton, Florida, 1990.

2. Bishop, Jim, "Waste Reduction," *Hazmat World*, October, 1988, pp. 56-61.

3. Freeman, Harry M., Ed., *Hazardous Waste Minimization*, McGraw-Hill Publishing Co., New York, 1990.

4. PRC Environmental Management, Inc., *Hazardous Waste Reduction in the Metal Finishing Industry*, Noyes Data Corporation, Park Ridge, NJ, 1989.

5. "Waste Minimization – A New Term for a Tested Practice," *Impact*, **6**(1), IT Corporation, Monroeville, PA, 1987, pp. 1-3.

6. Mueller Associates, Inc., *Waste Oil Reclaiming Technology, Utilization and Disposal*, Noyes Data Corporation, Park Ridge, NJ, 1989.

7. McCabe, Mark M., "Waste Oil," Section 4.1 in *Standard Handbook of Hazardous Waste Treatment and Disposal*, Harry M. Freeman, Ed., McGraw Hill, New York, 1989, pp. 4.3-4.12.

8. Breton, M., *Treatment Technologies for Solvent Containing Wastes*, Noyes Data Corporation, Park Ridge, NJ, 1988.

9. U.S. Environmental Protection Agency, ICF Consulting Associates, Inc., *Solvent Waste Reduction*, Noyes Data Corporation, Park Ridge, NJ, 1990.

10. Donahue, Bernard A., *Reclamation and Reprocessing of Spent Solvents*, Noyes Data Corporation, Park Ridge, NJ, 1989.

11. Holden, Tim, *How to Select Hazardous Waste Treatment Technologies for Soils and Sludges*, Noyes Data Corporation, Park Ridge, NJ, 1989.

12 Stowe, Elizabeth, "Flocculants: Flocculants Provide a Potpourri of Water and Industrial Wastewater Treatment Solutions," *Hazmat World*, March, 1990, 26-31.

13. "72 Wastes Are Added to the Land-Ban List," *Hazmat World*, February, 1989, pp. 16-17.

14. Porter, Mark C., Ed., *Handbook of Industrial Membrane Technology*, Noyes Data Corporation, Park Ridge, NJ, 1989.

15. Bridges, James S., and Dempsey, Clyde R., *Pesticide Waste Disposal Technology*, Noyes Data Corporation, Park Ridge, NJ, 1988.

16. "Handbook of Remedial Action at Waste Disposal Sites," EPA/625/6-85/006, U.S. Environmental Protection Agency, Hazardous Waste Engineering Research Laboratory, Cincinnati, Ohio, 1985, pp. 10-45–9-47.

17. Roy, Kimberly A., "Interox America," *Hazmat World*, March, 1990, 26-31.

18. Arienti, M., *Dioxin-Containing Wastes Treatment Technologies*, Noyes Data Corporation, Park Ridge, NJ, 1988.

19. Hanson, David J., "Hazardous Waste Management: Planning to Avoid Future Problems," *Chemical and Engineering News*, July 31, 1989, pp. 9-18.

20. Oppelt, E. Timothy, "Hazardous Waste Destruction," *Environmental Science and Technology*, **22**, 403–404 (1988).

21. Brunner, Calvin R., "Incineration: Today's Hot Option for Waste Disposal," *Chemical Engineering*, October 12, 1987, pp. 96-106.

22. Mournighan, Robert E., Marta K. Richards, and Howard Wall, "Incinerability Ranking of Hazardous Organic Compounds," presented at the 15th Annual Research Symposium on Remedial Action, Treatment, and Disposal of Hazardous Waste, Cincinnati, Ohio, April 10-12, 1989.

23. "On Site Engineering Report of Treatment Technology Performance and Operation for Wet Air Oxidation of F007 at Zimpro/Passavant, Incorporated in Rothschild, Wisconsin," Office of Solid Waste, U. S. Environmental Protection Agency, Washington, D. C., 1988.

24. Warner, H. Paul, "Destruction of Cyanides in Electroplating Wastewaters Using Wet Air Oxidation," presented at the 15th Annual Research Symposium on Remedial Action, Treatment, and Disposal of Hazardous Waste, Cincinnati, Ohio, April 10-12, 1989.

25. Peroxidation Systems, Inc., "Organic Oxidation On-Site," *Hazmat World*, May, 1989, p. 26.

26. Mueller, James, G., Peter J. Chapman, and P. Hap. Pritchard, "Creosote Contaminated Sites: Their Potential for Bioremediation," *Environmental Science and Technology*, **23**, 1197-1201 (1989).

27. J. M. Thomas and C. H. Ward. "*In Situ* Biorestoration of Organic Contaminants in the Subsurface," *Environmental Science and Technology*, **23**, 760-766 (1989).

28. Torpy, Michael F., Ed., *Anaerobic Treatment of Industrial Wastewaters*, Noyes Data Corporation, Park Ridge, NJ, 1988.

29. Burton, Dudley J., and K. Ravishankar, *Treatment of Hazardous Petrochemical and Petroleum Wastes*, Noyes Data Corporation, Park Ridge, NJ, 1989.

30. Phung, Tan, "Land Treatment of Hazardous Wastes," Section 9.4 in *Standard Handbook of Hazardous Waste Treatment and Disposal*, Freeman, Harry M., Ed., McGraw-Hill Book Company, New York, 1989, pp. 9.41-9.51.

31. Doyle, Richard C., Jenefir D. Isbister, George A. Anspach, David Renard, and Judith F. Kitchens, "Composting Explosives Contaminated Soil," in *Water and Residue Treatment*, Volume II, Hazardous Materials Control Research Institute, Silver Spring, Maryland, 1987, pp. 90-95.

32. Côté, Pierre, and Michael Gilliam, Eds., *Environmental Aspect of Stabilization and Solidification of Hazardous and Radioactive Wastes*, American Society for Testing and Materials, Philadelphia, PA, 1989.

33. Conner, Jesse R., *Chemical Fixation and Solidification of Hazardous Wastes*, Van Nostrand Reinhold, New York, 1990.

34. Arozerena, M. M., E. F. Barth, M. J. Cullinane, P. de Percin, M. Dosani, S. A. Hokanson, L. W. Jones, R. Kravitz, P. G. Malone, H. R. Maxey, C. A. Pryately, T. Whipple, and J. L. Zieleniewski, *Stabilization and Solidification of Hazardous Wastes*, Noyes Data Corporation, Park Ridge, NJ, 1990.

35. "Direct Waste Treatment," Section 10 in *Remedial Action at Waste Disposal Sites*, EPA/625/6-85/006, U.S. Environmental Protection Agency Hazardous Waste Engineering Research Laboratory, Cincinnati, Ohio, 1985, pp.10-1–10-151.

36. "Destruction and Immobilization of Metal-Contaminated PCB Sludges by Gasification on an Activated Char from Subbituminous Coal," Stanley E. Manahan, Shubhender Kapila, Chris Cady, and David Larsen, preprint extended abstract of papers presented before the Division of Environmental Chemistry, American Chemical Society, Dallas, Texas, April, 1989.

37. Manahan, Stanley E., "Humic Substances and the Fates of Hazardous Waste Chemicals," Chapter 6 in *Influence of Aquatic Humic Substances on Fate and Treatment of Pollutants*, American Chemical Society, Washington D.C., 1988.

38. Czupyrna, G., *In Situ Immobilization of Heavy-Metal Contaminated Soils*, Noyes Data Corporation, Park Ridge, NJ, 1989.

39. Raghavan, R., E. Coles, and D. Dietz, "Cleaning Excavated Soil Using Extraction Agents: A State-of-the-Art Review," EPA/600/S2-89/034, United States Environmental Protection Agency, Risk Reduction Engineering Laboratory, Cincinnati, Ohio 1990.

SUPPLEMENTARY REFERENCES

Lankford, Perry W., and W. Wesley Eckenfelder, Jr., Eds., *Toxicity Reduction in Industrial Effluents*, Van Nostrand Reinhold, New York, 1990.

Fleming, J. L., *Volatilization Technologies for Removing Organics from Water*, Noyes Data Corporation, Park Ridge, NJ, 1989.

Surprenant, N., *Halogenated-Organic Containing Wastes Treatment Technologies*, Noyes Data Corporation, Park Ridge, NJ, 1988.

Palmer, S. A. K., *Metal/Cyanide-Containing Wastes Treatment Technologies*, Noyes Data Corporation, Park Ridge, NJ, 1988.

Wilk, L., *Corrosive-Containing Wastes Treatment Technologies*, Noyes Data Corporation, Park Ridge, NJ, 1988.

Makofske, William J., and Michael R. Edelstein, Eds., *Radon and the Environment*, Noyes Data Corporation, Park Ridge, NJ, 1988.

Scholze, R. J.., *Biotechnology for Degradation of Toxic Chemicals in Hazardous Wastes* , Noyes Data Corporation, Park Ridge, NJ, 1988.

McArdle, J. L., *Treatment of Hazardous Waste Leachate*, Noyes Data Corporation, Park Ridge, NJ, 1988.

Glynn, W., *Mobile Waste Processing Systems and Treatment Technologies*, Noyes Data Corporation, Park Ridge, NJ, 1987.

Berger, Bernard B., *Control of Organic Substances in Water and Wastewater*, Noyes Data Corporation, Park Ridge, NJ, 1987.

Flick, Ernest W., *Fungicides, Biocides and Preservatives for Industrial Agricultural Applications*, Noyes Data Corporation, Park Ridge, NJ, 1987.

Grosse, Douglas, *Managing Hazardous Wastes Containing Heavy Metals*, SciTech Publishers, Matawan, NJ, 1990.

Allegri, Theodore C., *Handling and Management of Hazardous Materials and Wastes*, Chapman and Hall, New York, NY, 1986.

Evans, Jeffrey C., Ed., *Toxic and Hazardous Wastes: Proceedings of the Nineteenth Mid-Atlantic Industrial Waste Conference*, Technomic Publishing, Lancaster, PA, 1987.

Lave, Lester B., and Arthur C. Upton, Eds, *Toxic Chemicals, Health, and the Environment*, The Johns Hopkins University Press, Baltimore, MD, 1987.

Kokoszka, Leopold C., and Jared W. Flood, *Environmental Management Handbook: Toxic Chemical Materials and Wastes*, Marcel Dekker, New York, NY, 1989.

Knox, Peter J., Ed., *Resource Recovery of Municipal Solid Wastes*, American Institute of Chemical Engineers, New York, NY 1988.

Goldman, Benjamin A., James A. Hulme, and Cameron Johnson, *Hazardous Waste Management: Reducing the Risk*, Council on Economic Priorities., New York, 1986

Carter, Luther J., *Nuclear Imperatives and Public Trust: Dealing with Radioactive Waste*, Resources for the Future In., Baltimore, MD, 1987.

The Physical Properties of Liquid Metals. Takamichi Iida, Roderick I. L. Guthrie. Oxford University Press, New York, NY, 1988.

Reviews of Environmental Contamination and Toxicology. Vol. 103. George W. Ware, Ed., Springer-Verlag, New York, NY, 1988.

Freeman, Harry M., Ed., *Standard Handbook of Hazardous Waste Treatment and Disposal*, McGraw Hill, New York, 1989.

U.S. Environmental Protection Agency, *Extremely Hazardous Substances: Superfund Chemical Profiles*, Vols. 1 & 2, Noyes Publications, Park Ridge, NJ, 1989.

Palmer, S. A. K., *Metal/Cyanide Containing Wastes: Treatment Technologies*, Noyes Data Corp., Park Ridge, NJ, 1988.

Bridges, James S., and Clyde R. Dempsey, Eds., *Pesticide Waste Disposal Technology*, Noyes Publications, Park Ridge, NJ, 1988.

Bell, John M., Ed., *Proceedings of the 42nd Industrial Waste Conference*, May 12, 13, 14, 1987, Lewis Publishers, Chelsea, Michigan , 1988.

Scholze, R. J., *Biotechnology for Degradation of Toxic Chemicals in Hazardous Wastes*, Noyes Publications, Park Ridge, NJ, 1988.

Nunno, T., *Toxic Waste Minimization in the Printed Circuit Board Industry*, Noyes Data Corp., Park Ridge, NJ, 1989.

Cashman, John R., *Hazardous Materials Emergencies: Response and Control*, 2nd Ed. , Technomic Publishing Co., Lancaster, PA, 1988.

Halogenated-Organic-Containing Wastes: Treatment Technologies, Noyes Publications, Park Ridge, NJ, 1988.

Unterberg, W., *How to Respond to Hazardous Chemical Spills*, Noyes Publications, Park Ridge, NJ, 1988.

Konrad B. Krauskopf, *Radioactive Waste Disposal and Geology*, Chapman and Hall, New York, NY, 1988.

Cheremisinoff, Paul N., *Hazardous Materials Emergency Response Pocket Handbook*, Technomic Publishing, Lancaster, PA, 1989.

Secondary Reclamation of Plastics Waste. Research Report, Phase 1: Development of Techniques for Preparation and Formulation, Plastics Institute of America, Technomic Publishing Co., Lancaster, PA, 1987.

Park, Chris C., *Acid Rain: Rhetoric and Reality*, Methuen, New York, NY, 1988.

Licht, William, *Air Pollution Control Engineering*, 2nd Ed., Marcel Dekker, New York, NY, 1988.

Torpy, Ed., Michael F., *Anaerobic Treatment of Industrial Wastewaters*, Noyes Publications, Park Ridge, NY, 1988.

Benedict, Arthur H., Eliot Epstein, Joel Alpert, *Composting Municipal Sludge: A Technology Evaluation*, Noyes Publications, Park Ridge, NJ, 1988.

Nazaroff, William W., and Anthony V. Nero Jr., Eds., *Radon and Its Decay Products in Indoor Air*, John Wiley & Sons, New York, NY, 1988.

Tardiff, Robert G., Joseph V. Rodricks, Eds., *Toxic Substances and Human Risk*, Plenum Press, New York, NY, 1987.

Worobec, Mary Devine, and Girard Ordway, *Toxic Substances Controls Guide*, BNA Books Distribution Center, Edison, NJ, 1989.

Kittel, J. Howard, Ed., *Radioactive Waste Management Handbook, Vol. I: Near-Surface Land Disposal*, Harwood Academic Publishers, New York, NY, 1989.

Wentz, Charles A., *Hazardous Waste Management*, McGraw-Hill Book Co., New York, NY, 1989.

Salcedo, Rodolfo N., Frank L. Cross, Jr., and Randolph L. Chrismon, *Environmental Impacts of Hazardous Waste Treatment, Storage and Disposal Facilities*, Technomic Publishing Co., Lancaster, PA, 1989.

Hazardous Waste Management Facilities Directory: Treatment, Storage, Disposal and Recycling. U.S. EPA, Noyes Publications, Park Ridge, NJ, 1990.

Holden, Tim, *How to Select Hazardous Waste Treatment Technologies for Soils and Sludges: Alternative, Innovative and Emerging Technologies*, Noyes Publications, Park Ridge, NJ, 1989.

Arthur R. Tarrer, *Reclamation and Reprocessing of Spent Solvents*, Noyes Publications, Park Ridge, NJ, 1989.

Scholze, R. J., *Biotechnology for Degradation of Toxic Chemicals in Hazardous Wastes*, Noyes Publications, Park Ridge, NJ, 1988.

QUESTIONS AND PROBLEMS

1. Place the following hazardous waste management options in order of increasing desirability: (a) treat the waste to make it nonhazardous, (b) do not produce waste, (c) minimize quantities of waste produced, (d) dispose of the waste in a safe manner, (e) recycle the waste.

2. Match the waste recycling process or industry from the column on the left with the kind of material that can be recycled from the list on the right, below:

   1. Recycle as raw material to the generator
   2. Utilization for pollution control or waste treatment
   3. Materials with agricultural uses
   4. Organic substances
   5. Energy production

   (a) Waste alkali
   (b) Hydraulic and lubricating oils
   (c) Incinerable materials
   (d) Incompletely consumed feedstock material
   (e) Waste lime or phosphate-containing sludges

3. What material is recycled using hydrofinishing, treatment with clay, and filtration?

4. What is the "most important operation in solvent purification and recycle" that is used to separate solvents from impurities, water, and other solvents?

5. Dissolved air flotation (DAF) is used in the secondary treatment of wastes. What is the principle of this technique? For what kinds of hazardous waste substances is it most applicable?

6. Match the process or industry from the column on the left with its "phase of waste treatment" from the list on the right, below:

   1. Activated carbon sorption   (a) Primary treatment
   2. Precipitation   (b) Secondary treatment
   3. Reverse osmosis   (c) Polishing
   4. Emulsion breaking
   5. Slurrying

7. Distillation is used in treating and recycling a variety of wastes, including solvents, waste oil, aqueous phenolic wastes, and mixtures of ethylbenzene and styrene. What is the major hazardous waste problem that arises from the use of distillation for waste treatment?

8. Supercritical fluid technology has a great deal of potential for the treatment of hazardous wastes. What are the principles involved with the use of supercritical fluids for waste treatment? Why is this technique especially advantageous? Which substance is most likely to be used as a supercritical fluid in this application? For which kinds of wastes are supercritical fluids most useful?

9. What are some advantages of using acetic acid, compared, for example, to sulfuric acid, as a neutralizing agent for treating waste alkaline materials?

10. Which of the following would be **least likely** to be produced by, or used as a reagent for the removal of heavy metals by their precipitation from solution? (a) Na_2CO_3, (b) CdS, (e) $Cr(OH)_3$, (d) KNO_3, (e) $Ca(OH)_2$

11. Both $NaBH_4$ and Zn are used to remove metals from solution. How do these substances remove metals? What are the forms of the metal products?

12. Of the following, thermal treatment of wastes is **not** useful for (a) volume reduction, (b) destruction of heavy metals, (c) removal of volatile, combustible, mobile organic matter, (d) destruction of pathogenic materials, (e) destruction of toxic substances.

13. From the following, choose the waste liquid that is least amenable to incineration and explain why it is not readily incinerated: (a) methanol, (b) tetrachloroethylene, (c) acetonitrile, (d) toluene, (e) ethanol, (f) acetone.

14. Name and give the advantages of the process that is used to destroy more hazardous wastes by thermal means than are burned solely for the purpose of waste destruction.

15. What is the major advantage of fluidized-bed incinerators from the standpoint of controlling pollutant byproducts?

16. What is the best way to obtain microorganisms to be used in the treatment of hazardous wastes by biodegradation?

17. What are the principles of composting? How is it used to treat hazardous wastes?

18. How is Portland cement used in the treatment of hazardous wastes for disposal? What might be some disadvantages of such a use?

19. What are the advantages of aboveground disposal of hazardous wastes as opposed to burying wastes in landfills?

20. Describe and explain the best approach to managing leachate from hazardous waste disposal sites.

Toxicological Chemistry

20.1. BIOCHEMISTRY AND THE CELL

Ultimately, most pollutants and hazardous substances are of concern because of their toxic effects. **Biochemistry**, the science that deals with chemical processes and materials in living systems,[1] is summarized very briefly here preparatory to discussing the toxic effects of environmental pollutants.

The Cell

Figure 20.1 shows the major features of the **eukaryotic cell**, which is the basic structure in which biochemical processes occur in animals. These features are:

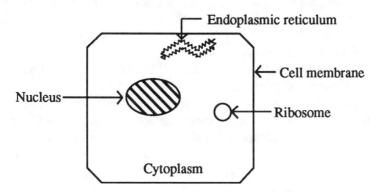

Figure 20.1. Some major features of the eukaryotic cell.

- **Cytoplasm**, which fills the cell and in which several important kinds of cell structures are contained
- **Mitochondria**, which mediate energy conversion and utilization in the cell
- **Ribosomes**, which participate in protein synthesis
- **Endoplasmic reticulum**, which is involved in the metabolism of some toxicants by enzymatic processes
- **Cell membrane**, which encloses the cell and regulates the passage of ions, nutrients, metabolic products, toxicants, and toxicant metabolites into and out of the cell interior; when its membrane is damaged by toxic sustances, a cell may not function properly and the organism may be harmed
- **Cell nucleus**, the "control center" of the cell which contains **deoxyribonucleic** acid through which the nucleus regulates cell division

20.2. PROTEINS, CARBOHYDRATES, AND LIPIDS

Proteins are the basic building blocks of biological material that constitute enzymes and most of the cytoplasm inside the cell. Proteins are composed primarily of biopolymers of **amino acids.**, which have the general structure shown in Figure 20.2, where R represents a group ranging from H atom (in the amino acid glycine) to moderately complex structures. The amino acids in proteins are joined at **peptide linkages** outlined by dashed lines in Figure 20.2.

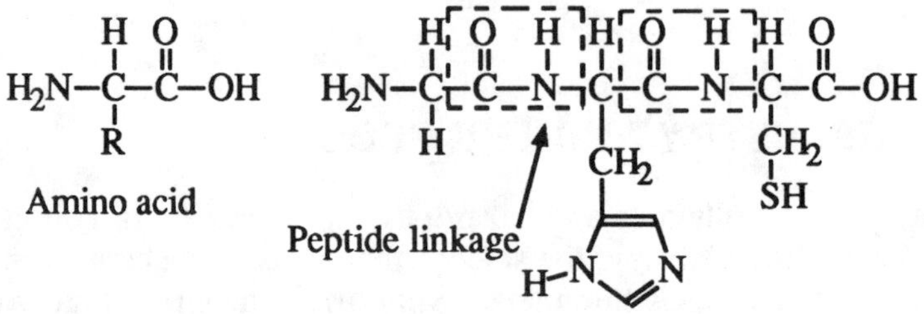

Figure 20.2. A tripeptide formed by the linking of three amino acids. The peptide linkages with which the amino acids are joined are outlined by dashed lines.

The **structures** of protein molecules determine the behavior of proteins in crucial areas such as the processes by which the body's immune system recognizes substances that are foreign to the body. Proteinaceous enzymes depend upon their structures for the very specific functions of the enzymes. The order of amino acids in the protein molecule determines its primary structure. Secondary and tertiary protein structures depend upon the ways in which the polypeptide molecules are bent and folded and by hydrogen bonding (see Section 2.5). The loss of a protein's secondary and tertiary structure is called **denaturation** and may be caused by heat or the action of foreign chemicals. Some corrosive poisons act by denaturing proteins.

Carbohydrates have the approximate simple formula CH_2O and include a diverse range of substances composed of simple sugars such as glucose:

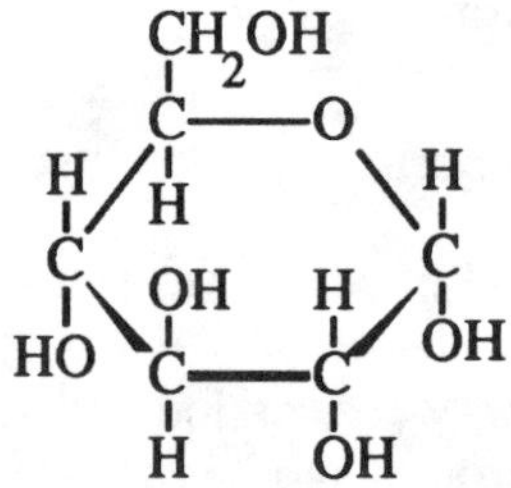

Glucose molecule

High-molecular-mass **polysaccharides**, such as starch and glycogen ("animal starch"), are biopolymers of simple sugars. The major functions of carbohydrates are to store and transfer energy, processes with which toxic substances may interfere.

Lipids are substances that can be extracted from plant or animal matter by organic solvents, such as chloroform, diethyl ether, or toluene. The most common lipids are fats and oils composed of **triglycerides** formed from the alcohol glycerol, $CH_2(OH)CH(OH)CH_2(OH)$ and a long-chain fatty acid such as stearic acid, $CH_3(CH_2)_{16}C(O)OH$ (Figure 20.3). Numerous other biological materials, including waxes, cholesterol, and some vitamins and hormones, are classified as lipids.

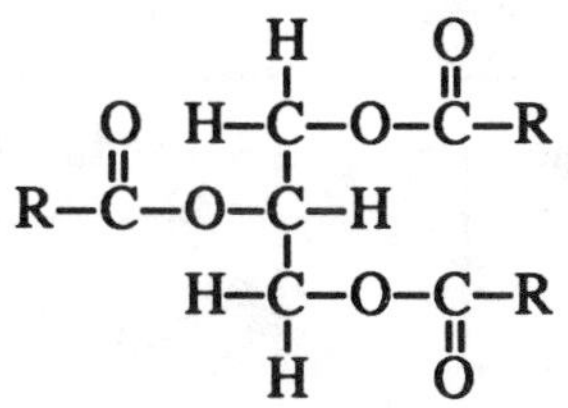

$$O \quad \begin{array}{c} H \quad O \\ | \quad || \\ H-C-O-C-R \\ | \\ R-C-O-C-H \\ || \quad | \\ \quad H-C-O-C-R \\ | \quad || \\ H \quad O \end{array}$$

Figure 20.3. General formula of triglycerides, which make up fats and oils. The R group is from a fatty acid and is a hydrocarbon chain, such as $-(CH_2)_{16}CH_3$.

Lipids are toxicologically important for several reasons. Some toxic substances interfere with lipid metabolism, leading to detrimental accumulation of lipids. Many toxic organic compounds are poorly soluble in water, but are lipid-soluble, so that bodies of lipids in organisms serve to dissolve and store toxicants.

20.3. ENZYMES

Enzymes are proteinaceous substances with highly specific structures that interact with particular substances or classes of substances called **substrates**. Enzymes act as catalysts to enable biochemical reactions to occur, after which they are regenerated intact to take part in additional reactions:

Enzyme + Substrate → Enzyme-substrate complex →

Products + Regenerated enzyme   (20.3.1)

Enzymes are named for what they do. As examples **lipase** enzymes cause lipid triglycerides to dissociate and form glycerol and fatty acids.

Some substances, such as cyanide, heavy metals, or insecticidal parathion, alter or destroy enzymes so that they function improperly or not at all. Toxicants can affect enzymes in several ways. Parathion, for example, bonds covalently to the nerve enzyme acetylcholinesterase, which can then no longer serve to stop nerve impulses. Heavy metals tend to bind to sulfur atoms in enzymes, thereby altering the shape and function of the enzyme.

Enzyme-Catalyzed Reactions of Xenobiotic Substances

The processes by which organisms metabolize **xenobiotic** species (synthetic substances that are foreign to living systems) are enzyme-catalyzed phase I and phase II reactions,[2] which are described briefly here.

Phase I Reactions

Lipophilic xenobiotic species in the body tend to undergo **phase I reactions** that make them more water-soluble and reactive by the attachment of polar functional groups, such as –OH (Figure 20.4). Most Phase I processes are "microsomal mixed-function oxidase" reactions catalyzed by the cytochrome P-450 enzyme system associated with the **endoplasmic reticulum** of the cell and occurring most abundantly in the livers of vertebrates.

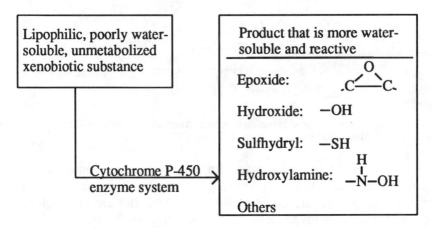

Figure 20.4. Illustration of phase I reactions.

Phase II Reactions

The polar functional groups attached to a xenobiotic compound in a Phase I reaction provide reaction sites for **Phase II** reactions. These are **conjugation reactions** in which enzymes attach **conjugating agents** to xenobiotics, their phase I reaction products, and non-xenobiotic compounds (Figure 20.5). The **conjugation product** of such a reaction is usually less toxic than the original xenobiotic compound, less lipid-soluble, more water-soluble, and more readily eliminated from the body. The major conjugating agents and the enzymes that catalyze their phase II reactions are glucuronide (UDP glucuronyltransferase enzyme), glutathione (glutathionetrans-ferase enzyme), sulfate (sulfotransferase enzyme), and acetyl (acetylation by acetyl-transferase enzymes). The most abundant conjugation products are glucuronides. A glucuronide conjugate is illustrated in Figure 20.6, where -X-R represents a xenobiotic species conjugated to glucuronide and R is an organic moiety. For example, if the xenobiotic compound conjugated is phenol, HXR is HOC_6H_5, X is the O atom, and R represents the phenyl group, C_6H_5.

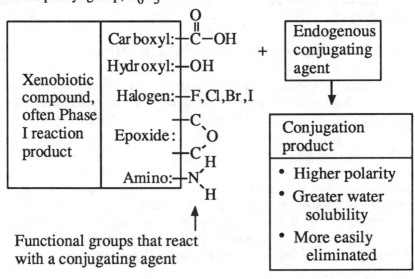

Figure 20.5. Illustration of phase II reactions.

20.4. TOXICOLOGY AND TOXICOLOGICAL CHEMISTRY

Toxicology is the science that deals with the effects of poisons upon living organisms. A **poison** or **toxicant**, is a substance that, above a certain level of exposure or dose, has detrimental effects on tissues, organs, or biological processes. Many toxicants are xenobiotic materials, which were defined and discussed in the preceding section. Among the important aspects of toxicology are the relationship between the demonstrated presence of a chemical or its metabolites in the body and observed symptoms of poisoning, mechanisms by which toxicants are transformed to other species by biochemical processes, the processes by which toxicants and their metabolites are eliminated from an organism, and treatment of poisoning with antidotes.

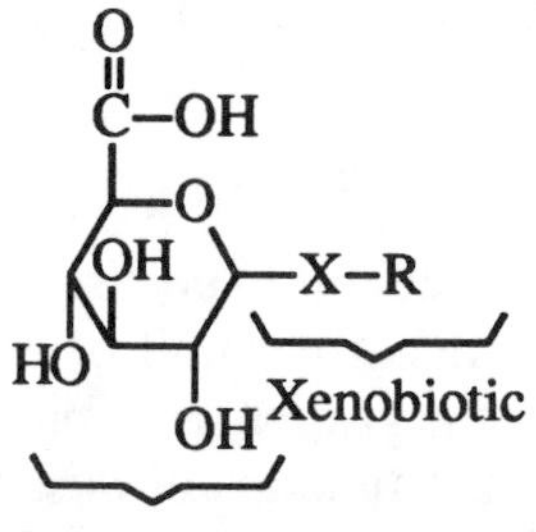

$$
\begin{array}{c}
\text{O} \\
\parallel \\
\text{C-OH} \\
\text{O} \\
\text{OH} \quad \text{X-R} \\
\text{HO} \\
\text{OH} \quad \text{Xenobiotic}
\end{array}
$$

Glucuronide

Figure 20.6. Glucuronide conjugate formed from a xenobiotic, HX-R.

Toxicological Chemistry

Toxicological chemistry is the science that deals with the chemical nature and reactions of toxic substances, including their origins, uses, and chemical aspects of exposure, fates, and disposal.[3] Toxicological chemistry addresses the relationships between the chemical properties and molecular structures of molecules and their toxicological effects. Figure 20.7 outlines the terms discussed above and the relationships among them.

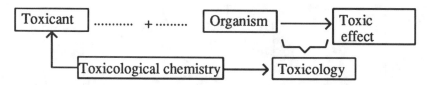

Figure 20.7. Toxicology is the science of poisons. Toxicological chemistry relates toxicology to the chemical nature of toxicants.

Toxicities

The major variables in the ways in which toxicants affect organisms are (1) toxicity of the same substance to different organisms, (2) toxicity of different substances to the same organism, (3) minimum levels for observable toxic effects, (4) sensitivity to small increments of toxicant, (5) levels at which most organisms experience the ultimate effect, particularly death, (6) reversibility of toxic effect, and (7) acute or chronic effect.

When a toxicant leaves no permanent effect, either through the action of the organism's natural defense mechanisms or the administration of substances to counteract the toxicant's action (antidotes), it is said to act in a **reversible** manner. Effects that last after the toxicant is eliminated, such as the scar from a sulfuric acid burn on the skin, are termed **irreversible**.

Acute toxicity refers to responses that are observed soon after exposure to a toxic substance. **Chronic toxicity** deals with effects that take a long time to be manifested. Chronic responses to toxicants may have latency periods as long as several decades in humans. Acute effects normally result from brief exposures to relatively high levels of toxicants and are comparatively easy to observe and relate to exposure to a poison. Chronic effects are often obscured by normal background maladies and tend to result from low exposures to a toxicant over relatively long periods of time. Chronic effects are much more difficult to study, but are of greater importance in dealing with hazardous wastes and pollutants.

Dose-Response Relationship

Dose is defined as the degree of exposure of an organism to a toxicant, commonly in units of mass of toxicant per unit of body mass of the organism. The observed effect of the toxicant is the **response**. A plot of the percentage of organisms that exhibit a particular response as a function of dose is a **dose-response** curve.[4] The statistical estimate of the dose that would cause death in 50 percent of the subjects is the inflection point of the S-shaped dose-response curve, shown in Figure 20.8, and is designated as LD_{50}. The figure shows that LD_{50} values and the slopes of the dose-response curves may differ substantially.

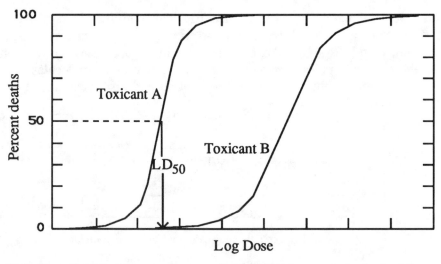

Figure 20.8. Illustration of a dose-response curve in which the response is the death of the organism. The cumulative percentage of deaths of organisms is plotted on the Y axis.

A response to a very low level of toxicant is known as **hypersensitivity**, an allergic reaction, which is an exaggerated response of the body's immune system, whereas response to only extremely high levels is called **hyposensitivity**. In some cases hypersensitivity develops as an extreme reaction to a toxicant after one or more doses of it.

Toxicity Ratings

Substances may be assigned **toxicity ratings** ranging from 1 to 6 as follows: (1) Practically non-toxic, >15 g/kg mass of toxicant per unit body mass; (2) slightly toxic, 5–15 g/kg; (3) moderately toxic, 0.5–5 g/kg; (4) very toxic, 50–500 mg/kg; (5) extremely toxic, 5-50 mg/kg; (6) supertoxic, <5 mg/kg.

Toxicants in the Body

The major routes and sites of absorption, metabolism, binding, and excretion of toxic substances in the body as illustrated in Figure 20.9. Toxicants in the body are metabolized, transported, and excreted; they have adverse biochemical effects; and they cause manifestations of poisoning. In is convenient to divide these processes into two major phases, a kinetic phase and a dynamic phase.

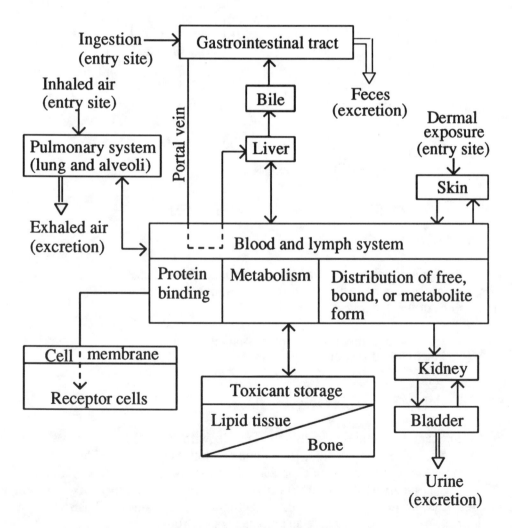

Figure 20.9. Major routes and sites of absorption, metabolism, binding, and excretion of toxic substances in the body.

Kinetic Phase

In the **kinetic phase** a toxicant or the metabolic precursor of a toxic substance (**protoxicant**) may undergo absorption, metabolism, temporary storage, distribution, and excretion, as illustrated in Figure 20.10. A toxicant that is absorbed may be passed through the kinetic phase unchanged as an **active parent compound**, metabolized to a **detoxified metabolite** that is excreted, or converted to a toxic **active metabolite**.

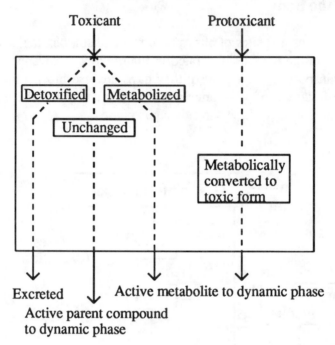

Figure 20.10. Processes involving toxicants or protoxicants in the kinetic phase.

Dynamic Phase

In the **dynamic phase** (Figure 20.11) a toxicant or toxic metabolite interacts with cells, tissues, or organs in the body to cause some toxic response. The three major subdivisions of the dynamic phase are the following:

- **Primary reaction** with a receptor or target organ
- A **biochemical response**
- Observable effects

Biochemical Effects in the Dynamic Phase

A toxicant or an active metabolite reacts with a receptor to cause a toxic response. Such a reaction occurs, for example, when benzene epoxide (see Section 20.12), forms an adduct with a nucleic acid unit in DNA resulting in alteration of the DNA. This is an example of an **irreversible** reaction between a toxicant and a receptor. A **reversible** reaction that can result in a toxic response is illustrated by the binding between carbon monoxide and oxygen-transporting hemoglobin in blood:

$$OHb + CO \longleftrightarrow CoHb + O_2 \tag{20.4.1}$$

Toxicant or toxic metabolite

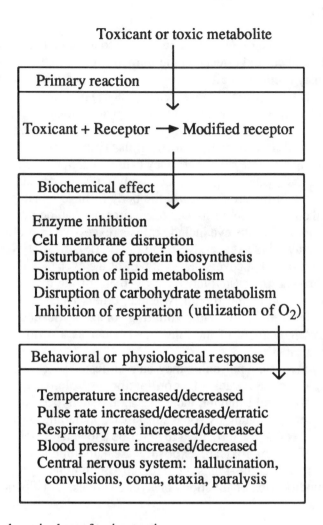

Figure 20.11. The dynamic phase of toxicant action.

The binding of a toxicant to a receptor may result in some kind of biochemical effect. The major ones of these are the following:

- Impairment of enzyme function by binding to the enzyme, coenzymes, metal activators of enzymes, or enzyme substrates

- Alteration of cell membrane or carriers in cell membranes

- Interference with carbohydrate metabolism

- Interference with lipid metabolism resulting in excess lipid accumulation ("fatty liver")

- Interference with respiration, the overall process by which electrons are transferred to molecular oxygen in the biological oxidation of energy-yielding substrates

- Stopping or interfering with protein biosynthesis by the action of toxicants on DNA

- Interference with regulatory processes mediated by hormones or enzymes.

Responses to Toxicants

Prominent among the more chronic responses to toxicant exposure are mutations, cancer, and birth defects and effects on the immune system. Other observable effects, some of which may occur soon after exposure, include, gastrointestinal illness, cardiovascular disease, hepatic (liver) disease, renal (kidney) malfunction, neurologic symptoms (central and peripheral nervous systems), skin abnormalities (rash, dermatitis).

Among the more immediate and readily observed manifestations of poisoning are alterations in the **vital signs** of **temperature, pulse rate, respiratory rate,** and **blood pressure**. Poisoning by some substances may cause an abnormal skin color (jaundiced yellow skin from CCl_4 poisoning) or excessively moist or dry skin. Toxic levels of some materials or their metabolites cause the body to have unnatural **odors**, such as the bitter almond odor of HCN in tissues of victims of cyanide poisoning. Symptoms of poisoning manifested in the eye include **miosis** (excessive or prolonged contraction of the eye pupil), **mydriasis** (excessive pupil dilation), **conjunctivitis** (inflammation of the mucus membrane that covers the front part of the eyeball and the inner lining of the eyelids) and **nystagmus** (involuntary movement of the eyeballs). Some poisons cause a moist condition of the mouth, whereas others cause a dry mouth. Gastrointestinal tract effects including pain, vomiting, or paralytic ileus (stoppage of the normal paristalsis movement of the intestines) occur as a result of poisoning by a number of toxic substances.

Central nervous system poisoning may be manifested by **convulsions, paralysis, hallucinations,** and **ataxia** (lack of coordination of voluntary movements of the body), as well as abnormal behavior, including agitation, hyperactivity, disorientation, and delirium. Severe poisoning by some substances, including organophosphates and carbamates, causes **coma,** the term used to describe a lowered level of consciousness.

Often the effects of toxicant exposure are subclinical in nature. The most common of these are some kinds of damage to immune system, chromosomal abnormalities, modification of functions of liver enzymes, and slowing of conduction of nerve impulses

20.5. TERATOGENESIS, MUTAGENESIS, CARCINOGENESIS, AND IMMUNE SYSTEM EFFECTS

Teratogenesis

Teratogens are chemical species that cause birth defects. These usually arise from damage to embryonic or fetal cells. However, mutations in germ cells (egg or sperm cells) may cause birth defects, such as Down's syndrome.

The biochemical mechanisms of teratogenesis are varied. These include enzyme inhibition by xenobiotics; deprivation of the fetus of essential substrates, such as vitamins; interference with energy supply; or alteration of the permeability of the placental membrane.

Mutagenesis

Mutagens alter DNA to produce inheritable traits. Although mutation is a natural process that occurs even in the absence of xenobiotic substances, most mutations are harmful. The mechanisms of mutagenicity are similar to those of carcinogenicity and mutagens often cause birth defects as well. Therefore, mutagenic hazardous substances are of major toxicological concern.

Carcinogenesis

Chemical carcinogenesis is the term that applies to the role of substances foreign to the body in causing the uncontrolled cell replication commonly known as cancer. In the public eye chemical carcinogenesis is the aspect of toxicology most commonly associated with hazardous substances.

The two major steps in the overall processes by which xenobiotic chemicals cause cancer are an **initiation stage** followed by a **promotional stage**.[5] Chemical carcinogens usually have the ability to form covalent bonds with macromolecular life molecules, especially DNA.[6] This can alter the DNA in a manner such that the cells replicate uncontrollably and form cancerous tissue. Many chemical carcinogens are **alkylating agents**, which act to attach alkyl groups — such as methyl (CH_3) or ethyl (C_2H_5) — or **arylating agents**, which act to attach aryl moieties, such as the phenyl group, C_6H_5, to DNA. Attachment is made through the N and O atoms in the nitrogenous bases that are contained in DNA.

Primary Carcinogens and Procarcinogens

Chemical substances that cause cancer directly are called **primary** or **direct-acting carcinogens**. Most xenobiotics involved in causing cancer are **precarcinogens** or **procarcinogens**. These species require metabolic activation by phase I or phase II reactions to produce **ultimate carcinogens**, which are actually responsible for carcinogenesis.[7]

Only a few chemicals have definitely been established as human carcinogens. A well documented example is vinyl chloride, $CH_2=CHCl$, which is known to have caused a rare form of liver cancer (angiosarcoma) in individuals who cleaned autoclaves in the polyvinylchloride fabrication industry. Animal tests are used to infer chemical carcinogenicity in humans.

Bruce Ames Test

Mutagenicity used to infer carcinogenicity is the basis of the **Bruce Ames** test, in which observations are made of the reversion of mutant histidine-requiring *Salmonella* bacteria back to a form that can synthesize its own histidine.[8] The test makes use of enzymes in homogenized liver tissue to convert potential procarcinogens to ultimate carcinogens. Histidine-requiring *Salmonella* bacteria are inoculated onto a medium that does not contain histidine, and those that mutate back to a form that can synthesize histidine establish visible colonies that are assayed to indicate mutagenicity.

According to Bruce Ames, the developer of the test by the same name, animal tests for carcinogens that make uses of massive doses of chemicals may give results that cannot be accurately extrapolated to assess cancer risks from smaller doses of chemicals.[9] This is because the huge doses of chemicals used kill large numbers of cells, which the organism's body attempts to replace with new cells. Rapidly dividing cells greatly increase the likelihood of mutations that result in cancer simply as the result of rapid cell proliferation, not genotoxicity.

Immune System Response

The **immune system** acts as the body's natural defense system to protect it from xenobiotic chemicals; infectious agents, such as viruses or bacteria; and neoplastic cells, which give rise to cancerous tissue. Adverse effects on the body's immune system are being increasingly recognized as important consequences of exposure to hazardous substances. Toxicants can cause **immunosuppression**, which is the impairment of the body's natural defense mechanisms. Xenobiotics can also cause the immune system to lose its ability to control cell proliferation, resulting in leukemia or lymphoma.

Another major toxic response of the immune system is **allergy** or **hypersensitivity**. This kind of condition results when the immune system overreacts to the presence of a foreign agent or its metabolites in a self-destructive manner. Among the xenobiotic materials that can cause such reactions are beryllium, chromium, nickel, formaldehyde, pesticides, resins, and plasticizers.

20.6. HEALTH HAZARDS

In recent years attention in toxicology has shifted from readily recognized, usually severe, acute maladies that developed on a short time scale as a result of brief, intense exposure to toxicants to delayed, chronic, often less severe illnesses caused by long-term exposure to low levels of toxicants. Although the total impact of the latter kinds of health effects may be substantial, their assessment is very difficult because of factors such as uncertainties in exposure, low occurrence above background levels of disease, and long latency periods.

Assessment of Potential Exposure

A critical step in assessing exposure to toxic substances, such as those from hazardous waste sites is evaluation of potentially exposed populations. The most direct approach to this is to determine chemicals or their metabolic products in organisms. For inorganic species this is most readily done for heavy metals, radionuclides, and some minerals, such as asbestos. Symptoms associated with exposure to particular chemicals may also be evaluated. Examples of such effects include skin rashes or subclinical effects, such as chromosomal damage.

Epidemiologic Evidence

Epidemiologic studies applied to hazardous wastes attempt to correlate observations of particular illnesses with probable exposure to such wastes. There are two major approaches to such studies. One approach is to look for diseases known to be caused by particular agents in areas where exposure is likely from such agents in hazardous wastes. A second approach is to look for **clusters** consisting of an abnormally large number of cases of a particular disease in a limited geographic area, then attempt to locate sources of exposure to hazardous wastes that may be responsible. The most common types of maladies observed in clusters are spontaneous abortions, birth defects, and particular types of cancer.

Epidemiologic studies are complicated by long latency periods from exposure to onset of disease, lack of specificity in the correlation between exposure to a particular waste and the occurrence of a disease, and background levels of a disease in the absence of exposure to a hazardous waste capable of causing the disease.

Estimation of Health Effects Risks

An important part of estimating the risks of adverse health effects from exposure to toxicants involves extrapolation from experimentally observable data. Usually the end result needed is an estimate of a low occurrence of a disease in humans after a long latency period resulting from low-level exposure to a toxicant for a long period of time. The data available are almost always taken from animals exposed at high levels of the substance for a relatively short period of time. Extrapolation is then made using linear or curvilinear projections to estimate the risk to human populations. There are, of course, very substantial uncertainties in this kind of approach.

Risk Assessment

Toxicological considerations are very important in estimating potential dangers of pollutants and hazardous waste chemicals. One of the major ways in which toxicology interfaces with the area of hazardous wastes is in **health risk assessment**, providing guidance for risk management, cleanup, or regulation needed at a hazardous waste site based upon knowledge about the site and the chemical and toxicological properties of wastes in it. Risk assessment includes the factors of site characteristics; substances present, including indicator species; potential receptors; potential exposure pathways; and uncertainty analysis. It may be divided into the following major components:[10]

- Identification of hazard
- Dose-response assessment
- Exposure assessment
- Risk characterization

20.7. TOXIC ELEMENTS AND ELEMENTAL FORMS

This section discusses toxicological aspects of elements (particularly heavy metals) whose presence in a compound frequently means that the compound is toxic, as well as the toxicities of some commonly used elemental forms, such as the chemically uncombined elemental halogens.

Ozone

Ozone (O_3, see Chapters 9, 13, and 14) has several toxic effects. Air containing 1 ppm by volume ozone has a distinct odor. Inhalation of ozone at this level causes severe irritation and headache. Ozone irritates the eyes, upper respiratory system, and lungs. Inhalation of ozone can cause sometimes fatal pulmonary edema. Chromosomal damage has been observed in subjects exposed to ozone.

Ozone generates free radicals in tissue. These reactive species can cause lipid peroxidation, oxidation of sulfhydryl (–SH) groups, and other destructive oxidation processes. Compounds that protect organisms from the effects of ozone include radical scavengers, antioxidants, and compounds containing sulfhydryl groups.

White Phosphorus

Elemental white phosphorus can enter the body by inhalation, by skin contact, or orally. It is a systemic poison, that is, one that is transported through the body to sites remote from its entry site. White phosphorus causes anemia, gastrointestinal system dysfunction, bone brittleness, and eye damage. Exposure also causes **phossy jaw**, a condition in which the jawbone deteriorates and becomes fractured.

Elemental Halogens

Elemental **fluorine** (F_2) is a pale yellow highly reactive gas that is a strong oxidant. It is a toxic irritant and attacks skin and the mucous membranes of the nose and eyes.

Chlorine (Cl_2) gas reacts in water to produce a strongly oxidizing solution. This reaction is responsible for some of the damage caused to the moist tissue lining the respiratory tract when the tissue is exposed to chlorine. The respiratory tract is rapidly irritated by exposure to 10-20 ppm of chlorine gas in air, causing acute discomfort that warns of the presence of the toxicant. Even brief exposure to 1,000 ppm of Cl_2 can be fatal.

Bromine (Br_2) is a volatile dark red liquid that is toxic when inhaled or ingested. Like chlorine and fluorine, it is strongly irritating to the mucous tissue of the respiratory tract and eyes and may cause pulmonary edema. The toxicological hazard of bromine is limited somewhat because its irritating odor elicits a withdrawal response.

Elemental **iodine** (I_2), a solid, is irritating to the lungs much like bromine or chlorine. However, the relatively low vapor pressure of iodine limits exposure to I_2 vapor.

Heavy Metals

Heavy metals (see Section 7.3) are particularly toxic in their chemically combined forms and some, notably mercury, are toxic in the elemental form. The toxic properties of some of the most hazardous heavy metals and metalloids are discussed here.

Although not truly a heavy metal, **beryllium** (atomic mass 9.01) is one of the more hazardous toxic elements. Its most serious toxic effect is berylliosis, a condition manifested by lung fibrosis and pneumonitis, which may develop after a latency period of 5-20 years. Beryllium exposure also causes skin granulomas and ulcerated skin and is a hypersensitizing agent.

Cadmium adversely affects several important enzymes; it can also cause painful osteomalacia (bone disease) and kidney damage. Inhalation of cadmium oxide dusts and fumes results in cadmium pneumonitis characterized by edema and pulmonary epithelium necrosis.

Lead, widely distributed as metallic lead, inorganic compounds, and organometallic compounds, has a number of toxic effects, including inhibition of the synthesis of hemoglobin. It also adversely affects the central and peripheral nervous systems and the kidneys.

Arsenic is a metalloid which forms a number of toxic compounds. The toxic +3 oxide, As_2O_3, is absorbed through the lungs and intestines. Biochemically, arsenic acts to coagulate proteins, forms complexes with coenzymes, and inhibits the production of adenosine triphosphate (ATP) in essential metabolic processes.

Elemental **mercury** vapor can enter the body through inhalation and be carried by the bloodstream to the brain where it penetrates the blood-brain barrier. It disrupts metabolic processes in the brain causing tremor and psychopathological symptoms such as shyness, insomnia, depression, and irritability. Divalent ionic mercury, Hg^{2+}, damages the kidney. Organometallic mercury compounds such as dimethylmercury, $Hg(CH_3)_2$, are also very toxic.

20.8. TOXIC INORGANIC COMPOUNDS

Cyanide

Both **hydrogen cyanide (HCN)** and **cyanide salts** (which contain CN^- ion) are rapidly acting poisons; a dose of only 60–90 mg is sufficient to kill a human. Uses of hydrogen cyanide as a pesticidal fumigant and cyanide salt solutions in chemical synthesis and metal processing pose risks of human exposure.

Metabolically, cyanide bonds to iron(III) in iron-containing ferricytochrome oxidase enzyme (see enzymes, Section 20.3), preventing its reduction to iron(II) in the oxidative phosphorylation process by which the body utilizes O_2. The crucial enzyme is inhibited because ferrouscytochrome oxidase, which is required to react with O_2, is not formed and utilization of oxygen in cells is prevented so that metabolic processes cease.

Carbon Monoxide

Carbon monoxide, CO, is a common cause of accidental poisonings. At CO levels in air of 10 parts per million (ppm) impairment of judgement and visual perception occur; exposure to 100 ppm causes dizziness, headache, and weariness; loss of consciousness occurs at 250 ppm; and inhalation of 1,000 ppm results in rapid death. Chronic long-term exposures to low-levels of carbon monoxide are suspected of causing disorders of the respiratory system and the heart.

After entering the blood stream through the lungs, carbon monoxide reacts with hemoglobin (Hb) to convert oxyhemoglobin (O_2Hb) to carboxyhemoglobin (COHb):

$$O_2Hb + CO \rightarrow COHB + O_2 \qquad (20.8.1)$$

Carboxyhemoglobin is much more stable than oxyhemoglobin so that its formation prevents hemoglobin from carrying oxygen to body tissues.

Nitrogen Oxides

The two most common toxic oxides of nitrogen are NO and NO_2, of which the latter is regarded as the more toxic. Nitrogen dioxide causes severe irritation of the innermost parts of the lungs resulting in pulmonary edema. In cases of severe exposures, fatal bronchiolitis fibrosa obliterans may develop approximately three weeks after exposure to NO_2. Fatalities may result from even brief periods of inhalation of air containing 200–700 ppm of NO_2. Biochemically, NO_2 disrupts lactic dehydrogenase and some other enzyme systems, possibly acting much like ozone, a stronger oxidant discussed later in this chapter. Free radicals, particularly HO·, are likely formed in the body by the action of nitrogen dioxide and the compound probably causes **lipid peroxidation** in which the C=C double bonds in unsaturated body lipids are attacked by free radicals and undergo chain reactions in the presence of O_2, resulting in their oxidative destruction.

Nitrous oxide, N_2O is used as an oxidant gas and in dental surgery as a general anesthetic. This gas was once known as "laughing gas," and was used in the late 1800s as a "recreational gas" at parties held by some of our not-so-staid Victorian ancestors. Nitrous oxide is a central nervous system depressant and can act as an asphyxiant.

Hydrogen Halides

Hydrogen halides (general formula HX, where X is F, Cl, Br, or I) are relatively toxic gases. The most widely used of these gases are HF and HCl; their toxicities are discussed here.

Hydrogen Fluoride

Hydrogen fluoride, (HF, mp -83.1°C, bp 19.5°C) is used as a clear, colorless liquid or gas or as a 30–60% aqueous solution of **hydrofluoric acid**, both referred to here as HF. Both are extreme irritants to any part of the body that they contact, causing ulcers in affected areas of the upper respiratory tract. Lesions caused by contact with HF heal poorly, and tend to develop gangrene.

Fluoride ion, F^-, is toxic in soluble fluoride salts, such as NaF, causing **fluorosis**, a condition characterized by bone abnormalities and mottled, soft teeth. Livestock is especially susceptible to poisoning from fluoride fallout on grazing land; severely afflicted animals become lame and even die. Industrial pollution has been a common source of toxic levels of fluoride. However, about 1 ppm of fluoride used in some drinking water supplies prevents tooth decay.

Hydrogen Chloride

Gaseous **hydrogen chloride** and its aqueous solution, called **hydrochloric acid**, both denoted as HCl are much less toxic than HF. Hydrochloric acid is a natural physiological fluid present as a dilute solution in the stomachs of humans and other animals. However, inhalation of HCl vapor can cause spasms of the larynx as well as pulmonary edema and even death at high levels. The high water affinity of hydrogen chloride vapor tends to dehydrate eye and respiratory tract tissue.

Interhalogen Compounds and Halogen Oxides

Interhalogen compounds, including ClF, BrCl, and BrF_3, are extremely reactive and are potent oxidants. They react with water to produce hydrohalic acid solutions (HF, HCl) and nascent oxygen {O}. Too reactive to enter biological systems in their original chemical state, interhalogen compounds tend to be powerful corrosive irritants that acidify, oxidize, and dehydrate tissue, much like those of the elemental forms of the elements from which they are composed. Because of these effects skin, eyes, and mucous membranes of the mouth, throat, and pulmonary systems are especially susceptible to attack.

Major halogen oxides, including fluorine monoxide (OF_2), chlorine monoxide (Cl_2O), chlorine dioxide (ClO_2), chlorine heptoxide (Cl_2O_7), and bromine monoxide (Br_2O), tend to be unstable, highly reactive, and toxic compounds that pose hazards similar to those of the interhalogen compounds discussed previously in this section. Chlorine dioxide, the most commonly used halogen oxide is employed for odor control and bleaching wood pulp. As a substitute for chlorine in water disinfection, it produces fewer undesirable chemical byproducts, particularly trihalomethanes.

The most important of the oxyacids and their salts formed by halogens are hypochlorous acid, HOCl, and hypochlorites, such as NaOCl, used for bleaching and disinfection. The hypochlorites irritate eye, skin, and mucous membrane tissue because they react to produce active (nascent) oxygen ($\{O\}$) and acid as shown by the reaction below:

$$HClO \rightarrow H^+ + Cl^- + \{O\} \tag{20.8.2}$$

Inorganic Compounds of Silicon

Silica (SiO_2, quartz) occurs in a variety of minerals such as sand, sandstone, and diatomaceous earth. **Silicosis** resulting from human exposure to silica dust from construction materials, sand blasting, and other sources has been a common occupational disease. A type of pulmonary fibrosis that causes lung nodules and makes victims more susceptible to pneumonia and other lung diseases, silicosis is one of the most common disabling conditions resulting from industrial exposure to hazardous substances. It can cause death from insufficient oxygen or from heart failure in severe cases.

Silane, SiH_4, and disilane, H_3SiSiH_3, are examples of inorganic **silanes**, which have H-Si bonds. Numerous organic ("organometallic") silanes exist in which alkyl moieties are substituted for H. Little information is available regarding the toxicities of silanes.

Silicon tetrachloride, $SiCl_4$, is the only industrially significant of the **silicon tetrahalides**, a group of compounds with the general formula SiX_4, where X is a halogen. The two commercially produced **silicon halohydrides**, general formula $H_{4-x}SiX_x$, are dichlorosilane (SiH_2Cl_2) and trichlorosilane, ($SiHCl_3$). These compounds are used as intermediates in the synthesis of organosilicon compounds and in the production of high-purity silicon for semiconductors. Silicon tetrachloride and trichlorosilane, fuming liquids which react with water to give off HCl vapor, have suffocating odors and are irritants to eye, nasal, and lung tissue.

Asbestos

Asbestos is the name given to a group of fibrous silicate minerals, typically those of the serpentine group, for which the approximate formula is $Mg_3P(Si_2O_5)(OH)_4$. Asbestos has been widely used in structural materials, brake linings, insulation, and pipe manufacture.[11] Inhalation of asbestos may cause asbestosis (a pneumonia condition), mesothelioma (tumor of the mesothelial tissue lining the chest cavity adjacent to the lungs), and bronchogenic carcinoma (cancer originating with the air passages in the lungs)[12] so that uses of asbestos have been severely curtailed and widespread programs have been undertaken to remove the material from buildings.

Inorganic Phosphorus Compounds

Phosphine (PH_3), a colorless gas that undergoes autoignition at 100°C, is used for the synthesis of organophosphorus compounds and is sometimes inadvertently produced in chemical syntheses involving other phosphorus compounds. It is a potential hazard in industrial processes and in the laboratory. Symptoms of poisoning from potentially fatal phosphine gas include pulmonary tract irritation, central nervous system depression, fatigue, vomiting, and difficult, painful breathing.

Phosphorus pentoxide, H_3PO_4, is produced as a fluffy white powder from the combustion of elemental phosphorus and reacts with water from air to form syrupy orthophosphoric acid, Because of the formation of acid by this reaction and its dehydrating action, P_4O_{10} is a corrosive irritant to skin, eyes and mucous membranes.

The most important of the **phosphorus halides**, general formulas PX_3 and PX_5, is phosphorus pentachloride used as a catalyst in organic synthesis, as a chlorinating agent and as a raw material to make phosphorus oxychloride ($POCl_3$). Because they react violently with water to produce the corresponding hydrogen halides and oxo phosphorus acids,

$$PCl_5 + 4H_2O \rightarrow H_3PO_4 + 5HCl \qquad\qquad (20.8.3)$$

the phosphorus halides are strong irritants to eyes, skin, and mucous membranes.

The major **phosphorus oxyhalide** in commercial use is phosphorus oxychloride ($POCl_3$) a faintly yellow fuming liquid. Reacting with water to form toxic vapors of hydrochloric acid and phosphonic acid (H_3PO_3), phosphorus oxyhalide is a strong irritant to the eyes, skin, and mucous membranes.

Inorganic Compounds of Sulfur

A colorless gas with a foul rotten-egg odor, **hydrogen sulfide** is very toxic. In some cases inhalation of H_2S kills faster than even hydrogen cyanide; rapid death ensues from exposure to air containing more than about 1000 ppm H_2S due to asphyxiation from respiratory system paralysis. Lower doses cause symptoms that include headache, dizziness, and excitement because of damage to the central nervous system. General debility is one of the numerous effects of chronic H_2S poisoning.

Sulfur dioxide, SO_2, dissolves in water, to produce sulfurous acid, H_2SO_3; hydrogen sulfite ion, HSO_3^-; and sulfite ion, SO_3^{2-}. Because of its water solubility, sulfur dioxide is largely removed in the upper respiratory tract. It is an irritant to the eyes, skin, mucous membranes and respiratory tract. Some individuals are hypersensitive to sodium sulfite (Na_2SO_3), which has been used as a chemical food preservative. These uses were further severely restricted in the U.S. in early 1990.

Number one in synthetic chemical production, **sulfuric acid** (H_2SO_4) is a severely corrosive poison and dehydrating agent in the concentrated liquid form; it readily penetrates skin to reach subcutaneous tissue causing tissue necrosis with effects resembling those of severe thermal burns. Sulfuric acid fumes and mists irritate eye and respiratory tract tissue and industrial exposure has caused tooth erosion in workers.

The more important halides, oxides and oxyhalides of sulfur are listed in Table 20.1. The major toxic effects of these compounds are given in the table.

Organometallic Compounds

The toxicological properties of some organometallic compounds — pharmaceutical organoarsenicals, organomercury fungicides, and tetraethyllead anti-knock gasoline additives — that have been used for many years are well known.

However, toxicological experience is lacking for many relatively new organometallic compounds that are now being used in semiconductors, as catalysis, and for chemical synthesis, so they should be treated with great caution until proven safe.

Table 20.1. Inorganic Sulfur Compounds

Compound name	Formula	Properties
Sulfur		
Monofluoride	S_2F_2	Colorless gas, mp -104°C, bp 99°C, toxicity similar to HF
Tetrafluoride	SF_4	Gas, bp -40°C, mp -124°C, powerful irritant
Hexafluoride	SF_6	Colorless gas, mp -51°C, surprisingly nontoxic when pure, but often contaminated with toxic lower fluorides
Monochloride	S_2Cl_2	Oily, fuming orange liquid, mp -80°C, bp 138°C, strong irritant to eyes, skin, and lungs
Tetrachloride	SCl_4	Brownish/yellow liquid/gas, mp -30°C, Decom. below 0°C, irritant
Trioxide	SO_3	Solid anhydride of sulfuric acid reacts with moisture or steam to produce sulfuric acid
Sulfuryl chloride	SO_2Cl_2	Colorless liquid, mp -54°C, bp 69°C, used for organic synthesis, corrosive toxic irritant
Thionyl chloride	$SOCl_2$	Colorless-to-orange fuming liquid, mp -105°C, bp 79°C, toxic corrosive irritant
Carbon oxysulfide	COS	Volatile liquid byproduct of natural gas or petroleum refining, toxic narcotic
Carbon disulfide	CS_2	Colorless liquid, industrial chemical, narcotic and central nervous system anesthetic

Organometallic compounds often behave in the body in ways totally unlike the inorganic forms of the metals that they contain. This is due in large part to the fact that, compared to inorganic forms, organometallic compounds have an organic nature and higher lipid solubility.

Organolead Compounds

Perhaps the most notable toxic organometallic compound is tetraethyllead, $Pb(C_2H_5)$, a colorless, oily liquid that was widely used as an octane-boosting gasoline additive. Tetraethyllead has a strong affinity for lipids and can enter the body by inhalation, ingestion, and absorption through the skin. Acting differently from inorganic compounds in the body, it affects the central nervous system with symptoms such as fatigue, weakness, restlessness, ataxia, psychosis, and convulsions. Recovery from severe lead poisoning tends to be slow. In cases of fatal tetraethyllead poisoning, death has occurred as soon as one or two days after exposure.

Organotin Compounds

The greatest number of organometallic compounds in commercial use are those of tin — tributyltin chloride and related tributyltin (TBT) compounds. These compounds have bactericidal, fungicidal, and insecticidal properties and have particular environmental significance because of their increasing applications as industrial biocides.[13,14] Organotin compounds are readily absorbed through the skin, sometimes causing a skin rash. They probably bind with sulfur groups on proteins and appear to interfere with mitochondrial function.

Carbonyls

Metal carbonyls regarded as extremely hazardous because of their toxicities include nickel carbonyl ($Ni(CO)_4$), cobalt carbonyl, and iron pentacarbonyl. Some of the hazardous carbonyls are volatile and readily taken into the body through the respiratory tract or through the skin. The carbonyls affect tissue directly and they break down to toxic carbon monoxide and products of the metal, which have additional toxic effects.

Reaction Products of Organometallic Compounds

An example of the production of a toxic substance from the burning of an organometallic compound is provided by the oxidation of diethylzinc:

$$Zn(C_2H_5)_2 + 7O_2 \rightarrow ZnO(s) + 5H_2O(g) + 4CO_2(g) \qquad (20.8.4)$$

Zinc oxide is used as a healing agent and food additive. However, inhalation of zinc oxide fume particles produced by the combustion of zinc organometallic compounds causes zinc **metal fume fever**. This is an uncomfortable condition characterized by elevated temperature and "chills."

20.9. TOXICOLOGY OF ORGANIC COMPOUNDS

Alkane Hydrocarbons

Methane, ethane, *n*-butane, and isobutane (both C_4H_{10}) are regarded as **simple asphyxiants** which deprive air of sufficient oxygen to support respiration. The most common toxicological occupational problem associated with the use of hydrocarbon liquids in the workplace is dermatitis caused by dissolution of the fat portions of the skin and characterized by inflamed, dry, scaly skin. Inhalation of volatile liquid 5–8 carbon *n*-alkanes and branched-chain alkanes may cause central nervous system depression manifested by dizziness and loss of coordination. Exposure to *n*-hexane and cyclohexane results in loss of myelin (a fatty substance constituting a sheath around certain nerve fibers) and degeneration of axons (part of a nerve cell through which nerve impulses are transferred out of the cell). This has resulted in multiple disorders of the nervous system (**polyneuropathy**) including muscle weakness and impaired sensory function of the hands and feet.

Alkene and Alkyne Hydrocarbons

Ethylene, a widely used colorless gas with a somewhat sweet odor, acts as a simple asphyxiant and anesthetic to animals and is phytotoxic (toxic to plants). The toxicological properties of propylene (C_3H_6) are very similar to those of ethylene. Colorless, odorless gaseous 1,3-butadiene is an irritant to eyes and respiratory system mucous membranes; at higher levels it can cause unconsciousness and even death. Acetylene, H-C≡C-H, is a colorless gas with an odor resembling garlic. It acts as an asphyxiant and narcotic, causing headache, dizziness, and gastric disturbances. Some of these effects may be due to the presence of impurities in the commercial product.

Benzene and Aromatic Hydrocarbons

Inhaled benzene is readily absorbed by blood, from which it is strongly taken up by fatty tissues. For the non-metabolized compound, the process is reversible and benzene is excreted through the lungs. As shown in Figure 20.12, benzene is converted to phenol by a Phase I oxidation reaction (see Section 20.3) in the liver. The benzene epoxide intermediate in this reaction is probably responsible for the unique toxicity of benzene, which involves damage to bone marrow.

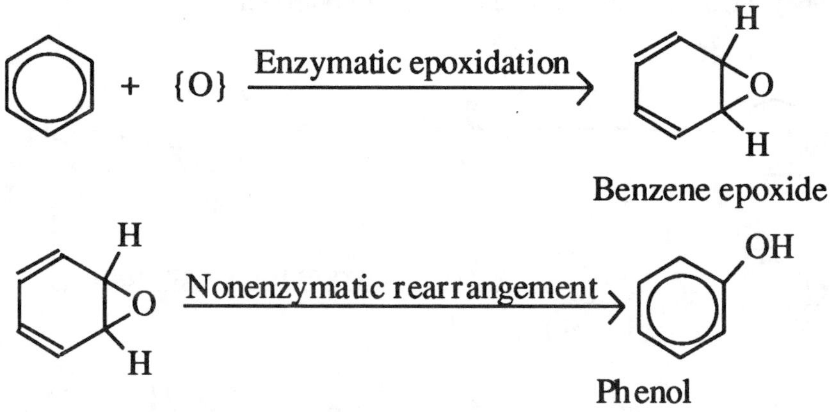

Figure 20.12. Conversion of benzene to phenol in the body.

Benzene is a skin irritant, and progressively higher local exposures can cause skin redness (erythema), burning sensations, fluid accumulation (edema) and blistering. Inhalation of air containing about 7 g/m^3 of benzene causes acute poisoning within an hour, because of a narcotic effect upon the central nervous system manifested progressively by excitation, depression, respiratory system failure, and death. Inhalation of air containing more than about 60 g/m^3 of benzene can be fatal within a few minutes.

Long-term exposures to lower levels of benzene cause non-specific symptoms, including fatigue, headache, and appetite loss. Chronic benzene poisoning causes blood abnormalities, including a lowered white cell count, an abnormal increase in blood lymphocytes (colorless corpuscles introduced to the blood from the lymph glands), anemia, a decrease in the number of blood platelets required for clotting (thrombocytopenia), and damage to bone marrow. It is thought that preleukemia, leukemia, or cancer may result.

Toluene

Toluene, a colorless liquid boiling at 101.4°C, is classified as moderately toxic through inhalation or ingestion; it has a low toxicity by dermal exposure. Toluene can be tolerated without noticeable ill effects in ambient air up to 200 ppm. Exposure to 500 ppm may cause headache, nausea, lassitude, and impaired coordination without detectable physiological effects. Massive exposure to toluene has a narcotic effect, which can lead to coma. Because it possesses an aliphatic side-chain that can be oxidized enzymatically leading to products that are readily excreted from the body (see the metabolic reaction scheme in Figure 20.13), toluene is much less toxic than benzene.

Naphthalene

As is the case with benzene, **naphthalene** undergoes a Phase I oxidation reaction that places an epoxide group on the aromatic ring. This process is followed by Phase II conjugation reactions to yield products that can be eliminated from the body.

Exposure to naphthalene can cause anemia and marked reductions in red cell count, hemoglobin, and hematocrit in genetically susceptible individuals. Naphthalene causes skin irritation or severe dermatitis in sensitized individuals. Headaches, confusion, and vomiting may result from inhalation or ingestion of naphthalene. Death from kidney failure occurs in severe instances of poisoning.

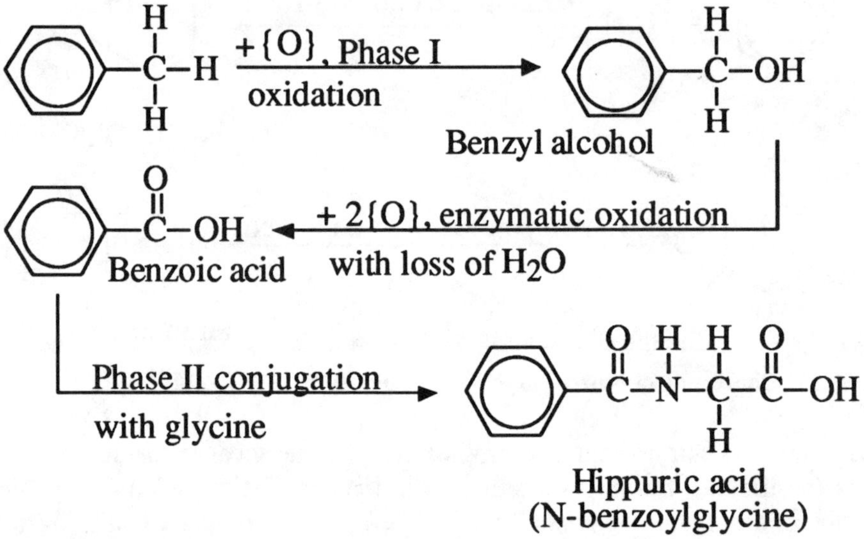

Figure 20.13. Metabolic oxidation of toluene with conjugation to hippuric acid, which is excreted with urine.

Polycyclic Aromatic Hydrocarbons

Benzo(a)pyrene (see Section 12.4) is the most studied of the polycyclic aromatic hydrocarbons (PAHs). Some metabolites of PAH compounds, particularly the 7,8-diol-9,10 epoxide of benzo(a)pyrene shown in Figure 20.14 are known to cause cancer. There are two stereoisomers of this metabolite, both of which are known to be potent mutagens and presumably can cause cancer.

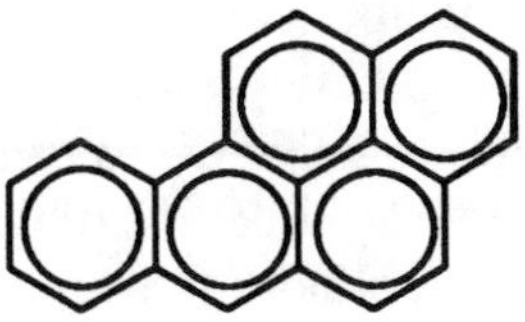

Benzo(a)pyrene

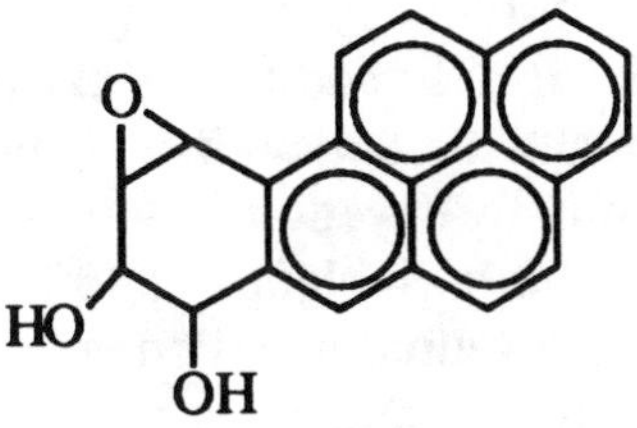

7,8-Diol-9,10-epoxide
of benzo(a)pyrene

Figure 20.14. Benzo(a)pyrene and its carcinogenic metabolic product.

Oxygen-Containing Organic Compounds

Oxides

Hydrocarbon **oxides** such as ethylene oxide and propylene oxide,

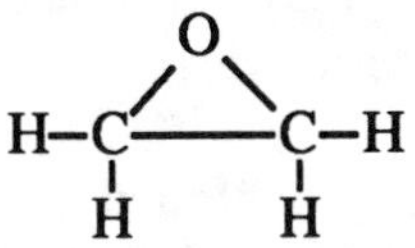

Ethylene oxide 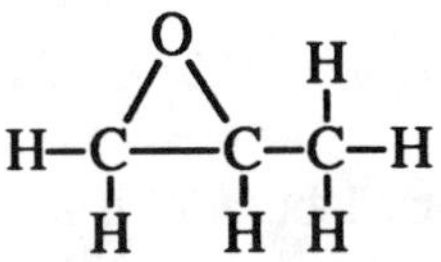   Propylene oxide

which are characterized by an **epoxide** functional group bridging oxygen between two adjacent C atoms, are significant for both their uses and their toxic effects. Ethylene oxide, a gaseous colorless, sweet-smelling, flammable, explosive gas used as a chemical intermediate, sterilant, and fumigant, has a moderate to high toxicity, is a mutagen, and is carcinogenic to experimental animals. Inhalation of relatively low levels of this gas results in respiratory tract irritation, headache, drowsiness, and dyspnea, whereas exposure to higher levels causes cyanosis, pulmonary edema, kidney damage, peripheral nerve damage, and even death. Propylene oxide is a colorless, reactive, volatile liquid (bp 34°C) with uses similar to those of ethylene oxide and similar, though less severe, toxic effects. The toxicity of 1,2,3,4-butadiene epoxide, the oxidation product of 1,3-butadiene, is notable in that it is a direct–acting (primary) carcinogen.

Alcohols

Human exposure to the three light alcohols shown in Figure 20.15 is common because they are widely used industrially and in consumer products.

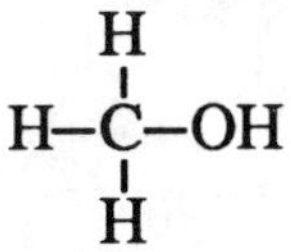

Methanol 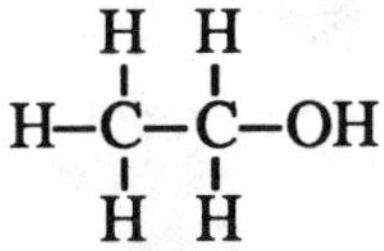   Ethanol 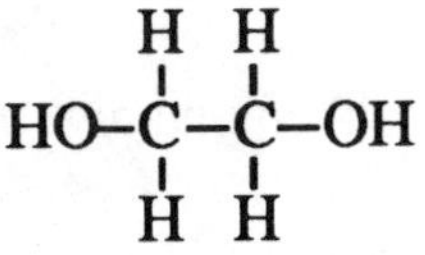   Ethylene glycol

Figure 20.15. **Alcohols** such as these three compounds are oxygenated compounds in which the hydroxyl functional group is attached to an aliphatic or olefinic hydrocarbon skeleton.

Methanol, which has caused many fatalities when ingested accidentally or consumed as a substitute for beverage ethanol, is metabolically oxidized to formaldehyde and formic acid. In addition to causing acidosis, these products affect the central nervous system and the optic nerve. Acute exposure to lethal doses causes an initially mild inebriation, followed in about 10–20 hours by unconsciousness, cardiac depression, and death. Sublethal exposures can cause blindness from deterioration of the optic nerve and retinal ganglion cells. Inhalation of methanol fumes may result in chronic, low level exposure.

Ethanol is usually ingested through the gastrointestinal tract, but can be absorbed as vapor by the alveoli of the lungs. Ethanol is oxidized metabolically more rapidly than methanol, first to acetaldehyde (discussed later in this section), then to CO_2. Ethanol has numerous acute effects resulting from central nervous system depression. These range from decreased inhibitions and slowed reaction times at 0.05% blood ethanol, through intoxication, stupor and — at more than 0.5% blood ethanol — death.

Despite its widespread use in automobile cooling systems, exposure to ethylene glycol is limited by its low vapor pressure. However, inhalation of droplets of ethylene glycol can be very dangerous. In the body, ethylene glycol initially stimulates the central nervous system, then depresses it. Glycolic acid, $HOCH_2CO_2H$, formed as an intermediate metabolite in the metabolism of ethylene glycol, may cause acedemia and oxalic acid produced by further oxidation may precipitate as kidney-damaging calcium oxalate, CaC_2O_4.

Of the higher alcohols, 1-butanol is an irritant, but its toxicity is limited by its low vapor pressure. Unsaturated (alkenyl) allyl alcohol, $CH_2=CHCH_2OH$, has a pungent odor and is strongly irritating to eyes, mouth, and lungs.

Phenols

Figure 20.16 shows some of the more important phenolic compounds, aryl analogs of alcohols which have properties much different from those of the aliphatic

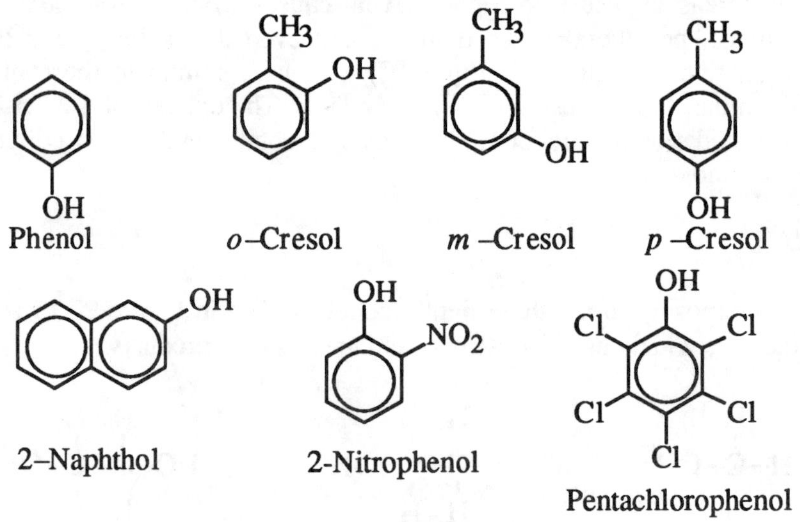

Figure 20.16. Some phenols and phenolic compounds.

and olefinic alcohols. Nitro groups ($-NO_2$) and halogen atoms (particularly Cl) bonded to the aromatic rings strongly affect the chemical and toxicological behavior of phenolic compounds.

Although the first antiseptic used on wounds and in surgery, phenol is a protoplasmic poison that damages all kinds of cells and is alleged to have caused "an astonishing number of poisonings" since it came into general use.[15] The acute toxicological effects of phenol are predominantly upon the central nervous system and death can occur as soon as one-half hour after exposure. Acute poisoning by phenol can cause severe gastrointestinal disturbances, kidney malfunction, circulatory system failure, lung edema, and convulsions. Fatal doses of phenol may be absorbed through the skin. Key organs damaged by chronic exposure to phenol include the spleen, pancreas, and kidneys. The toxic effects of other phenols resemble those of phenol.

Aldehydes and Ketones

Aldehydes and ketones are compounds that contain the carbonyl (C=O) group, as shown by the examples in Figure 20.17.

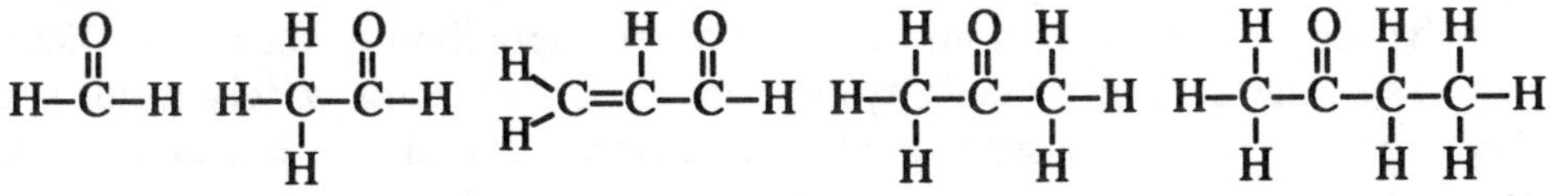

Figure 20.17. Commercially and toxicologically significant aldehydes and ketones.

Formaldehyde is uniquely important because of its widespread use and toxicity. In the pure form formaldehyde is a colorless gas with a pungent, suffocating odor and **formalin** is a 37–50% aqueous solution of formaldehyde containing some methanol. Exposure to inhaled formaldehyde via the respiratory tract is usually to molecular formaldehyde vapor, whereas exposure by other routes is usually to formalin. Prolonged, continuous exposure to formaldehyde can cause hypersensitivity. A severe irritant to the mucous membrane linings of both the respiratory and alimentary tracts, formaldehyde reacts strongly with functional groups in molecules. Formaldehyde has been shown to be a lung carcinogen in experimental animals. The toxicity of formaldehyde is largely due to its metabolic oxidation product, formic acid (see below).

The lower aldehydes are relatively water-soluble and intensely irritating. These compounds attack exposed moist tissue, particularly the eyes and mucous membranes of the upper respiratory tract. (Some of the irritating properties of photochemical smog, Chapter 13, are due to the presence of aldehydes.) However, aldehydes that are relatively less soluble can penetrate further into the respiratory tract and affect the lungs. Colorless, liquid acetaldehyde is relatively less toxic than acrolein and acts as an irritant and systemically as a narcotic to the central nervous system. Extremely irritating, lachrimating acrolein vapor has a choking odor and inhalation of it can cause severe damage to respiratory tract membranes. Tissue exposed to acrolein may undergo severe necrosis, and direct contact with the eye can be especially hazardous.

The ketones shown in Figure 20.17 are relatively less toxic than the aldehydes. Pleasant smelling acetone can act as a narcotic and causes dermatitis by dissolving fats from skin. Not many toxic effects have been attributed to methylethyl ketone exposure. It is suspected of having caused neuropathic disorders in shoe factory workers.

Carboxylic Acids

Formic acid, HCO_2H, is a relatively strong acid that is corrosive to tissue. In Europe, decalcifier formulations for removing mineral scale that contain about 75% formic acid are sold and children ingesting these solutions have suffered corrosive lesions to mouth and esophageal tissue. Although acetic acid as a 4–6% solution in vinegar is an ingredient of many foods, pure acetic acid (glacial acetic acid) is extremely corrosive to tissue that it contacts. Ingestion of, or skin contact with acrylic acid can cause severe damage to tissues.

Ethers

The common ethers have relatively low toxicities because of the low reactivity of the C–O–C functional group which has very strong carbon-oxygen bonds. Exposure to volatile diethyl ether is usually by inhalation and about 80% of this compound that gets into the body is eliminated unmetabolized as the vapor through the lungs. Diethyl ether depresses the central nervous system and is a depressant widely used as an anesthetic for surgery. Low doses of diethyl ether causes drowsiness, intoxication, and stupor, whereas higher exposures cause unconsciousness and even death.

Acid Anhydrides

Strong smelling, intensely lachrimating **acetic anhydride**,

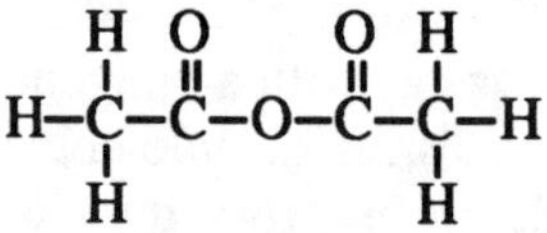

$$
\begin{array}{ccccc}
H & O & & O & H \\
| & \| & & \| & | \\
H-C-&C&-O-&C&-C-H \\
| & & & & | \\
H & & & & H
\end{array}
\quad \text{Acetic anhydride}
$$

is a systemic poison. It is especially corrosive to the skin, eyes, and upper respiratory tract, causing blisters and burns that heal only slowly. Levels in the air should not exceed 0.04 mg/m^3 and adverse effects to the eyes have been observed at about 0.4 mg/m^3.

Esters

Many esters (Figure 20.18) have relatively high volatilities so that the pulmonary system is a major route of exposure. Because of their generally good solvent properties, esters penetrate tissues and tend to dissolve body lipids. For example, vinyl acetate acts as a skin defatting agent. Because they hydrolyze in water, ester toxicities tend to be the same as the toxicities of the acids and alcohols from which they were formed. Many volatile esters exhibit asphyxiant and narcotic action. Whereas many of the naturally occurring esters have insignificant toxicities at low doses, allyl acetate and some of the other synthetic esters are relatively toxic.

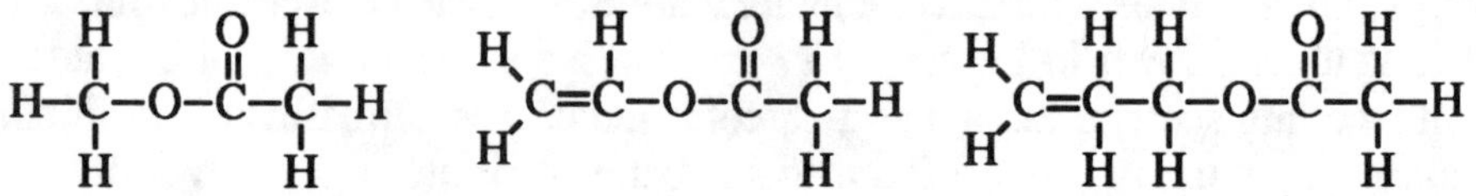

Methyl acetate   Vinyl acetate   Allyl acetate

Figure 20.18. Examples of esters.

Organonitrogen Compounds

Organonitrogen compounds constitute a large group of compounds with diverse toxicities. Examples of several of the kinds of organonitrogen compounds discussed here are given in Figure 20.19.

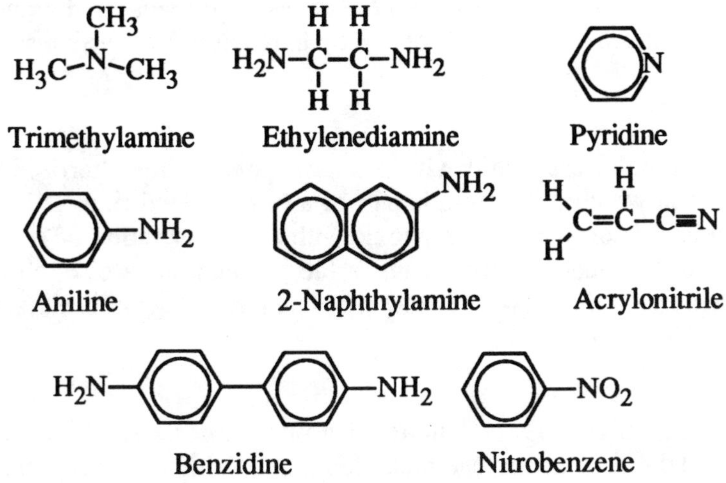

Figure 20.19. Some toxicological significant organonitrogen compounds.

Aliphatic Amines

The lower amines, such as the methylamines, are rapidly and easily taken into the body by all common exposure routes. They are basic and react with water in tissue,

$$R_3N + H_2O \rightarrow R_3NH^+ + OH^- \tag{20.9.1}$$

raising the pH of the tissue to harmful levels, acting as corrosive poisons (especially to sensitive eye tissue), and causing tissue necrosis at the point of contact. Among the systemic effects of amines are necrosis of the liver and kidneys, lung hemorrhage and edema, and sensitization of the immune system. The lower amines are among the more toxic substances in routine, large-scale use.

Ethylenediamine is the most common of the **alkyl polyamines**, compounds in which two or more amino groups are bonded to alkane moieties. Its toxicity rating is only 3, but it is a strong skin sensitizer and can damage eye tissue.

Carbocyclic Aromatic Amines

Aniline is a widely used industrial chemical and is the simplest of the **carbocyclic aromatic amines**, a class of compounds in which at least one substituent group is an aromatic hydrocarbon ring bonded directly to the amino group. There are numerous compounds with many industrial uses in this class of amines. Some of the carbocyclic aromatic amines have been shown to cause cancer in the human bladder, ureter, and pelvis, and are suspected of being lung, liver, and prostate carcinogens. A very toxic colorless liquid with an oily consistency and distinct odor, aniline readily enters the body by inhalation, ingestion, and through the skin. Metabolically, aniline converts iron(II) in hemoglobin to iron(III). This causes a condition called **methem-**

oglobinemia, characterized by cyanosis and a brown-black color of the blood, in which the hemoglobin can no longer transport oxygen in the body. This condition is not reversed by oxygen therapy.

Both **1-naphthylamine** (alpha-naphthylamine) and **2-naphthylamine** (beta-naphthylamine) are proven human bladder carcinogens. In addition to being a proven human carcinogen, **benzidine**, *p*-aminodiphenyl, is highly toxic and has systemic effects that include blood hemolysis, bone marrow depression, and kidney and liver damage. It can be taken into the body orally, by inhalation, and by skin sorption.

Pyridine

Pyridine, a colorless liquid with a sharp, penetrating, "terrible"odor, is an aromatic amine in which an N atom is part of a 6-membered ring. This widely used industrial chemical is only moderately toxic with a toxicity rating of 3. Symptoms of pyridine poisoning include anorexia, nausea, fatigue, and, in cases of chronic poisoning, mental depression. In a few rare cases pyridine poisoning has been fatal.

Nitriles

Nitriles contain the -C≡N functional group. Colorless, liquid **acetonitrile**, CH_3CN, is widely used in the chemical industry. With a toxicity rating of 3–4, acetonitrile is considered relatively safe, although it has caused human deaths, perhaps by metabolic release of cyanide. **Acrylonitrile**, a colorless liquid with a peach-seed (cyanide) odor, is highly reactive because it contains both nitrile and C=C groups. Ingested, absorbed through the skin, or inhaled as vapor, acrylonitrile metabolizes to release deadly HCN, which it resembles toxicologically.

Nitro Compounds

The simplest of the **nitro compounds, nitromethane** H_3CNO_3, is an oily liquid that causes anorexia, diarrhea, nausea, and vomiting and damages the kidneys and liver. **Nitrobenzene**, a pale yellow oily liquid with an odor of bitter almonds or shoe polish, can enter the body by all routes. It has a toxic action much like that of aniline, converting hemoglobin to methemoglobin, which cannot carry oxygen to body tissue. Nitrobenzene poisoning is manifested by cyanosis.

Nitrosamines

N-nitroso compounds (**nitrosamines**), which are characterized by the -N–N=O functional group, have been found in a variety of materials to which humans may be exposed, including beer, whiskey, and cutting oils used in machining. Cancer may result from exposure to a single large dose or from chronic exposure to relatively small doses of some nitrosamines. Once widely used as an industrial solvent and known to cause liver damage and jaundice in exposed workers, dimethylnitrosamine was shown to be carcinogenic from studies starting in the 1950s.

$$CH_3$$
$$|$$
$$N-N=O$$
$$|$$
$$CH_3$$

Dimethylnitrosamine
(N-nitrosodimethylamine)

Isocyanates and Methyl Isocyanate

Compounds with the general formula R–N=C=O, **isocyanates** are widely used industrial chemicals noted for the high chemical and metabolic reactivity of their characteristic functional group. **Methyl isocyanate**, H_3C–N=C=O, was the toxic agent involved in the catastrophic industrial poisoning in Bhopal, India on December 2, 1984, the worst industrial accident in history. In this incident several tons of methyl isocyanate were released, killing 2,000 people and affecting about 100,000. The lungs of victims were attacked; survivors suffered long-term shortness of breath and weakness from lung damage as well as numerous other toxic effects including nausea and bodily pain.[16]

Organonitrogen Pesticides

Pesticidal *carbamates* are characterized by the structural skeleton of carbamic acid outlined by the dashed box in the structural formula of carbaryl in Figure 20.20. Widely used on lawns and gardens, insecticidal **carbaryl** has a low toxicity to mammals. Highly water-soluble **carbofuran** is taken up by the roots and leaves of plants and poisons insects that feed on the leaves. The toxic effects to animals of carbamates are due to the fact that they inhibit acetylcholinesterase directly without the need to first undergo biotransformation. This effect is relatively reversible because of metabolic hydrolysis of the carbamate ester.

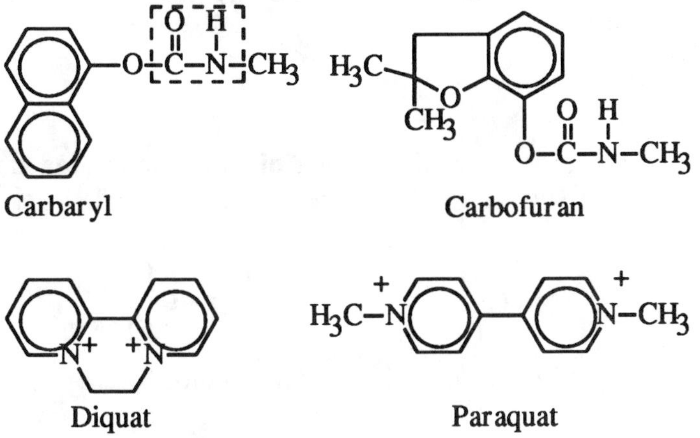

Carbaryl   Carbofuran

Diquat   Paraquat

Figure 20.20. Examples of organonitrogen pesticides.

Reputed to have "been responsible for hundreds of human deaths,"[17] herbicidal **paraquat** has a toxicity rating of 5. Dangerous or even fatal acute exposures can occur by all pathways, including inhalation of spray, skin contact, and ingestion. Paraquat is a systemic poison that affects enzyme activity and is devastating to a number of organs. Pulmonary fibrosis results in animals that have inhaled paraquat aerosols, and the lungs are also adversely affected by non-pulmonary exposure. Acute exposure may cause variations in the levels of catecholamine, glucose, and insulin. The most prominent initial symptom of poisoning is vomiting, followed within a few days by dyspnea, cyanosis, and evidence of impairment of the kidneys, liver, and heart. Pulmonary fibrosis, often accompanied by pulmonary edema and hemorrhaging, is observed in fatal cases.

Organohalide Compounds

Alkyl Halides

The toxicities of alkyl halides, such as carbon tetrachloride, CCl_4, vary a great deal with the compound. Most of these compounds cause depression of the central nervous system, and individual compounds exhibit specific toxic effects.

During its many years of use as a consumer product, carbon tetrachloride compiled a grim record of toxic effects which led the U. S. Food and Drug Administration (FDA) to prohibit its household use in 1970. It is a systemic poison that affects the nervous system when inhaled, and the gastrointestinal tract, liver, and kidneys when ingested. The biochemical mechanism of carbon tetrachloride toxicity involves reactive radical species including,

Unpaired
electrons

that react with biomolecules, such as proteins and DNA. The most damaging such reaction occurs in the liver as **lipid peroxidation**, consisting of the attack of free radicals on unsaturated lipid molecules, followed by oxidation of the lipids through a free radical mechanism.

Alkenyl Halides

The most significant **alkenyl** or **olefinic organohalides** are the lighter chlorinated compounds, such as vinyl chloride and tetrachloroethylene:

Vinyl chloride   Tetrachloroethylene

Because of their widespread use and disposal in the environment, the numerous acute and chronic toxic effects of the alkenyl halides are of considerable concern.

The central nervous system, respiratory system, liver, and blood and lymph systems are all affected by vinyl chloride exposure, which has been widespread because of this compound's use in polyvinylchloride manufacture. Most notably, vinyl chloride is carcinogenic, causing a rare angiosarcoma of the liver. This deadly form of cancer has been observed in workers chronically exposed to vinyl chloride while cleaning autoclaves in the polyvinylchloride fabrication industry. The alkenyl organohalide, 1,1-dichloroethylene, is a suspect human carcinogen based upon animal studies and its structural similarity to vinyl chloride The toxicities of both 1,2-dichloroethylene isomers are relatively low. These compounds act in different ways in that the *cis* isomer is an irritant and narcotic, whereas the *trans* isomer affects both the central nervous system and the gastrointestinal tract, causing weakness, tremors, cramps, and nausea. A suspect human carcinogen, trichloroethylene has caused liver carcinoma in experimental animals and is known to affect numerous body organs. Like other organohalide solvents, trichloroethylene causes skin dermatitis from dissolution of skin lipids and it can affect the central nervous and respiratory systems, liver, kidneys, and heart. Symptoms of exposure include disturbed vision, headaches, nausea, cardiac arrhythmias, and burning/tingling sensations in the nerves (paresthesia). Tetrachloroethylene damages the liver, kidneys, and central nervous system. It is a suspect human carcinogen.

Aryl Halides

Individuals exposed to irritant monochlorobenzene by inhalation or skin contact suffer symptoms to the respiratory system, liver, skin, and eyes. Ingestion of this compound causes effects similar to those of toxic aniline, including incoordination, pallor, cyanosis, and eventual collapse.

The dichlorobenzenes are irritants that affect the same organs as mono-chlorobenzene; the 1,4- isomer has been known to cause profuse rhinitis (running nose), nausea, jaundice, liver cirrhosis, and weight loss associated with anorexia. *Para*-dichlorobenzene (1,2-dichlorobenzene), a chemical used in air fresheners and mothballs, has become the center of a controversy regarding the evaluation of carcinogenicity. Based upon animal tests that involved subjecting rats to large amounts of the chemical, then extrapolating to humans with adjustments for differences in body size and many orders of magnitude in dose, the U. S. Department of Health and Human Service's National Toxicology Program has classified 1,2-dichlorobenzene as a potential cancer-causing substance. Some industry groups, including the Synthetic Organic Chemicals Association, have filed suit contending that the procedures used are inaccurate and out of date. The suit is viewed as a general challenge to the current use of animal studies to establish possible carcinogenicity of chemicals.[18] Specifically, the suit asks that the report acknowledge that the protein through which the chemical acts in rats does not exist in humans. Many authorities contend that more sophisticated alternatives to animal studies should be used to predict human carcinogenicity of chemicals, especially biochemical investigations of the interactions of suspect chemicals and their metabolites in the body, pathways through the body, and chemical (structural and functional) similarities to known carcinogens.

Because of their once widespread use in electrical equipment, as hydraulic fluids, and in many other applications, polychlorinated biphenyls (PCBs, see Section 7.10) became widespread, extremely persistent environmental pollutants.[19] PCBs have a strong tendency to undergo bioaccumulation in lipid tissue. Polybrominated biphenyl analogs (PBBs) were much less widely used and distributed. However, PBBs were involved in one major incident that resulted in catastrophic agricultural losses when livestock feed contaminated with PBB flame retardant caused massive livestock poisoning in Michigan in 1973.

Organohalide Insecticides

Exhibiting a wide range of kind and degree of toxic effects, many organohalide insecticides (see Section 7.10) affect the central nervous system, causing symptoms such as tremor, irregular jerking of the eyes, changes in personality, and loss of memory. Such symptoms are characteristic of acute DDT poisoning. However, the acute toxicity of DDT to humans is very low and it was used for the control of typhus and malaria in World War II by large scale direct application to people. The chlorinated cyclodiene insecticides — aldrin, dieldrin, endrin, chlordane, heptachlor, endosulfan, and isodrin — act on the brain, releasing betaine esters and causing headaches, dizziness, nausea, vomiting, jerking muscles, and convulsions. Dieldrin, chlordane, and heptachlor have caused liver cancer in test animals and some chlorinated cyclodiene insecticides are teratogenic or fetotoxic. Because of these effects, aldrin, dieldrin, heptachlor, and — more recently — chlordane have been prohibited from use in the U. S.

The major **chlorophenoxy** herbicides are 2,4-dichlorophenoxyacetic acid (2,4-D), 2,4,5-trichlorophenoxyaceticacid (2,4,5-T or Agent Orange), and Silvex. Large doses of 2,4-dichlorophenoxyacetic acid have been shown to cause nerve damage, (peripheral neuropathy), convulsions, and brain damage. According to a National Cancer Institute study,[20] Kansas farmers who had handled 2,4-D extensively have suffered 6 to 8 times the incidence of non-Hodgkins lymphoma as comparable unexposed populations. With a toxicity somewhat less than that of 2,4-D, Silvex is largely excreted unchanged in the urine. The toxic effects of 2,4,5-T (used as a herbicidal warfare chemical called "Agent Orange") have resulted from the presence of 2,3,7,8-tetrachloro-p-dioxin (TCDD, commonly known as "dioxin", discussed below), a manufacturing by-product. Autopsied carcasses of sheep poisoned by this herbicide have exhibited nephritis, hepatitis, and enteritis.

TCDD

Polychlorinated dibenzodioxins, which have the same basic structure as that of TCDD (2,3,7,8-tetrachlorodibenzo-p-dioxin),

TCDD (2,3,7,8-tetrachloro-dibenzo-p-dioxin)

but different numbers and arrangements of chlorine atoms on the ring structure. Extremely toxic to some animals, the toxicity of TCDD to humans is rather uncertain; it is known to cause a skin condition called chloracne. TCDD has been a manufacturing byproduct of some commercial products (see the discussion of 2,4,5-T, above), contaminant identified in some municipal incineration emissions, and widespread environmental pollutant from improper waste disposal. This compound has been released in a number of industrial accidents, the most massive of which exposed several tens of thousands of people to a cloud of chemical emissions spread over an approximately 3-square-mile area at the Givaudan-La Roche Icmesa manufacturing plant near Seveso, Italy, in 1976. On an encouraging note from a toxicological perspective, no abnormal occurrences of major malformations were found in a study of 15,291 children born in the area within 6 years after the release.[21]

Chlorinated Phenols

The chlorinated phenols used in largest quantities have been **pentachlorophenol** (Chapter 7) and the trichlorophenol isomers used as wood preservatives. Although exposure to these compounds has been correlated with liver malfunction and dermatitis, contaminant polychlorinated dibenzodioxins may have caused some of the observed effects.

Organosulfur compounds

Despite the high toxicity of H_2S, not all organosulfur compounds are particularly toxic. Their hazards are often reduced by their strong, offensive odors that warn of their presence.

Inhalation of even very low concentrations of the alkyl **thiols**, such as methanethiol, H_3CSH, can cause nausea and headaches; higher levels can cause increased pulse rate, cold hands and feet, and cyanosis. In extreme cases, unconsciousness, coma, and death occur. Like H_2S, the alkyl thiols are precursors to cytochrome oxidase poisons.

An oily water-soluble liquid, **methylsulfuric acid** is a strong irritant to skin, eyes, and mucous tissue. Colorless, odorless **dimethylsulfate** is highly toxic and is a

$$H_3C-O-\overset{\overset{\displaystyle O}{\|}}{\underset{\underset{\displaystyle O}{\|}}{S}}-OH \quad \text{Methylsulfuric acid} \qquad H_3C-O-\overset{\overset{\displaystyle O}{\|}}{\underset{\underset{\displaystyle O}{\|}}{S}}-O-CH_3 \quad \text{Dimethyl- sulfate}$$

primary carcinogen. Skin or mucous membranes exposed to dimethylsulfate develop conjunctivitis and inflammation of nasal tissue and respiratory tract mucous membranes following an initial latent period during which few symptoms are observed. Damage to the liver and kidney, pulmonary edema, cloudiness of the cornea, and death within 3–4 days can result from heavier exposures.

Sulfur Mustards

A typical example of deadly **sulfur mustards**, compounds used as military poisons, or "poison gases," is mustard oil (bis(2-chloroethyl)sulfide):[22]

$$\text{Cl}-\overset{\overset{\displaystyle H}{|}}{\underset{\underset{\displaystyle H}{|}}{C}}-\overset{\overset{\displaystyle H}{|}}{\underset{\underset{\displaystyle H}{|}}{C}}-S-\overset{\overset{\displaystyle H}{|}}{\underset{\underset{\displaystyle H}{|}}{C}}-\overset{\overset{\displaystyle H}{|}}{\underset{\underset{\displaystyle H}{|}}{C}}-\text{Cl} \quad \text{Mustard oil}$$

An experimental mutagen and primary carcinogen, mustard oil produces vapors that penetrate deep within tissue, resulting in destruction and damage at some depth from the point of contact; penetration is very rapid, so that efforts to remove the toxic agent from the exposed area are ineffective after 30 minutes. This military "blistering gas" poison, causes tissue to become severely inflamed with lesions that often become infected. These lesions in the lung can cause death.

Organophosphorous Compounds

Organophosphorus compounds have varying degrees of toxicity. Some of these compounds, such as the "nerve gases" produced as industrial poisons, are deadly in minute quantities. The toxicities of major classes of organophosphate compounds are discussed in this section.

Organophosphate Esters

Some organophosphate esters are shown in Figure 20.21. **Trimethylphosphate** is probably moderately toxic when ingested or absorbed through the skin, whereas moderately toxic **triethylphosphate**, $(C_2H_5O)_3PO$, damages nerves and inhibits acetylcholinesterase. Notoriously toxic **tri-*o*-cresylphosphate, TOCP,** apparently is metabolized to products that inhibit acetylcholinesterase. Exposure to TOCP causes degeneration of the neurons in the body's central and peripheral nervous systems with early symptoms of nausea, vomiting, and diarrhea accompanied by severe abdominal pain. About 1–3 weeks after these symptoms have subsided, peripheral paralysis develops manifested by "wrist drop" and "foot drop," followed by slow recovery, which may be complete or leave a permanent partial paralysis.

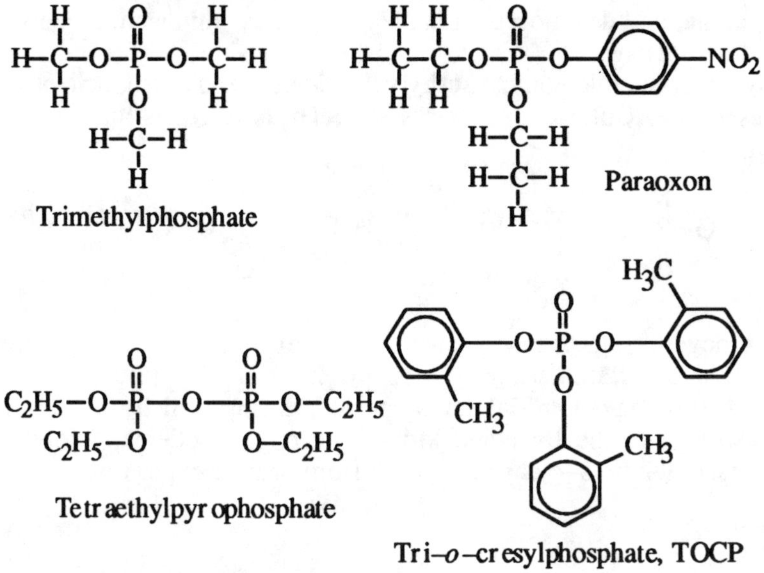

Figure 20.21. Some organophosphate esters.

Briefly used in Germany as a substitute for insecticidal nicotine, **tetraethylpyrophosphate, TEPP,** is a very potent acetylcholinesterase inhibitor. With a toxicity rating of 6 (supertoxic), TEPP is deadly to humans and other mammals.

Phosphorothionate and Phosphorodithioate Ester Insecticides

Because esters containing the P=S (thiono) group are resistant to non-enzymatic hydrolysis and are not as effective as P=O compounds in inhibiting acetylcholinesterase, they exhibit higher insect:mammal toxicity ratios than their non-sulfur analogs. Therefore, **phosphorothionate** and **phosphorodithioate** esters (Figure 20.22) are widely used as insecticides. The insecticidal activity of these compounds requires metabolic conversion of P=S to P=O (oxidative desulfuration). Environmentally, organophosphate insecticides are superior to many of the organochlorine insecticides because the organophosphates readily undergo biodegradation and do not bioaccumulate.

The first commercially successful phosphorothionate/phosphorodithioate ester insecticide was **parathion,** *O,O*-diethyl-*O-p*-nitrophenylphosphorothionate, first licensed for use in 1944. This insecticide has a toxicity rating of 6 (supertoxic). Since its use began, several hundred people have been killed by parathion, including 17 of 79 people exposed to contaminated flour in Jamaica in 1976. As little as 120 mg of parathion has been known to kill an adult human and a dose of 2 mg has been fatal to a child. Most accidental poisonings have occurred by absorption through the skin. Methylparathion (a closely related compound with methyl groups instead of ethyl groups) is regarded as extremely toxic.

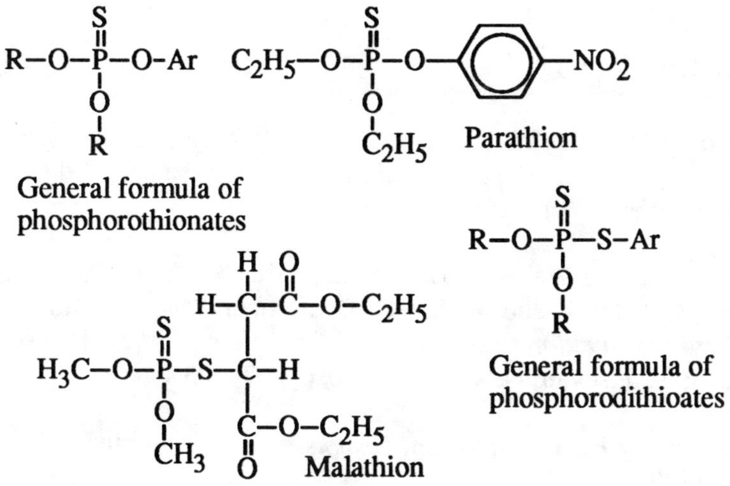

Figure 20.22. Phosphorothionate and phosphorodithioate ester insecticides. Malathion contains hydrolyzable carboxyester linkages.

In order for parathion to have a toxic effect, it must be converted metabolically to paraoxon (Figure 20.21), which is a potent inhibitor of acetylcholinesterase. Because of the time required for this conversion, symptoms develop several hours after exposure, whereas the toxic effects of TEPP or paraoxon develop much more rapidly. Humans poisoned by parathion exhibit skin twitching and respiratory distress. In fatal cases, respiratory failure occurs due to central nervous system paralysis.

Malathion is the best known of the phosphorodithioate insecticides. It has a relatively high insect:mammal toxicity ratio because of its two carboxyester linkages which are hydrolyzable by carboxylase enzymes (possessed by mammals, but not insects) to relatively non-toxic products. For example, although malathion is a very effective insecticide, its LD_{50} for adult male rats is about 100 times that of parathion.

Organophosphorus Military Poisons

Powerful inhibitors of acetylcholinesterase enzyme, organophosphorus "nerve gas" military poisons such as **Sarin** and **VX**. (The possibility that military poisons such as these might be used in war was a major concern during the 1991 mid-East conflict, which, fortunately, ended without their being employed.) A systemic poison to the central nervous system that is readily absorbed as a liquid through the skin, Sarin may be lethal at doses as low as about 0.01 mg/kg; a single drop can kill a human.

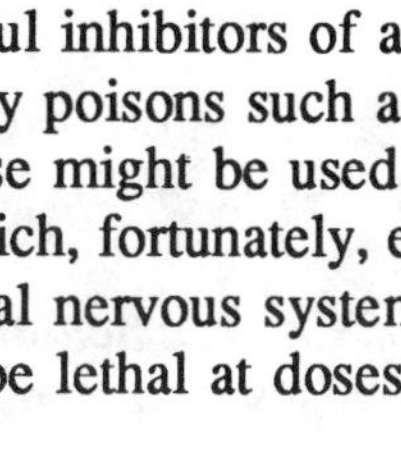

Sarin 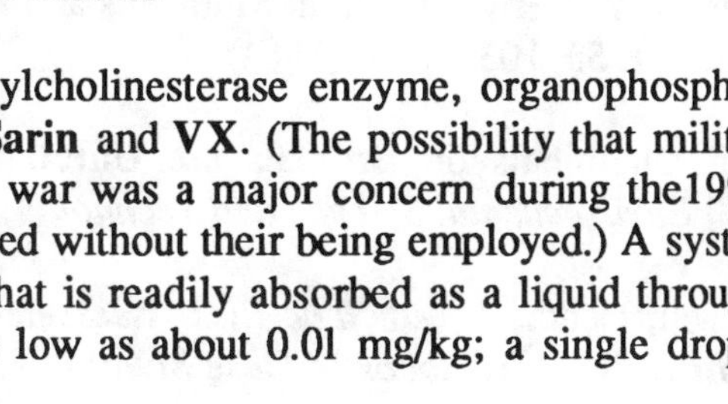   VX

LITERATURE CITED

1. Feigl, Dorothy M., John W. Hill, and Erwin Boschman, *Foundations of Life: An Introduction to General, Organic, and Biological Chemistry*, 3rd Ed., Macmillan Publishing Co., New York, 1991.

2. Hodgson, Ernest, and Walter C. Dauterman, "Metabolism of Toxicants — Phase I Reactions," Chapter 4, and Walter C. Dauterman, "Metabolism of Toxicants," Chapter 5, in *Introduction to Biochemical Toxicology*, Ernest Hodgson and Frank E. Guthrie, Eds., Elsevier, New York, 1980, pp. 67–105.4.

3. Manahan, Stanley E., *Toxicological Chemistry*, Lewis Publishers, Inc., Chelsea, Michigan, 1989.

4. James, Robert C., "General Principles of Toxicology," Chapter 2 in *Industrial Toxicology*, Philip L. Williams and James L. Burson, Eds., Van Nostrand Reinhold Company, New York, 1985.

5. Diamond, Leila, "Tumor Promoters and Cell Transformation," Chapter 3 in *Mechanisms of Cellular Transformation by Carcinogenic Agents*, D. Grunberger and S. P. Goff, Eds., Pergamon Press, New York, 1987, pp. 73–132.

6. Singer, B., and D. Grunberger *Molecular Biology of Mutagens and Carcinogens*, Plenum Press, New York, 1983.

7. Levi, Patricia E., "Toxic Action," Chapter 6 in *Modern Toxicology*, Ernest Hodgson and Patricia E. Levi, Eds., Elsevier, New York, 1987, pp. 133–184.

8. "The Detection of Environmental Mutagens and Potential Carcinogens," Bruce N. Ames, *Cancer*, **53**, 1034-1040 (1984).

9. "Tests of Chemicals on Animals are Unreliable as Predictors of Cancer in Humans," *Environmental Science and Technology* **24**, 1990 (1990).

10. Paustenbach, Dennis J., Ed., *The Risk Assessment of Environmental and Human Health Hazards: A Textbook of Case Studies*, John Wiley and Sons, New York, 1988.

11. Steinway, Daniel M., "Scope and Numbers of Regulations for Asbestos-Containing Materials, Abatement Continue to Grow," *Hazmat World*, April, 1990, pp. 32-58.

12. Fisher, Gerald L., and Michael A. Gallo, Eds., *Asbestos Toxicity*, Marcel Dekker, New York, 1988.

13. Seligman, Peter F., et al., "Distribution and Fate of Tributyltin in the Marine Environment," *American Chemical Society Division of Environmental Chemistry Preprint Extended Abstracts*, **28**, 573–579 (1988).

14. Clark, Elizabeth M., Robert M. Sterritt, and John N. Lester, "The Fate of Tributyltin in the Aquatic Environment," *Environmental Science and Technology*, **22**, 600–604 (1988).

15. Gosselin, Robert E., Roger P. Smith, and Harold C. Hodge, "Phenol," in *Clinical Toxicology of Commercial Products*, 5th ed., Williams and Wilkins, Baltimore/London, 1984, pp. III-344–III-348.

16. Lepowski, Wil, "Methyl Isocyanate: Studies Point to Systemic Effects," *Chemical and Engineering News*, June 13, 1988, p. 6.

17. Gosselin, Robert E., Roger P. Smith, and Harold C. Hodge, "Paraquat," in *Clinical Toxicology of Commercial Products,*5th ed., Williams and Wilkins, Baltimore/London, 1984, pp. III-328–III-336.

18. Shabecoff, Philip, "Industry Fights Use of Animal Tests to Assess Cancer Risk," *New York Times*, July 25, 1989, p. 20.

19. Safe, S., Ed., *Polychlorinated Biphenyls (PCBs): Mammalian and Environmental Toxicology*, Springer-Verlag, New York, 1987.

20. Silberner, J., "Common Herbicide Linked to Cancer," *Science News*, **130**(11), 167–174 (1986).

21. "Dioxin is Found Not to Increase Birth Defects," *New York Times*, March 18, 1988, p. 12.

22. "Global Experts Offer Advice on Chemical Weapons Treaty," *Chemical and Engineering News*, July 27, 1987, pp. 16-27.

SUPPLEMENTARY REFERENCES

Hallenbeck, William H., and Kathleen M. Cunningham-Burns, *Quantitative Risk Assessment for Environmental and Occupational Health*, Lewis Publishers, Chelsea, MI, 1986.

Fawell, J. K., and S. Hunt, *Environmental Toxicology: Organic Pollutants*, John Wiley & Sons, New York, N. Y., 1988.

Stryer, Lubert, *Biochemistry*, 3rd Ed., W. H. Freeman & Co., New York, N. Y., 1988.

Kneip, Theodore J., and John V. Crable, Eds., *Methods for Biological Monitoring: A Manual for Assessing Human Exposure to Hazardous Substances*, American Public Health Association, Washington, D.C., 1988.

Fisher, Gerald L., and Michael A. Gallo, Eds., *Asbestos Toxicity*, Marcel Dekker, New York, NY, 1988.

Elements of Toxicology and Chemical Risk Assessment, Revised Ed., Environ Corp., Washington, DC, 1988.

Proctor, Nick H., James P. Hughes, and Michael L. Fischman, *Chemical Hazards of the Workplace*, 2nd Ed., J. B. Lippincott Co., Philadelphia, PA, 1988.

Schmal, Dietrich, *Combination Effects in Chemical Carcinogenesis*, VCH Publishers, New York, NY, 1988.

Lioy, Paul J., and Joan M. Daisey, Eds., *Toxic Air Pollution: A Comprehensive Study of Non-Criteria Air Pollutants*, Lewis Publishers, Chelsea, MI, 1987.

Ram, Neil M., Edward J. Calabrese, and Russell F. Christman, Eds., *Organic Carcinogens in Drinking Water: Detection, Treatment, and Risk Assessment*, John Wiley and Sons, New York, NY, 1986.

McLachlan, John A., Robert M. Pratt, Clement L. Markert, Eds., *Developmental Toxicology: Mechanisms and Risk*, Cold Spring Harbor Laboratory, Cold Spring Harbor, NY, 1988.

Tardiff, Robert G., Joseph V. Rodricks, Eds., *Toxic Substances and Human Risk*, Plenum Press, New York, NY, 1987.

Worobec, Mary Devine, and Girard Ordway, *Toxic Substances Controls Guide*, BNA Books Distribution Center, Edison, NJ, 1989.

Tardiff, Robert G., and Joseph V. Rodricks, Eds., *Toxic Substances and Human Risk: Principles of Data Interpretaion*, Plenum Publishing, New York, NY, 1988.

Paddock, Todd, *Dioxins and Furans: Questions and Answers*, Academy of Natural Sciences, Philadelphia, PA, 1989.

Cairns, John, Jr., and James R. Pratt, Eds., *Functional Testing of Aquatic Biota for Estimating Hazards of Chemicals*, ASTM, Philadelphia, PA, 1989.

R. Colin Garner and Jan Hradec, Eds., *Biochemistry of Chemical Carcinogenesis*, Plenum, New York, 1990.

Carcinogenic, Mutagenic, and Teratogenic Marine Pollutants. Impact on Human Health and the Environment, Gulf, Houston, TX, 1990.

Cohen, Gary, and John O'Connor, *Fighting Toxics. A Manual for Protecting your Family, Community, and Workplace*, Island Press, Washington, DC, 1990.

Poole, A., and G. B. Leslie, *A Practical Approach to Toxicological Investigations*, Cambridge University Press, New York, 1990.

Cooper, C. S., and P. L. Gover, Eds., *Chemical Carcinogenesis and Mutagensis II*, Springer-Verlag, New York, 1990.

Mathews, Christopher K., and K. E. van Holde, Biochemistry. Benjamin/Cummings, Redwood City, CA, 1990.

Marks, Dawn B., *Biochemistry*, Williams and Wilkins, Baltimore, 1990.

Ragsdale, Nancy N., and Robert E. Menzer, Eds., *Carcinogenicity and Pesticides. Principles, Issues, and Relationships*, American Chemical Society, Washington, DC, 1989.

Nichol, John, *Bites and Stings. The World of Venomous Animals*, Facts on File, New York, 1989.

Baker, Scott R., and Chris F. Wilkinson, Eds., *The Effects of Pesticides on Human Health*, Princeton Scientific, Princeton, NJ, 1990.

Sandhu, Shahbeg S., Ed., *In Situ Evaluation of Biological Hazards of Environmental Pollutants*, Plenum, New York, 1990.

Feo, Francesco, Eds., *Chemical Carcinogenesis. Models and Mechanisms*, Plenum, New York, 1988.

Clarke, Lee, *Acceptable Risk? Making Decisions in a Toxic Environment*, University of California Press, Berkeley, 1989.

Liberman, Daniel F., and Judith G. Gordon, Eds., *Biohazards Management Handbook*, Dekker, New York, 1989.

QUESTIONS AND PROBLEMS

1. How are conjugating agents and Phase II reactions involved with some toxicants?

2. What is the toxicological importance of proteins, particularly as related to protein structure?

3. What is the toxicological importance of lipids? How do lipids related to hydrophobic ("water-disliking") pollutants and toxicants?

4. What is the function of a hydrolase enzyme?

5. What are Phase I reactions? What enzyme system carries them out? Where is this enzyme system located in the cell?

6. Name and describe the science that deals with the chemical nature and reactions of toxic substances, including their origins, uses, and chemical aspects of exposure, fates, and disposal.

7. What is a dose-response curve?

8. What is meant by a toxicity rating of 6?

9. What are the three major subdivisions of the *dynamic phase* of toxicity, and what happens in each?

10. Match the cell structure on the left with its function on the right, below:

    1. Mitochondria   (a) Toxicant metabolism
    2. Endoplasmic reticulum   (b) Fills the cell
    3. Cell membrane   (c) Deoxyribonucleic acid
    4. Cytoplasm   (d) Mediate energy conversion and utilization
    5. Cell nucleus   (e) Encloses the cell and regulates the passage of materials into and out of the cell interior

11. Characterize the toxic effect of carbon monoxide in the body. Is its effect reversible or irreversible? Does it act on an enzyme system?

12. Of the following, choose the one that is **not** a biochemical effect of a toxic substance: (a) impairment of enzyme function by binding to the enzyme, (b) alteration of cell membrane or carriers in cell membranes, (c) change in vital signs, (d) interference with lipid metabolism, (e) interference with respiration.

13. Distinguish among teratogenesis, mutagenesis, carcinogenesis, and immune system effects. Are their ways in which they are related?

14. As far as environmental toxicants are concerned, compare the relative importance of acute and chronic toxic effects and discuss the difficulties and uncertainties involved in studying each.

15. What are some of the factors that complicate epidemiologic studies of toxicants?

16. List and discuss two elements that are invariably toxic in their elemental forms. For another element, list and discuss two elemental forms, one of which is quite toxic and the other of which is essential for the body. In what sense is even the toxic form of this element "essential for life?"

17. What is a toxic substance that bonds to iron(III) in iron-containing ferricytochrome oxidase enzyme, preventing its reduction to iron(II) in the oxidative phosphorylation process by which the body utilizes O_2?

18. What are interhalogen compounds, and which elemental forms do their toxic effects most closely resemble?

19. Name and describe the three health conditions that maybe caused by inhalation of asbestos.

20. Why might tetraethyllead be classified as "the most notable toxic organometallic compound"?

21. What is the most common toxic effect commonly attributed to low-molecular-mass alkanes?

22. Although benzene and toluene have a number of chemical similarities, their metabolisms and toxic effects are quite different. Explain.

23. Information about the toxicities of many substances to humans is lacking because of limited data on direct human exposure. (Volunteers to study human health effects of toxicants are in notably short supply.) However, there is a great deal of information available about human exposure to phenol and the adverse effects of such exposure. Explain.

24. What are neuropathic disorders? Why are organic solvents frequently the cause of such disorders?

25. What is a major metabolic effect of aniline? What is this effect called? How is it manifested?

26. What are the organic compounds characterized by the –H–N=O functional group? What is their major health effect?

27. What structural group is characteristic of carbamates? For what purpose are these compounds commonly used? What are their major advantages in such an application?

28. What is lipid peroxidation? Which common toxic substance is known to cause lipid peroxidation?

29. Biochemically, what do organophosphate esters such as parathion do that could classify them as "nerve poisons"?

30. Comment on the toxicity of the compound below:

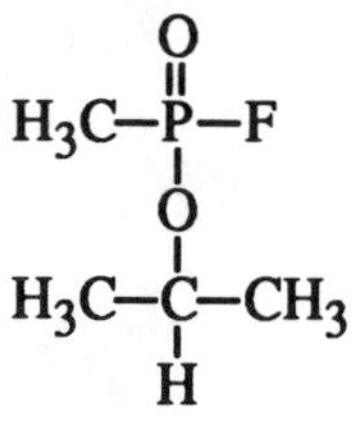

$$
\begin{array}{c}
\quad\quad\; O \\
\quad\quad\; \| \\
H_3C-P-F \\
\quad\quad\; | \\
\quad\quad\; O \\
\quad\quad\; | \\
H_3C-C-CH_3 \\
\quad\quad\; | \\
\quad\quad\; H
\end{array}
$$

Resources and Energy

21.1. THE NATURAL RESOURCES-ENERGY-ENVIRONMENT TRIANGLE

Natural resources, energy, and the environment are intimately related (Fig. 21.1). Perturbations in one usually cause perturbations in the other two. For example, reductions in automotive exhaust pollutant levels with the use of catalytic devices, discussed in Chapter 13, have resulted in increased demand for platinum metal, a scarce natural resource, and greater gasoline consumption than would be the case if exhaust emissions were not controlled at all. The availability of many metals depends upon the quantity of energy used and the amount of environmental damage tolerated in the extraction of low-grade ores. Many other such examples could be cited. Because of these intimate interrelationships, resources and energy must be discussed along with environmental chemistry.

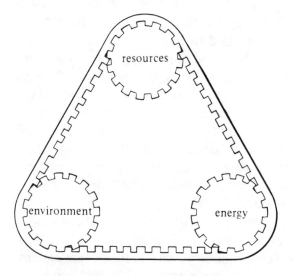

Figure 21.1. Strong connections exist among natural resources, energy, and environment.

Problems of energy, natural resources, and environment came to a head in the mid-1970s, particularly during the 1973-74 "energy crisis." This crisis was accompanied by worldwide shortages of some foods and minerals, followed in some cases by surpluses, such as the surplus wheat resulting from increased planting and a copper surplus resulting from the efforts of copper-producing nations to acquire foreign currency by copper export. These should be regarded as only temporary, so long as a growing world population and increasing affluence continue to place demands upon limited resources.

It is beyond the scope of one chapter to discuss natural resources and energy in anything like a comprehensive manner. Instead, this chapter touches upon the major aspects of natural resources and energy resources as they relate to environmental chemistry.

In discussing minerals and fossil fuels, two terms related to available quantities are used which should be defined. The first of these is **resources**, which refers to quantities that are estimated to be *ultimately* available. The second term is **reserves**, which refers to well-identified resources that can be profitably utilized with existing technology.

21.2. METALS

With an adequate supply of all of the important elements and energy, almost any needed material can be manufactured. Most of the elements, including practically all of those likely to be in short supply, are metals. Some of these are virtually unavailable in the U.S., which imports almost all its aluminum, chromium, cobalt, manganese, palladium, platinum, and titanium, and the majority of its bismuth, cadmium, mercury, nickel, tungsten and zinc.

Some metals are considered especially crucial because of their importance to industrialized societies, uncertain sources of supply, and price volatility in world markets. One of these is antimony, used in auto batteries, fire-resistant fabrics, and rubber. Chromium, another crucial metal, is used to manufacture stainless steel (especially for parts exposed to high temperatures and corrosive gases), jet aircraft, automobiles, hospital equipment, and mining equipment. The U.S. imports 91% of its chromium, largely from the Republic of South Africa and the U.S.S.R., and obtains the remainder by recycling. Some limited U.S. resources of chromium exist in the Stillwater Complex of Montana and in Oregon beach sands. Cadmium, nickel, and zinc may be substituted for chromium for corrosion protection (plastic and rubber "cushions" have now replaced chromium-plated automobile bumpers), whereas substitutes for iron alloys of chromium may be made with nickel, cobalt, molybdenum, and vanadium. The U.S. imports 93% of its cobalt, the bulk of which comes from Zaire, and obtains 7% from recycling. U.S. domestic production stopped in 1979. Cobalt is employed in high-temperature alloys in jet engines, magnets, catalysts, and other uses. Nickel can be substituted for cobalt in many applications, although the product is generally inferior. Manganese is employed in steel making as ferromanganese. U.S. imports of manganese come largely from Gabon, Brazil, and the Republic of South Africa. More than 80% of identified world resources of this metal are found in the Republic of South Africa and the U.S.S.R. The platinum-group metals (platinum, palladium, iridium, rhodium) are used as catalysts in the chemical industry, in petroleum refining, and in automobile exhaust antipollution devices. The U.S. imports 87% of these metals and receives the remainder from recycling. The Republic of South Africa and the U.S.S.R are the major import sources. Substitutes in the electrical and electronic industries include gold, silver, and tungsten. Nickel, vanadium, titanium, and rare earths can substitute in catalytic applications. Titanium is used to manufacture jet engines, steel and other alloys, spacecraft and missile components, and equipment in the chemical processing industry. Titanium dioxide is widely employed as a nontoxic white paint pigment. The U.S. imports 100% of its titanium metal, largely from Australia, but produces most TiO_2 domestically. Adequate domestic supplies of the titanium ore, ilmenite, are available. Vanadium is used to make iron and steel alloys, in the production of titanium alloys, and as a catalyst in sulfuric acid production. Domestic sources can supply U.S. needs, although imported vanadium is cheaper. Another critical metal is germanium, used in the electronics industry.

Mining and processing of metal ores involve major environmental concerns, including disturbance of land, air pollution from dust and smelter emissions, and water pollution from disrupted aquifers. This problem is aggravated by the fact that the general trend in mining involves utilization of less rich ores. This is illustrated in Figure 21.2, showing the average percentage of copper in copper ore mined since 1900. The average percentage of copper in ore mined in 1900 was about 4%, but by 1982 it was about 0.6% in domestic ores and 1.4% in richer foreign ores. Ores as low as 0.1% copper may eventually be processed. Increased demand for a particular metal, coupled with the necessity to utilize lower grade ores, has a vicious multiplying effect upon the amount of ore that must be mined and processed, and accompanying environmental consequences.

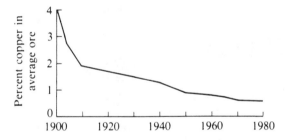

Figure 21.2. Average percentage of copper in ore that has been mined.

Discussion of all the major environmental manifestations of the mining of each metal would require a chapter much longer than this one. Instead, these are summarized in Table 21.1.

The sea bottom may serve to supply minerals. One such source consists of manganese nodules found at depths of from 2,500 to 6,000 meters; these bodies contain iron, copper, cobalt, and nickel, as well as manganese. Oceanic sulfide deposits may turn out to be sources of about 20 minerals that are strategically short in the U.S. These deposits are associated with areas where the seafloor is spreading, many of which occur within U.S. territorial waters in the Pacific. The deposits are formed by the upwelling of the Earth's molten silicate magma core from which metals are leached by superheated ocean water infiltrating fractures in the ocean floor. As this mineral-laden water rises through the Earth's crust to the ocean's bottom, it deposits metal sulfide ores, as well as ferric hydroxide.

Table 21.1. Worldwide and Domestic Metal Resources

Metal	Properties[1]	Major uses	Ores and aspects of resources[2]
Aluminum	mp 660 °C, bp 2467 °C, sg 2.70, malleable, ductile	Metal products, including autos, aircraft, electrical equipment. Conducts electricity better than copper per unit weight and is used in electrical transmission lines.	From bauxite ore containing 35–55% Al_2O_3. About 60 million metric tons of bauxite produced worldwide annually, about 27% used in the U.S., which produces about 1.5 million metric tons of bauxite per year. U.S. resources of bauxite are 40 million metric tons, world resources are 15 billion metric tons.
Chromium	mp 1903 °C, bp 2642 °C, sg 7.14, hard, silvery color	Metal plating, stainless steel, wear-resistant and cutting-tool alloys, chromium chemicals, including chromate used as an anticorrosive and cooling-water additive.	From chromite having the general formula [Mg(II), Fe(II)][Cr(III), Al(III), Fe(III)]$_2O_4$. Resources of 1 billion metric tons in South Africa and Rhodesia, large deposits in Russia, virtually none in the U.S.
Cobalt	mp 1495 °C, bp 2880 °C, sg 8.71, bright, silvery	Manufacture of hard, heat-resistant alloys such as stellite, permanent magnet alloys, driers, pigments, glazes, catalysts, animal-feed additive.	From a variety of minerals, such as linnaeite, Co_3S_4, and as a by-product of other metals. World consumption is 25,000 metric tons Co per year, 30% used in the U.S. U.S. imports 80% of its cobalt. Abundant global and U.S. resources.
Copper	mp 1083 °C, bp 2582 °C, sg 8.96, dense, ductile, malleable	Electrical conductors, alloys, chemicals. Many uses.	Occurs in low percentages (see Fig. 17.2) as sulfides, oxides, and carbonates in other minerals. U.S. consumption is 1.5 million metric tons per year. World resources of 344 million metric tons, including 78 million in U.S.
Gold	mp 1063 °C, bp 2660 °C, sg 19.3	Jewelry, basis of currency, electronics, increasing industrial uses.	In various minerals at a very low 10 ppm for ores currently being processed in the U.S.; by-product of copper refining. World resources of 1 billion oz, 80 million oz in U.S.

Iron	mp 1535 °C, bp 2885 °C, sg 7.86, silvery metal in (rare) pure form	By far the most widely produced metal, usually as steel, a high-tensile-strength material containing 0.3–1.7% C. Made into many alloys for special purposes.	Occurs as hematite (Fe_2O_3), goethite ($Fe_2O_4 \cdot H_2O$), and magnetite (Fe_3O_4) in abundant supply globally and in the U.S.
Lead	mp 327 °C, bp 1750 °C, sg 11.35, silvery color	Fifth most widely used metal. Storage batteries, gasoline additives, pigments, ammunition.	Major source is galena, PbS. Worldwide consumption of lead is 3.5 million metric tons, about 1/3 in U.S. (not including recycled scrap constituting 40% of use). Global reserves of 140 million metric tons, 39 million metric tons in U.S.
Manganese	mp 1244 °C, bp 2040 °C, sg 7.3, hard, brittle, gray-white	Sulfur and oxygen scavenger in steel, manufacture of alloys, dry cells, chemicals.	Found in a variety of minerals, primarily oxides, in manganese nodules on the ocean floor. About 20 million metric tons of manganese ore produced globally each year, 2 million tons consumed in the U.S., no domestic production. World reserves of manganese are 6.5 billion metric tons.
Mercury	mp −38 °C, bp 357 °C, sg 13.6, shiny liquid metal	Instruments, electronic apparatus, electrodes, chemical compounds (such as fungicides and slimicides).	From cinnabar, HgS. Annual world production of 11,500 metric tons, 1/3 of which is used in the U.S. World resources of 275,000 metric tons, only 6600 metric tons in the U.S.
Molybdenum	mp 2620 °C, bp 4825 °C, sg 9.01, ductile, silvery-gray	Alloys, pigments, catalysts, chemicals, and lubricants.	Molybdenite (MoS_2) and wulfenite ($PbMoO_4$) are major ores. About two-thirds of global molybdenum production is in the U.S.; global resources are quite large.
Nickel	mp 1455 °C, bp 2835 °C, sg 8.90, silvery color	Alloys, coins, storage batteries, catalysts (e.g., for hydrogenation of vegetable oil).	Found in ores associated with iron. U.S. consumption of nickel is 150,000 metric tons per year, of which less than 10% is produced domestically. Large domestic reserves of low-grade ore are available.

Metal	Properties[1]	Major uses	Ores and aspects of resources[2]
Silver	mp 961 °C, bp 2193 °C, sg 10.5, shiny metal	Photographic materials, electronics, sterling ware, jewelry, bearings, dentistry.	Found with sulfide minerals, by-product of copper, lead, and zinc smelting. Annual U.S. consumption of 150 million troy ounces will soon exhaust known resources.
Tin	mp 232 °C, bp 2687 °C, sg 7.31	Coatings, solders, bearing alloys, bronze, chemicals.	Found in many compounds associated with granitic rocks and chrysolites. World consumption of 190,000 metric tons/year, U.S. consumption of 60,000 metric tons/year, world resources of 10 million metric tons.
Titanium	mp 1677 °C, bp 3277 °C, sg 4.5, silvery color	Strong, corrosion-resistant, used in aircraft and their engines, valves, pumps, paint pigments.	Ranks ninth in elemental abundance, commonly as TiO_2; no shortages of titanium are likely.
Tungsten	mp 3380 °C, bp 5530 °C, sg 19.3, gray	Very strong, high boiling point, used in alloys, drill bits, turbines, nuclear reactors, tungsten carbide.	Found as tungstates, such as scheelite ($CaWO_4$); U.S. has 7% of world reserves, China 60%. Could be in short supply in U.S. by year 2000.
Vanadium	mp 1917 °C, bp 3375 °C, sg 5.87, gray	Used to make strong steel alloys.	Occurs in igneous rocks, primarily as V(III), primarily a by-product of other metals, U.S. consumption of 5,000 metric tons/year equals production.
Zinc	mp 420 °C, bp 907 °C, sg 7.14, bluish-white	Widely used in alloys (brass), galvanization, paint pigments, chemicals. Fourth in metal production worldwide.	Found in many ore minerals, including sulfides, oxides, and silicates. World production is 5 million metric tons/year (10% from U.S.) and annual U.S. consumption is 1.5 million metric tons. World resources are 235 million metric tons, 20% in the U.S.

[1] Mp, melting point; bp, boiling point; sg, specific gravity.
[2] All figures are approximate; quantities of minerals considered available depend upon price, technology, recent discoveries, and other factors, so that these quantities are subject to fluctuation.

21.3   NONMETAL MINERAL RESOURCES

A number of minerals other than those used to produce metals are important resources. There are so many of these that it is impossible to discuss them all in this chapter; however, mention will be made of the major ones. As with metals, the environmental aspects of mining many of these minerals are quite important. Typically, even the extraction of ordinary rock and gravel can have important environmental effects.

About 1.2 million metric tons of barite, composed of finely ground $BaSO_4$, are employed annually in the U.S. to make drilling "mud", which is used in rotary drilling rigs to seal the walls, cool the drill bit, and lubricate the drill stem. World resources of barite exceed 300 million metric tons with about 100 million metric tons in the U.S.

Clays have been discussed as suspended and sedimentary matter in water (Chapter 5) and as secondary minerals in soil (Chapter 15). Various clays are also used for clarifying oils, as catalysts in petroleum processing, as fillers and coatings for paper, and in the manufacture of firebrick, pottery, sewer pipe, and floor tile. Major types of clays that have industrial uses are shown in Table 21.2. U.S. production of clay is about 60 million metric tons per year, and global and domestic resources are abundant.

Table 21.2.   Major Types of Clays and Their Uses in the U.S.

Type of clay	Percent use	Composition	Uses
Miscellaneous	72	variable	filler, brick, tile, portland cement, many others
Fireclay	12	variable; can be fired at high temperatures without warping	refractories, pottery, sewer pipe, tile, brick
Kaolin	8	$Al_2(OH)_4Si_2O_5$; is white and can be fired without losing shape or color	paper filler, refractories, pottery, dinnerware, petroleum-cracking catalyst
Bentonite and fuller's earth	7	variable	drilling muds, petroleum catalyst, carriers for pesticides, sealers, clarifying oils
Ball clay	1	variable, very plastic	refractories, tile, whiteware

Fluorine compounds are widely used in industry, with fluorspar, CaF_2, being used in large quantities as a flux in steel manufacture. Synthetic and natural cryolite, Na_3AlF_6, is used as a solvent for aluminum oxide in the electrolytic preparation of aluminum metal. Freon-12, difluorodichloromethane, is widely used as a refrigeration and air-conditioning fluid and is considered to be a major stratospheric pollutant (see Section 11.8). Sodium fluoride is used for water fluoridation. World reserves of high-grade fluorspar are around 190 million metric tons, about 13% of which is in the United States. This is sufficient for several decades at projected rates of use. A great deal of byproduct fluorine is recovered from the processing of fluorapatite, $Ca_5(PO_4)_3F$, used as a source of phosphorus.

Micas are complex aluminum silicate minerals which are transparent, tough, flexible, and elastic. Muscovite, $K_2O \cdot 3Al_2O_3 \cdot 6\ SiO_2 \cdot 2\ H_2O$, is major type of mica. Better grades of mica are cut into sheets and used in electronic apparatus, capacitors, generators, transformers, and motors. Finely divided mica is widely used in roofing, paint, welding rods, and many other applications. Sheet mica is imported into the United States largely because of high labor costs in the U.S. Domestic production of finely divided "scrap" mica is in excess of demand, so that shortages of this mineral are unlikely.

Phosphorus, along with nitrogen and potassium, is one of the major fertilizer elements (see Chapter 16). Many soils are deficient in phosphate. Its non-fertilizer applications include supplementation of animal feeds, synthesis of detergent builders, and preparation of chemicals such as pesticides and medicines.

The major phosphate minerals are the apatites, having the general formula $Ca_5(PO_4, CO_3)_3(F, OH, Cl)$. The most common of these are fluorapatite, $Ca_5(PO_4)_3F$ and hydroxyapatite, $Ca_5(PO_4)_3(OH)$. Ions of Na, Sr, Th, and U are found substituted for calcium in apatite minerals. Small amounts of PO_4^{3-} may be replaced by SO_4^{2-}, VO_4^{3-}, or AsO_4^{3-}. Because of the latter ion, traces of arsenic are sometimes found in phosphate products. Byproduct vanadium in phosphate minerals may have economic value.

Approximately 17% of world phosphate production is from igneous minerals, primarily fluorapatites. About three-fourths of world phosphate production is from sedimentary deposits, generally of marine origin. Vast deposits of phosphate, accounting for approximately 5% of world phosphate production, are derived from guano droppings of seabirds and bats.

Current U.S. production of phosphate rock is around 40 million metric tons per year, most of it from Florida. Tennessee and several of the western states are also major producers of phosphate. Reserves of phosphate minerals in the United States amount to 10.5 billion metric tons, containing approximately 1.4 billion metric tons of phosphorus. Identified world reserves of phosphate rock are approximately 6 billion metric tons.

Pigments and fillers of various kinds are used in large quantities. The only naturally occurring pigments still in wide use are those containing iron. These minerals are colored by limonite, an amorphous brown-yellow compound with the formula $2\ Fe_2O_3 \cdot 3H_2O$, and hematite, composed of gray-black Fe_2O_3. Along with varying quantities of clay and manganese oxides, these compounds are found in ocher, sienna, and umber. Manufactured pigments include carbon black, titanium dioxide, and zinc pigments. About 1.5 million metric tons of carbon black, manufactured by the partial combustion of natural gas, are used in the U.S. each year, primarily as a reinforcing agent in tire rubber.

Over 7 million metric tons of minerals are used in the U.S. each year as fillers for paper, rubber, roofing, battery boxes, and many other products. Among the minerals used as fillers are asbestos, carbon black, diatomite, barite, fuller's earth, kaolin, mica, limestone, pyrophyllite, and wollastonite $(CaSiO_3)$.

Although sand and gravel are the cheapest of mineral commodities per ton, the average annual dollar value of these materials is greater than all but a few mineral products because of the huge quantities involved. In tonnage, sand and gravel production is exceeded only by that of fossil fuels.

At present, old river channels and glacial deposits are used as sources of sand and gravel. Many valuable deposits of sand and gravel are covered by construction and lost to development. Transportation and distance from source to use are especially crucial for this resource. Environmental problems involved with defacing land can be severe, although bodies of water used for fishing and other recreational activities frequently are formed by removal of sand and gravel.

The biggest single use for sulfur is in the manufacture of sulfuric acid. However, the element is employed in a wide variety of other industrial and agricultural products. Current consumption of sulfur amounts to approximately 10 million metric tons per year in the United States.

Sulfur can exist in many forms and undergoes a complicated geobiochemical cycle (Figure 21.3). Sulfur may be expelled from volcanoes as sulfur dioxide or as hydrogen sulfide, which is oxidized to sulfur dioxide and sulfates in the atmosphere. Highly insoluble metal sulfides are oxidized upon exposure to atmospheric oxygen to relatively soluble metal sulfates. In the ground and in water, sulfate is converted to organic sulfur by plants and bacteria. Bacteria mediate transitions among sulfate, elemental sulfur, organic sulfur, and hydrogen sulfide. Sulfur (-II) may either escape to the atmosphere as H_2S or precipitate as sulfides of metals, primarily iron.

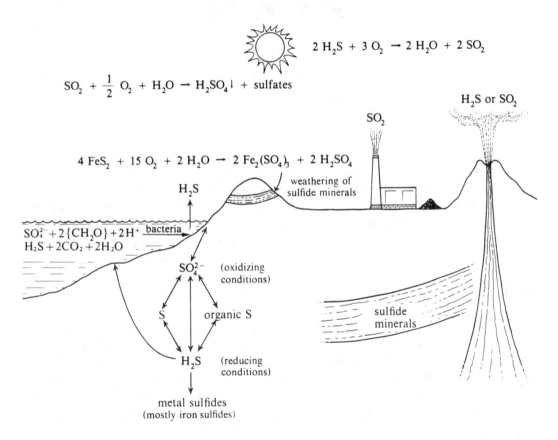

Figure 21.3. Biogeochemical sulfur cycles.

The four most important sources of sulfur are (in decreasing order): deposits of elemental sulfur; H_2S recovered from sour natural gas; organic sulfur recovered from petroleum; and pyrite (FeS_2). Supply of sulfur is no problem either in the United States or worldwide. The United States has abundant deposits of elemental sulfur, and sulfur recovery from fossil fuels as a pollution control measure could even result in surpluses of this element.

21.4. WOOD — A MAJOR RENEWABLE RESOURCE

Fortunately, one of the major natural resources in the world, wood, is a renewable resource. Production of wood and wood products is the fifth largest industry in the United States and forests cover one third of the United States surface area. Wood ranks first worldwide as a raw material for the manufacture of other products, including lumber, plywood, particle board, cellophane, rayon, paper, methanol, plastics, and turpentine.

Chemically, wood is a complicated substance consisting of long cells having thick walls composed of polysaccharides such as cellulose,

Cellulose polymer

The polysaccharides in cell walls account for approximately three-fourths of *solid wood*, wood from which extractable materials have been removed by an alcohol-benzene mixture. Wood typically contains a few tenths of a percent ash.

A wide variety of organic compounds can be extracted from wood by water, alcohol-benzene, ether, and steam distillation. These compounds include tannins, pigments, sugars, starch, cyclitols, gums, mucilages, pectins, galactans, terpenes, hydrocarbons, acids, esters, fats, fatty acids, aldehydes, resins, sterols, and waxes. Substantial amounts of methanol (sometimes called *wood alcohol*) are extracted from wood. Methanol, once a major source of liquid fuel, is again being considered for that use.

A major use of wood is in paper manufacture. The widespread use of paper is a mark of an industrialized society. The manufacture of paper is a highly advanced technology. Paper consists essentially of cellulosic fibers tightly pressed together. The lignin fraction must first be removed from the wood, leaving the cellulosic fraction. Both the sulfite and alkaline processes for accomplishing this separation have resulted in severe water and air pollution problems, although substantial progress has been made in alleviating these.

Wood fibers and particles can be used for making fiberboard, paper-base laminates (layers of paper held together by a resin and formed into the desired structures at high temperatures and pressures), particle board (consisting of wood particles bonded together by a phenol-formaldehyde or urea-formaldehyde resin) and nonwoven textile substitutes consisting of wood fibers held together by adhesives. Chemical processing of wood enables the manufacture of many useful products, including methanol (a gasoline substitute) and sugar. Sugar and methanol are potential major products from the 60 million metric tons of wood wastes produced in the U.S. each year.

21.5. THE ENERGY PROBLEM

Since the first "energy crisis" of 1973-74, much has been said and written, many learned predictions have gone awry, and some concrete action has even taken place. Prophecies of catastrophic economic disruption, people "freezing in the dark," and freeways given over to bicycles (though perhaps not a bad idea) have not been fulfilled. However, the several-fold increase in crude oil prices since 1973 has extacted a toll. In the U.S. and other industrialized nations, the economy has been plagued by inflation, recession, unemployment, and obsolescence of industrial equipment. The economies of some petroleum-deficient developing countries have been devastated by energy prices.

The solutions to energy problems are strongly tied to environmental considerations. For example, a massive shift of the energy base to coal in nations that now rely largely on petroleum for energy would involve much more strip mining, potential production of acid mine water, use of scrubbers, and release of greenhouse gases (carbon dioxide from coal combustion and methane from coal mining). Similar examples could be cited for most other energy alternatives.

Clearly, chemists must be involved in developing alternative energy sources. Chemical processes are used in the conversion of coal to gaseous and liquid fuels. New materials developed through the applications of chemistry will be employed to capture solar energy and convert it to electricity. The environmental chemist has a key role to play in making alternative energy sources environmentally acceptable. The energy problem poses both questions and opportunities for students entering a career in chemistry. It is important, therefore, that these students know the basics of energy resources and alternative energy resources as well as their environmental aspects.

21.6. WORLD ENERGY RESOURCES

At present, most of the energy consumed by humans is produced from fossil fuels. Estimates of the amounts of fossil fuels available differ. Because of undiscovered deposits of petroleum and natural gas, there may be considerably more fossil fuel available than is commonly realized (or less than the more optimistic estimates).

Estimates of the quantities of recoverable fossil fuels in the world before 1800 are given in Figure 21.4. By far the greatest recoverable fossil fuel is in the form of coal and lignite. Furthermore, only a small percentage of this energy source has been utilized to date, whereas much of the recoverable petroleum and natural gas has already been consumed. Projected use of these latter resources indicates rapid depletion.

Although it is not the purpose of this chapter to give a detailed discussion of energy resources and projected rates of use, some of the statistics in these areas are interesting as well as sobering. In the 48 coterminous United States, the peak of petroleum production has been reached. Alaskan oil can help the petroleum supply only temporarily. Worldwide petroleum resources are approximately 10 times those of the United States, and peak world production will be reached within a few years. Despite the vast amounts of oil that remain in tar sand deposits and oil shales, the economic costs of their utilization may be prohibitive.

Although world coal resources are enormous and potentially can fill energy needs for a century or two, their utilization is limited by environmental disruption from mining and emissions of carbon dioxide and sulfur dioxide. These would become intolerable long before coal resources were exhausted. Assuming only uranium-235 as

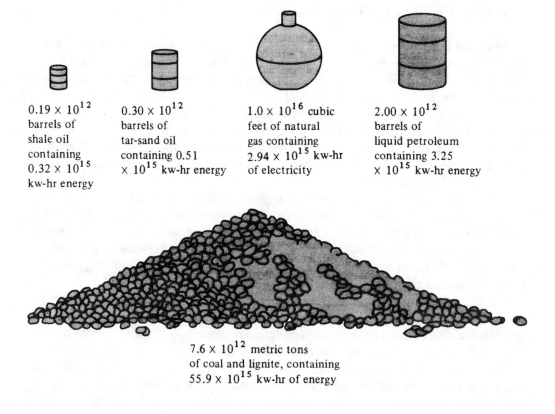

0.19 × 10^{12}
barrels of
shale oil
containing
0.32 × 10^{15}
kw-hr energy

0.30 × 10^{12}
barrels of
tar-sand oil
containing 0.51
× 10^{15} kw-hr energy

1.0 × 10^{16} cubic
feet of natural
gas containing
2.94 × 10^{15} kw-hr
of electricity

2.00 × 10^{12}
barrels of
liquid petroleum
containing 3.25
× 10^{15} kw-hr energy

7.6 × 10^{12} metric tons
of coal and lignite, containing
55.9 × 10^{15} kw-hr of energy

Figure 21.4. Original amounts of the world's recoverable fossil fuels (quantities in thermal kilowatt hours of energy based upon data taken from M. K. Hubbert, "The Energy Resources of the Earth," in *Energy and Power*, W. H. Freeman and Co., San Francisco, 1971).

a fission fuel source, total recoverable reserves of nuclear fuel are roughly about the same as fossil fuel reserves. These are many orders of magnitude higher if the use of breeder reactors is assumed. Extraction of only 2% of the deuterium present in the Earth's oceans would yield about a billion times as much energy by controlled nuclear fusion as was originally present in fossil fuels! This prospect is tempered by the lack of success in developing a controlled nuclear fusion reactor. Hopes for such a process soared with the announcement of a "cold fusion" process in 1989, but the world still awaits an operational cold fusion reactor. Geothermal power, currently utilized in northern California, Italy, and New Zealand, has the potential for providing a high percentage of energy worldwide. The same limited potential is characteristic of several renewable energy resources, including hydroelectric energy, tidal energy, and wind power. All of these will continue to contribute significant, but relatively small, amounts of energy. Renewable, nonpolluting solar energy comes as close to being an ideal energy source as any available. It almost certainly has a bright future.

Globally, therefore, the energy picture is both hopeful and grim. In the U.S., certainly, liquid petroleum and natural gas are on the way out as major producers of energy. The prospects for massive utilization of shale oil deposits appear to be very dim. Nuclear power may provide an increasing share of energy production, though difficult political, safety, technical, and environmental problems will continue to delay its rate of development. Miscellaneous sources such as wind, hydroelectric, tidal, and geothermal power will make useful but limited contributions to total energy production. Were it not for the associated environmental problems, coal, an abundant and versatile fossil fuel, could fill the energy gap until more exotic sources can be developed. Within a few decades, widespread utilization may be made of nuclear fusion or solar power, each of which promises to be an abundant, relatively nonpolluting source of energy.

21.7 ENERGY CONSERVATION

From about 1960 until 1973, energy production increased in the United States slightly over 4% per year, a growth rate with a doubling time of approximately 16 years. Such a growth rate can be used or misused to support many different theories. It can be projected to show, for example, that within a few decades facilities for the production of power must be greatly expanded to provide for energy needs. It can also be used to show that devastating environmental damage will occur as a result of the exploitation of energy resources. It can even be used to calculate the exact date upon which the whole nation will glow in the dark from exponentially increasing waste heat radiated out into space. What is wrong with such projections is that they need not, would not, and indeed cannot come true.

Any consideration of energy needs and production must take energy conservation into consideration. This does not have to mean cold classrooms with thermostats set at 60°F in mid-winter, nor swelteringly hot homes with no air-conditioning, nor total reliance on the bicycle for transportation, although these, and even more severe, conditions are routine in many countries. The fact remains that the United States has wasted energy at a deplorable rate. For example, U.S. energy consumption is significantly higher per capita than that of some other countries that have equal, or significantly better, living standards. Obviously, a great deal of potential exists for energy conservation that will ease the energy problem.

Transportation is the economic sector with the greatest potential for increased efficiencies. The private auto and airplane are only about one-third as efficient as buses or trains for transportation. Transportation of freight by truck requires about 3800 Btu/ton-mile, compared ton only 670 Btu/ton-mile for a train. It is terribly inefficient compared to rail transport (as well as dangerous, labor-intensive, and environmentally disruptive). Major shifts in current modes of transportation in the U.S. will not come without anguish, but energy conservation dictates that they be made.

Household and commercial uses of energy are relatively efficient. Here again, appreciable savings can be made. The all-electric home requires much more energy (considering the percentage wasted in generating electricity) than a home heated with fossil fuels. The sprawling ranch-house style home uses much more energy per person than does an apartment unit or row house. Improved insulation, sealing around the windows, and other measures can conserve a great deal of energy. Electric generating plants centrally located in cities can provide waste heat for commercial and residential heating and cooling and, with proper pollution control equipment, can use refuse for a significant fraction of fuel

As scientists and engineers undertake the crucial task of developing alternative energy sources to replace dwindling petroleum and natural gas supplies, energy conservation must receive proper emphasis. In fact, zero energy-use growth, at least on a per capita basis, is a worthwhile and achievable goal. Such a policy would go a long way toward solving many environmental problems. With ingenuity, planning, and proper management, it could be achieved while increasing the standard of living and quality of life.

Despite the importance of energy conservation, it should be pointed out that there are pitfalls involved in a total reliance on it as a solution to the energy problem. David Lilienthal, the first chairman of the Atomic Energy Commission, stated eloquently in his book on atomic energy:[1] "proponents of energy conservation who have put it forward as *the answer* to the 'energy crisis' invite the charge that this is an 'elitist' doctrine and a form of the 'no growth' concept..... To offer it as a basic social doctrine of 'less is better' or as a slogan to rally opposition to nuclear energy expansion and the excesses of industrialization and consumerism will antagonize rather than persuade the general American public and a world in which poverty closely associated with energy shortages in endemic."

In the broadest sense a form of energy conservation, waste–to–energy processes hold the prospect of meeting some energy needs while disposing of potential pollutants. For example, almost half of the 1.4 million tons of solid waste generated in the State of Maine is burned in facilities with energy-generating capabilities.[2]

21.8. ENERGY CONVERSION PROCESSES

As shown in Figure 21.5, energy occurs in several forms and must be converted to other forms. The efficiencies of conversion vary over a wide range. Conversion of electrical energy to radiant energy by incandescent light bulbs is very inefficient — less than 5% of the energy is converted to visible light and the remainder is wasted as

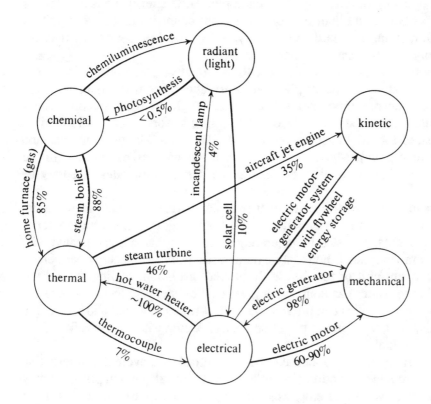

Figure 21.5. Kinds of energy and examples of conversion between them, with conversion efficiency percentages.

heat. At the other end of the scale, a large electrical generator is around 80% efficient in converting mechanical energy to electrical energy. The once much-publicized Wankel rotary engine converts chemical to mechanical energy with an efficiency of about 18%, compared to 25% for a gasoline-powered piston engine and about 37% for a diesel engine. A modern coal-fired steam-generating power plant converts chemical energy to electrical energy with an overall efficiency of about 40%.

One of the most significant energy conversion processes is that of thermal energy to mechanical energy in a heat engine such as a steam turbine. The Carnot equation,

$$\text{Percent efficiency} = \frac{T_1 - T_2}{T_1} \times 100 \tag{21.8.1}$$

states that the percent efficiency is given by a fraction involving the inlet temperature (for example, of steam), T_1, and the outlet temperature, T_2. These temperatures are expressed in Kelvin (°C + 273). Typically, a steam turbine engine operates with approximately 810 K inlet temperature and 330 K outlet temperature. These temperatures substituted into Equation 21.8.1 give a maximum theoretical efficiency of 59%. However, because it is not possible to maintain the incoming steam at the maximum temperature and because mechanical energy losses occur, overall efficiency of conversion of thermal energy to mechanical energy in a modern steam power plant is approximately 47%. Taking into account losses from conversion of chemical to thermal energy in the boiler, the total efficiency is about 40%.

Some of the greatest efficiency advances in the conversion of chemical to mechanical or electrical energy have been made by increasing the peak inlet temperature in heat engines. The use of superheated steam has raised T_1 in a steam power plant from around 550 K in 1900 to about 850 K at present. Improved materials and engineering design, therefore, have resulted in large energy savings.

The efficiency of nuclear power plants is limited by the maximum temperatures attainable. Reactor cores would be damaged by the high temperatures used in fossil-fuel-fired boilers and have a maximum temperature of approximately 620 K. Because of this limitation, the overall efficiency of conversion of nuclear energy to electricity is about 30%.

Most of the 60% of energy from fossil-fuel-fired power plants and 70% of energy from nuclear power plants that is not converted to electricity is dissipated as heat, either to the atmosphere or to bodies of water and streams. The latter is thermal pollution, which may either harm aquatic life or, in some cases, actually increase bio-activity in the water to the benefit of some species. This waste heat is potentially very useful in applications like home heating, water desalination, and aquaculture (growth of plants in water). Increased production of waste heat by human activity could eventually result in marked changes in climate.

Some devices for the conversion of energy are shown in Figure 21.6. Substantial advance have been made in energy conversion technology over many decades and more can be projected for the future. Through the use of higher temperatures and larger generating units, the overall efficiency of fossil-fueled electrical power generation has increased approximately ten-fold since 1900, from less than 4% to a maximum of around 40%. An approximately four-fold increase in the energy-use efficiency of rail transport occurred during the 1940s and 1950s with the replacement of steam locomotives with diesel locomotive. During the coming decades, increased efficiency can be anticipated from such techniques as combined power cycles in

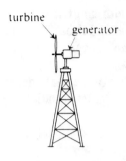

(1) Turbine for conversion of kinetic or potential energy of a fluid to mechanical and electrical energy.

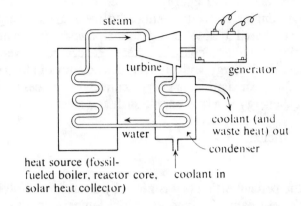

(2) Steam power plant in which high-energy fluid is produced by vaporizing water.

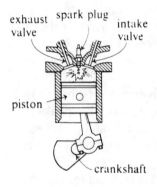

(3) Reciprocating internal combustion engine.

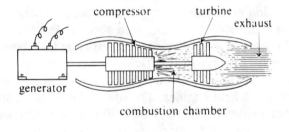

(4) Gas turbine engine. Kinetic energy of hot exhaust gases may be used to propel aircraft.

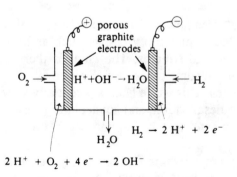

(5) Fuel cell for the direct conversion of chemical energy to electrical energy.

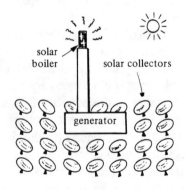

(6) Solar thermal electric conversion.

Figure 21.6. Some energy conversion devices.

connection with magnetohydrodynamics (Figure 21.8) probably will be developed as a very efficient energy source used in combination with conventional steam generation. Entirely new devices such as thermonuclear reactors for the direct conversion of nuclear fusion energy to electricity will very likely be developed.

21.9. PETROLEUM AND NATURAL GAS

Since its first commercial oil well in 1859, somewhat more than 100 billion barrels of oil have been produced in the United States, most of it in recent years. In 1990 world petroleum consumption was at a rate of about 65 million barrels per day.

Liquid petroleum is found in rock formations ranging in porosity from 10 to 30%. Up to half of the pore space is occupied by water. The oil in these formations must flow over long distances to an approximately 6-inch diameter well from which it is pumped. The rate of flow depends on the permeability of the rock formation, the viscosity of the oil, the driving pressure behind the oil, and other factors. Because of limitations in these factors **primary recovery** of oil yields an average of about 30% of the oil in the formation, although it is sometimes as little as 15%. More oil can be obtained using **secondary recovery** techniques, which involve forcing water under pressure into the oil-bearing formation to drive the oil out. Primary and secondary recovery together typically extract somewhat less than 50% of the oil from a formation. Finally, **tertiary recovery** techniques can be used to extract even more oil. This normally uses injection of pressurized carbon dioxide, which forms a mobile solution with the oil and allows it to flow more easily to the well. Other chemicals, such as detergents, may be used to aid in tertiary recovery. Currently, about 300 billion barrels of U.S. oil is not available through primary recovery alone. A recovery efficiency of 60% through secondary or tertiary techniques could double the amount of available petroleum. Much of this would come from fields which have already been abandoned or essentially exhausted using primary recovery techniques. Advanced recovery techniques are now being more widely practiced as increased petroleum prices have made them more feasible.

Shale oil is a possible substitute for liquid petroleum. Shale oil is a pyrolysis product of oil shale, a rock containing organic carbon in a complex structure called kerogen. It is believed that approximately 1.8. trillion barrels of shale oil could be recovered from deposits of oil shale in Colorado, Wyoming, and Utah. In the Colorado Piceance Creek basin alone, more than 100 billion barrels of oil could be recovered from prime shale deposits.

Shale oil may be recovered from the parent mineral by retorting the mined shale in a surface retort or by burning the shale underground with an *in situ* process. Both processes present environmental problems. Surface retorting requires the mining of enormous quantities of mineral and disposal of the spent shale, which has a volume greater than the original mineral. *In situ* retorting limits the control available over infiltration of underground water and resulting water pollution. Water passing through spent shale become quite saline, so there is major potential for saltwater pollution.

During the late 1970s and early 1980s, several corporations began building facilities for shale oil extraction in northwestern Colorado. The most ambitious of these was the Colony project, a joint construction project of Exxon and Tosco. This plant, situated about 15 miles north of Parachute, Colorado, was to produce 50,000 barrels of synthetic crude oil per day starting around 1987. The amounts of money involved were staggering. Exxon spent $300 million to purchase a 60% share of the project from Tosco, and the two companies spend an additional $400 million on the project. By 1982 the total cost of the project was estimated to be $6 billion or more. The Synthetic Fuels Corporation, an agency set up by the U.S. government, had lent $80 million on the project. In the face of escalating construction costs and a "soft" crude oil market, Exxon withdrew from the project in May, 1982, stopping construction on the plant, and shale oil extraction is still not practiced commercially.

An even larger synthetic fuels project that failed was the $13.1 billion (Canadian) Alsands oil sands project located in Alberta. This project would have produced 137,000 barrels of oil per day from the mining of tar sand and extraction of heavy oil from it. In April, 1982, both Shell Canada Ltd. and Gulf Canada Ltd. withdrew from the project, the cost of which had escalated from an initial estimate of $5 billion.

Natural gas, consisting almost entirely of methane, has become more attractive as an energy source. This is because of uncertainties regarding natural gas availability, coupled with the potential for the discovery and development of truly enormous new sources of this premium fuel. In 1968, discoveries of natural gas deposits in the U.S. fell below annual consumption for the first time. This trend continued during the following decade, except for 1970, when gas was discovered in Alaska. During the record-cold winter of 1976-77, severe natural gas shortages occurred in parts of the U.S. Price incentives resulting from the passage of the 1978 Gas Policy Act have tended to reverse the adverse trend of natural gas discoveries versus consumption.

In addition to its use as a fuel, natural gas can be converted to many other hydrocarbon materials. It can be used as a raw material for the Fischer-Tropsch synthesis of gasoline. The discovery and development of truly massive sources of natural gas, such as may exist in geopressurized zones, could provide abundant energy reserves for the U.S., though at substantially increased prices.

21.10. COAL

From Civil War times until World War II, coal was the dominant energy source behind U.S. industrial expansion. However, the greater convenience of lower-cost petroleum resulted in a decrease in the use of coal for U.S. energy requirements after World War II. Annual coal production fell by about one-third, reaching a low of approximately 400 million tons in 1958. Since that time U.S. production has increased substantially, setting records of 981 million tons in 1989 and 1.02 billion tons in 1990.[3]

The general term *coal* describes a large range of solid fossil fuels derived from partial degradation of plants. Table 21.3 shows the characteristics of the major classes of coal found in the U.S., differentiated largely by percentage of fixed carbon, percentage of volatile matter, and heating value (*coal rank*). Chemically, coal is a very complex material and is by no means pure carbon. For example, a chemical formula for Illinois No. 6 coal, a type of bituminous coal, would be something like $C_{100}H_{85}S_{2.1}N_{1.5}O_{9.5}$.

Anthracite, a hard, clean-burning, low-sulfur coal, is the most desirable of all coals. Approximately half of the anthracite originally present in the United States has been mined. Bituminous coal found in the Appalachian and north central coal fields is the most widely used. It is an excellent fuel with a high heating value. Unfortunately, most bituminous coals have a high percentage of sulfur (an average of 2-3%), so the use of this fuel presents environmental problems. Huge reserves of virtually untouched subbituminous and lignite coals are found in the Rocky Mountain states and in the northern plains of Dakotas, Montana, and Wyoming. Subbituminous and lignite coals have a relatively high oxygen content, largely in the form of carboxyl, phenolic hydroxyl, and carbonyl groups in a humic acid fraction of the fuel. Some of the lower-grade lignites are almost pure humic acid (see Chapter 3). These fuels have the advantage of being low in sulfur content and are finding increasing use in power plants that have to meet SO_2 emission standards. They have a number of disadvantages, including low heat content and high moisture contents. Also, they are generally found great distances from areas having the greatest need for fossil fuels. Lignite, in particular, tends to lose moisture and crumble when transported to form "bug dust." Despite these disadvantages, the low sulfur content and ease of mining these low-grade fuels is resulting in a rapid increase in their use.

Table 21.3. Major Types of Coal Found in the United States

| Type of coal | Proximate analysis, percent[1] | | | | Range of heating value (Btu/pound) |
	Fixed carbon	Volatile matter	Moisture	Ash	
Anthracite	82	5	4	9	13,000–16,000
Bituminous					
Low-volatile	66	20	2	12	11,000–15,000
Medium-volatile	64	23	3	10	11,000–15,000
High-volatile	46	44	6	4	11,000–15,000
Subbituminous	40	32	19	9	8,000–12,000
Lignite	30	28	37	5	5,500–8,000

[1] These values may vary considerably with the source of coal.

Some geographical areas have very large coal resources. Huge reserves of subbituminous coal are found in the Rocky Mountain states, particularly Montana and Wyoming. An especially striking example is the approximately 350 billion tons of virtually untouched lignite in North Dakota, of which 16-20 billion tons may be mined at present prices. Figure 21.7 shows areas in the U.S. with major coal reserves.

The extent to which coal can be used as a fuel depends upon solutions to several problems, including (1) minimizing the environmental impact of coal mining; (2) removing ash and sulfur from coal prior to combustion; (3) removing ash and sulfur dioxide from stack gas after combustion; (4) conversion of coal to liquid and gaseous fuels free of ash and sulfur (see Section 21.11); and, most important, (5) whether or not the impact of increased carbon dioxide emissions upon global climate can be tolerated. Progress is being made on minimizing the environmental impact of mining. As more is learned about the processes by which acid mine water is formed, measures can be taken to minimize the production of this water pollutant. Particularly on flatter lands, strip-mined areas can be reclaimed with relative success. Inevitably, some environmental damage will result from increased coal mining, but the environmental impact can be reduced by various control measures. Washing, flotation, and chemical processes can be used to remove some of the ash and sulfur prior to burning. Approximately half of the sulfur in the average coal occurs as pyrite, FeS_2, and half as organic sulfur. Although little can be done to remove the latter, much of the pyrite can be separated from most coals by physical and chemical processes.

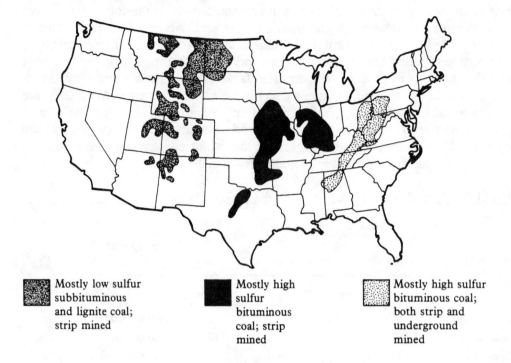

Mostly low sulfur
subbituminous
and lignite coal;
strip mined

Mostly high
sulfur
bituminous
coal; strip
mined

Mostly high sulfur
bituminous coal;
both strip and
underground
mined

Figure 21.7. Areas with major coal reserves in the coterminous United States.

The maintenance of air pollution emission standards requires the removal of
sulfur dioxide from stack gas in coal-fired power plants. Stack gas desulfurization
presents some economic and technological problems; the major processes available
for it are summarized in Section 11.5.

Magnetohydrodynamic power combined with conventional steam generating
units has the potential for a major breakthrough in the efficiency of coal utilization. A
schematic diagram of magnetohydrodynamic (MHD) generator is shown in Figure
21.8. This device uses a plasma of ionized gas at around 2400°C blasting through a
very strong magnetic field of at least 50,000 gauss to generate direct current. The

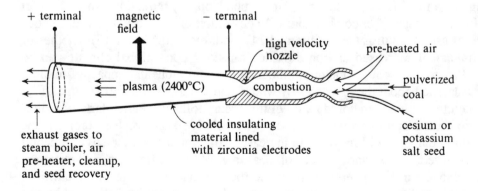

Figure 21.8. A magnetohydrodynamic power generator.

ionization of the gas is accomplished by injecting a "seed" of cesium or potassium salts. In an MHD generator, the ultra-high-temperature gas issuing through a supersonic nozzle contains ash, sulfur dioxide, and nitrogen oxides which severely erode and corrode the materials used. This hot gas is used to generate steam for a conventional steam power plant, thus increasing the overall efficiency of the process. The seed salts combine with sulfur dioxide and are recovered along with ash in the exhaust. Pollutant emissions are low. Despite some severe technological difficulties, there is a chance that MHD power could become feasible on a large scale, and an experimental MHD generator was tied to a working power grid in the U.S.S.R. for several years. Most important, the overall efficiency of combined MHD-steam power plants should reach 60%, one and one-half times the maximum of present steam-only plants.

21.11. COAL CONVERSION

Coal can be converted to gaseous, liquid, or low-sulfur, low-ash fuels. All of these are less polluting than coal, and the gases and liquids can be used with distribution systems and equipment designed for natural gas or petroleum. A major advantage of coal conversion is that it enables use of high-sulfur coal which otherwise could not be burned without intolerable pollution or expensive stack gas cleanup.

Coal conversion is an old idea; a house belonging to William Murdock at Redruth, Cornwall, England, was illuminated with coal gas in 1792. The first municipal coal gas system was employed to light Pall Mall in London in 1807. The coal-gas industry began in the U.S. in 1816. The early coal-gas plants used coal pyrolysis (heating in the absence of air) to produce a hydrocarbon-rich product particularly useful for illumination. Later in the 1800s the water-gas process was developed, in which steam was added to hot coal to produce a mixture consisting primarily of H_2 and CO. It was necessary to add volatile hydrocarbons to this "carbureted" water-gas to bring its illuminating power up to that of gas prepared by coal pyrolysis. The U.S. had 11,000 coal gasifiers operating in the 1920s. At the peak of its use in 1947, the water-gas method accounted for 57% of U.S.-manufactured gas. The gas was made in low-pressure, low-capacity gasifiers which by today's standards would be inefficient and environmentally unacceptable (several locations of these old plants have been designated as hazardous waste sites because of residues of coal tar and other wastes). During World War II, Germany developed a major synthetic petroleum industry based on coal, which reached a peak capacity of 100,000 barrels per day in 1944. A plant now operating in Sasol, South Africa, converts several tens of thousands tons of coal per day to synthetic petroleum.

The major routes for coal conversion are shown in Figure 21.9. The more promising coal-derived fuels are the following:

1. Solvent-refined coal (SRC), a solid high-Btu product with low sulfur, ash, and water content

2. Low-sulfur boiler fuels that are liquid at elevated temperatures

3. Liquid hydrocarbon fuels, including gasoline, diesel fuel, and naphtha

4. Synthetic natural gas (SNG), essentially pure methane

5. Low-sulfur, low-Btu gas for industrial use

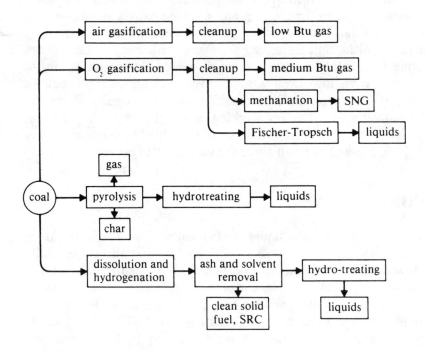

Figure 21.9. Routes to coal conversion.

High-heat-content (high-Btu) synthetic natural gas, SNG, can be produced by any of several processes. The steps in a typical process for SNG production are:

1. Steam and oxygen react with coal in a gasifier, producing a low-Btu gas consisting of carbon monoxide, carbon dioxide, hydrogen, and some methane. The major reactions, simplifying coal as C, are:

$$2C + O_2 \rightarrow 2CO \qquad \Delta H = -94.1 \text{ kcal} \qquad (21.11.1)$$

(per mole of carbon at 25°C)

$$2CO + O_2 \rightarrow 2CO_2 \qquad\qquad (21.11.2)$$

$$C + H_2O \rightarrow CO + H_2 \qquad \Delta H = +31.4 \text{ kcal} \qquad (21.11.3)$$

$$C + 2H_2 \rightarrow CH_4 \qquad \Delta H = -17.9 \text{ kcal} \qquad (21.11.4)$$

2. The gas product is freed from tar and dust by scrubbing with water.

3. The shift reaction,

$$H_2O + CO \rightarrow H_2 + CO_2 \qquad\qquad (21.11.5)$$

over a shift catalyst increases the molar ratio of H_2 to CO to an optimum value of approximately 3.5:1.

4. Acid gases (H_2S, COS, CO_2) are removed in an alkaline scrubber. Complete sulfur removal is necessary to avoid poisoning the methanation catalyst.

5. The catalytic hydrogenation of CO (methanation step) produces methane gas:

$$CO + 3H_2 \rightarrow CH_4 + H_2O \tag{21.11.6}$$

   Probably the most promising route for coal gasification is the **Texaco process**, which gasifies a water slurry of coal at temperatures of 1250°C to 1500°C and pressures of 350 to 1200 pounds per square inch. Under these severe conditions, no tar by-products are produced. The process is very efficient, in part because the gas coming from the gasifier is cooled by raising steam in a heat exchanger, and the steam is employed to drive a steam turbine. After particles, NH_3, and H_2S are removed from the gas, it is used to fire a gas turbine engine. The heat from the engine exhaust is used to raise steam to drive another steam turbine.

   Chemical addition of hydrogen to coal can liquify it and produce a synthetic petroleum product. This can be done with a hydrogen donor solvent, which is recycled and itself hydrogenated with H_2 during part of the cycle. Such a process forms the basis of the successful **Exxon Donor Solvent process**, which has been used in a 250 ton/day pilot plant.

   A number of environmental implications are involved in the widespread use of coal gasification. To produce significant quantities of gas from coal would require increased coal mining. Strip mining would have to increase and large quantities of water would be required. Scarcity of water is a particularly severe constraint on gasification processes in the Powder River basin of Wyoming and Montana, which is located in a region where the average rainfall is only 15-30 cm per year. Indigenous water supplies could not support massive coal gasification operations. Another consideration involves the thermal efficiency of the conversion of coal to gas, which has been estimated at approximately 65%. Therefore, huge quantities of heat would be produced by widespread gasification, causing thermal pollution. Since coal gasification is carried out in closed systems, however, there is little potential for air pollution.

   Methanol, CH_3OH, is a convenient liquid fuel which can be produced from coal. Methanol is produced on a small scale by the destructive distillation of wood (see Section 21.4). It was widely used for heating and lighting in mid-19th century France and for automobile fuel in gasoline-short countries during World War II. On a large scale, methanol is produced by the reaction of carbon monoxide and hydrogen at a lower H_2/CO ratio than that required for the production of methane:

$$CO + 2H_2 \rightarrow CH_3OH \tag{21.11.7}$$

This reaction can be carried out using a copper-based catalyst at 50 atm and 250°C. The carbon monoxide and hydrogen used for methanol production are both produced from coal, oxygen, and steam. At levels up to 15%, methanol makes an excellent additive for gasoline. It has a high octane number (106) and improves fuel economy and acceleration time in automobiles. It also reduces the emissions of practically all automotive pollutants. A significant advantage of methanol is that it can be substituted for present liquid fuels with minimum disruption in existing processing, transportation, and end-use facilities.

21.12. NUCLEAR FISSION POWER

The awesome power of the atom revealed at the end of World War II held out enormous promise for the production of abundant, cheap energy. This promise has never really come to fruition, although nuclear energy currently provides a significant percentage of electric energy in many countries.

Nuclear power reactors currently in use depend upon the fission of uranium-235 nuclei by reactions such as

$$\ce{^{235}_{92}U} + \ce{^{1}_{0}n} \longrightarrow \ce{^{133}_{51}Sb} + \ce{^{99}_{41}Nb} + 4\ce{^{1}_{0}n} \qquad (21.12.1)$$

to produce two radioactive fission products, an average of 2.5 neutrons, and an average of 200 MeV of energy per fission. The neutrons, initially released as fast-moving, highly energetic particles, are slowed to thermal energies in a moderator medium. For a reactor operating at a steady state, exactly one of the neutron products from each fission is used to induce another fission reaction in a chain reaction:

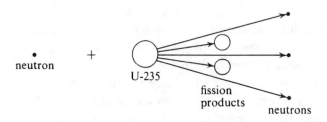

The energy from these nuclear reactions is used to heat water in the reactor core and produce steam to drive a steam turbine, as shown in Figure 21.10.

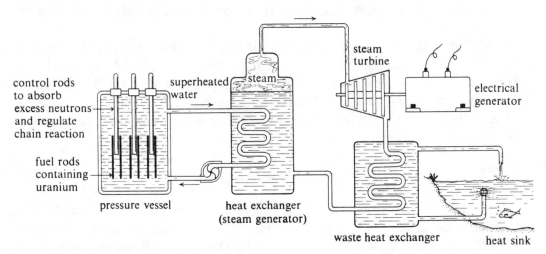

Figure 21.10. A typical nuclear fission power plant.

Because of limitations of structures and materials, nuclear reactors operate at a maximum temperature of around 620 K, compared to approximately 800 K in a fossil-fuel power plant. The thermal efficiency of nuclear power generation is limited by the Carnot relationship and is therefore inherently low. The overall efficiency for production of electricity in a nuclear-fission plant does not exceed 30%, leaving 70% of the energy to be disposed of in the environment.

The course of nuclear power development was permanently altered by the now-famous reactor accident at Three Mile Island (TMI) in Pennsylvania. The incident began on March 28, 1979 with a partial loss of coolant water from the Metropolitan Edison Company's nuclear reactor located on Three Mile Island in the Susquehanna River, 28 miles outside of Harrisburg, Pennsylvania. The loss resulted in loss of control, overheating, and partial disintegration of the reactor core. Some radioactive xenon and krypton gas were released and some radioactive water was dumped into the Susquehanna River. A much worse accident occurred at Chernobyl in the Soviet Union in April of 1986 when a reactor blew up spreading radioactive debris over a wide area and killing a number of people (officially 31, but probably many more). Thousands of people were evacuated and the entire reactor structure had to be entombed in concrete. Food was seriously contaminated as far away as northern Scandanavia.

A limitation of fission reactors is the fact that only 0.71% of natural uranium is fissionable uranium-235. This situation could be improved by the development of **breeder reactors**, which convert uranium-238 (natural abundance 99.28%) to fissionable plutonium-239.

A major consideration in the widespread use of nuclear fission power is the production of large quantities of highly radioactive waste products. These remain lethal for thousands of years. They must either be stored in a safe place or disposed of permanently in a safe manner. At the present time, spent fuel elements are being stored under water at the reactor sites. Eventually, the wastes from this fuel will have to be buried.

Another problem to be faced with nuclear fission reactors is their eventual decommissioning. There are three possible solutions. One is dismantling soon after shutdown, in which the fuel elements are removed, various components are flushed with cleaning fluids, and the reactor is cut up by remote control and buried. "Safe storage" involves letting the reactor stand 30-100 years to allow for radioactive decay, followed by dismantling. The third alternative is entombment, encasing the reactor in a concrete structure.

21.13. NUCLEAR FUSION POWER

The two main reactions by which energy can be produced from the fusion of two light nuclei into a heavier nucleus are the deuterium-deuterium reaction,

$$\ce{^2_1H} + \ce{^2_1H} \rightarrow \ce{^3_2He} + \ce{^1_0n} + 1\text{MeV} \tag{21.13.1}$$

and the deuterium-tritium reaction:

$$\ce{^2_1H} + \ce{^3_1H} \rightarrow \ce{^4_2He} + \ce{^1_0n} + 17.6\text{MeV} \tag{21.13.2}$$

The second reaction is more feasible because less energy is required to fuse the two nuclei than to fuse two deuterium nuclei. The total energy from deuterium-tritium fusion is limited by the availability of tritium, which is made from nuclear reactions of lithium-6 (natural abundance, 7.4%). The supply of deuterium, however, is essentially unlimited; one out of every 6700 atoms of hydrogen is the deuterium isotope. The ^{3}He by-product of Reaction 21.13.1. reacts with neutrons, which are abundant in a nuclear fusion reactor, to produce tritium required for Reaction 21.13.2.

The power of nuclear fusion has not yet been harnessed in a sustained, controlled reaction of appreciable duration. Most approaches have emphasized "squeezing" a plasma (ionized gas) of fusionable nuclei in a strong magnetic field. Another approach to obtaining the high temperatures required for fusion involves the use of a number of high powered lasers focused on a pellet of fusionable material. A great flurry of excitement over the discovery of a cheap, safe, simple fusion power source was generated by an announcement from the University of Utah in 1989 of the attainment of "cold fusion" in the electrolysis of deuterium oxide (heavy water). Funding was appropriated and laboratories around the world were thrown into frenetic activity in an effort to duplicate the reported results. Some investigators reported evidence, particularly the generation of anomalously large amounts of heat, to support the idea of cold fusion, whereas others scoffed at the idea. To date the promise of cold fusion has not been realized.

Controlled nuclear fusion processes would produce almost no radioactive waste products. However, tritium is very difficult to contain, and some release of the isotope would occur. The deuterium-deuterium reaction promises an unlimited source of energy. Either of these reactions would be preferable to fission in terms of environmental considerations. Therefore, despite the possibility of insurmountable technical problems involved in harnessing fusion energy, the promise of this abundant, relatively nonpolluting energy source makes its pursuit well worth a massive effort.

21.14. GEOTHERMAL ENERGY

Underground heat in the form of steam, hot water, or hot rock used to produce steam is already being used as an energy resource. This energy was first harnessed for the generation of electricity at Larderello, Italy, in 1904, and has since been developed in Japan, Russia, New Zealand, the Phillipines, and at the Geysers in northern California.

Underground dry steam is relatively rare but is the most desirable from the standpoint of power generation. More commonly, energy reaches the surface as superheated water and steam. In some cases, the water is so pure that it can be used for irrigation and livestock; in other cases, it is loaded with corrosive, scale-forming salts. Utilization of the heat from contaminated geothermal water generally requires that the water be reinjected into the hot formation after heat removal to prevent contamination of surface water.

The utilization of hot rocks for energy requires fracturing of the hot formation, followed by injection of water and withdrawal of steam. This technology is still in the experimental state, but promises approximately ten time as much energy production as steam and hot water sources.

Land subsidence and seismic effects are environmental factors that may hinder the development of geothermal power. However, this energy source holds considerable promise, and its development continues.

21.15. THE SUN: AN IDEAL ENERGY SOURCE

The recipe for an ideal energy source calls for one that is unlimited in supply, widely available, and inexpensive; it should not add to the Earth's total heat burden or produce chemical air and water pollutants. Solar energy fulfills all of these criteria.

Solar energy does not add excess heat to that which must be radiated from the Earth. On a global basis, utilization of only a small fraction of solar energy reaching the Earth could provide for all energy needs. In the United States, for example, with conversion efficiencies ranging from 10-30%, it would only require collectors ranging in area from one-tenth down to one-thirtieth that of the state of Arizona to satisfy present U.S. energy needs. (This is still an enormous amount of land, and there are economic and environmental problems related to the use of even a fraction of this amount of land for solar energy collection. Certainly, many residents of Arizona would not be pleased at having so much of the state devoted to solar collectors, and some environmental groups would protest the resultant shading of rattlesnake habitat.)

Solar power cells for the direct conversion of sunlight to electricity have been developed and are widely used for energy in space vehicles. With present technology, however, they remain too expensive for large-scale generation of electricity. Therefore, most schemes for the utilization of solar power depend upon the collection of thermal energy, followed by conversion to electrical energy. The simplest such approach involves focusing sunlight on a steam-generating boiler. Parabolic reflectors can be used to focus sunlight on pipes containing heat-transporting fluids. Selective coatings on these pipes can be used so that only a small percentage of the incident energy is reradiated from the pipes.

A major disadvantage of solar energy is its intermittent nature. However, flexibility inherent in an electric power grid would enable it to accept up to 15% of its total power input from solar energy units without special provision for energy storage. Existing hydroelectric facilities may be used for pumped-water energy storage in conjunction with solar electricity generation. Heat or cold can be stored in water, in a latent form in water (ice) or eutectic salts, or in beds of rock. Enormous amounts of heat can be stored in water as a supercritical fluid contained at high temperatures and very high pressures deep underground. Mechanical energy can be stored with compressed air or flywheels.

Hydrogen gas, H_2, is an ideal chemical fuel that may serve as a storage medium for solar energy. Electricity generated by solar means can be used to electrolyze a salt solution containing an anion that is very difficult to oxidize, so that oxygen is released at the anode and hydrogen is produced at the cathode. The net reaction is

$$4H_2O + \text{electrical energy} \rightarrow H_2(g) + O_2(g) \qquad (21.15.1)$$

Hydrogen, and even oxygen, can be piped some distance and the hydrogen burned without pollution or used in a fuel cell (Figure 21.6). This may, in fact, make possible a "hydrogen economy." Disadvantages of hydrogen uses as a fuel include the fact that it has a heating value per unit volume of about one-third that of natural gas and that it is explosive over a wide range of mixtures with air.

No really insurmountable barriers exist to block the development of solar energy, such as might be the case with fusion power. In fact, the installation of solar space and water heaters became widespread in the late 1970s and research on solar energy was well supported in the U.S. until after 1980, when it became fashionable to believe that free market forces had solved the "energy crisis." With the installation of more heating devices and the probable development of some cheap, direct solar electrical generating capacity, it is likely that during the coming century solar energy will be providing an appreciable percentage of energy needs in areas receiving abundant sunlight.

21.16. ENERGY FROM BIOMASS

All fossil fuels originally came from photosynthetic processes. Photosynthesis does hold some promise of producing combustible chemicals to be used for energy production and could certainly produce all needed organic raw materials. It suffers from the disadvantage of being a very inefficient means of solar energy collection (a collection efficiency of only several hundredths of a percent by photosynthesis is typical of most common plants). However, the overall energy conversion efficiency of several plants, such as sugarcane, is around 0.6%. Furthermore, some plants, such as *Euphorbia lathyrus* (gopher plant), a small bush growing wild in California, produce hydrocarbon emulsions directly. The fruit of the Philippine plant, *Pittsosporum reiniferum*, can be burned for illumination due to its high content of hydrocarbon terpenes (see Section 12.2), primarily α-pinene and myrcene. Conversion of agricultural plant residues to energy could be employed to provide some of the energy required for agricultural production. Indeed, until about 70 years ago, virtually all of the energy required in agriculture — hay and oats for horses, home-grown food for laborers, and wood for home heating — originated from plant materials produced on the land. (An interesting exercise is to calculate the number of horses required to provide the energy used for transportation at the present time in the Los Angeles basin. It can be shown that such a large number of horses would fill the entire basin with manure at a rate of several feet per day.)

Annual world production of biomass is estimated at 146 billion metric tons, mostly from uncontrolled plant growth. Many farm crops and trees can produce 10-20 metric tons per acre per year of dry biomass and some algae and grasses can produce as much as 50 metric tons per acre per year. The heating value of this biomass is 5000-8000 Btu/lb for a fuel having virtually no ash or sulfur (compare heating values of various coals in Table 21.3). Current world demand for oil and gas could be met with about 6% of the global production of biomass. Meeting U.S. demands for oil and gas would require that about 6-8% of the land area of the coterminous 48 states be cultivated intensively for biomass production. Another advantage of this source of energy is that use of biomass for fuel would not add any net carbon dioxide to the atmosphere.

As it has been throughout history, biomass is significant as heating fuel, and in some parts of the world is the fuel most widely used for cooking. Air pollution from wood burning stoves and furnaces is a growing problem in some areas. Currently wood provides about 8% of world energy needs. This percentage could increase through the development of "energy plantations" consisting of trees grown solely for their energy content.

Seed oils show promise as fuels, particularly for use in diesel engines. The most common plants producing seed oils are sunflowers and peanuts. More exotic species include the buffalo gourd, cucurbits, and Chinese tallow tree.

Biomass could be used to replace much of the 50-60 million metric tons of petroleum and natural gas currently consumed in the manufacture of primary chemicals in the U.S. each year. Among the sources of biomass that could be used for chemical production are grains and sugar crops (for ethanol manufacture), oilseeds, animal byproducts, manure, and sewage (the last two for methane generation). The biggest potential source of chemicals is the lignocellulose making up the bulk of most plant material. For example, both phenol and benzene might be produced directly from lignin. Brazil has maintained a program for the production of chemicals from fermentation-produced ethanol.

21.17. GASOHOL

A major option for converting photosynthetically-produced biochemical energy to form suitable for internal combustion engines is the production of either methanol or ethanol. Methanol is created by the destructive distillation of wood (Section 21.4) or from synthesis gas manufactured from coal or natural gas (Section 21.11). Ethanol is most commonly manufactured by fermentation of carbohydrates. Either one of these chemicals can be used by itself as fuel in a suitably designed internal combustion engine. More commonly, these alcohols are blended in proportions of up to 20% with gasoline to give **gasohol**, a fuel that can be used in existing internal combustion engines with little or no adjustment.

Gasohol offers a number of advantages. A high octane rating is obtained without adding tetraethyl lead. The reduction in exhaust emissions can be substantial, up to 50% for carbon monoxide and NO_x. Most important, because of its photosynthetic origin, alcohol may be considered a renewable resource rather than a depletable fossil fuel. The manufacture of alcohol can be accomplished by the fermentation of sugar obtained from the hydrolysis of cellulose in wood wastes (see Section 21.4) and crop wastes. Fermentation of these waste products offers an excellent opportunity for recycling.

Brazil has been a leader in the manufacture of ethanol for fuel uses, with 4 billion liters produced in 1982. At one time Brazil had over 450,000 automobiles that could run on pure alcohol, although many of these were converted back to gasoline during the era of relatively low petroleum prices in the 1980s. The extensive involvement of Brazil with ethanol production stems from the fact that it has few fossil resources but does have ideal conditions for the growth of large quantities of biomass. To date, most of Brazil's alcohol has come from the fermentation of sugarcane. However, a potentially much greater source of fermentable biomass is casava, or manioc, a root crop growing abundantly throughout the country. Significant amounts of gasoline in the United States is are supplemented with ethanol, more as an octane-ratings booster, rather than as a fuel supplement.

Methanol, another potential ingredient of gasohol, can also be made from biomass. This can be accomplished by converting biomass, such as wood, to CO and H_2, and synthesizing methanol from these gases.

21.18. FUTURE ENERGY SOURCES

As discussed in this chapter, a number of options are available for the supply of energy in the future. The major possibilities are summarized in Table 21.4 (next page).

LITERATURE CITED

1.Lillienthal, David E., *Atomic Energy: A New Start*, Harper and Row Publishers, Inc., New York, 1980.

2."Trash–to–Electricity Disposal Runs into Problems in Maine," *New York Times*, January 22, 1991, p. A13.

3."Coal Output Near a Record," *New York Times*, December 17, 1990, p. C-2.

Table 21.4. Possible Future Sources of Energy.

Source	Principles
Coal conversion	Manufacture of gas, hydrocarbon liquids, alcohol, or solvent-refined coal from coal.
Oil shale	Retorting petroleum-like fuel from oil shale.
Geothermal	Utilization of underground heat.
Gas-turbine topping cycle	Utilization of hot combustion gases in a turbine, followed by steam generation.
MHD	Electrical generation by passing a hot gas plasma through a magnetic field.
Thermionics	Electricity generated across a thermal gradient.
Fuel cells	Conversion of chemical to electrical energy.
Solar heating and cooling	Direct use of solar energy for heating and cooling through the application of solar collectors.
Solar cells	Use of silicon semiconductor sheets for the direct generation of electricity from sunlight.
Solar thermal electric	Conversion of solar energy to heat followed by conversion to electricity.
Wind	Conversion of wind energy to electricity.
Ocean thermal electric	Use of ocean thermal gradients to convert heat energy to electricity.
Nuclear fission	Conversion of energy released from fission of heavy nuclei to electricity.
Breeder reactors	Nuclear fission combined with conversion of nonfissionable nuclei to fissionable nuclei.
Nuclear fusion	Conversion of energy released by the fusion of light nuclei to electricity.
Bottoming cycles	Utilization of waste heat from power generation for various purposes.
Solid waste	Combustion of trash to produce heat and electricity.
Photosynthesis	Use of plants for the conversion of solar energy to other forms by a biomass intermediate.
Hydrogen	Generation of H_2 by thermochemical means for use as an energy-transporting medium.

Supplementary References

Medvedev, Zhores A., *The Legacy of Chernobyl*, Norton, New York, 1990.

Ogden, Joan M., and Robert H. Williams, *Solar Hydrogen. Moving Beyond Fossil Fuels*, World Resources Institute, Washington, DC, 1989.

World Resources, 1990-91, Oxford University Press, New York, 1990.

Cohen, Bernard L., *The Nuclear Energy Option. An Alternative for the 90s*, Plenum, New York, 1990.

Cassedy, Edward S., and Peter Z. Grossman, *Introduction to Energy. Resources, Technology, and Society*, Cambridge University Press, New York, 1990.

Van der Voort, E., and G. Grassi, Eds., *Wind Energy–1*, Harwood, New York, 1988.

Van der Voort, E., and G. Grassi, Eds., *Wind Energy–2*, Harwood, New York, 1988.

World Resources Institute and International Institute for Environment and Development, *World Resources, 1988-89*, Basic Books, New York, 1988.

Berlin, Robert E., and Catherine C. Stanton, *Radioactive Waste Management*, Wiley-Interscience, New York, 1989.

Rush to Burn. Solving America's Garbage Crisis?, Island Press, Washington, DC, 1989.

Oppenheimer, Ernest J., *Natural Gas, the Best Energy Choice*, Pen & Podium, Inc., New York, N.Y. 1989.

Sperling, Daniel, *New Transportation Fuels: A Strategic Approach to Technological Change*, University of California Press, Berkeley, CA, 1989.

Carter, Luther J., *Nuclear Imperatives and Public Trust: Dealing with Radioactive Waste*, Resources for the Future In., Baltimore, MD, 1987.

Klass, Donald L., *Energy From Biomass and Wastes IX*, Institute of Gas Technology, Chicago, Illinois, 1986.

Krauskopf, Konrad B., *Radioactive Waste Disposal and Geology*, Chapman and Hall, New York, 1988.

Burnes, Ed., Michael E., *Low-Level Radioactive Waste Regulation: Science, Politics, and Fear*, Lewis Publishers, Chelsea, MI, 1988.

Bibby, D. M., *Methane Conversion*, Elsevier Science Publishers, New York, NY, 1988.

Kittel, Ed., Howard, *Radioactive Waste Management Handbook, Vol. I: Near-Surface Land Disposal*, J. Harwood Academic Publishers, New York, NY, 1989.

Park, Chris C., *Chernobyl: The Long Shadow*, Routledge, London and New York, 1989.

Speight, James G., Ed., *Fuel Science and Technology Handbook*, Marcel Dekker, New York, NY, 1990.

QUESTIONS AND PROBLEMS

1. What pollution control measures may produce a shortage of platinum metals?

2. List and discuss some of the major environmental concerns related to the mining and utilization of metal ores.

3. What are the major phosphate minerals?

4. Arrange the following energy conversion processes in order from the least to the most efficient: (a) electric hot water heater, (b) photosynthesis, (c) solar cell, (d) electric generator, (e) aircraft jet engine.

5. Considering the Carnot equation and common means for energy conversion, what might be the role of improved materials (metal alloys, ceramics) in increasing energy conversion efficiency?

6. Why is shale oil, a possible substitute for petroleum in some parts of the world, considered to be a pyrolysis product?

7. List some coal ranks and describe what is meant by coal rank.

8. Why was it necessary to add hydrocarbons to gas produced by reacting steam with hot carbon from coal in order to make a useful gas product?

9. What is the principle of the Exxon Donor Solvent process for producing liquid hydrocarbons from coal?

10. As it is now used, what is the principle or basis for the production of energy from uranium by nuclear fission? Is this process actually used for energy production? What are some of its environmental disadavanges? What is one major advantage?

11. What would be at least two highly desirable features of nuclear fusion power if it could ever be achieved in a controllable fashion on a large scale?

12. Justify describing the sun as "an ideal energy sources." What are two big disadvantages of solar energy?

13. What are some of the greater implications of the use of biomass for energy? How might such widespread use affect greenhouse warming? How might it affect agricultural production of food?

14. Describe how gasohol is related to energy from biomass.

15. Using some specific examples, describe what is meant by the "resources–energy–environment triangle."

16. How does the trend toward utilization of less rich ores affect the environment? What does it have to do with energy utilization?

17. Of the resources listed in Chapter 21, list and discuss those that are largely from byproduct sources.

18. Why is the total dollar value of "cheap" sand and gravel so high? What does this fact imply for environmental protection?

INDEX

ABS detergents, 163
Abstraction reactions
  (atmospheric chemical), 322
Acetaldehyde, 51, 304, 515
Acetic anhydride, 516
Acetone, 304, 515
Acetonitrile, 518
Acetylchlolinesterase, 524
Acetylene, 301, 511
Acid deposition, 343
  fog, 250
  hydrolysis, 374
  mine water, 34, 138
Acid
  precipitation, 343
  rain, 284, 343
Acid/base neutralization, 462
Acidic solutions, 6
Acidity, 32, 39, 146, 158
Acids, 6
Acridine, 261
Acrolein, 304, 515
Acrylonitrile, 518
Activated
  carbon, 201, 460
  adsorption, 190
Activated sludge process, 187
Active metabolite, 498
Active parent compound, 498
Activity, 175
Acute toxicity, 496
Addition reactions, 322
Adsorption by aerosol particles,
  439
Advanced design incinerators, 469
Advanced waste treatment, 189
Aeration zone, 375

Aerobic
  degradation, 432
  processes, 446
  waste treatment, 471
Aerosol, 251
Agency for Toxic Substances
  and Disease Registry (ATSDR), 420
Agent orange, 522
Aggregation, 435
  of particles, 99
Aggressive water, 195
Air parcel, 349
Airglow, 239
Alar, 397
Albedo, 359
Alcohols, 14, 306
Aldehydes, 303, 515
Aldicarb, 166
Aldrin-dieldrin, 166
Algae, 114
Algal nutrients, 146, 156
Aliphatic alcohols, 305
Alkaline soils, 31, 32, 42, 146, 158, 159,
  383
Alkanes Hydrocarbons, 8, 14, 510
Alkenes, 9, 297
Alkenyl halides, 308
Alkoxides, 465
Alkoxyl radicals, 329
Alkyl
  benzene sulfonate, 163
  nitrates, 330
  nitrites, 330
  peroxyl radicals, 330
  radical, 298
Alkylating agents, 501
Alkynes, 14, 297
Allergy, 502